2150

PRINCIPLES OF
Paleontology

An assemblage of trilobites of Devonian age: *Phacops rana milleri* Stewart (×1.3). Reprinted from *Trilobites: A Photographic Atlas* by Riccardo Levi-Setti, by permission of The University of Chicago Press. Copyright © 1975 by The University of Chicago.

Second Edition

PRINCIPLES OF Paleontology

David M. Raup
UNIVERSITY OF ROCHESTER

Steven M. Stanley
THE JOHNS HOPKINS UNIVERSITY

W. H. FREEMAN AND COMPANY
San Francisco

COVER DESIGN
Based on illustration from J. H. Callomon, Sexual dimorphism in Jurassic ammonites, *Trans. Leicester Lit. Philos. Soc.,* **57:** 21–56 (1963).

Library of Congress Cataloging in Publication Data

Raup, David M.
Principles of paleontology.

Bibliography: p.
Includes index.
1. Paleontology. I. Stanley, Steven M., joint author. II. Title
QE711.2.R37 1978 560 77–17443
ISBN 0–7167–0022–0

Copyright © 1971, 1978 by W. H. Freeman and Company

No part of this book may be reproduced by any mechanical, photographic, or electronic process, or in the form of a phonographic recording, nor may it be stored in a retrieval system, transmitted, or otherwise copied for public or private use, without written permission from the publisher.

Printed in the United States of America

2 3 4 5 6 7 8 9

Contents

Preface ix

PART I DESCRIPTION AND CLASSIFICATION OF FOSSILS

1. Preservation and the Fossil Record 1

    The Nature of the Sample 3
    Kinds of Organisms 10
    Numbers of Individuals 13
    Preservability 14
        Biologic Destruction 14
        Mechanical Destruction 16
        Chemical Destruction 20
    Biologic Structures Most Likely To Be Preserved 20
    Environment and Preservability 21
    Post-mortem Transport 24
    Conclusion 25

2. Describing a Single Specimen 25

    The Pictorial Description 28
        Photographic or Mechanical Images 28
        The Line Drawing 34
    Descriptive Terminology 35
    Description by Measurement 37
    Describing Internal Structures 44

3. Ontogenetic Variation 45

    Types of Growth 46
        Accretion of Existing Parts 46
        Addition of New Parts 47
        Molting 49

Modification 51
Mixed Growth Strategies 51
Describing Ontogenetic Change 55
Growth Rates 59
Reasons for Anisometric Growth 63
Ontogeny of Colonial Organisms 66

4. The Population as a Unit 69

Populations in Biology 69
Individual Variation within Populations 71
Ontogenetic Differences 71
Genetic Differences 72
Nongenetic Differences 74
Fossil Populations 75
Population Dynamics 78
Procedures in Describing Variation 85
Variation within Assemblages 85
Variation between Assemblages 87
Multivariate Analysis 90

5. The Species as a Unit 101

The Biologic Species Definition 102
The Origin of Species 103
Natural Selection 103
Geographic Speciation 104
The Subspecies 105
Clines and Ring Species 105
Addition of Species without Geographic Isolation 106
Rates of Evolution and Speciation 106
Biologic Methods of Species Discrimination 107
The Species Problem in Paleontology 109
Formal Naming and Description of Species 114
The Format of a Description 116
Diagnosis and Description 117
Type Specimens 125
Changing Species Names 126
Perspective 127

6. Grouping of Species into Higher Categories 129

Classification versus Evolution 130
Taxonomic Categories 131
Criteria for Definition of Higher Categories 138
Numerical Taxonomy 142
Cladistic Taxonomy 146

7. Identification of Fossils 151

Normal Procedures 152
Keys 155

Automatic Methods of Identification 158
Presentation of Results 158
Bibliographic Sources 161

PART II THE USES OF PALEONTOLOGIC DATA 613

8. Adaptation and Functional Morphology 165

The Nonrandomness of Adaptation 166
Theoretical Morphology 170
Methods of Functional Morphologic Analysis 178
Examples of Functional Analysis 180
Vision in Trilobites 180
Biomechanics of Pterosaurs 182
Evolution of Jaw Mechanics 188
Life-habit Transitions in Mussels 192
Conclusion 195

9. Biostratigraphy 197

Rock-stratigraphic Units 198
Biostratigraphic Units—The Biozone 203
Correlation with Fossils 207
Stratigraphic Ranges and Zones 207
Percentage of Common Taxa 209
Index Fossils 213
Morphologic Features 213
Ecologic Patterns 215
Quantitative Correlation Methods 216
Time and Time-rock Units 222
Accuracy of Correlation 225

10. Paleoecology 231

Fundamental Ecologic Principles 232
The Marine Ecosystem 235
Life Habits 239
Direct Evidence through Preservation 239
Homology 243
Functional Morphology 243
Evidence of Biologic Activity 246
Limiting Factors 248
Temperature 248
Oxygen 248
Water Depth 255
Salinity 261
Substratum 267
Food 273
Spatial Distribution of Populations 274
Fossil Communities 278

Limiting Factors and Species Interactions 278
Organic Reef Communities 282
Soft-bottom Communities 287
Terrestrial Communities 293
Post-mortem Information Loss 296

11. Evolution and the Fossil Record 303

Extinction 304
Rates of Evolution and Extinction 307
Rate of Progressive Evolution 307
Taxonomic Frequency Rates 316
The Distribution of Evolutionary Rates in Phylogeny 324
Rates of Extinction above the Species Level 330
Mass Extinction 332
Diversity Controls 333
The Nature of Trends 333
Phyletic Trends 335
Phylogenetic Trends 348
Patterns of Evolution and Extinction 356
Adaptive Radiation 357
Evolutionary Convergence and Parallelism 363
Ecologic Replacement 365
Determining Phylogenetic Relationships 369

12. Biogeography 379

The Earth's Climate 380
Winds and Ocean Currents 380
Types of Climate 385
Biotic Distributions 387
Biomes and Provinces 387
Latitudinal Gradients in Taxonomic Diversity 391
Climates of the Past 393
Botanical Evidence 393
Sedimentary Evidence 398
Data From Marine Life 400
Determining Geographic Ranges of Extinct Taxa 404
Barriers to Dispersal 404
The Nature of Barriers 406
Island Biogeography 408
Evolution of Taxa Occupying Island-like Habitats 413
Plate Tectonics and Physiographic Change 416
The New Plate Tectonics Framework 417
Biologic and Paleobiologic Data 423
Mass Extinction 438

Bibliography 449
Author Index 469
Subject Index 473

Preface

In the preface to the first edition of this book, we noted that our goal was to write a book that would present the principles of paleontology at a level suitable for an undergraduate course. What was true then, in 1971, is no less true of the present edition. This book is not designed to provide the entire content of such a course; rather, it is meant to provide a conceptual background for the course and essential information and ideas that may be only partially presented—or presented from a different point of view—in lectures and laboratories. Many teachers of undergraduate paleontology courses follow a phylum-by-phylum taxonomic format in their lectures, which we believe makes assignment of extensive reading on principles and approaches all the more important.

There are twelve chapters in our volume; an average assignment of one chapter per week allows for coverage of the entire book in a typical semester.

In Part I of the text, "Description and Classification of Fossils," we give some general information about the fossil record in an introductory chapter. In subsequent chapters we discuss the ways in which fossils are studied as specimens within species, as species, and as hierarchical groups of species. Part I might have been called "Taxonomy" or "Systematics," but is slightly different in scope from these broad disciplines.

A major addition to Part I in this edition is an introduction to the philosophy and methodology of multivariate analysis, which has become a fundamental tool of the paleontologic trade.

We explain in Part II, "The Uses of Paleontologic Data," ways in which information derived from fossil study is applied to various geologic and biologic problems.

Increasingly after 1971, Part II of the first edition began to cry out for change. That our textbook would require revision in order to accommodate these changes after only seven years pays an indirect compliment to our profession. Paleontology has grown rapidly in exciting new directions, just as it had in the few years preceding publication of the first edition. In Part II of the present edition, we discuss many of the results of this rapid new growth.

Some of the modifications we have made and the motives for making them are as follows: In the past few years, functional morphology has attained new levels of sophistication; of necessity, case studies presented in the revised Chapter 8 have been chosen arbitrarily from a host of potentially useful published examples. Paleoecology, which during the 1960's occupied center stage in the paleontologic theater, has matured as a subdiscipline but has also lost some of its luster; appreciation of the incompleteness of the invertebrate fossil record has led to a general narrowing of goals in the study of ancient marine communities. After more than two decades of relative dormancy following the publication of George Gaylord Simpson's *Major Features of Evolution,* large-scale evolution has moved into the limelight, and new developments have led to a major restructuring of the discussion of "Evolution and the Fossil Record," which is now Chapter 11. Plate tectonics has provided a fresh and, one must hope, proper backdrop for the study of biogeography; the final chapter of the present edition, "Biogeography," is designed to guide the student into this exciting area. A brief discussion of plate tectonic theory is included for readers, such as some students of biology, who have not encountered the subject elsewhere.

We express special thanks to our friends Robert Linsley and R. Peter Richards for offering numerous suggestions that have substantially improved our revision. For any errors or weaknesses that remain, we accept full responsibility. We also thank those readers of the first edition who wrote to offer suggestions or point out errors, though, unfortunately, every suggestion could not be heeded. Nevertheless, we request the same kind of response from users of the second edition.

December 1977

David M. Raup
Steven M. Stanley

PRINCIPLES OF
Paleontology

PART I

DESCRIPTION AND CLASSIFICATION OF FOSSILS

CHAPTER 1

Preservation and the Fossil Record

The fossil record is only a small sample of past life. The sample is not random: it is highly biased by a variety of biologic and geologic factors. Therefore, any study of fossils or use of paleontologic data must be based on a clear understanding of the strengths and weaknesses of the record. We must learn what can be done with fossils and what cannot.

The Nature of the Sample

The fossil record contains about 250,000 species of plants and animals. The number is not precisely known because of the complexity of the world paleontological literature and because there is some redundancy in naming species. It is clear however, that the sample is a very large one and that it is steadily growing as new geologic areas are explored and old ones are more thoroughly worked. From this sample we can learn much about the history of life on the planet. When looked at from another viewpoint, the sample is actually very small, because it contains but a tiny fraction of past life. We do not actually know how many species existed in the geologic past, but we can make some reasonably good guesses based on the size of the present biota and on our knowledge of evolutionary turnover—extinction and origination of species.

About 1,500,000 different species of plants and animals are known to be living today. Since this number is based only on species that have actually been found, described, and classified, the real number is probably considerably higher. Some authorities have estimated that when the job of description of living species is complete, as many as 4,500,000 species of plants and animals will be known. Thus, the number of fossil species (250,000) is about 5 percent of the *probable total* for living organisms. The comparison is particularly striking when one considers that the fossil record covers several billions of years, whereas living flora and fauna represent only an instant of geologic time. If overall preservation of fossils were even reasonably good, we would expect the number of fossil species to be far greater than the number of living species.

Actually, the fossil sample is not quite as impoverished as the foregoing figures would suggest. About three-quarters of a million living species are insects, and though insects are occasionally preserved in the fossil record (about 8,000 species are known), their terrestrial habitat and delicate skeletons mean that fossilization is a rare event. If insects and a few other hard-to-preserve groups are ignored, the fossil record appears to be a considerably better sample of past life. This reasoning cannot be carried far, however, because it leads to the absurd conclusion that the fossil record is excellent if one considers only those organisms that are excellently preserved! It is conventional, nevertheless, to evaluate the completeness of the fossil record in terms of a few large groups which (a) are readily preserved under most common geologic conditions and (b) include a large enough number of species to have a significant impact on the fossil record as a whole. These groups are the well-skeletonized marine representatives of the following nine phyla: Protozoa, Archaeocyatha (extinct), Porifera, Coelenterata, Ectoprocta, Brachiopoda, Mollusca, Arthropoda, and Echinodermata. Thus excluded are insects and several other large groups, such as the worms. Also excluded are vertebrates and plants because, despite their evolutionary importance, they usually do not make up a significant fraction of fossil assemblages.

Various zoologists and paleontologists have made estimates of the number of living species in the nine readily fossilizable groups listed in the preceding paragraph. For example, Valentine (1970) estimated that about 100,000 living species have been discovered and perhaps another 50,000 are yet to be found. In the fossil record, about 180,000 species in the nine groups have been found (Raup, 1976a). In order to translate these figures into an estimate of the completeness of the fossil record, we must know two more things: (a) the evolutionary turnover rate and (b) the history of standing diversity (number of coexisting species) over geologic time.

Estimates of the average rate of species turnover have been made. One is that of Simpson (1952), who estimated that the average life span of a species is 2.75 million years. He acknowledged, however, that the true average may be as low as 0.5 million years or as high as 5.0 million years. Some of the estimates by other people have been higher: in a recent analysis, Valentine (1970) used average species durations of from 5 to 10 million years. Regardless of which figure is used, however, it is clear that there has been complete species replacement many times during geologic history.

Estimates of changes in diversity over geologic time are hard to obtain. The fossil record extends back about 3.5 billion years but nearly all fossils are found in rocks of Cambrian or younger age (0.6 billion years). The rise in diversity seen in the Cambrian is almost certainly real but it is difficult to say when the number of species reached its present count. It *appears* that diversity does increase through time after the Cambrian but, as we shall see, part of this may be due to the fact that sampling is better in younger rocks.

Valentine (1970), Durham (1967), and several other paleontologists have used various estimates and assumptions to calculate the number of species that have lived in the nine readily preservable groups. Durham concluded that only about one in forty-four has been found, or 2.3 percent of the preservable species that have lived since the beginning of the Cambrian. Valentine's figures yield a range of from 4.5 percent to 13.6 percent. Regardless of which estimates we accept, we must agree that the fossil record yields but a tiny fraction of the species that have lived and that even within phyla whose members are relatively easy to preserve, fossilization, and ultimate discovery by paleontologists, are rare events.

Because preservation is such an improbable event, the slightest advantage enjoyed by a particular type of organism or a particular habitat can mean that a favored organism or habitat is represented in the record, yet a less favored one is not. One of the purposes of this chapter will be to develop an awareness of some general principles governing the preservation process and of the biases that inevitably exist.

The quality of a fossil record as a sample of past life varies with what is known as *taxonomic level.* All organisms, fossil and living, are classified not only into species but also into higher groupings or categories. At the higher end of the classificatory scale are taxonomic groups such as *phylum* and *class.* At the lower end of the scale are groups such as *family* and *genus.* For reasons stemming from sampling theory, our knowledge of past life is far better at high taxonomic levels than at low taxonomic levels. In any sample of organisms we need to look at many fewer individuals to find all phyla than to find all species or genera. This is because—by logic—there are more individuals in a phylum than in any of its genera or species and thus there is a greater chance of finding at least one representative of the phylum than of any one of its species. This can be illustrated by an actual example from the fossil record. Table 1-1 shows data on the diversity found in a well sample from some Tertiary sediments in Denmark. In this case, the paleontologist was concerned only with molluscan fossils; therefore, from his point of view at least, the sample contained representatives of but one phylum (Mollusca). Within the Mollusca he found 2,954 specimens which were distributed among three molluscan classes (Bivalvia, Scaphopoda, and Gastropoda). As is almost always so in the classification system, each lower taxonomic level yielded a larger number of taxa (see Table 1-1).

Had the basic sample from the Danish well been larger (that is, had it yielded more than 2,954 specimens), the numbers in Table 1-1 would most probably have been larger but the differences would have been concentrated at the lower taxonomic levels. The probability of finding additional species would be much greater than the probability of finding additional orders or classes. At all taxonomic levels, "diminishing returns" take over at large sample sizes: more and more specimens are added to already recognized taxa.

If fewer than 2,954 specimens had been collected and examined, fewer taxa would have been found and the effect would also have been concentrated at the lower taxonomic levels. This aspect of the sampling phenomenon is shown graphically in Figure 1-1. The number of specimens in the Danish sample is plotted against number

TABLE 1-1
Numbers of Taxa Found in a Molluscan Sample from Arnum Formation (Miocene of Denmark).

	2,954 specimens
Phyla	1
Classes	3
Orders	12
Families	44
Genera	64
Species	86

SOURCE: Sorgenfrei, 1958.

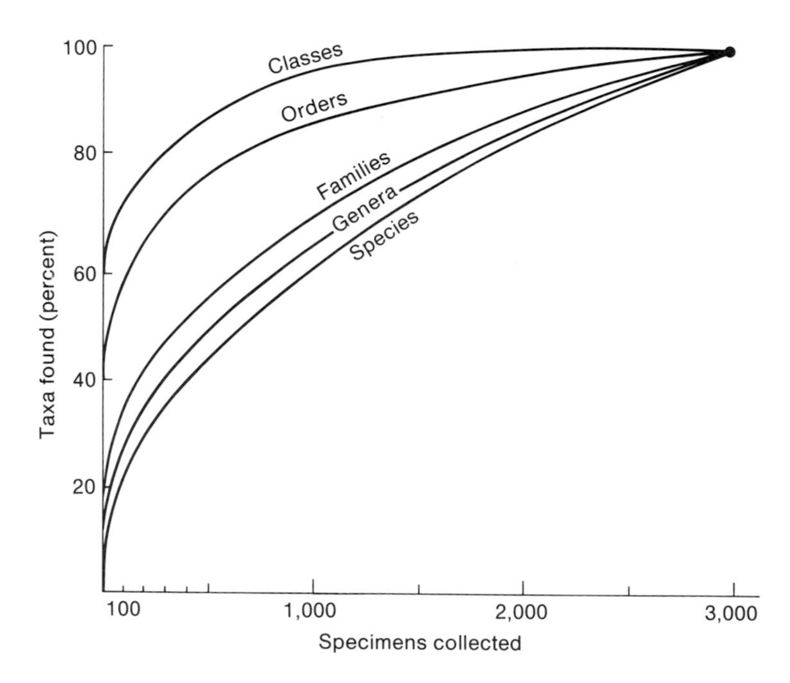

FIGURE 1-1
Rarefaction curves for molluscan fossils found in a well sample of Miocene age in Denmark (based on data from Sorgenfrei, 1958). The point at the upper right represents the actual sample. The curves show how many taxa would have been found had the sample been smaller.

BOX 1-A Rarefaction Method

In order to compute a rarefaction curve such as the species curve in Figure 1-1, it is necessary to know the number of individuals in the sample (N), the number of species (S), and the number of individuals found for each species (N_i, where $i = 1, 2, \ldots, S$). With this information, selected points on the rarefaction curve may be computed from the following equation:

$$E(S_n) = \sum_{i=1}^{S}\left[1 - \frac{\binom{N - N_i}{n}}{\binom{N}{n}}\right].$$

The quantity $E(S_n)$ is the estimated number of species that would be found in a sample of n individuals. For mathematical details, see Raup (1975a). In practice, the computations are carried out for an arbitrary series of n values (all less than N) and each $E(S_n)$ thus produced yields one point on the rarefaction curve. Or if several samples are to be compared at a standard n, that n can be used for a single computation for each sample. It is also possible to compute the uncertainty attached to the estimated species numbers, that is, the variance of $E(S_n)$.

In the case of the species rarefaction in Figure 1-1, N was equal to 2,954 and S was 86. The most common species was represented by 818 individuals; forty of the species had only one specimen each. A few of the computed values of $E(S_n)$ and its variance are given below:

Number of Specimens (n)	*Estimated number of species,* $E(S_n)$	*Variance of* $E(S_n)$
2,500	79.66	5.42
2,000	71.93	9.89
1,500	63.14	12.59
1,000	52.52	13.59
500	38.56	12.23
100	19.05	6.64
50	14.05	5.05
10	6.24	2.88

If rarefaction is to be done at higher taxonomic levels (as in Figure 1-1), data for genera, families, and higher categories are simply substituted for the species data and the same equation is used. The use of rarefaction at higher taxonomic levels is described by Raup (1975a).

of taxa at various levels, with the scale for the latter being expressed as a percentage of the number actually found. The large point on the right of the graph (where the several curves converge) is the actual sample; that is, 2,954 specimens and 100 percent of all taxonomic categories. The curves extending back toward the left are calculated by a statistical technique called ***rarefaction*** and enable one to estimate how many taxa *would have been found* had the sample been smaller (see Box 1-A). In this case, had the sample contained only 2,000 specimens, all three classes (100 percent), but probably only about 80 percent of the species, would still have been found. Had only 100 specimens been collected, the rarefaction method estimates that about two-thirds of the classes would still have been discovered (the two most common in the total assemblage), but only about 20 percent of the species would have been found.

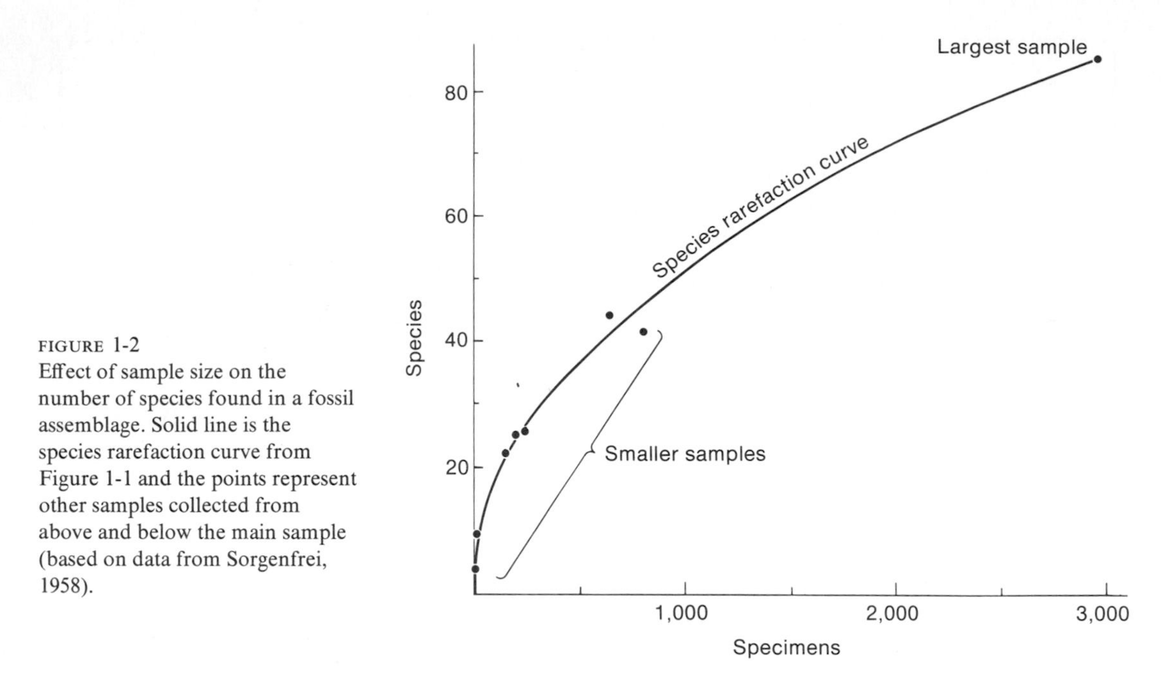

FIGURE 1-2
Effect of sample size on the number of species found in a fossil assemblage. Solid line is the species rarefaction curve from Figure 1-1 and the points represent other samples collected from above and below the main sample (based on data from Sorgenfrei, 1958).

In Figure 1-2, the effect of sample size on species number is shown in another way for the Danish well data. The sample of 2,954 individuals was the largest of eight samples taken from one sedimentary formation (the Arnum Formation of Miocene age). The other seven samples came from different stratigraphic intervals above and below the large sample. The numbers of specimens and species for each are plotted in Figure 1-2 along with the rarefaction curve for species in the large sample (taken from Figure 1-1). The points for the small samples fall very close to the theoretical sampling curve—suggesting that the differences in numbers of species are just the result of sample size.

It follows from the foregoing reasoning that in regions of the world or in parts of the geologic column where preservation is best (that is, where a greater number of individual organisms are preserved), our taxonomic knowledge of the fossil record will be more complete. As a general rule, time periods late in geologic history are better represented, both geologically and paleontologically, than periods further back in time. There are two main reasons for this. First, younger rocks are less likely to be covered or obscured by other rocks; they are in effect near the "top of the pile." It is characteristic that the younger time periods are represented by greater thicknesses of known sediments over larger parts of the earth's surface. Second, younger rocks are less likely to have been eroded away, or to have had their fossil record obscured by metamorphism.

The results of both factors are illustrated in Figure 1-3. The diagram at the left as one examines Figure 1-3 is a plot of what is known as *sediment survival* through

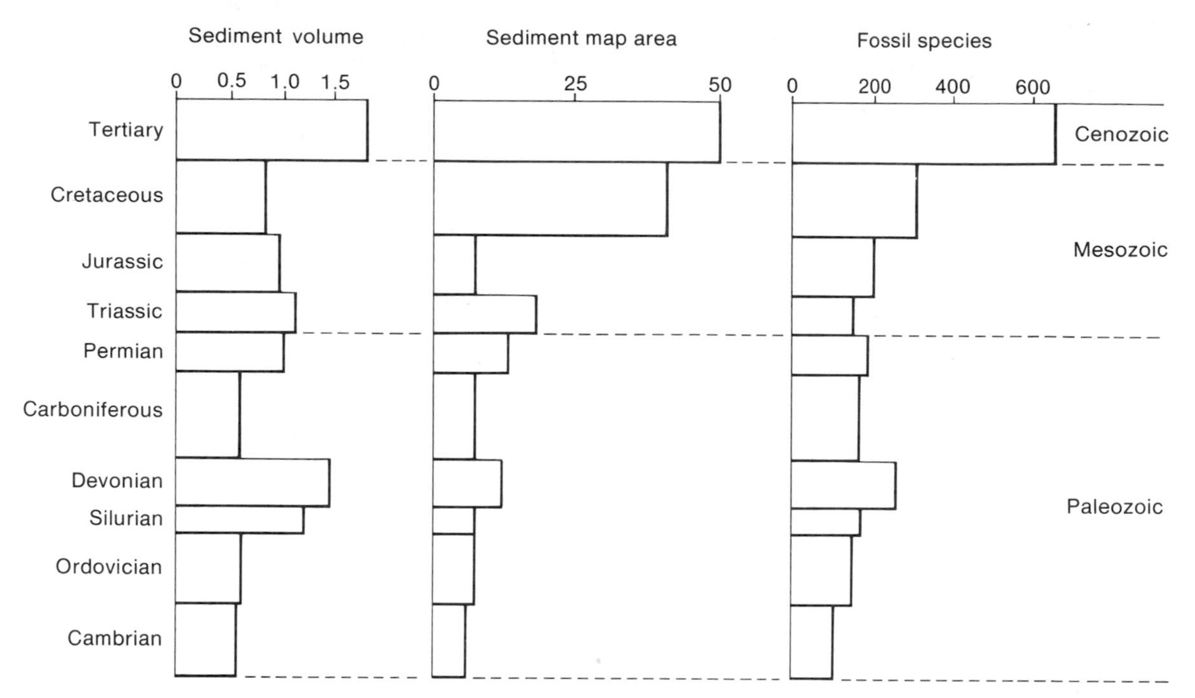

FIGURE 1-3
Sedimentary rocks and fossil diversity through geologic time. The diagram at the left shows estimates of the volumes of sedimentary rock for the periods of the Phanerozoic, adjusted for the durations of those periods (data from Gregor, 1970). The middle diagram shows comparable estimates for the areas of geologic maps occupied by sedimentary rocks (data from Blatt and Jones, 1975). The diagram at the right is an estimate of the numbers of invertebrate species per million years that have been discovered and named since 1900 (Raup, 1976a). Pleistocene and Recent data are ignored in all three diagrams. All data were compiled from worldwide sampling. Time scale is adjusted to match that of Figure 1-4.

geologic time. Sediment survival is commonly expressed as the volume of sedimentary rocks preserved and recognized per million years of geologic time. As can be seen in Figure 1-3, volume increases in an irregular fashion toward the present. This does not mean that rates of sedimentary accumulation have increased, but rather that the likelihood of sediments being preserved and observed by geologists is greater as the present is approached. Figure 1-3 also shows another way of looking at sediment survival (middle diagram). Here, what has been plotted is the land area covered by sedimentary rocks of various ages. The data were gathered on a worldwide basis from geologic maps. The basic pattern is the same as that for sediment volume except that the increase toward the Recent is more striking. This stems from the fact that large areas of the world are now covered by a relatively thin veneer of young rocks (Cretaceous and Tertiary coastal plain sequences, for example).

Figure 1-3 also shows trends in the number of fossil species known from the fossil record. The diagram at the right is based on counts of published records of about 70,000 invertebrate species—a sample which represents a large fraction of the total species that have been found. The pattern of change in number of species is remarkably like the patterns for sediment volume and area. In view of our analysis of sampling, it is certainly plausible that some or all of the increase in the number of species through time is the result not of true biologic change but rather of an increase in the quality of the sample (the number of individuals preserved, found, and classified). If the number of phyla or classes were plotted in the same fashion, the correspondence with sediment survival would not be close. For example, after an initial burst in the beginning of the Cambrian, the number of phyla in the fossil record remains nearly constant. This probably means that the number of specimens per phylum was great enough, even in the early parts of the record, to have yielded all or nearly all groups that could reasonably be expected to be preserved.

It is possible, of course, that the similarity of the curves in Figure 1-3 is just a coincidence or that it is due to other interrelationships between quantities of sedimentary rocks and fossils. For example, it is known that areas actively receiving marine sediment were larger at some times in the past than at others. At times when the areas were extensive, there would have been a larger habitable area for marine organisms. It is conceivable, therefore, that species diversity was actually greater in the Devonian than in the Carboniferous, for example, because of greater habitable area and that this larger area was also responsible for increased volume and area of sediment accumulation. This and other possibilities are still open questions. Some of them will be explored more fully in later chapters.

Kinds of Organisms

Several sources contain tabulations of the taxonomic distribution of species. Among them are Easton (1960), Grant (1963), Van Valen (1973b), and Raup (1976a). The numbers of species in various biologic groups are estimates based on surveys of the taxonomic literature and, as such, are subject to considerable uncertainty. Clearly, however, the numbers of species are not randomly distributed among the various phyla and classes. It is usually true that a given phylum or class contains many more living species than fossil species, but certain groups are more heavily represented in the fossil record than among living organisms. For example, there are about 10,000 fossil cephalopod species known, but only 400 living cephalopod species are known.

A basic question concerning all tabulations of species abundance is: To what extent is the fossil record biased by preferential preservation? Only about 8,000 fossil insect species are known but where we do find a well-preserved insect fauna, it is diverse, containing insects that show a wide variety of structural adaptations to a wide variety of environments. In an authoritative review of the insect fossil record, Carpenter (1953) made the following general comments:

> From a survey of the Carboniferous fauna it is apparent that the insects had acquired surprising diversity. . . . I am convinced that we have not yet begun to appreciate the extent of the Upper Carboniferous insect fauna. . . . If the same number of living species were collected at a few isolated localities over the world, we could not expect to obtain from them a good idea of the complexity of the world fauna as it exists today. It is not beyond the realm of possibility, therefore, that the extinct orders of Carboniferous insects were in their time comparable in extent to the major orders now living.

We may conclude, therefore, that the sparse fossil record of insects is due primarily to the unlikeliness of their preservation. The virtual nonexistence of a Cretaceous insect record in all probability stems from a lack—by pure chance—of insect environments suitable for preservation during the Cretaceous.

As noted earlier, known fossil species actually outnumber living species in a few biologic groups. This is most striking when we consider cephalopods and crinoids but is also true for brachiopods, bivalves, and echinoids. Clearly, there has been an evolutionary *decrease* in diversity of cephalopods in the course of time. One of the most common fossils in some parts of the record (particularly in the Mesozoic of Western Europe) is the shelled cephalopod. Throughout much of the Mesozoic, ammonoid and nautiloid cephalopods constitute a large part of the fossil record in terms of number of species, morphologic types, and numbers of individuals. Yet comparable cephalopods are represented in modern seas by only one genus containing four species. Without question the Mesozoic was truly an "age of cephalopods."

Similarly, marine environments of the Paleozoic era supported thousands of species of crinoids and brachiopods. The present-day sparsity of these groups can be explained only by evolutionary decline (extinction).

The bivalves and the echinoids show a somewhat different pattern. The fossil record of both of these groups exhibits a progressive *increase* in number of fossil species through time and suggests that the number of species living today should exceed the number living *at any single point in time* in the geologic past. The fact that the *total* number of fossil species for these groups exceeds the number of living species simply reflects the fact that the fossil record is a composite of a long period of time.

Another important point concerning the groups in which the number of fossil species exceeds the number of living species is that these are groups that, by the nature of their skeletons, have a relatively high probability of being preserved. They are also groups that contain many species that lived in environments that were relatively favorable for fossil preservation.

Figure 1-4 shows the observed diversity through geologic time for selected animal groups. We have already discussed the decline in brachiopod diversity from the mid-Paleozoic onward. Data for diversity of echinoderms show a peak during the Paleozoic, a pronounced reduction in the Triassic, and a steady increase between Jurassic and Recent. The bimodal pattern reflects the composite nature of the sample: the Paleozoic peak stems from pelmatozoan echinoderms (crinoids, in particular); the post-Triassic peak documents the evolutionary radiation of eleutherozoan echinoderms (e.g., echinoids, starfish).

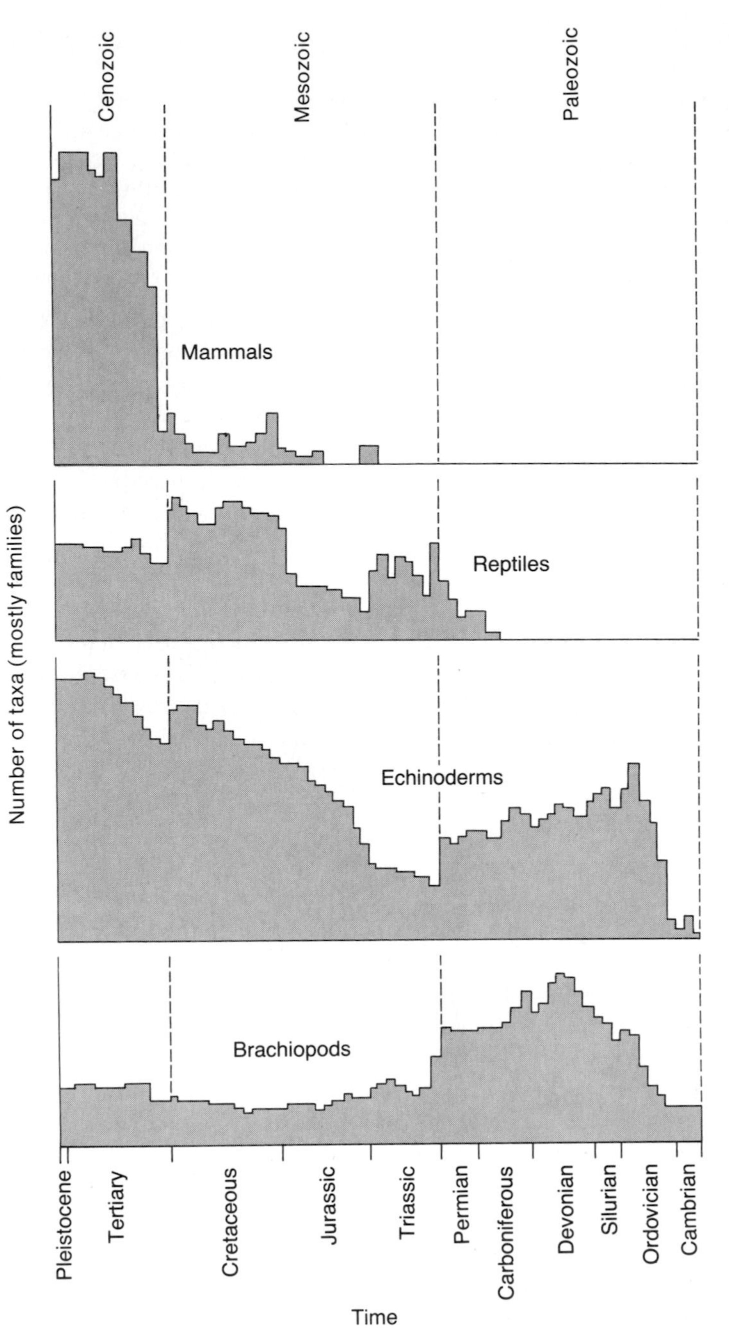

FIGURE 1-4
Diversity throughout time for four animal groups. The time scale has been adjusted in order to show data within each period clearly. (Data from Harland et al., 1967.)

The reptile diversity pattern in Figure 1-4 shows a high in the late Paleozoic and another in the Cretaceous. The abrupt drop at the end of the Cretaceous corresponds to the extinction of the dinosaurs and complements the dramatic rise in mammal diversity following the Cretaceous.

For a given biologic group, a fossil record that is meager may mean either that preservation of the organisms was poor or that they expressed little evolutionary diversity. A rich fossil record may result from better-than-average preservability, decrease in evolutionary diversity through time, or the simple fact that the fossil record is the composite of a long period of evolutionary turnover. One of the most important areas of research open to the paleontologist is the further evaluation of the fossil record as a sample of former life.

Numbers of Individuals

Our interpretation of the fossil record may be assisted by our considering living plants and animals as potential fossils. As we do this we should take into account numbers of individuals (abundance) as well as numbers and kinds of species (diversity). The number of individuals per species ranges from a few hundred to many millions or even billions. Each is a potential fossil.

From the biologist we get an occasional glimpse of the magnitude of the numbers involved. Figure 1-5, for example, shows the larger invertebrate organisms typically found in one-quarter of a square meter of sea bottom off the Kii Peninsula, Japan, at

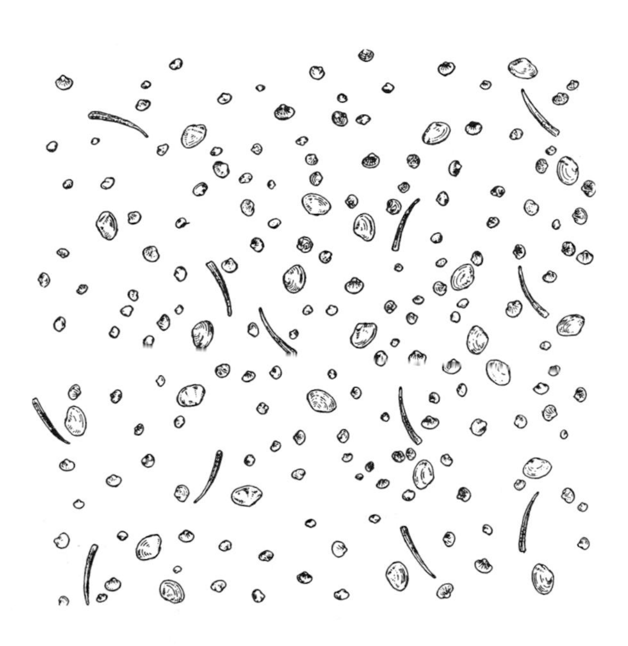

FIGURE 1-5
Animals obtained by dredging ¼ square meter of level sea bottom off the Kii Peninsula, Japan. Species are the bivalves *Macoma incongrua* (larger) and *Cardium hungerfordi* (smaller) and the scaphopod *Dentalium octangulatum* (saber-tooth shapes). (From Thorson, 1957.)

depths from 10 to 60 fathoms. This assemblage represents the "standing crop" of individuals of three species: two bivalves and one scaphopod. There are about 25 individuals of the larger bivalve (*Macoma incongrua*), about 160 of the smaller bivalve (*Cardium hungerfordi*), and 12 of the scaphopod (*Dentalium octangulatum*). All this in one-quarter of a square meter! The average age of the specimens is approximately two years. This rate of production would yield 1,000 potential fossils in ten years, or one hundred million in a million years. If these calculations were extended to include a larger geographic area and longer periods, the number of specimens—and the tonnage of potential fossils—would be truly staggering. In just the tiny area of sea bottom that we have been discussing would probably be produced in a million years more individuals than the total number of fossil specimens (of all species) that have ever been studied.

The sample calculation just carried out is, of course, subject to many errors and many tacit assumptions. The density of life varies greatly from place to place and from time to time. Many sites have higher densities than the sample area we used; many sites have lower. The calculation does give us some impression, however, that the fossils we have to work with represent an almost infinitesimal fraction of the total life of the geologic past. If the fossil record were a random sample of the plants and animals that have lived, there would be less difficulty. But it clearly is not random. As one of the biases in the fossil record, we may note that those species represented by many individuals are more likely to be preserved and found (other things being equal) than those species having fewer individuals.

Preservability

To be a part of the fossil record, all or part of an organism or some trace of its activity must be preserved in the rock. Tracks and trails of animals often form part of the fossil record and may give us information not given by the preserved body or skeleton. If an organism or some trace of its activity is preserved, the wide variety of destructive processes that operate on dead organisms and their surroundings have been, at least partly, unsuccessful.

BIOLOGIC DESTRUCTION

Biologic agents of destruction are present in nearly all environments. Predators and scavengers are ubiquitous in the biologic world. Some are larger than the organisms they feed on; some are much smaller. Hardly a biologic structure fails to attract scavengers or other biologic agents of destruction. We tend to think, for example, that the shell of an oyster is almost a fossil as soon as it is formed by the animal. The shell is quite sturdy and is made up largely of calcium carbonate. The structure of the oyster

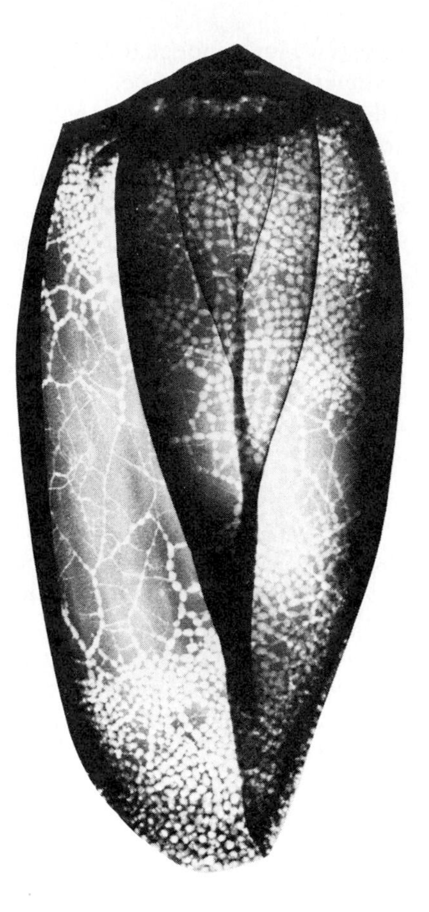

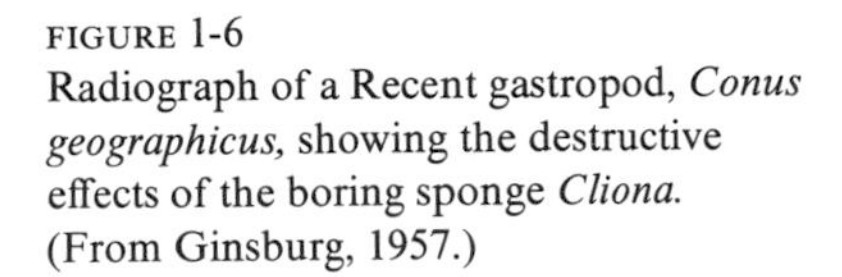

FIGURE 1-6
Radiograph of a Recent gastropod, *Conus geographicus,* showing the destructive effects of the boring sponge *Cliona.* (From Ginsburg, 1957.)

shell is not continuous or solid, however, but rather is constructed of tiny needles or lamellae of calcite held together by a network of organic tissue commonly referred to as the organic matrix. The strength of the shell is thus partly dependent on the integrity of the organic matrix.

As soon as an oyster or other mollusc dies, its shell is subject to deterioration resulting from attack by a great variety of boring organisms, including worms, sponges, other molluscs, and algae (see Figure 1-6). Most sea bottoms on which living shelled organisms are abundant have surprisingly few intact, empty shells.

If an organism is buried by sedimentation shortly after death, it is partially insulated from destructive biologic processes. The importance of this insulating effect is often exaggerated, however, for the unconsolidated sediment immediately below the sediment-water interface in a normal aquatic environment is anything but biologically inert. In fact, much of the bacterial decay of biologic tissue is concentrated in the upper few inches of the sediment. A shell may survive long enough to be buried, only to be destroyed or fragmented beneath the sediment-water interface.

Although biologic destruction of potential fossils is generally recognized as an important factor in limiting the fossil record, our knowledge of the process is woefully small. Little research has been done on the process, particularly on bacterial activity and effects of bacteria on the chemistry of aqueous solutions in sediments. Yet understanding biologic destruction is vitally important if we are to understand fully the preservation process and its effect on the fossil record.

MECHANICAL DESTRUCTION

Several important studies have been made on mechanical breakage and abrasion of skeletal material of potential fossils. As has been known for a long time, organisms whose early post-mortem history takes place in a high-energy environment may be abraded beyond recognition or completely destroyed by the action of wind, waves, and currents. Also, some kinds of skeletons are known to be more susceptible to mechanical destruction than others, which of course contributes to bias in the fossil record.

One of the simplest and most meaningful studies of mechanical destruction was carried out by Chave (1964). Particle-against-particle abrasion was tested by placing shells and other skeletal parts of various marine invertebrates with chert pebbles in tumbling barrels. The time required for various degrees of destruction by abrasion was carefully noted. Some of the results are shown in Figure 1-7. In this illustration, time is shown on the horizontal axis and the percentage of the sample larger than 4 millimeters in diameter is indicated on the vertical axis. This particular experiment was performed on skeletons of gastropods, corals, echinoids, and bryozoans. The differences in durability are striking. After more than a hundred hours of tumbling, more than 60 percent of the material of one gastropod species still remained as particles larger than 4 millimeters and, therefore, as potentially recognizable fragments. By contrast, all of the bryozoans and calcareous algae were gone after one hour of tumbling. In view of this, it is not surprising that bryozoans make up a relatively small fraction of the fossil record and gastropods a relatively large fraction.

Figure 1-8 shows the results of another tumbling barrel experiment carried out by Chave. Six different kinds of bivalve shells were used, and the variation in durability among them was almost as great as that among the groups of organisms in the first experiment. The relative durability of most of these shells can be interpreted in terms of such differences as their size, thickness, and internal fabric.

Studies such as Chave's can be used to evaluate preservability bias in the fossil record. They also are of value in providing material for comparison with actual fossils.

A further example of the way in which mechanical destruction may bias the fossil record is shown in Figure 1-9, which is also drawn from Chave's experimental work. A

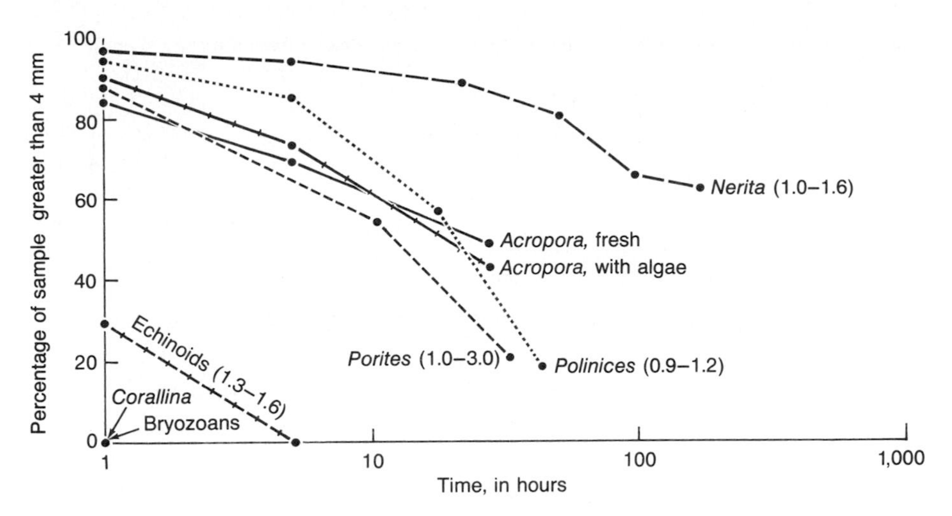

FIGURE 1-7
Experimental abrasion of skeletal material by tumbling with chert pebbles. The gastropod *Nerita* is most durable; the corals *Acropora* and *Porites* and the gastropod *Polinices* are intermediate; the calcareous alga *Corallina,* the bryozoans, and the echinoids are least durable under these conditions. The numbers in parentheses after the names refer to the original size, in inches, of the specimens. (From Chave, 1964.)

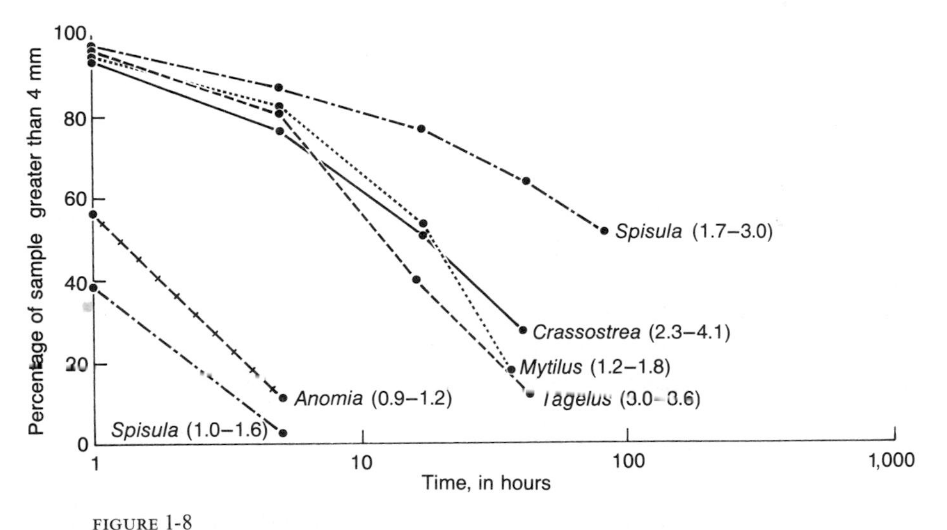

FIGURE 1-8
Experimental abrasion of skeletal material by tumbling with chert pebbles. Species are all marine bivalves. (From Chave, 1964.)

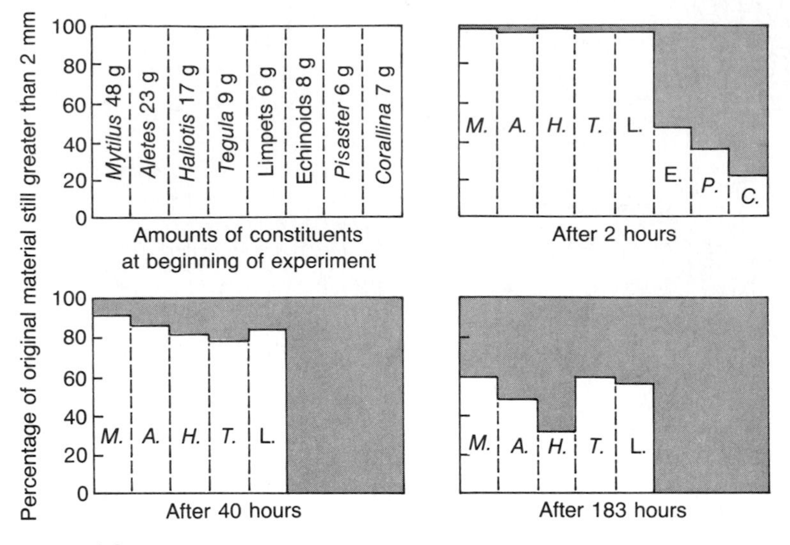

FIGURE 1-9
Experimental abrasion of skeletal material. Starting sample (upper left) contained fresh specimens of the bivalve *Mytilus,* the gastropods *Aletes, Haliotis,* and *Tegula,* various species of limpets and echinoids, the starfish *Pisaster,* and the calcareous alga *Corallina.* The series of diagrams shows selective destruction of the assemblage by tumbling. (From Chave, 1964.)

representative assemblage of shells from Corona del Mar, California, was placed with sand in a tumbling barrel. The illustration shows the effect on the composition of the assemblage; the bryozoans and calcareous algae were completely removed from the assemblage, in the sense that particle size was reduced to less than 2 millimeters, and the relative abundance of the remaining forms was considerably altered.

Figure 1-10 shows a photograph of a coarse sandstone slab containing an assemblage of molluscan fossils that Chave suggests may have survived mechanical attack similar to that to which his tumbled Corona del Mar sample was subjected. This emphasizes the fact that when we look at a fossil assemblage in a rock we must attempt to look beyond the assemblage, even though the general level of preservation may seem to be good, in order to deduce what organisms may have lived at the site but have not been preserved.

FIGURE 1-10
Slab of fossiliferous sandstone from the Pliocene of California. Condition and composition of the assemblage are comparable to the results of the tumbling experiments illustrated in Figure 1-9. The assemblage is dominated by worn specimens of gastropods and bivalves. (From Woodring et al., 1940.)

CHEMICAL DESTRUCTION

The skeleton of an animal may withstand biologic and mechanical destructive processes yet not become part of the permanent fossil record. Simple chemical solution is one of the most important reasons why we do not have more identifiable fossils than we do. Chemical solution can take place at any time after the death of the animal—even after its skeleton has been a fossil for a long time. Solution may occur not only on the sea bottom but also in soft sediments. Solution of an organism's remains by ground water millions of years after its burial is common.

The ability of a fossil to survive solution depends on its chemical composition and on the composition and physical characteristics of the waters to which it is exposed. The chemistry of local waters may, in turn, be influenced strongly by the biologic environment, particularly by its bacterial activity.

If a shell is disintegrated by chemical solution after being imbedded in rock, the chances are good that the cavity may remain as a fossil. For many paleontologic purposes such a fossil, which gives evidence about morphology, may be adequate.

Biologic Structures Most Likely to be Preserved

Skeletons containing a high percentage of mineral matter are most readily preserved; soft tissue not intimately connected with skeletal parts is least likely to be preserved. This means that the fossil record contains a biased selection not only of types of organisms but also of parts of organisms. Much of our paleontologic knowledge about mammals, for example, is based on teeth alone, the teeth being much more durable than other parts of the skeleton.

A striking effect of the difference in preservability of so-called hard parts and soft parts is that only in rare instances do we have any knowledge of the color of extinct organisms. Pigments, except in rare situations, are simply not preserved with fossils.

Evolutionary change may greatly alter the fossil record of a group. If, for example, an evolving animal group developed resistant skeletal elements, its preservability would suddenly have increased, and it would have begun to make a greater impact on the fossil record. Arthropods probably took such an evolutionary step. Nearly all arthropods have a skeleton. The chemical composition of their skeletons varies widely, however, as does their strength and resistance to decay, solution, and abrasion. The skeleton of a crab, for example, is less fully calcified than that of a trilobite. The trilobite skeleton when it was part of a living organism was made up of a denser, firmer structure containing more pure calcium carbonate and proportionately less organic material than that of the crab. The trilobite was thus more likely to be fossilized, and, indeed, the fossil record of trilobites is much more complete than that of crabs. Many paleontologists have suggested that the rather abrupt beginning of the fossil record of trilobites and those of other organisms in the Cambrian stems not from the sudden evolution of many forms but rather from the sudden evolution of calcification.

Environment and Preservability

A few sedimentary formations display a quantity and quality of preservation orders of magnitude better than the average fossiliferous rock. One of the most famous of these is the Solnhofen Limestone (Jurassic) of southern Bavaria. In the Solnhofen there are, for example, about 180 species of insects in an excellent state of preservation. Also, a variety of marine invertebrates are found with features rarely seen elsewhere in the fossil record. Tracks and trails abound, as well as impact marks showing where (and in what orientation) animals such as cephalopods and crustaceans landed on the sedimentary surface. The Solnhofen is particularly well known for a few specimens of the first known bird, *Archaeopteryx,* in which the feathers are clearly seen as impressions on the rock surface (Figure 1-11).

Other dramatic examples of preservation include the Hunsrück Shale of the Devonian of Germany (see Figure 1-12), the Burgess Shale (Cambrian of British Columbia), the La Brea tar deposits (Pleistocene of southern California), the Mazon Creek Shale (Pennsylvanian of Illinois), and the Baltic amber (Oligocene of Germany). These deposits and a few others like them are called ***Lagerstätten*** (German for "mother lode"). A significant fraction of our knowledge of past life comes from Lagerstätten and many first or only occurrences of rare taxa are known from them (*Archaeopteryx,* for example).

The examples just described are all different—in time, geologic setting, and environment. So there is no simple explanation for their occurrence. Obviously, an environment in which burial is rapid is conducive to fossilization, but other factors must be involved. The Solnhofen Limestone is an extremely fine-grained, well-bedded limestone showing none of the features that a geologist normally associates with rapid sedimentation. The origin of the limestone is not known for certain, but in all probability it was formed by very slow accumulation of fine particles. It has been suggested that minor slumping or movement of sediment by density currents may have buried the organisms quickly without leaving clear evidence of the burial event, but this has not been definitely proven. A striking feature of the Solnhofen fossil assemblage is that very few of the organisms are bottom-dwelling forms. They are mostly swimming or flying organisms, or organisms that show evidence of having drifted or floated to the site of deposition. In other words, there is little evidence that animals actually lived on the bottom of the Solnhofen sea. This goes a long way toward explaining why the preservation is so good: if the site of deposition could not support life, the chances of destruction by scavengers and by decay were greatly reduced.

The Solnhofen story suggests a general explanation of Lagerstatten: perhaps the probability of preservation is enhanced if potential fossils are transported away from their normal habitat to a place or situation where biological activity is low or completely absent. The spectacular preservation of vertebrates in the tar pits of southern California is another example. A tar pit is anything but the normal habitat of animals and those that are preserved in the pits apparently stumbled into them by accident. The preservation of insects in amber is similar in many respects to that of vertebrates in the tar pits.

FIGURE 1-11
Fossils from the Solnhofen Limestone (Jurassic, Bavaria.) A: *Mesolimulus walchi* ($\times \frac{1}{5}$). Its final walking tracks are preserved (arrow *t*); the curved grooves below were made by the animal sliding upside-down over the sediment before reaching its final resting place. B: *Archaeopteryx lithographica* ($\times \frac{1}{2}$). Marks made by its feathers are preserved. (A from Leich, 1965; B courtesy of the U.S. National Museum.)

B

FIGURE 1-12
Radiograph of a trilobite from the Hünsruck Shale (Devonian of Germany). The specimen is unusual in showing an almost complete array of appendages. (X-ray taken by W. Stürmer, Erlangen, West Germany.)

One common denominator in examples of excellent fossil preservation is thus an environment that was biologically inert. In such an environment, the remains of animals and plants are protected from the normal biologic agents of destruction as well as from many physical and chemical agents. The implication of this is that it may be somewhat more likely to find fossils where they did not live than where they did.

In addition to the foregoing, an organism's habitat in life influences the probability of its preservation. Habitat greatly influences preservability of an organism. A mountain sheep is less likely to be preserved than a hippopotamus because the sheep normally lives farther from an area of sedimentation. Likewise, within the marine environment, large differences exist. One of the most important from a paleontologic viewpoint is the distinction between those animals that live *in* the sediment on the sea bottom and those that live *on* the surface of the sediment or swim or float in the overlying water. For the organism living in the sediment, sedimentation rates may be less important because the organism is wholly buried in the sediment even when it is alive and is thus somewhat protected against scavengers and to a large extent against mechanical abrasion and breakage.

Post-mortem Transport

The remains of different organisms are transported in many different ways; furthermore, similar organisms are transported in different ways in different environments. One of the simplest forms of transport is illustrated by free-swimming organisms that upon dying sink to the sea bottom—an entirely different environment, which may in fact be biologically inert. This possibility has led to one of the most widely accepted interpretations of the excellent preservation of fossils in the Solnhofen.

Some living pelagic organisms, for example most crustaceans, remain suspended in the water by their own swimming activity; others are suspended in the water or float at the surface because of their buoyancy. Usually when organisms of the first sort die, their remains settle to the bottom. With the second sort of organism, there may be considerable transport of the buoyant cadaver by currents and wind. The coiled cephalopod furnishes a prime example. In life, the gas-filled chambers of the shell provide buoyancy. Upon death this buoyancy may actually increase because decay of the soft body of the organism makes the shell lighter. The shell may float in the surface water for days, weeks, or even months before becoming so damaged that it sinks to the bottom. Post-mortem transport is probably the rule rather than the exception for coiled cephalopods and similar organisms.

Most post-mortem transport of bottom-dwelling marine organisms depends on currents. In relatively shallow water in which currents are strong, the skeleton of an organism may be carried a considerable distance to its final site of burial. The possibility also exists that bottom-dwelling organisms may float after death because the decay of the soft tissue produces gas that may become trapped inside the skeleton. Unfortunately it is not known how common this phenomenon is. Certainly fish and other soft-bodied animals do become bloated during decay and rise to the surface. Occurrences of the same phenomenon have been recorded for rather heavy-shelled invertebrates but not enough research has been done to assess this possibility as a major factor in fossil preservation.

Pollen and spores of terrestrial plants, which make up an important part of the fossil record, can be carried long distances by wind. Also, rivers and streams carry a variety of plant and animal fragments. Very commonly, terrestrial organisms, particularly plants, are carried to the ocean and float for hundreds of miles before sinking to the bottom. An interesting result of this is that several sea urchin species living in very deep water off the coast of New Guinea depend, for food, upon terrestrial plant material brought into the ocean by the rivers of the New Guinea coast.

It is exceedingly difficult to say just how important a role is played by post-mortem transport. Physical deterioration of fossils or their discovery where the organisms are known not to have lived gives clear evidence of post-mortem transport. On the other hand some fossils show no evidence of having been disturbed after death.

The significance of post-mortem transport as a biasing factor in the fossil record depends entirely on the use to which the fossil record is put. If the concern is with tracing the evolutionary development of a biologic group on a broad scale, which is done by studying comparative morphology and distribution in time, movement of a few miles or even a few hundreds of miles is negligible. Similarly, if fossil evidence is to be used to reconstruct regional climate, movement of a few miles or tens of miles may be insignificant. On the other hand, movement of a short distance may be vitally important when the problem is to reconstruct local environmental conditions.

Conclusion

We have seen that preservation of an organism's remains or of evidence of its activity is a rare event. The more we investigate the difficulties of fossil preservation, the more surprised we become that the fossil record is as good as it is. But the number of potentially fossilizable plants and animals is so enormous that even such an unlikely event as preservation becomes a relatively common phenomenon. Although we have a generally accurate understanding of some of the basic problems and processes in fossil preservation there still are enormous areas of ignorance. For example, it has been suggested in this chapter that geologically unusual or even catastrophic conditions contribute to the preservation of fossils. But to what degree? We do not have enough information yet to answer this question.

The best way for us to proceed is never to accept a fossil assemblage at face value; it probably represents a strongly biased picture of past life. A corollary is that lack of fossils in given rocks should not be taken to indicate that animals or plants were definitely absent.

Although the fossil record is limited in many ways, it still contains an extraordinary amount of information. There are perhaps two reasons why the rather poor overall sample of past life in the fossil record is often adequate for the types of study that characterize paleontology:

1. In *large-scale* studies of rates, trends, and patterns of evolution and of evolutionary relationships within major plant and animal groups, paleontologists tend to restrict themselves to groups that have relatively good fossil records—fossil records that are adequate statistical samples.

2. In interpreting fossil faunas and floras of *local* rock units, the paleontologist attempts to deal with rocks in which a large proportion of potentially fossilizable species have been preserved.

Thus, to undertake a particular research project, an attempt is made to choose certain fossil groups or rocks units that lend themselves to a particular approach. Our main purpose in the remainder of this book will be to explore the methods by which fossil information, carefully evaluated in light of preservational biases, can be interpreted and applied to geologic and biologic problems.

Supplementary Reading

Durham, J.W. (1967) The incompleteness of our knowledge of the fossil record. *Jour. Paleont.*,**41**:559–565. (A presidential address to the Paleontological Society on the nature of the fossil record, particularly with regard to marine invertebrates.)

Harland, W.B., et al. (1967) *The Fossil Record.* London, Geological Society of London, 827 p. (The most complete compilation of data about the diversity of the fossil record thus far published.)

Newell, N.D. (1959) Adequacy of the fossil record. *Jour. Paleont.*, **33**:488–499. (A general discussion of the size of the fossil record in relation to its study by paleontologists.)

Raup, D.M. (1972) Taxonomic diversity during the Phanerozoic. *Science,* **177**:1065–1071. (An analysis of the effects of bias in preservation.)

Raup, D.M. (1976) Species diversity in the Phanerozoic: a tabulation. *Paleobiol.*, **2**:279–288.

Richardson, E.S., Jr., ed. (1971) *Extraordinary Fossils.* Proceedings of the North American Paleontological Convention, Part I. Lawrence, Kans., Allen Press, p.1153–1269. (An important collection of papers describing unusual preservational circumstances.)

Schäfer, W. (1972) *Ecology and Paleoecology of Marine Environments.* Chicago, University of Chicago Press (English translation), 568 p. (Includes the best treatment in English of the processes of fossil preservation.)

Simpson, G.G. (1952) How many species? *Evolution,* **6**:342. (Calculated estimates of the number of species that have lived on the earth.)

Simpson, G.G. (1960) The history of life. *In* Tax, S., ed. *Evolution after Darwin.* Chicago, University of Chicago Press, vol. 1, p.117–180. (A general discussion of biases in the fossil record.)

Teichert, C. (1956) How many fossil species? *Jour. Paleont.*, **30**:967–969.

Valentine, J.W. (1970) How many marine invertebrate fossil species? A new approximation. *Jour. Paleont.*, **44**:410-415.

Valentine, J.W. (1973) *Evolutionary Paleoecology of the Marine Biosphere.* Englewood Cliffs, N.J., Prentice-Hall, 511 p. (A synthesis of the ecology of the marine fossil record from an evolutionary viewpoint.)

CHAPTER 2

Describing a Single Specimen

Description of fossils is fundamental to nearly all paleontologic research. The question for a given study is what to describe and how to describe it. If the objective is to define and classify a new fossil species, the description should be as complete as possible, but at the same time be such that comparison with other fossils is facilitated. Comparison is usually based on selected attributes. For any fossil, there are literally thousands that *could* be described. How then are attributes chosen? Once they have been chosen, a practical and meaningful scheme of expressing differences in them must be developed. How is this done?

Problems of description are not limited to the definition and classification of new species. Let us suppose that a paleontologist wishes to test the evolutionary hypothesis that *size* in a group of organisms increases with time. This seems simple enough, but several problems soon arise. What is the best measure of size? Should he use total weight, total volume, maximum length or width, surface area, or some combination of these? Should the whole organism be considered, or can parts of the organism be assumed to reflect overall size? Choosing attributes is in part a biologic problem, but in part a uniquely paleontologic one. Many attributes that the biologist would measure are not available to the paleontologist, who must operate within the restrictions of fossilized material. For example, characteristics such as weight and volume are greatly affected by fossilization. The paleontologist must arrive at a selection of attributes that is both biologically sound and paleontologically reasonable.

There is no single format that can be followed to produce a perfect or all-purpose description. In this chapter we will explore some of the general problems of selection of attributes and methods of description. We will assume that we have at hand fossils that are collected, prepared, and ready to be described, bypassing the important problems of fossil collection and preparation. (There are references at the end of the chapter to several comprehensive works on collection and preparation.)

The Pictorial Description

PHOTOGRAPHIC OR MECHANICAL IMAGES

In some ways, a good photograph of a fossil is the best description. It is both objective and comprehensive. For this reason, the photograph has become an indispensable part of the formal definition of a fossil species. The photograph cannot, of course, record certain attributes like chemical composition and some minute or internal structures.

A high point in the paleontologic use of photography is illustrated by the pairs of stereophotographs that are reproduced as Figure 2-1. Various special techniques are commonly used to show structures that would otherwise be virtually invisible to the camera, as they are to the naked eye.

Photographic methods other than the conventional ones that use visible light may be used to bring out characteristics that are not ordinarily visible. Figure 2-2 shows several examples. X-radiography shows many internal structures not otherwise apparent. The stereographic pair of X-ray photographs of a brachiopod in Figure 2-2 shows the internal lophophore support structure characteristic of this group of brachiopods. The spiral lophophore is completely embedded in solid sediment that fills the shell. The X-ray photograph shows the lophophore with a clarity otherwise attainable only with elaborate sectioning techniques and reconstruction of the structure from a series of sections. This illustrates another advantage of photographic description: it is essentially nondestructive.

Figure 2-2 also shows both an example of the effectiveness of ultraviolet photography, which depends upon the fluorescence of small amounts of organic materials to outline skeletal structures not evident in visible light, and an example of photography with infrared film. Differential absorption of heat by infrared film brings out structures not otherwise visible. Electron microscopy has become important in recent years for the examination and photography of ultramicrofossils and of the fine details of structure in larger fossils. Examples are shown in Figure 2-3.

It is also possible to treat photographic information numerically—by computer. In order to do this, the information about shape or structure must be put into "machine-

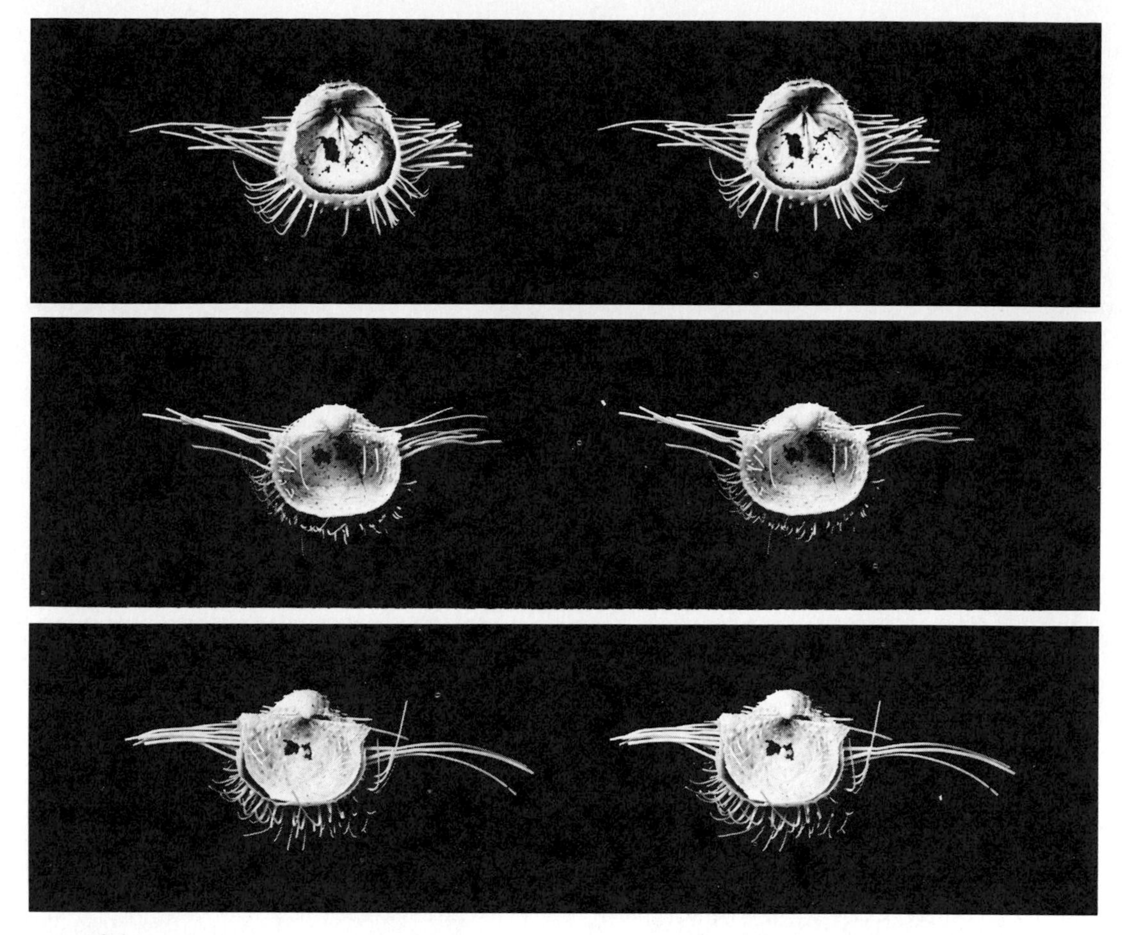

FIGURE 2-1
Stereophotographs of the brachiopod *Echinauris opuntia* from the Permian of Pakistan. Three-dimensional images can be seen when the pairs of photographs are viewed through a stereoscope. (From Grant, 1968.)

recognizable form." One of the simplest methods is called ***digitizing.*** The usual procedure is to superimpose on a photograph a conventional orthogonal coordinate system (that is, a grid in *X* and *Y*). The *X*- and *Y*-coordinate values of points on the photograph are noted and recorded on punched cards, punched paper tape, or magnetic tape that can be read into the memory of the computer. The photographic quality of the image recorded depends only on the number of points digitized.

This digitizing method assumes that we are dealing with a two-dimensional structure or a structure that can be expressed as a plane projection. For the computer

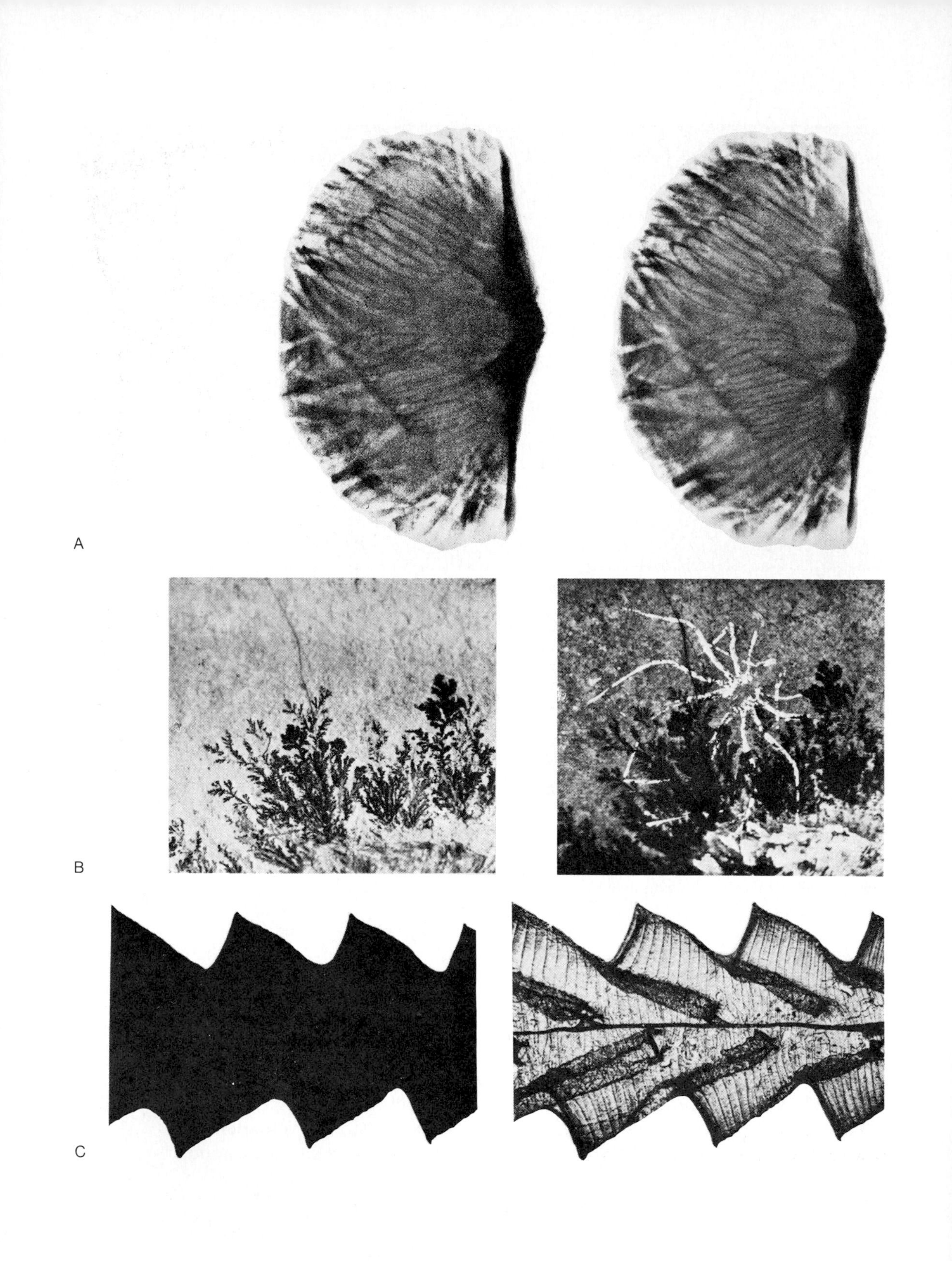
A
B
C

FIGURE 2-2
A: X-ray photographs (stereo pair) of the brachiopod *Spinocyrtia euruteines* from the Devonian of Ohio, showing the internal lophophore support structure. B: Bedding surface of Solnhofen Limestone in visible light (left) showing dendrites but no fossils and in ultraviolet light (right) showing larva of a decapod crustacean. C: Graptolite *Diplograptus gracilis* in visible light (left) and in infrared light (right). (A from Zangerl, 1965; B from Leon, 1933; C from Kraft, 1932.)

study of most fossils we need instead a method of three-dimensional digitizing. The most straightforward method is to digitize a stereographic pair of photographs. Each point on the fossil is read into the computer as two pairs of *X*-, *Y*-coordinates representing the point as observed from two slightly different viewpoints. Trigonometric treatment of the data in the computer recreates the three-dimensional image. Alternatively, for simple shapes, the digitizing can be done in three dimensions by locating points in an *X*-, *Y*-, *Z*-coordinate system on the fossil itself rather than working from photographs.

Technology has been developed to automate the job of digitizing. Commonly, a photograph or drawing is placed on a drafting surface and the operator moves a stylus over those areas of the photograph that he wishes to digitize (see Figure 2-4). When a point to be digitized is reached by the stylus, the operator pushes a button or a foot pedal to record automatically the coordinates of that point. A more fully automatic form of digitizing involves *scanning*. In this method, a light-sensitive instrument scans a photograph in a series of traverses. The variations in darkness that compose the photographic image are recorded in some machine-recognizable form. The quality of the reproduction depends on the number of traverses made across the original photograph and the sensitivity of the scanning instrument.

The reading of photographic information into a computer does not in itself solve any problems of fossil description, except that it may facilitate the communication of descriptive information. We are still faced with the problem of what to do with a photograph of a fossil: what attributes to choose for a particular paleontologic problem, and how to express them. This emphasizes a basic disadvantage of photographic description. Because it is objective, it includes too much information, the irrelevant as well as the relevant. When we look at the photograph of a fossil, we see almost all its attributes. Some are biologic in origin. Others are geologic (relating to the organism's post-mortem history). Even if we are successful in separating out a set of attributes of particular interest, the set may include more attributes than the human mind can absorb and manipulate. It is common for a student presented with photographs of two fossils to recognize immediately that they are different, but then to be unable to identify their differences and similarities. Description demands simplification: reducing the attributes observed and described to a manageable number and, in particular, eliminating those likely to be irrelevant or redundant. A certain degree of subjectivity is essential for effective fossil description. The perfect photograph so completely lacks subjectivity that other methods of description are often preferred.

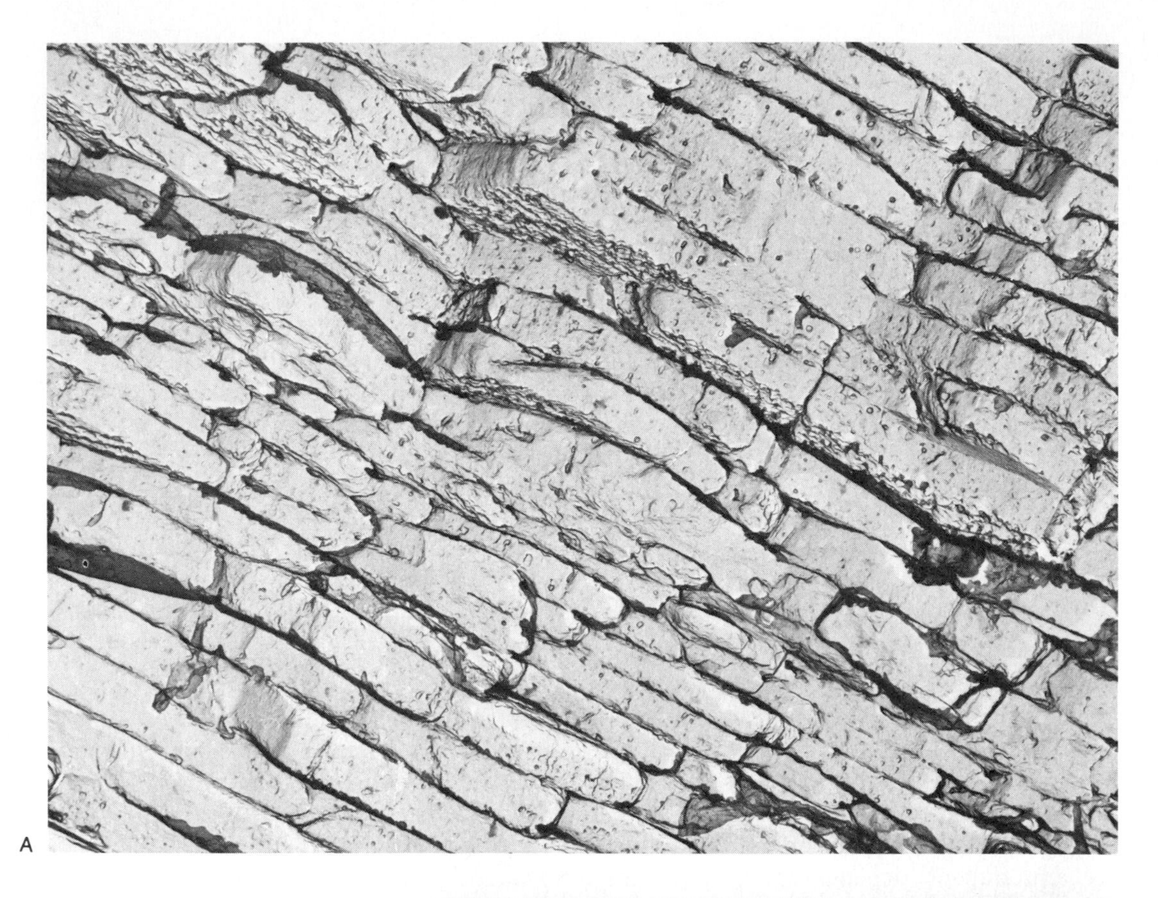
A

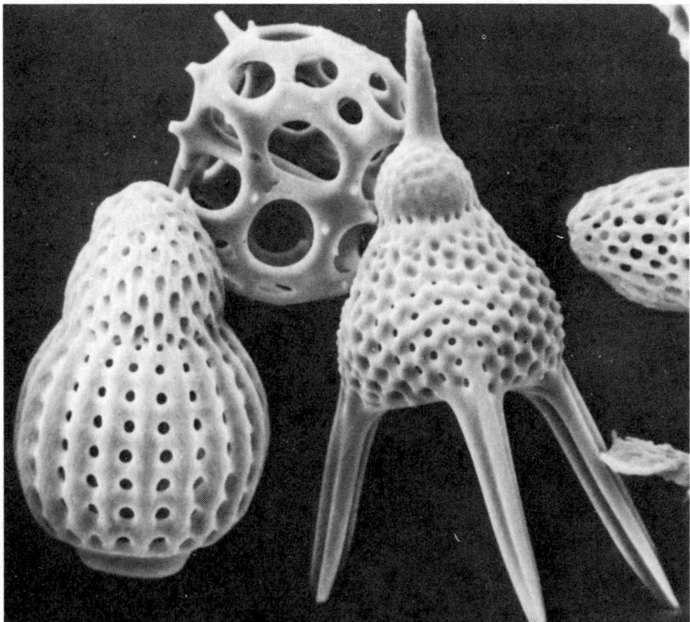
B

FIGURE 2-3

Electron microscopy of fossil and Recent invertebrate skeletons. A: Shell structure of the Jurassic bivalve *Praemytilus* enlarged 7,140 times. B: Radiolarians from the Eocene of Barbados at a magnification of 425. C and D: Surface of the test of the Recent ostracode *Carinocythereis* aff. *carinata* from the Bay of Naples, Italy at magnifications of 110 and 1,000. (A from Hudson, 1968; B–D provided by W. W. Hay and P. A. Sandberg.)

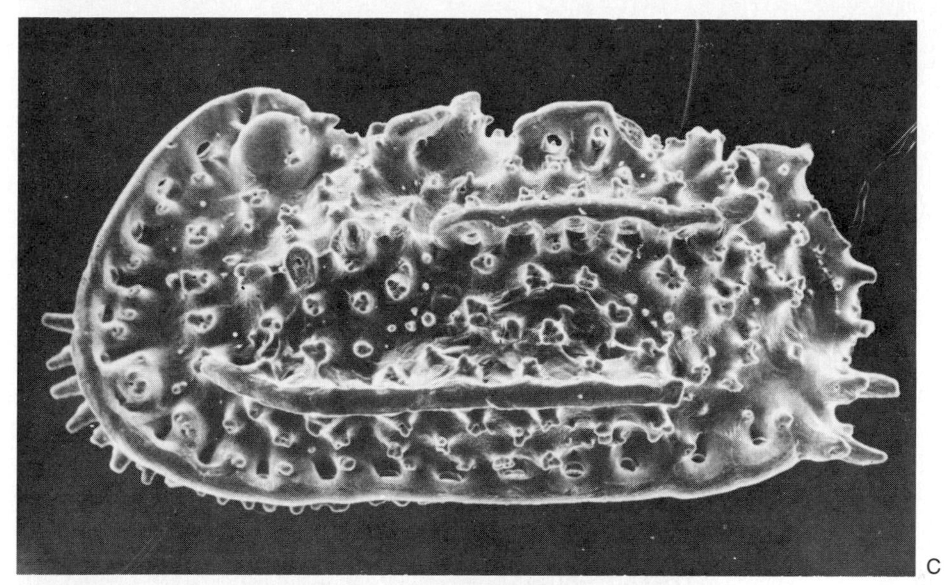

C

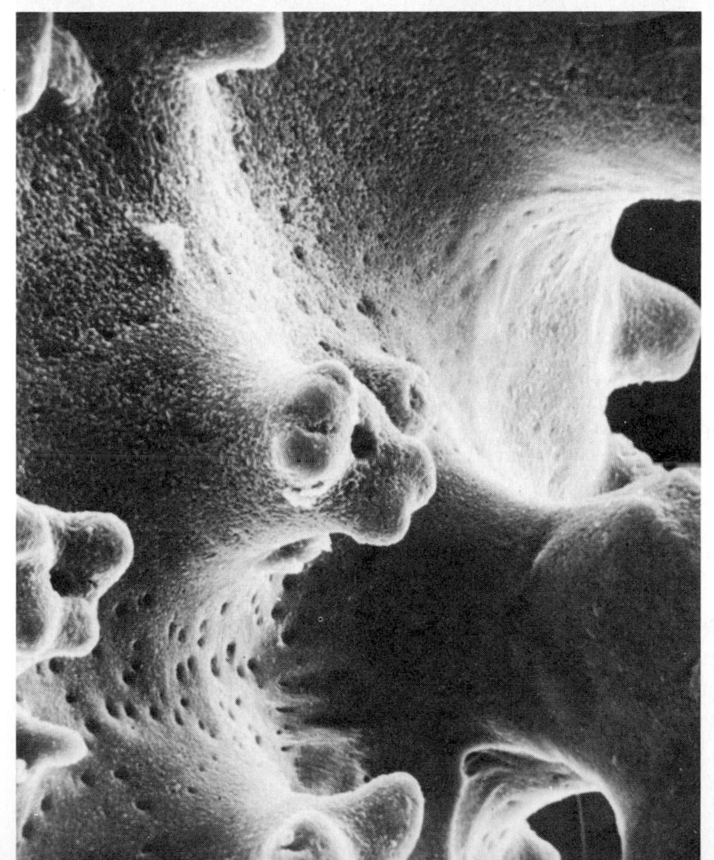

D

FIGURE 2-4
Automatic digitizing machine. A drawing of ammonite sutures is being digitized. A suture line is tracked by a stylus in the arm crossing the drawing board and *X*-, *Y*-coordinates of points on the line are recorded automatically. (Photograph provided by G. E. G. Westermann.)

THE LINE DRAWING

The most common and often the most effective step toward limiting description is to draw a sketch of the salient aspects of a fossil structure, illustrating certain attributes and ignoring others.

Some discussion is in order concerning the techniques of drawing. Sloppiness should be avoided not only because it introduces unnecessary "noise," but also because it introduces *uncontrolled* subjectivity. An accurate drawing is a rigorously executed replica. Because a drawing is usually two-dimensional, it is most appropriately made from a two-dimensional projection of the entire object. For this reason, drawings are commonly traced from photographs. As an alternative, a variety of instruments have been developed for projecting photographs or optical images onto a drafting surface. The camera lucida is the simplest such instrument.

The line drawing is more readily converted into machine-recognizable form than is the photograph because it contains less information. A scanning system need only distinguish between black and white to transcribe sketched information. By the same token, computer processing of such information is easier because less information is involved. For some paleontologic purposes a line drawing is still too complex because it contains too much information.

Descriptive Terminology

By far the most common medium for describing a structure is the word, or a combination of words. An extremely powerful and economical tool, a word is in some instances worth a thousand pictures. All fossil horses may be described as being either one-, three-, or four-toed (with respect to the forefoot). By the use of these simple words, fossil horses can be divided into descriptively valid groups. The significance attached to these words is, of course, highly subjective. By using them we *imply* that a major difference (involving many attributes) exists among groups of horses that can be expressed in terms of the number of toes on the forefoot.

The use of descriptive terminology to describe form and structure has many obvious advantages. Most important, it reduces a great deal of information to a single word or relatively few words. If carefully chosen, the words are self-explanatory or relatively easy to learn. The terminology soon becomes a natural part of the vocabulary of the experienced biologist or paleontologist. Descriptive terms are easily codified and thus can be put into machine-recognizable form, and they are readily manipulated by a computer.

As an example of the effective use of descriptive terminology, the formal description (from Sohl, 1960) of a gastropod species follows. Photographs of a specimen of the species are shown in Figure 2-5.

> Shell small, trochiform, phaneromphalous with nacreous inner shell layer; holotype with about 7¼ rapidly expanding whorls. Protoconch smooth on early whorl with coarse axial costae appearing at slightly more than one whorl, followed almost immediately by fine spiral lirae; suture impressed. Whorl sides slope less steeply than general slope of spire, giving an outline interrupted by overhang of periphery of preceding whorls;

FIGURE 2-5
Cretaceous gastropod *Calliomphalus conanti.* See text for the formal description of this species. (From Sohl, 1960.)

periphery subround to subangular; whorl side slopes steeply below periphery to broadly rounded base. Sculpture of axial and spiral elements same size; 8 spiral lirae on upper slope possess subdued tubercles where overridden by somewhat coarser and closer spaced axial cords; base covered by about 10 unequally spaced spirals with poorly defined tubercles and numerous axial lirae. Umbilicus narrow, bordered by a margin bearing low nodes. Aperture incompletely known, subcircular, slightly wider than high and reflexed slightly at junction of inner lip and umbilical margin.

To a person unfamiliar with gastropod morphology and its descriptive terminology, this description may be nearly unintelligible. For the person acquainted with the subject, however, the description should provide a convincing sketch of a group of specimens.

Any system of descriptive terminology, though, can lead to difficulty. Note the word "small" in the first line of the description. This implies a size comparison with other organisms—but what other organisms? Certainly all gastropods are small when compared with elephants but large whan compared with foraminiferans or plant spores. In this description, the word "small" means small in comparison with other gastropods of this same general type, or at most with all gastropods. By convention, terms such as small, large, wide, and narrow imply comparison only with closely related organisms. The dimension that conventionally denotes the general size of gastropods is the maximum height of the shell. The use of the descriptive term "small" is, therefore, a shorthand form of expressing a fairly rigorous description. Once a worker has learned the basic vocabulary, he finds this mode of description is remarkably efficient.

Notice the term "trochiform" in the first line of the description. It refers to a general shape common among gastropods, which is illustrated together with other common shapes in Figure 2-6. The origin of such terms, some of which are several centuries old, is quite varied. Some terms are derived from descriptive words: "turbinate" comes from the Latin *turbinatus,* meaning top-shaped. Others are from geometry: conical, biconical, obconical, and so on. Still others are derived from the names of particular organisms (genera or species), which display the form well or are common enough to be familiar.

A common problem encountered in using descriptive terminology is that of deciding where one category leaves off and another begins. What do we do, for example, if we are faced with a gastropod shell that is somewhere between naticiform and turbinate? One way of solving such problems is simply to add more terms for the intermediate shapes. Thus we have such terms in gastropod description as high turbinate and low turbinate. Inevitably—however many categories we may have—some fossils will be found that are best described as being somewhere between two or more of the categories. A system should have few enough categories to be efficient, yet enough that the whole body of material may be subdivided meaningfully.

One reason that intermediate forms do not present more problems than they do, is that form in the animal and plant worlds is not randomly distributed. Certain shapes are much more common than others. If a system of descriptive terminology accurately

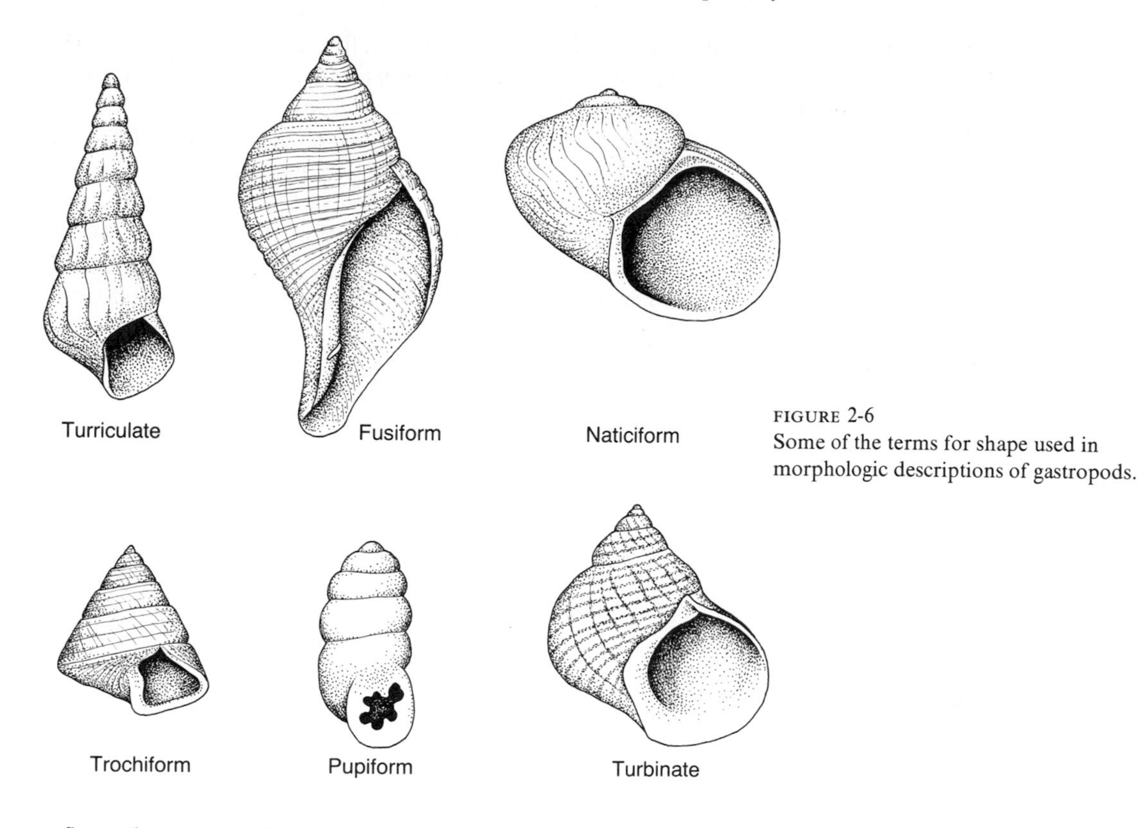

FIGURE 2-6
Some of the terms for shape used in morphologic descriptions of gastropods.

reflects the natural clusters of predominant forms, then relatively few specimens that are between two categories will be found. Thus, the establishment of a system of descriptive terminology may be an important scientific contribution in itself, representing a fundamental interpretation of the natural world.

The risk, of course, is that errors or misinterpretations will be promulgated. Many completely inappropriate systems of descriptive terminology have become fixed in paleontologic literature. Because they do not represent natural groupings, they have obstructed and delayed development of meaningful interpretations. The risk must be looked upon as one we knowingly take when we move beyond the photograph as a means of description.

Description by Measurement

If precise measurements of fossil form are used rather than descriptive terminology, some problems are avoided but others arise. As long as enough precision is used in measurement, the problem of intermediate forms does not exist. Also, measurements are machine-recognizable data and can be manipulated directly and efficiently.

Manipulation of quantitative data enjoys a good reputation. It is generally thought that the quantitative approach in science is the only truly objective approach. In actual fact, measurement is often the most subjective of the descriptive methods.

Consider the measurements indicated in Figure 2-7, in which length and height of a bivalve shell are defined by linear dimensions. Length is defined as the maximum linear dimension in an antero-posterior direction. Height is defined as the maximum dimension perpendicular to the length. These two dimensions yield quite a bit of information. They not only tell us something about the overall size of the shell but also tell us something about shape. The length-to-height ratio is a measure of shape. As the shell beomes more equidimensional, this ratio approaches one. As the shell becomes more elongate, the ratio increases.

The three bivalve shells in Figure 2-7 have nearly identical length-to-height ratios, but even a child could see the great difference in shape displayed by the specimens. Here, we see one of the greatest drawbacks of a system of description based solely on

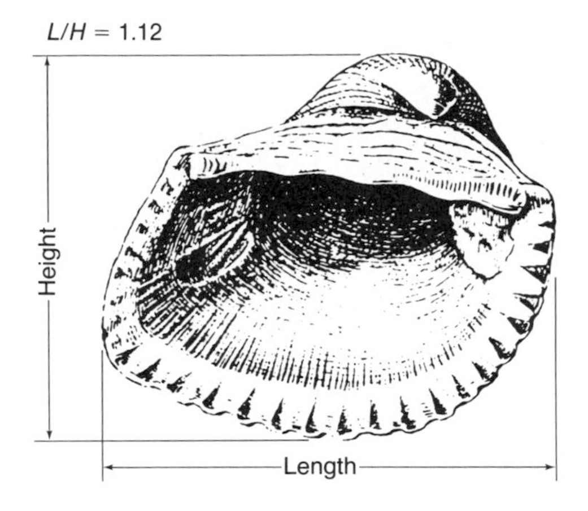

FIGURE 2-7
Interior views of three bivalve shells. These forms all have approximately the same ratio of the two measurements selected as principal dimensions. (From Vokes, 1957.)

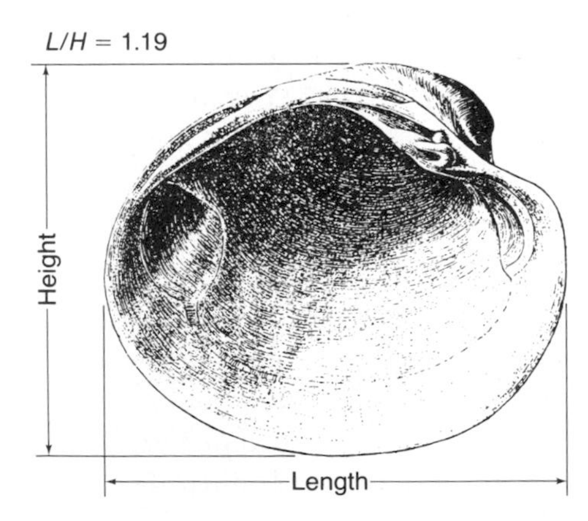

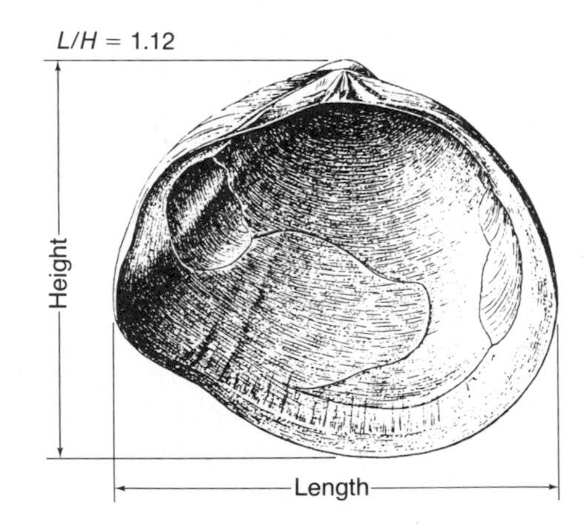

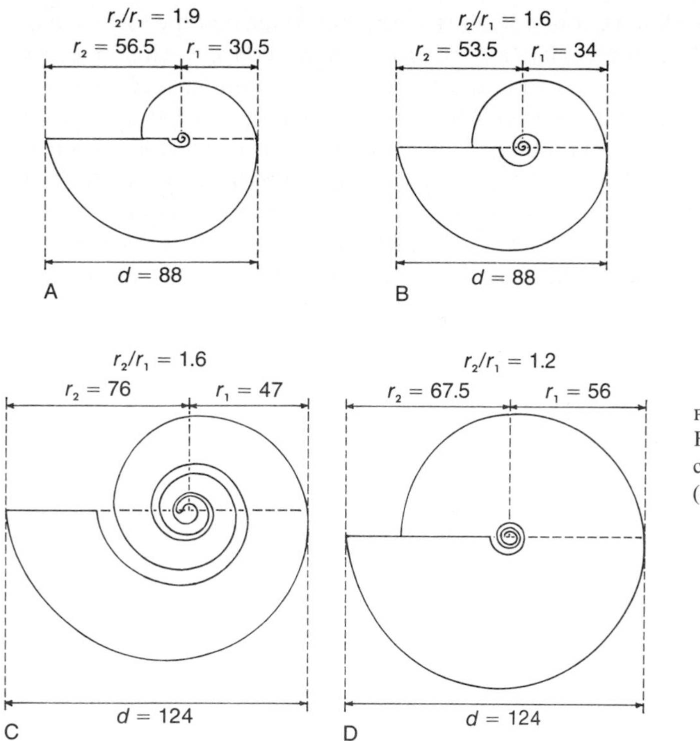

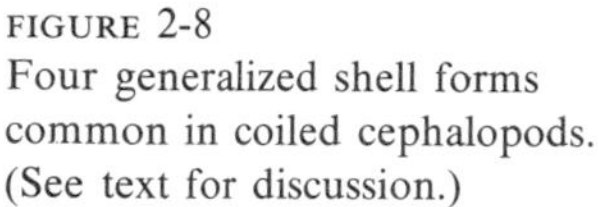
FIGURE 2-8
Four generalized shell forms common in coiled cephalopods. (See text for discussion.)

measurement. The obvious differences in these shapes are not reflected by their length-to-height ratios. If we describe shape by giving two dimensions perpendicular to each other, *we are tacitly assuming that the shape is rectangular.* If we knew in advance that all bivalve shells were rectangular, there would be no problem; the length and height measurements or the length-to-height ratio would be completely descriptive. The important thing to remember here is that *when we choose to define shape by a set of dimensions, we are assuming a model. We are assuming that there is an ideal geometric form, and we can only measure or observe differences in shape with reference to this form.*

Four hypothetical coiled cephalopods are shown in Figure 2-8. For each, the largest diameter, *d,* is indicated, which distinguishes the large forms from the small. The diameter, however, tells us little about shape. The model tacitly assumed is a circle. Any number of quite different spiral cephalopods can be inscribed in the same circle, and thus the diameter gives us little indication of anything except size. For each of the shapes in Figure 2-8, a pair of unequal radii (r_1 and r_2), each originating in the morphologic center of the spiral, gives us more information. Notice that the ratio of the two radii in A is larger than their ratio in D, and that this, indeed, reflects some of the differences between the two shapes. (Note that the sum of the two radii is slightly less than the diameter.) The use of radii in describing coiled shapes can be carried much farther. It has been known for one hundred and fifty years that the form of many spiral invertebrate shells is mathematically rigorous. In the coiled cephalopod, radii measured at equal angular intervals around the shell usually maintain a constant ratio to each other. This is illustrated in Figure 2-9, in which the ratio of radii separated by 180 degrees is always the same. With this in mind, we can look at Figure 2-8 again and describe each of the cephalopod shell forms by the ratio of radii separated by half a revolution. Using this method of description, we can successfully distinguish three cephalopod forms. The differences between B and C are evident in the drawings, but the two forms are indistinguishable from each other by the method of comparing ratios of radii.

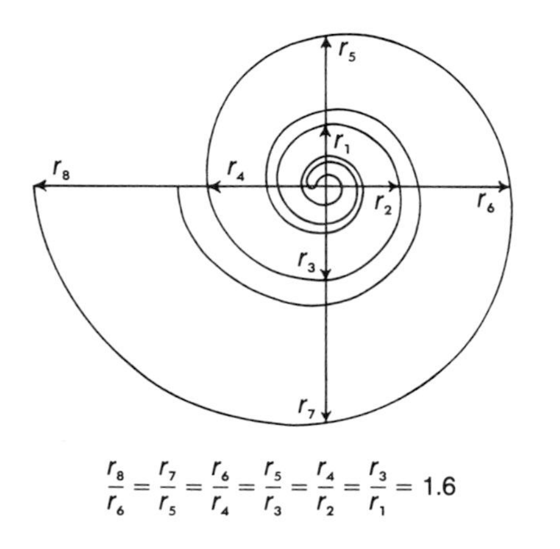

$$\frac{r_8}{r_6} = \frac{r_7}{r_5} = \frac{r_6}{r_4} = \frac{r_5}{r_3} = \frac{r_4}{r_2} = \frac{r_3}{r_1} = 1.6$$

FIGURE 2-9
Generalized cephalopod shell form illustrating the constancy of the ratios of radii.

In description by measurement, the number of measurements and their precision are generally less important than the *choice* of what is to be measured. We could have employed a wide variety of other linear dimensions of the cephalopods to describe their form, but probably no other set would be as economical or direct as the ratios of radii (though it should be kept in mind that more dimensions would be necessary to distinguish B and C in Figure 2-8). At best, measurement involves a selection of information. That is, only a part of the information that would be included in a photograph is used, and thus morphology is generalized and simplified. Whether this is justified in a given inquiry depends entirely upon the problem and upon the scientific questions being asked. There is no unique set of dimensions that most satisfactorily describes a given organism. However, it is fairly common for a basic set of dimensions to be adopted by most paleontologists for a particular fossil group.

In some cases, dimensions cannot be defined on any reasonable biologic basis. This happens when the form is relatively featureless, with no obvious structures to serve as markers, or when a simple geometric model, such as a rectangle or spiral, is not evident. In such instances, techniques of curve fittings are often helpful. The most common is harmonic analysis, and its utility stems from the fact that any simple curve can be described mathematically as the sum of a series of sine and cosine curves. Structural geologists, for example, use harmonic analysis (also known as Fourier analysis) to describe the geometry of folded rocks.

Figure 2-10 illustrates the use of harmonic analysis in the description of morphology. The shape in question is the outline of the carapace of the ostracode *Xestoleberis reniformis.* Measurements were made by Kaesler and Waters (1972) on the following basis: a "center" was chosen at the dorsal adductor muscle scar and a series of radii were measured at equal angular intervals from this point to the margin of the carapace. Using this basic information, equations were fit to the data. The fitting process can be done at any level of precision. At one extreme, the equation can be just that of a circle having a radius equal to the average radius of the ostracode. This has the advantage of simplicity but suffers from the fact that a circle is a poor description of the shape. To increase precision, more terms are added to the equation; the new terms contain sines and/or cosines and are called harmonics: first harmonic, second harmonic, and so on. As each new harmonic term is added, the graph of the equation comes closer to passing through the original measured points. Ultimately, an equation can be formed that reproduces the original shape almost perfectly.

Figure 2-10 shows part of the sequence of curve fitting for the ostracode. (See Box 2-A for more detail on the method.) The reconstruction at the left center of Figure 2-10

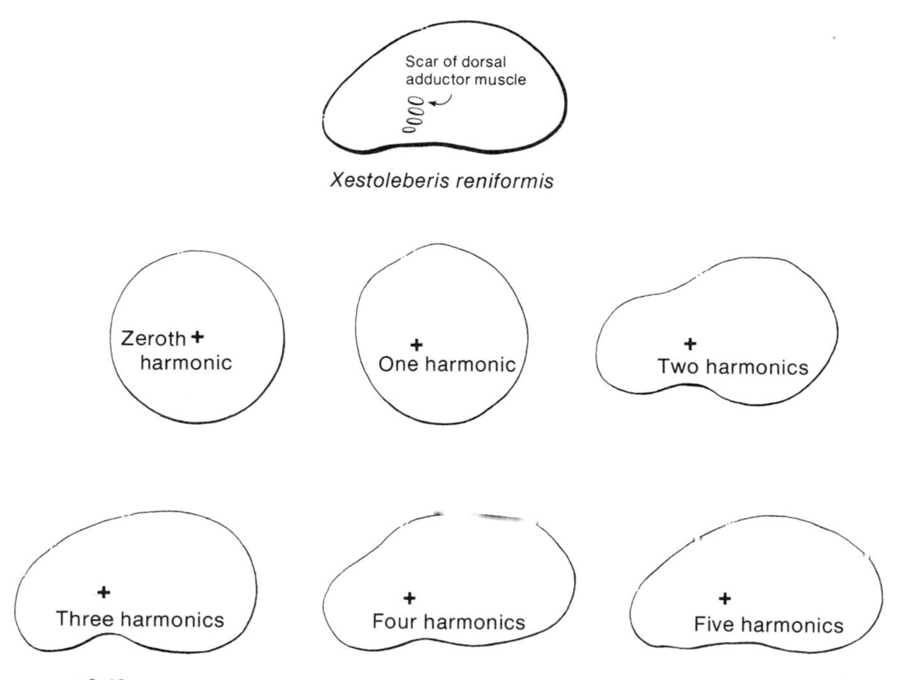

FIGURE 2-10
Reconstruction of the outline of an ostracode by harmonic analysis. When as many as five harmonic terms are used, the graph of the equation for the resulting outline (lower right) simulates the actual outline with nearly perfect accuracy.

BOX 2-A Harmonic Analysis of Shape

The curve fitting in Figure 2-10 was accomplished by using the following general equation, which is a conventional equation in the field of harmonic analysis:

$$r_\theta = \bar{r} + \sum_{i=1}^{\infty} (c_i \cos \phi_i \cos i\theta + c_i \sin \phi_i \sin i\theta).$$

This may be simplified to

$$r_\theta = \bar{r} + \sum_{i=1}^{\infty} c_i \cos (i\theta + \phi_i).$$

For details of this equation, see Kaesler and Waters (1972).

In these equations, r_θ is the radius to the margin of the ostracode at some angle θ, $\bar{r}$ is the average radius, c_i is the coefficient of the ith term, and ϕ_i is the angular position of the ith term (called the phase angle). The number of terms used in any harmonic analysis depends upon the complexity of the curve being described and upon the precision that is desired. In this case, the number of terms (harmonics) cannot exceed half the number of measurements (radii) that were made. The first few terms describe major (long period) features of the shape and successively later terms (higher harmonics) describe finer and finer detail.

The equation for the reconstruction at lower right in Figure 2-10 (five harmonics) is

$$r_\theta = 213 + \overbrace{65 \cos(\theta + 156)}^{\text{1st harmonic}} + 87 \cos(2\theta + 342)$$

$$+ 22 \cos(3\theta + 213) + 17 \cos(4\theta + 351) + \underbrace{\overset{\text{Coefficient}\ \searrow}{12} \cos(5\theta + \overset{\text{Phase angle}\ \searrow}{240})}_{\text{5th harmonic}}.$$

The other reconstructions in Figure 2-10 use the same equation but with successive terms truncated (from the right). The average radius (first term or zeroth harmonic) is 213.

A considerable amount of general information about shape can be gleaned from the array of coefficients and phase angles. For example, the coefficients of the cosine terms (c_i values: 65, 87, 22, 17, and 12 in the equation given above) are indicators of the contributions of the several harmonics. In this case, c_2 is 87 and is higher than the coefficients of the other four harmonics. This is characteristic of elliptical or elongate shapes. If the coefficient of the third harmonic predominates, the overall form has a triangular appearance. If the fourth harmonic is emphasized, the shape is quadrate, and so on. Hexagonal shapes, such as are common in colonial corals, have high sixth harmonics. Thus, the spectrum of amplitudes of the harmonic terms provides real information about shape; differences from specimen to specimen can be readily interpreted.

is the circle mentioned earlier and is called the zeroth harmonic. The second outline is based on the two-term equation; that is, it is the sum of the zeroth and first harmonics. Finally (lower right), the outline uses five harmonic terms and can be seen to simulate the original ostracode shape quite closely. This ostracode shape was simple enough that further terms would not significantly enhance the reconstruction.

Thus we see how harmonic analysis provides a mathematical fingerprint of a shape. The method makes few, if any, biological assumptions. Many shapes described in this way—perhaps representing several related species—can be compared by comparing the coefficients of the several harmonic terms. The method is used in paleontology not only for ostracodes but also for other kinds of shapes—for example, the polygonal shapes of morphological units within bryozoan and coral colonies.

The cephalopod shell illustrated in Figure 2-9 could be described by harmonic analysis, although the coordinate system would have to be modified. Such methodology would be clearly inappropriate, however, in view of the readily available and geometrically simple relationships of spiral shells that have already been noted. Harmonic analysis is something of a "brute force" technique and thus has a limited area of application. It is best used where simpler and more direct approaches are unavailable.

After it has been decided what dimensions will be measured for a given investigation, two problems remain: what method of measurement should be used, and what degree of precision should be sought. All measurements thus far discussed were of linear dimensions: that is, a curved surface projected onto a plane for measurement. Projection makes measurement easier, and is generally valid, but depends upon a simplification: a curved surface is assumed to be planar. The same assumption is made when calipers are used to measure the distance between two points on the curved surface of a specimen. Measurements made on photographs produce no serious distortions as long as the plane of the photograph represents the desired projection. Any optical projection of an image (including photography) produces some distortion, which should of course be minimized. What distortion is tolerable depends upon the desired precision and accuracy of the results.

No general rules can be made governing precision in measurement. What precision is appropriate and valuable depends entirely upon the problem being tackled. For example, if we wanted to distinguish between a trilobite and a dinosaur, measurement to the nearest foot would suffice. To measure to the nearest inch or hundredth of an inch would be superfluous. If, however, we wanted to distinguish between two very similar organisms, much greater precision would be necessary. The expert uses only as much precision as his problem demands.

The digitizing method discussed earlier is often applicable to the problem of measuring dimensions. The straight-line distance between two points is easy to calculate, given their coordinates. For a single measurement or a very few measurements on each specimen, the digitizing method is not time saving. However, if many measurements are to be made on each specimen, digitizing and automatic computation are of great benefit. The precision possible by this method depends only on the precision with which a point can be located in a coordinate system.

Describing Internal Structures

Most of the descriptive modes discussed thus far are usually applied to the surface of a fossil. Often equally or more important are internal structures such as the fabric of crystals making up a shell or the internal pore pattern in a mammal bone. The problems in describing internal structures are much the same as those for external form. But the techniques are different. We have seen that special types of photography, such as X-ray photography, yield information on internal structures. A specimen that is not amenable to such photographic investigation must usually be cut or sectioned to show its internal structures. The result may be a polished section that can be observed in reflected light or a thin section that can be observed in transmitted light.

The references at the end of this chapter include several basic summaries of the range of techniques available for paleontologic description. It should be kept in mind that new techniques are constantly being developed, many in fields far removed from paleontology, which have direct applicability to problems of describing internal structures in fossils. Historically, some of the greatest advances in paleontology have been made by those who have successfully adapted techniques from other disciplines.

Supplementary Reading

Brown, C. A. (1960) *Palynological Techniques.* Baton Rouge, La., C. A. Brown. 188 p. (A privately published handbook of techniques designed for work with fossil pollen and spores.)

Camp, C. L., and Hanna, G. D. (1937) *Methods in Paleontology.* Berkeley, University of California Press, 153 p. (A classic treatment of paleontological techniques.)

Kummel, B., and Raup, D. M., eds. (1965) *Handbook of Paleontological Techniques.* San Francisco, W. H. Freeman and Company, 852 p. (A collection of 86 articles written by specialists on various aspects of paleontological techniques, including collecting, preparing and illustrating; it also contains comprehensive bibliographies of techniques.)

McLean, J. D., Jr. (1959 to date) *Manual of Micropaleontological Techniques.* Alexandria, Va., McLean Paleontology Laboratory. (A continuing publication in looseleaf form covering many topics in the preparation and laboratory treatment of microfossils, particularly foraminiferans.)

Oxnard, C. (1973) *Form and Pattern in Human Evolution.* Chicago, University of Chicago Press, 218 p. (Contains many innovative methods of morphologic description, including optical data analysis.)

Scott, G. H. (1975) An automated coordinate recorder for biometry. *Lethaia,* **8**:49–52. (Description of special digitizer equipment for use with standard stereo microscopes. The instrument is especially good for collecting the raw data for harmonic analysis.)

CHAPTER 3

Ontogenetic Variation

This chapter is the first in a series dealing with the problems of describing and interpreting variation in fossils. Differences between fossil specimens result from innumerable biologic and geologic causes. Differences in age of individual organisms at the time of their deaths is one of the most important of the purely biologic factors. Two individuals may have been genetically identical and may have lived in identical environments, yet their fossilized remains may be strikingly different simply because one was older than the other at the time of death. The differences are not limited to size. More often than not, growth is accompanied by changes in form. Growth stages may in fact be so different that fossil specimens in different stages may not be recognizable as members of the same species, particularly with species whose ***ontogeny***, or normal life cycle, includes a metamorphosis, as does that of arthropods.

Ontogeny must be understood in order that fossil specimens which are members of the same species can be recognized and in order that the range of morphology displayed by a species can be assessed and interpreted properly. Ontogenetic change in morphology is just as important to the total description of an organism as is the adult form, especially in the study of organic evolution. To consider an evolutionary series as a sequence of adult forms is to oversimplify. Rather, evolution must be looked upon as a sequence of ontogenies.

Types of Growth

Organic growth is extremely complicated; it usually involves several types of change, among which are changes in cell size, number of cells, number of cell types, and relative positions of cells. During life, an organism may change in form abruptly (undergo metamorphosis) or it may change gradually.

All organisms that depend on a hard skeleton—for support, for protection, or for muscle attachment—must enlarge the skeletal structure to accommodate growth of the soft body. Postembryonic growth of skeletons is accomplished in four ways.

ACCRETION OF EXISTING PARTS

Most shelled molluscs increase their skeleton size simply by adding new material to the shell throughout life, which has the obvious advantage of permitting continued use of skeletal material deposited at earlier ontogenetic stages. It has the disadvantage, however, that the form of the juvenile shell must be incorporated as part of the adult shell. This type of growth is well illustrated by gastropods (see Figures 2-5 and 2-6). The shell of the coiled gastropod may be looked upon as a hollow tapering tube that is coiled about an axis. As the animal inside the shell grows larger, new material is added to the aperture, or opening, of the tube. Growth lines showing increments of growth can often be seen paralleling the outline of the aperture. Most adult gastropods occupy their entire shells and thus the shape of the soft body of a gastropod is, in effect, an internal mold of the shell.

Growth by simple accretion is found in many other animal groups, particularly among animals whose shell is external and serves primarily for protection and muscle attachment. Figure 3-1 illustrates growth in bivalved molluscs. The bivalve grows by accretion in much the same way that the gastropod or cephalopod does. In fact, as we shall see in subsequent chapters, the geometry of the bivalve is not substantially different from that of the other molluscan groups we have considered. We may look upon the individual valve as a hollow cone that tapers from a large, open end to a small umbo or "beak." The principal difference is that the "cone" shell of the bivalve is quite shallow and tapers much more sharply from opening to beak than does that of the gastropod or cephalopod. The cross-section of the shell (Figure 3-1) is marked by growth lines which can be observed in the shell as well as on the outer surface. New skeletal material is added not only at the leading edge of the shell, but also as a covering over most of the interior surface of the shell. Figure 3-1 includes an enlarged photograph of a series of growth lines. In this instance the shell was grown under

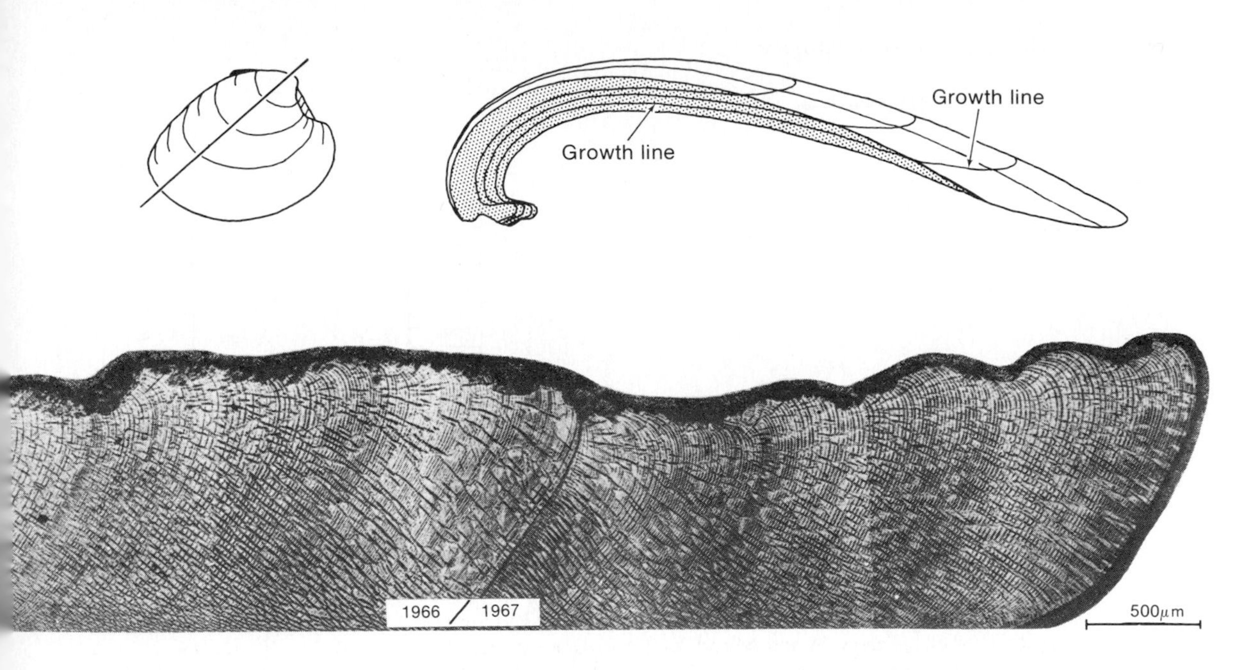

FIGURE 3-1
Accretionary growth in bivalved molluscs. Photograph shows approximately one year's growth in the species *Mercenaria mercenaria.* The leading edge of the shell is to the right. (From Pannella and MacClintock, 1968.)

controlled conditions and it is possible to establish that the growth lines are in fact daily. Tidal cycles (following the lunar month) can be identified in the pattern of line spacing.

ADDITION OF NEW PARTS

A common method of skeletal growth for those organisms whose skeletons consist of many parts, either tightly articulated or fitting loosely in the soft tissue, is addition of new skeletal parts. Most echinoids, for example, have a rigid skeleton made up of as many as several thousand tightly articulated calcite plates, arranged in a radial fashion, with columns of plates extending from one "pole" of the crudely spherical skeleton to the other. The continuous addition of new plates largely accounts for growth. An important fact about the growth of the echinoid skeleton is that new plates are always added at the same place, namely at the tops of the plate columns.

Figure 3-2 schematically shows some aspects of the growth of crinoid stems by addition of new columnals. In this type of growth, large columnals are added periodically at the base of the calyx (A and B in Figure 3-2). They may or may not grow appreciably by accretion after their formation. As each new columnal is added at

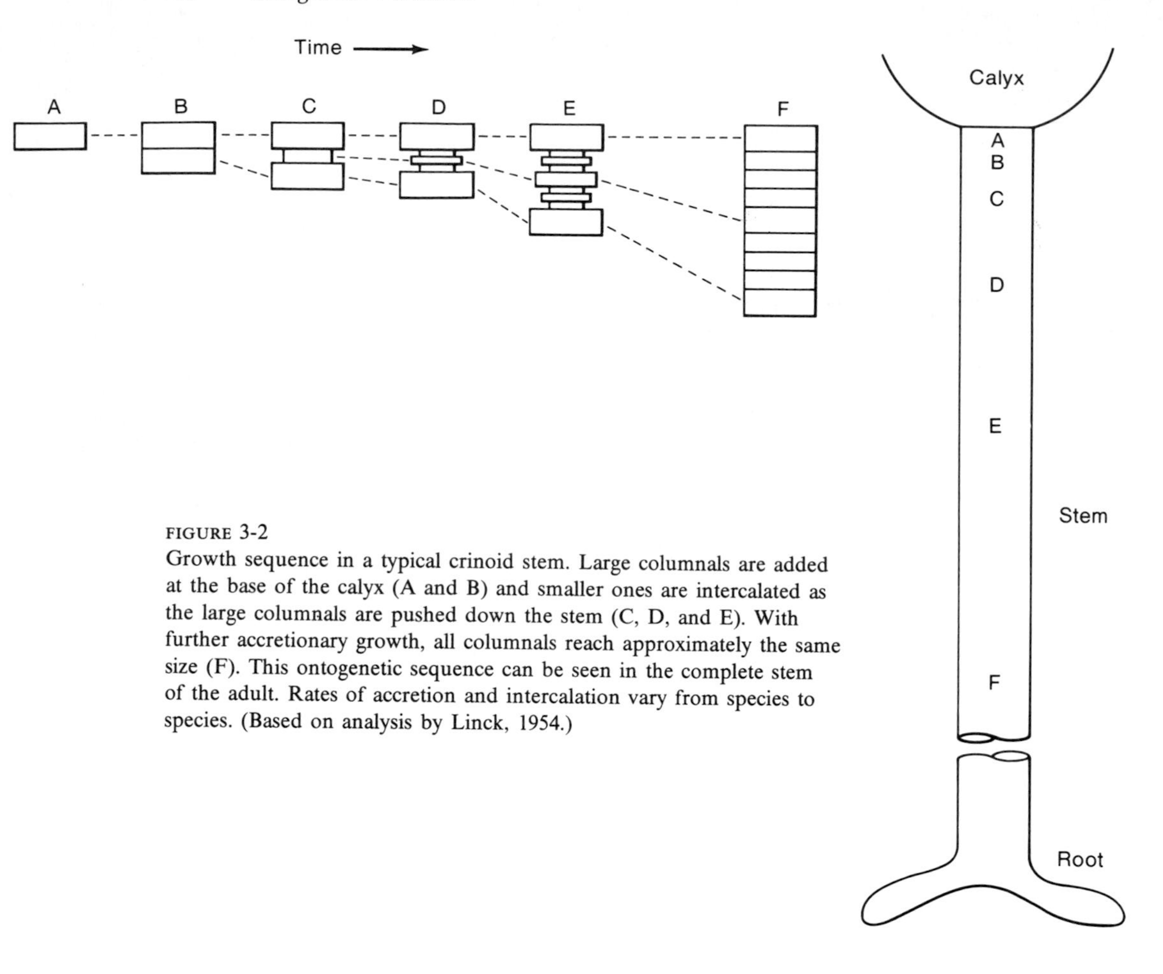

FIGURE 3-2
Growth sequence in a typical crinoid stem. Large columnals are added at the base of the calyx (A and B) and smaller ones are intercalated as the large columnals are pushed down the stem (C, D, and E). With further accretionary growth, all columnals reach approximately the same size (F). This ontogenetic sequence can be seen in the complete stem of the adult. Rates of accretion and intercalation vary from species to species. (Based on analysis by Linck, 1954.)

the calyx, the previous ones are inevitably displaced down the stem. At some distance away from the calyx, smaller columnals appear between the pre-existing ones. They grow by accretion and as they do, yet more small columnals are intercalated. The result of this process (E in Figure 3-2) is an orderly set of generations of intercalated columnals—with the generations being distinguishable by size. The sequence can be seen by observing one part of the stem as it grows over a period of time or, more conveniently, by analyzing the sequence of ontogenetic stages down the stem of an adult individual. In this type of growth, the intercalated columnals reach a terminal size nearly equivalent to the size of those originally formed at the base of the calyx. Thus, near the bottom of the stem, the columnals are all nearly the same size. It has been shown by Seilacher et al. (1968) that different crinoid species exhibit modifications of this basic growth model in order to accommodate different functional problems. For example, parts of the stem where large columnals alternate with small ones are much more flexible than where all columnals are the same size. Depending on the crinoid's mode of life, flexibility of the stem may be desirable or not, and flexibility may vary from one part of the stem to another.

Addition of new parts is an integral part of the growth of many other organisms. Figure 3-3 shows several stages in the ontogeny of a trilobite. One of the more obvious ontogenetic changes is the gradual addition of segments to the thorax.

MOLTING

The trilobite shown in Figure 3-3 uses another basic mechanism of growth, the periodic shedding of the entire skeleton and formation of a new one to accommodate the increase in size of the soft parts. Figure 3-4 shows a plot of cephalon length against

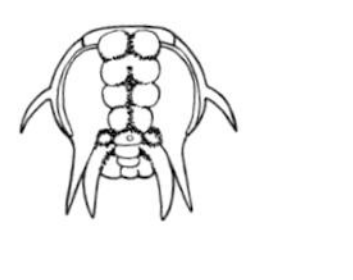

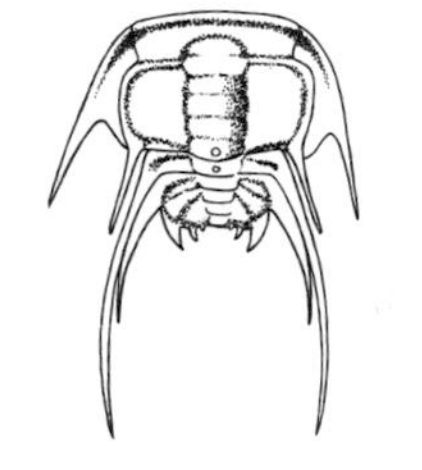

FIGURE 3-3
Four stages in ontogeny of the trilobite *Paradoxides*. Ontogeny is accompanied by radical changes in the form of existing parts and by the addition of new parts. (From Whittington, 1957.)

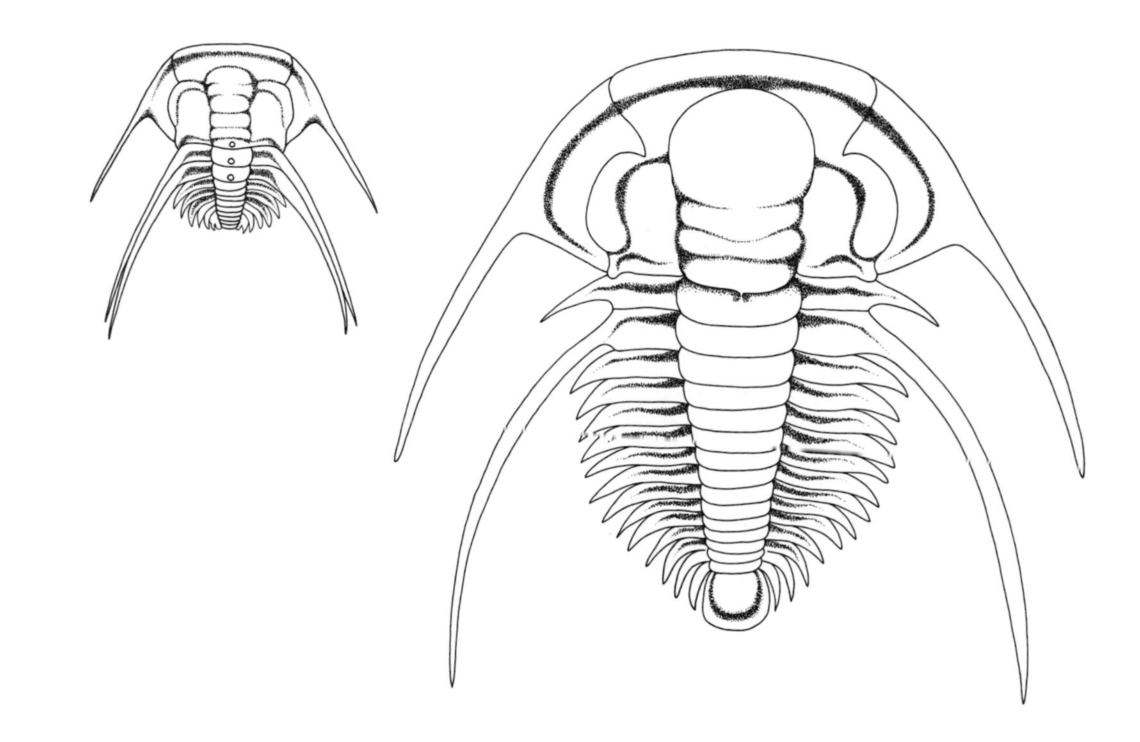

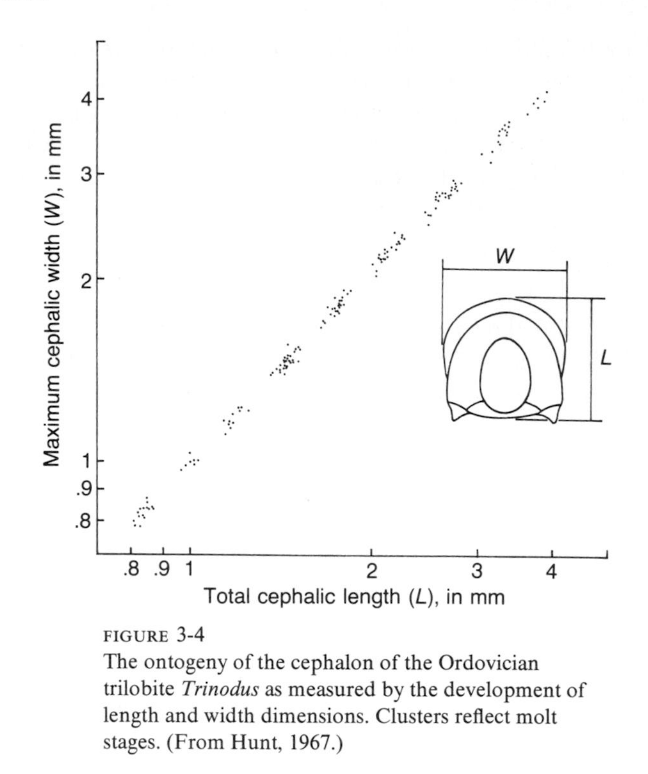

FIGURE 3-4
The ontogeny of the cephalon of the Ordovician trilobite *Trinodus* as measured by the development of length and width dimensions. Clusters reflect molt stages. (From Hunt, 1967.)

width in an assemblage of trilobites of one species. As can be seen from the graph, the measurements fall into clusters, each representing a molt stage. Differences between points in a cluster represent minor differences in size and shape among the individuals of that stage. Specimens represented by points intermediate between the clusters are rare.

Skeletal growth by molting has one clear advantage over both growth by accretion and growth by addition of parts. The skeleton of the juvenile individual need not form part of the adult skeleton, and the adult has much more freedom to change shape during growth. At the same time, growth by molting has decided disadvantages in that the organism must pass through rather perilous periods when it lacks a hard skeleton, and it must expend considerable metabolic energy in replacing its entire skeleton at intervals.

In terms of numbers of species, skeletal growth by molting is by far the most common mode of growth among organisms possessing a skeleton. This stems simply from the fact that arthropods employ this mode of growth, and arthropods make up about three-quarters of all living animal species. In other groups of organisms molting is occasionally encountered, but usually as part of a major metamorphosis following a larval stage.

MODIFICATION

This mode of growth is most common in the formation of certain bones of higher vertebrates. By an elaborate process of replacement and re-formation of skeletal materials, the form and structure of the bone changes as size increases. This mode of growth shares with the molting system the advantage that skeletal form at one stage of growth is largely independent of the previous stages but not the disadvantage that the organism is without a skeleton for certain periods.

MIXED GROWTH STRATEGIES

It is difficult to classify organisms rigidly according to mode of skeletal growth, primarily because many organisms employ a combination of the basic mechanisms just described. The crinoid stem example (Figure 3-2) combines addition of new parts and accretion of existing parts. Figure 3-5 shows the product of another such combination, a pair of plate columns from an echinoid. As we have seen, addition of new plates is a principal means of increasing skeletal size in echinoids. At the same time, however, the existing plates grow by peripheral accretion. Major periods of accretion are recorded by growth lines that trace the history of change in size and shape of the plates. The growth pattern is somewhat more complicated in the echinoid than in the crinoid stem but it is still amenable to description by a rather simple model relating several rates of accretion and new plate formation. This is illustrated in Figure 3-6 by a purely hypothetical computer simulation of echinoid growth. In order to make the simulation, the various rates of growth (including production of new plates) were translated into computer language. The simulation process serves among other purposes as a way to check one's understanding of the growth sequence. If the simulation does not match the pattern observed in nature, then the model used is incorrect. In this case, a comparison between Figures 3-5 and 3-6 shows differences in relative plate sizes at the lower end. The plate pattern produced by the computer is a common one among echinoids but does not match the particular species illustrated in Figure 3-5.

Another example of a combination of the same two modes of growth is shown in Figure 3-7. There, we have a coiled cephalopod in which the outer shell grows by accretion in much the same manner as in gastropods. But the two organisms are quite different biologically: the gastropod fills the entire shell whereas the body of the cephalopod is contained in what is known as the body chamber, which occupies the last half or three-quarters of a revolution about the coiling axis. The body chamber is separated from the rest of the shell (phragmocone) by a calcified wall called the septum. During ontogeny, the animal increases in size and moves outward in the shell so that it is always in a position near the opening. New partitions or septa are deposited periodically to accommodate the movement of the animal forward in the shell. A succession of septa is thereby deposited, representing an addition of parts. The

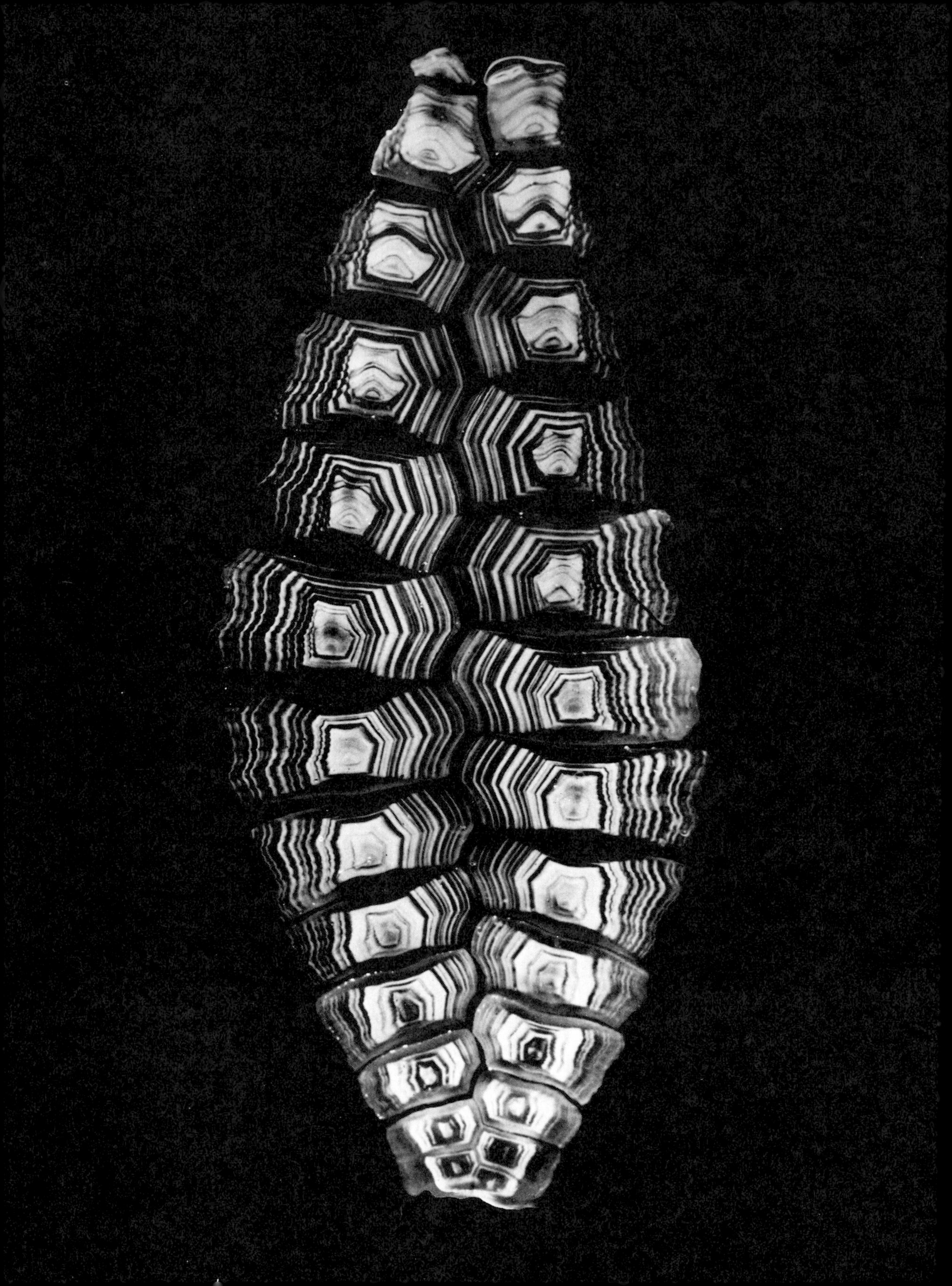

FIGURE 3-5
Growth by a combination of addition of new parts and accretion of existing parts. A pair of plate columns from the echinoid *Strongylocentrotus* is shown. Each plate contains a "nest" of growth lines reflecting its accretionary history. New plates are added always at the tops of columns. (From Raup, 1968.)

FIGURE 3-6
Computer simulation of plate growth in echinoids. Compare with the photograph of an actual specimen, on the facing page. (From Raup, 1968.)

FIGURE 3-7
X-ray photograph of the living cephalopod *Nautilus* illustrating growth by addition of parts and by accretion. During ontogeny, shell length is increased by accretion and number of chambers is increased by addition of septa. (Photograph by R. M. Eaton.)

differences in growth between the gastropod and the cephalopod are closely related to physiologic and habitat differences. The cephalopod depends on having a portion of its shell empty in order to make it buoyant for floating or swimming.

Describing Ontogenetic Change

For the biologist, the description of changes undergone by an organism during ontogeny is relatively simple. He can often observe the growing animal directly. Furthermore, he can observe precisely the time at which ontogenetic changes take place. The paleontologist, on the other hand, can only rarely deduce how old an organism whose remains he finds was at the time of its death, and he has no opportunity to observe its growth. Two approaches can be used, however, to obtain a considerable amount of ontogenetic information from the fossil record. First, a "growth series" of specimens may be assembled (as in Figure 3-3) and the change in shape with increasing size can be observed or measured. Second, the ontogeny of a single adult specimen may be described from features such as growth lines (like those in Figure 3-5), which really represent temporary cessation, or slowing, of growth.

The choice between these two methods depends largely upon the mode or modes of growth employed by the organism. If growth was by simple accretion with no resorption or modification of previous ontogenetic stages and if growth was recorded by recognizable growth lines, either approach may be used. The same is true if growth was solely by addition of parts. But if molting or modification was a part of the organism's ontogeny, the paleontologist must use a series of specimens to simulate that ontogeny. When either method may be used, the one employing a single adult individual is usually preferable because the difficulties of interpreting the variation between different specimens are avoided.

Figure 3-8 shows graphs of two brachiopod dimensions. The upper graph is based on measurements of approximately seventy-five fossil specimens of a species from a single outcrop and thus may be reasonably assumed to represent growth in a single population. The scatter of points forms a somewhat fan-shaped pattern that extends out from the origin. Notice that the scatter of points is slightly curved (convex toward the "width" axis). This may be interpreted as indicating that the rate of length growth increases relative to the rate of width growth as the animal becomes larger. To make this interpretation, we must assume that size increases as some function of time. The diagram contains no information, however, about the relationship between size increase and absolute time. All we can learn is something about the rate of increase in one dimension *relative* to the rate of increase in the other dimension.

The curved scatter of points implies a change in ratio between length and width, and therefore a change in shape during growth. A line drawn at 45° is included in the plots in Figure 3-8. All points below this line represent individuals whose width exceeds their length, while points above the line represent individuals whose length exceeds their width. The change in relative growth rates produces a gradual shift from a shell

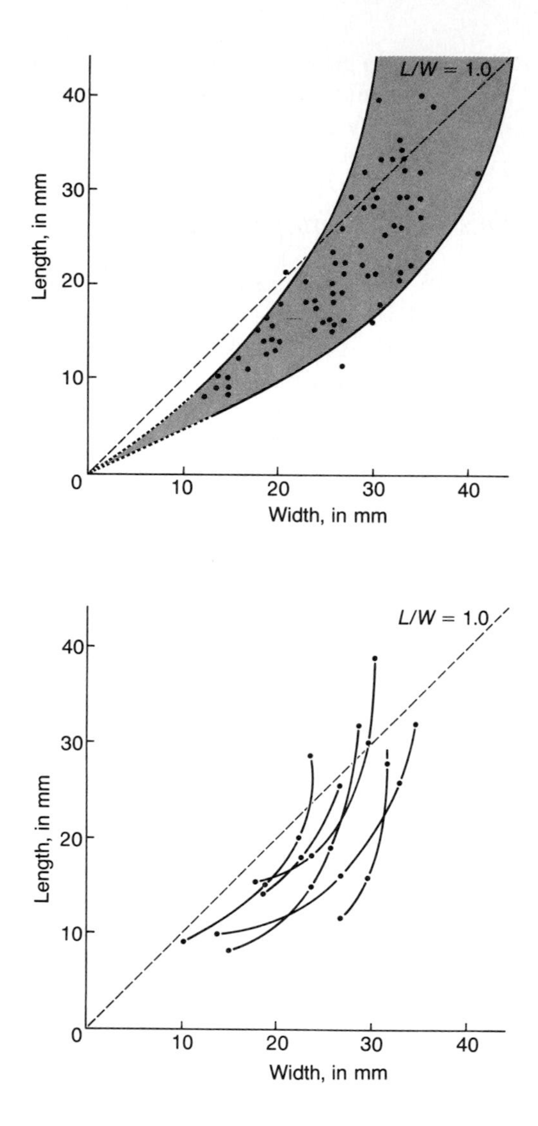

FIGURE 3-8
Ontogenetic change in shape in brachiopods. Plots are of measurements made on an assemblage of the Carboniferous brachiopod *Ectochoristites.* Length refers to the antero-posterior length of the pedicle valve; width is measured along the hinge of the pedicle valve. Points connected by curved lines in the lower graph are based on growth line measurements of single specimens. (From Campbell, 1957.)

that is wider than long to one that is longer than wide. We could refine this interpretation by fitting a curve to the scatter of data. We would then have a generalized picture of the change in shape during ontogeny for this species with respect to the two dimensions.

What is the explanation of the departure of many of the points from a perfect curve? There are several possible reasons. One is that in describing ontogeny from data drawn from a variety of individuals, the observer is also recording genetic differences between the individuals, variability unrelated to ontogeny. A second possibility is that some of the specimens were damaged by purely geologic causes after death and that the scatter is thus not biologic at all. It is also possible that the method of measurement used was not precise enough and that the specimens themselves fall much closer to the idealized line than would be indicated by the observed scatter.

The lower graph in Figure 3-8 shows several growth curves for the same brachiopod species. The points on each curve refer to measurements of growth lines of a single adult specimen. This plot confirms that the growth curve for the species is curvilinear, and illustrates the importance of being able to trace the ontogeny of a single individual. Both graphs in Figure 3-8 yield essentially the same information—but the lower one gives us a more definitive picture of the growth pattern and of the differences between individuals in the fossil assemblage.

Figure 3-9 shows results, plotted in graphic form, of ontogenetic reconstruction based on measurements made on a single adult specimen—the coiled cephalopod shown in Figure 3-7. The specimen was measured for a series of radii (see Chapter 2 for discussion of the descriptive significance of radii in coiled cephalopods). Graph A in Figure 3-9 relates the length of a radius to its angular position. Because we know that the outer shell of a cephalopod grows by accretion, each angular increment of shell must represent an increase in age. The curved pattern of points indicates that the rate of radius increase changes relative to the rate of angular movement. (The scatter

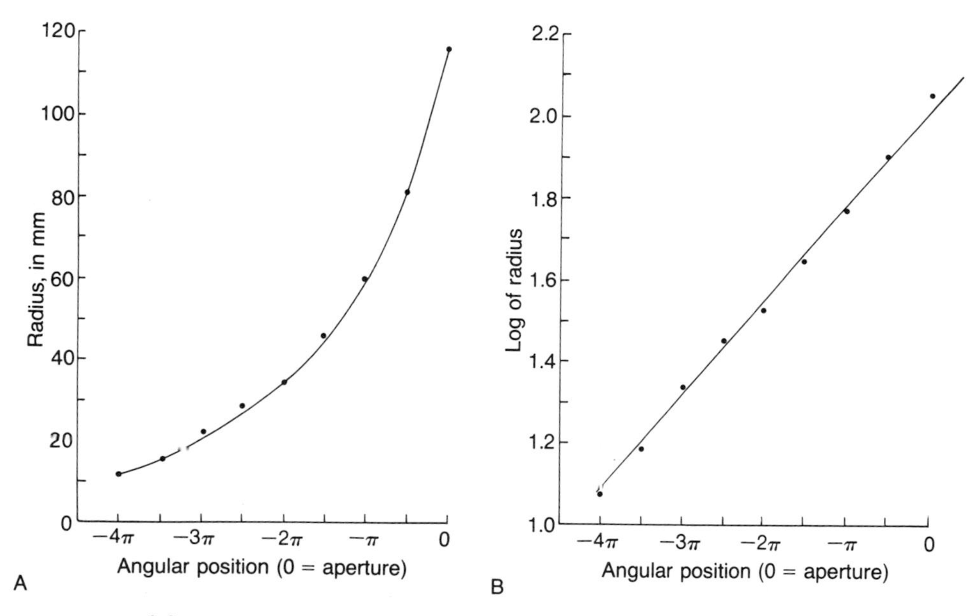

FIGURE 3-9
Relation of length of radius to its angular position during the course of cephalopod ontogeny. The measurements plotted here were made on the *Nautilus* specimen shown in Figure 3-7.

of points is much less than in the brachiopod graph because only one individual is plotted here.)

Does the curved pattern of points indicate a change of shape? We cannot answer this question directly because the plotted dimensions are not linear, as were the brachiopod dimensions. The ontogeny of this cephalopod can be explored somewhat further by plotting the angular position of the radius against the *logarithm* of the radius (graph B in Figure 3-9). Notice that the scatter of points in the new plot falls quite rigorously on the straight line. Thus the angular position of the radius increases with the logarithm of the radius. In other words, successive radii are in geometric progression. We thus have what is known as a logarithmic or equiangular spiral. One of the mathematical characteristics of this spiral is that its shape remains constant. In this sense, we can conclude that the shape of the cephalopod does not change during ontogeny. The shape does change, of course, in that as growth proceeds, there are more and more revolutions about the axis.

In summary it may be said that of the two possible approaches to ontogenetic studies of fossils, the one based on a single specimen is easier to handle and the results are generally more accurate. Nevertheless, much the same information can be deduced from a series of specimens if the nonontogenetic variability inherent in a heterogeneous group of individuals is not excessive.

The examples of ontogenetic analysis shown thus far were done by measuring two morphologic attributes and plotting them against each other to observe their relative rates of development. This can be expanded to include more traits, and often is. It is possible when working with purely graphical techniques to add a third attribute, thereby seeing the relationship between three growth rates, but to add a fourth or a fifth or a sixth moves beyond the practical limits of graphical analysis. For this, more sophisticated "multivariate" methods of statistical analysis must be employed. Such methods have been used in paleontology with considerable success.

Coordinate transformation is a quite different approach to the description of ontogenetic change and is particularly suited to describing and interpreting complex changes in a whole organism or an entire structure of an organism. In Figure 3-10, two trilobite growth stages are reproduced as sketches. Their scale has been adjusted so that the sizes of the two are comparable. This makes ontogenetic changes in form more evident. An arbitrary *X*-, *Y*-grid has been superimposed on the first stage. The same grid is indicated on the second stage but is deformed to express the change in shape of the grid necessary to accommodate the ontogenetic change in morphology.

Coordinate transformation was first proposed by D'Arcy Wentworth Thompson in his classic work *On Growth and Form.* Relatively little use was made of the concept, however, because it is very time-consuming and because subjectivity enters into the construction of deformed grids. The high-speed computer has changed things, however. Recently, computers have been combined with methods of "trend surface analysis" (developed in other areas of science), permitting the paleontologist to create Thompson's deformed grids with great speed and accuracy and at the same time to reason more rigorously about the precise mathematical character of the deformation.

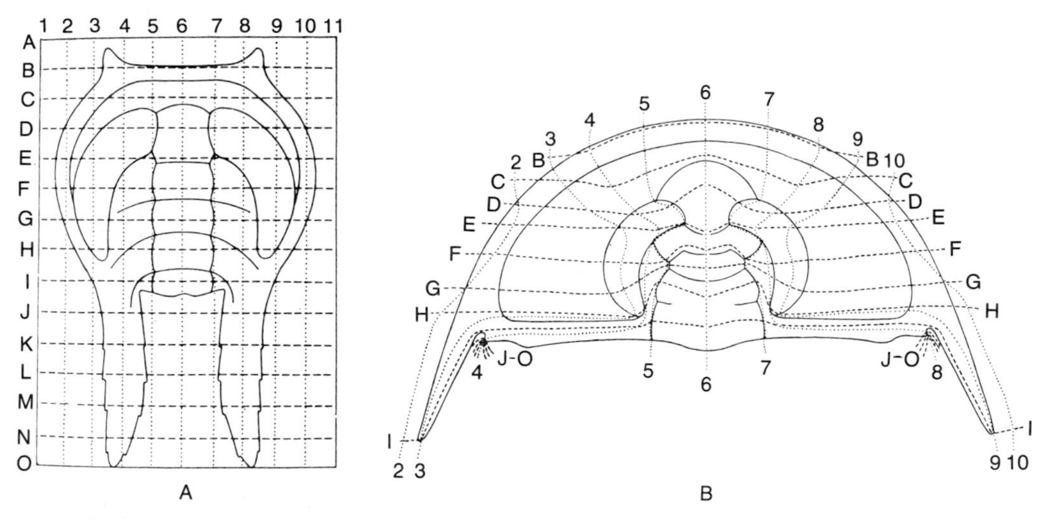

FIGURE 3-10
Trilobite ontogeny interpreted by coordinate transformation. An X-, Y-grid was superimposed arbitrarily on stage A and the morphologic positions of grid intersections were noted. After the grid intersections were located at morphologically homologous points in stage B, the deformed grid was drawn on it. The deformed grid thus indicates spatially the morphogenetic changes that transform A to B. (From Palmer, 1957.)

Growth Rates

Some plants and animals grow very rapidly; others grow very slowly. About twenty years are generally required, for example, to complete human growth. But a monkey completes its growth in about two years. Other organisms, such as insects, may go through a complete ontogeny in a matter of days or weeks. Such differences are largely hereditary; that is, they are under genetic control and have been produced by evolution. In all organisms, however, there is a certain amount of variability in growth rate that is not under genetic control but that depends upon the specific conditions under which the organism lives. Factors such as nutrition, crowding, and other conditions imposed by the physical and biologic environment control growth rates to a certain extent. The geneticist uses the term "norm of reaction" to express the amount of nongenetic variation possible in a morphologic feature or in the growth rate of that feature.

In nearly all organisms the rate of growth varies with time; that is, it changes during ontogeny. Most of this change in growth rate is genetically controlled although it, too, is subject to a norm of reaction. A typical pattern of growth with respect to time is shown in the first curve of Figure 3-11. Here, size in a hypothetical organism is plotted against time. Size increases slowly at first, then more rapidly, and finally much more

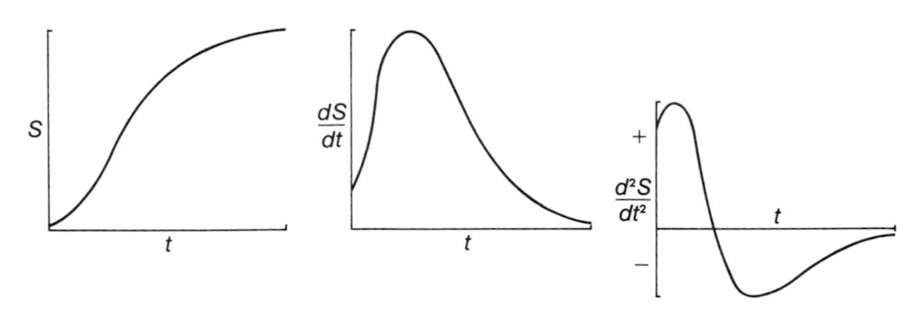

FIGURE 3-11
Generalized pattern of growth relating the size S of an organism to time t. (From Medawar, 1945.)

slowly, so that in the later stages, growth has nearly stopped. This variation in growth rate is shown more clearly in the second graph in Figure 3-11, where the *rate of growth* (the first derivative of size with respect to time) is plotted against time. In geometric terms, the second curve traces the change in slope of the first curve. The third graph shows the *rate of change* in slope of the first curve and is thus the second derivative of size with respect to time.

Some animals do not follow this general pattern of growth, but it is sufficiently widespread in the plant and animal worlds that we may use it as a general model. One of the principal differences between organisms lies in the form of the upper part of the curve in the first graph. The growth curve for most mammals, for example, actually levels off. The organism may continue to live for a considerable time after a plateau has been reached but growth will have stopped. The period after completion of growth is commonly referred to as *maturity,* or the true adult stage. In most plants, however, and in most invertebrate animals, growth continues throughout the life of the organism, although at a greatly reduced rate, making it difficult to define a true adult stage. At best we can define a stage at which most of the growth will have taken place, but growth will not have ceased.

When the growth of a real organism is measured, the data obtained seldom yield the perfect curve shown in Figure 3-11. This is because the norm of reaction for growth rates is generally quite large. If, during the development of an organism, a sudden improvement in nutrition takes place, the curve may be displaced upward, thus breaking the smooth pattern. Environmental variability often is so high that a clear pattern of growth through time cannot be deduced.

Regardless of environmental change during ontogeny, different parts of an organism typically grow at different rates. We have already seen an example of this in brachiopods (Figure 3-8) in which length and width grew at different rates. The different rates of growth of parts of an animal are well coordinated with respect to each other. If an environmental change causes a decrease in the rate of growth of one part, this is accompanied by a corresponding decrease in rate of growth of another part. The change in the relative size of the human head during ontogeny is well known. The head becomes *relatively* smaller as growth proceeds because it does not grow as rapidly as most of the rest of the body. The proportion of the size of the head to the size of other parts is always the same for a given point in ontogeny, although environmental factors may amplify or restrict the person's overall growth. From an evolutionary point of view, the advantages of coordinated growth are obvious. The functioning of the entire organism depends to a large extent on its maintaining a balance between sizes and proportions of various parts. If growth rates of various parts were under independent control, then normal changes in growth rate caused by the environment would produce an organism that would not be functional.

From the paleontologic viewpoint, the coordination between the growth of the various parts is extremely important. It permits us to neglect the absolute time factor and to study the development of an organism and the change in shape of the organism without reference to and without measuring time. Figures 3-8 and 3-9 illustrate this. In both examples it was possible to learn a considerable amount about the ontogeny of the organisms involved, but in neither did we have any knowledge of growth rates with respect to absolute time.

Nearly all paleontologic studies of ontogeny require the measurement of change in one morphologic attribute in relation to change in another. We can define two basically different types of growth: ***isometric*** growth and ***anisometric*** growth. If the ratio between the sizes of two parts of an organism does not change during ontogeny, we have isometric growth (Figure 3-12,A and B). If the ratio does change, we have anisometric growth (Figure 3-12,C and D). To say that growth is isometric implies that there is no change in shape during growth; if growth is anisometric, there must be change in shape. Two kinds of isometric growth are shown in Figure 3-12. In the pattern of isometric growth plotted on graph A, both parts grow at precisely the same rate. This yields a straight line at 45 degrees to either axis. In the pattern of isometric growth plotted on graph B, one part grows more rapidly than the other but the ratio between them is constant, giving a straight line that is separated from one axis by a smaller angle than from the other. In the pattern of anisometric growth plotted on graph C, the X part grows more rapidly than the Y part at first but this relationship is subsequently reversed. The plot of growth is a curve. The pattern of growth plotted on graph D is different from all of the other three patterns in that the plotted line would not pass through the origin if extended. In pattern D as in pattern C, shape changes with growth; that is, growth is anisometric. To summarize, growth is isometric if the plotted line is straight *and* passes through the origin (or would do so if extended). All other conditions produce anisometric growth.

Anisometric growth is far more common than isometric growth. In other words, *shape change during ontogeny is the rule rather than the exception.* Anisometric growth has been the subject of a tremendous amount of research in both biology and paleontology and some important generalizations have resulted. Principal among these is that anisometric growth very often may be approximated by the following

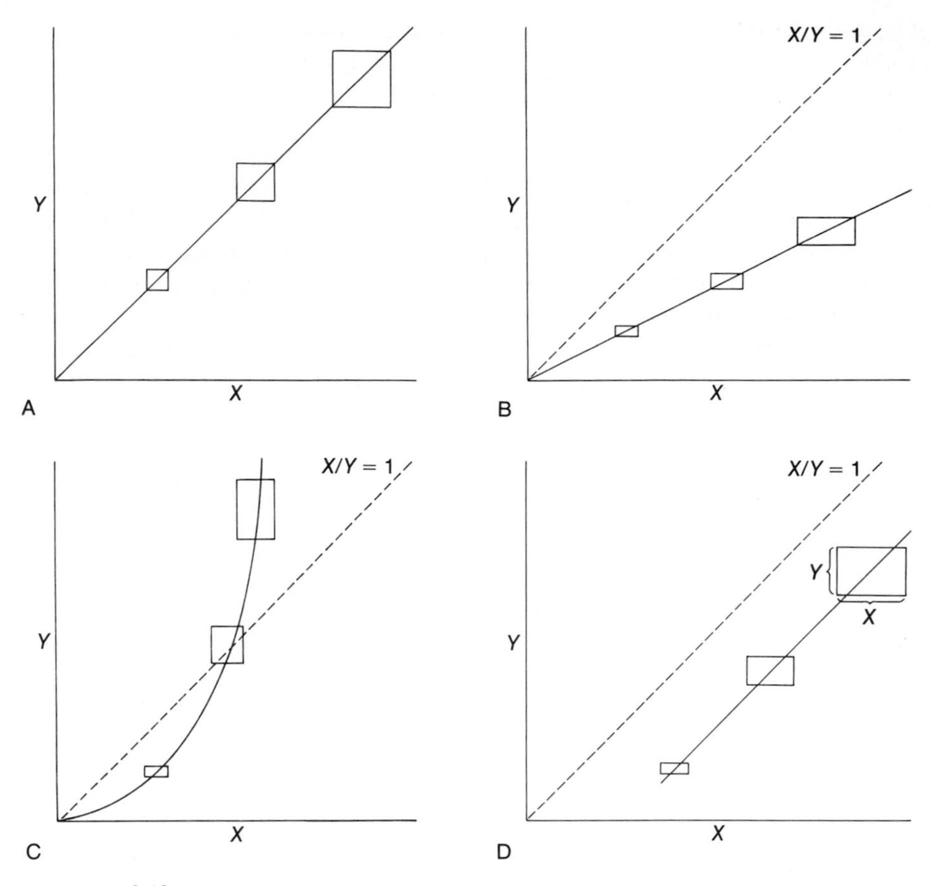

FIGURE 3-12
Typical patterns of growth observed when two morphologic dimensions are plotted against one another. In each graph, the shape of a hypothetical organism is shown by a square or rectangle at three ontogenetic stages. In A and B growth is isometric, and there is no change in shape; in C and D growth is anisometric, and shape is continually changing.

relationship between the two parts:

$$Y = bX^a,$$

where a and b are constants and $a \neq 1$. Growth that proceeds according to this relationship is called ***allometry.*** (It should be noted that isometric growth obtains when the constant a is one.) Allometric growth could be the subject of several chapters. The student should refer to the works of Huxley, Thompson, and Gould listed in the references at the end of this chapter. (Several authors have defined the term allometry more broadly to include all anisometric growth.)

Reasons for Anisometric Growth

So far we have considered anisometric growth and the resulting changes in shape during ontogeny simply as a biologic process and have concentrated on some of the problems of defining and describing it. Of greater interest, however, and of considerably greater importance is the *interpretation of change of shape* during ontogeny. The modern student of evolution subscribes to the idea that changes in form during ontogeny are not the result of whim, are not accidental, but rather have become a part of the hereditary material because of their functional value. We must search, therefore, for functional or adaptive explanations for the changes in shape of a structure or in number of structures that we observe. Why, for example, does the shape of the brachiopod shell change during growth?

Where ontogenetic change is abrupt, as in the metamorphosis of an insect, it can often be related to sudden change in mode of life. The frog develops through a succession of fairly gradual changes into an adult. The tadpole lives in and depends upon an aquatic environment: it extracts oxygen directly from the water. The adult frog, although partially dependent upon the proximity of water, is essentially a terrestrial organism. Some ontogenetic changes in frog anatomy are produced by addition of new structures and deletion of old; others are brought about by changes in relative rates of growth.

This sort of relation does not apply to the brachiopod. Immature and mature brachiopods occupy nearly the same environment. Both are aquatic. Both use basically the same methods of getting food, and in general have the same life activities. As far as we know the principal change during growth is simply in size. It is fruitful, therefore, to look more carefully into the consequences of size increase.

Consider a leg bone of a terrestrial vertebrate. The bone serves several functions, but one of the most important is to support the body. It must be strong enough to bear the weight of the body. The strength of a bone (or any other supporting structure, for that matter) is approximately proportional to its cross-sectional area. Thus, a thick bone will be stronger than a thin bone regardless of its length. Now consider that the bone must grow in order to accommodate an increase in size and mass of the body that is to be supported. Suppose that the growth brings about a doubling in all the linear dimensions of the animal. That is, the length of the body is doubled, and so on. We are thus assuming, for the moment, completely isometric growth.

The doubling of the linear dimensions produces an eight-fold increase in volume or weight. This stems from the simple geometric fact that as we increase the linear dimension of any structure (without changing shape), the volume increases as the cube of the linear dimension. If the linear dimensions of the bone that supports the weight have doubled, the cross-sectional area of the bone has increased by a factor of only four (area increases as the square of the linear dimension). It is thus inevitable that the volume of the animal supported by the bone increases more rapidly than the cross-sectional area of the bone. Therefore, the weight of the body increases more rapidly than the effective strength of the bones that support it. D'Arcy Wentworth Thompson named this rule of scaling inequality the ***principle of similitude.*** One way in

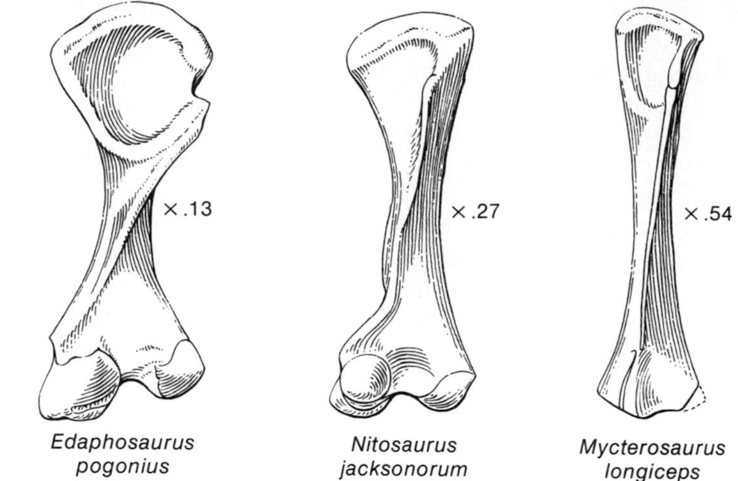

FIGURE 3-13
Femurs of three pelycosaurian reptiles, illustrating allometric change in proportions with increasing size. Actual size increases from right to left. (From Gould, 1967.)

which the organism can combat it is to increase the cross-sectional area of the bone more rapidly than would be the case if all linear dimensions of the bone were to grow in a constant ratio. This, in turn, demands a change in shape of the bone. That is, the bone must become relatively stouter in order to carry the increased weight of the body. This is in fact what occurs in the ontogenetic development of many terrestrial vertebrates.

Species that differ in size often show predictable differences in shape of supporting structures. Galileo was one of the first to recognize this phenomenon. He wrote: "I have sketched a bone whose natural length has been increased three times and whose thickness has been multiplied until, for a correspondingly large animal, it would perform the same function which the small bone performs for its small animal."

An actual case of the effect of scale on shape for leg bones is shown in Figure 3-13. Here, the femurs of three different pelycosaurian reptiles are shown with the scales of the drawings adjusted so that all bones appear to be the same length. Actually, the femur on the left is about four times longer than the one on the right. Size increase is predictably accompanied by an increase in the relative diameter of the shaft but in this particular case, even more modification occurs in the areas of muscle attachment at the ends of the bone. The reason for this is that pelycosaurs had a sprawling gait and body support depended more on muscle strength than bone strength.

We can extend the general line of reasoning to include a wide variety of structures. The efficiency of many respiratory devices, for example, depends upon the surface area of the respiratory surface. If the area of respiratory surfaces does not increase during general size increase, the surface-to-volume ratio becomes smaller and smaller and the respiratory system becomes less effective. Some organisms have evolved a common strategy for countering this problem: the respiratory surface becomes more irregular or convoluted during ontogeny, thus expanding surface area more rapidly relative to the organism's general increase in size. Another strategy evolved is simply a limit to maximal size; an optimal surface-to-volume ratio is then maintained.

A classic example of the relation between surface area and volume is found among insects. If we could so stimulate growth of a mosquito that it grew to be 10 feet long but was otherwise a perfect replica of a normal mosquito, it could not survive, because the relation between its surface area and volume would be unsuitable for temperature regulation and respiration. In order for a mosquito of such length to be viable, its morphology would have to be altered drastically.

A few kinds of animals increase the size of certain organs or of their entire bodies in a way that avoids the problems of similitude altogether. They grow not by maintaining geometric similarity, or some semblance of it, but by addition of identical structural units. Figure 3-14 illustrates this mode of growth for a sponge. A sponge is essentially an assemblage of flagellated chambers through which water is pumped for feeding and respiration. As the sponge grows, it does not increase the size of its chambers, but simply adds new chambers of the original size. In effect, each unit that is added is self-supporting, so any number can be added. Efficiency of feeding and respiration does not limit the size of a sponge. The vertebrate lung grows in the same fashion, and thus avoids problems of similitude.

Let us return to structural changes more likely to be represented in the fossil record. We have seen that the bivalve shell grows by accretionary thickening as well as by addition of material at the leading edge (Figure 3-1). The primary function of the thickening is to make the pre-existing shell (which was the entire shell of an earlier growth stage) thick enough and therefore strong enough for the adult animal. If material were not added to the inner surface, the shell would become larger relative to its thickness and would soon have insufficient strength. Gastropods and cephalopods generally do not employ a thickening process. They are able to make do with the originally deposited thickness because the shell surface has greater concavity and thus an inherent strength that is lacking in the shallower bivalve shell.

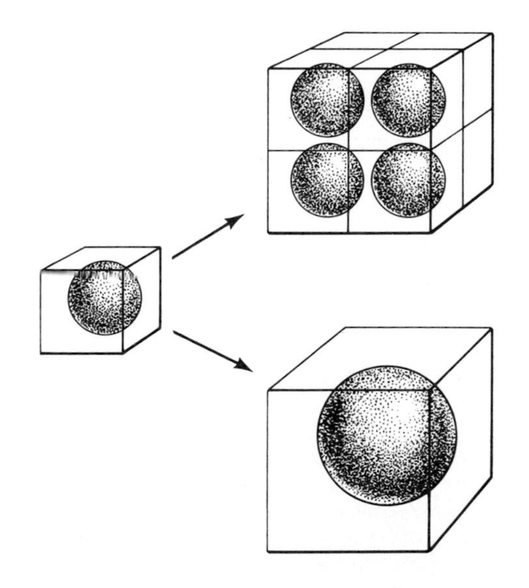

FIGURE 3-14
Diagrammatic illustration of the manner in which a sponge grows. The sponge starts out as a single chamber in a block of "tissue." Instead of maintaining geometric similarity as it grows larger (lower right), it grows larger by adding units like the original one (upper right, where chambers are shown for only the front four blocks). In real sponges, canals connect chambers with the external environment.

Many other examples of the effect of size on ontogenetic variation could be cited for groups of animals common in the fossil record. Many gaps in our knowledge exist, however. We cannot, for example, offer a convincing explanation for the change in shape of the brachiopod shells discussed earlier. Perhaps the change is size related. There are many fossil groups for which we do not have sufficient knowledge of biologic functions to be able to interpret the observed ontogenetic variation. If such knowledge could be obtained it would be applicable not only to interpreting differences between growth stages of a single species, but also to the understanding and interpretation of differences between species.

In summary, ontogenetic change in shape accompanies two other sorts of change: change in mode of life and change in size. Change in size is usually gradual, as is the change in shape accompanying it. Changes in mode of life may be sudden or gradual. The ontogenetic change in shape observed in the brachiopod may correspond to the gradual cumulative change in size or it may correspond to a rather subtle change in mode of life that has not yet been recognized.

Ontogeny of Colonial Organisms

So far, we have been concerned with organisms that live as autonomous individuals, not intimately affected by or dependent upon others. As we study the ontogeny of colonial organisms several special problems arise. Some aspects of colonial growth are very similar to those encountered in studying the ontogeny of single organisms that have many growing parts, such as echinoids. Topologically and geometrically, the many plates of the echinoid skeleton develop as if they were individuals in a colony.

In colonial corals, the individuals, or corallites, are often bound together by an integrated skeleton. Obvious effects of this mode of life are found when we compare an individual near the center of a colony with one on the margin. The relative lack of crowding on the margin of the colony may produce an ontogeny quite distinct from that of an individual completely surrounded by others. (A comparable effect is sometimes noticed in noncolonial organisms, such as oysters, which in certain environments grow so closely together that the growth of one affects the shape of its neighbors.)

Among colonial corals, the first member of a colony must attach to some solid substrate, such as a pebble, an empty shell of another organism, or even a dead coral. As the colony develops, new individuals attach to older parts of the colony. The ontogeny of a given individual varies considerably with its place in the ontogeny of the colony as a whole.

Figure 3-15 illustrates some of the complexities as well as the elegance often found

FIGURE 3-15
Ontogeny of the central region of a branch of the Ordovician trepostome bryozoan *Rhombotrypa quadrata.* A: Unrolled tangential peel of circumference of branch (after outer growing surface has been removed). B: Longitudinal section. C: Transverse section. D: Schematic drawing of the central region of a branch showing cycles in the development of quadrate zooecia. E: Detail of the transition from one growth cycle to the next younger cycle. Photographs B and C are ten times actual size; A is five times actual size. (From Boardman and McKinney, 1976.)

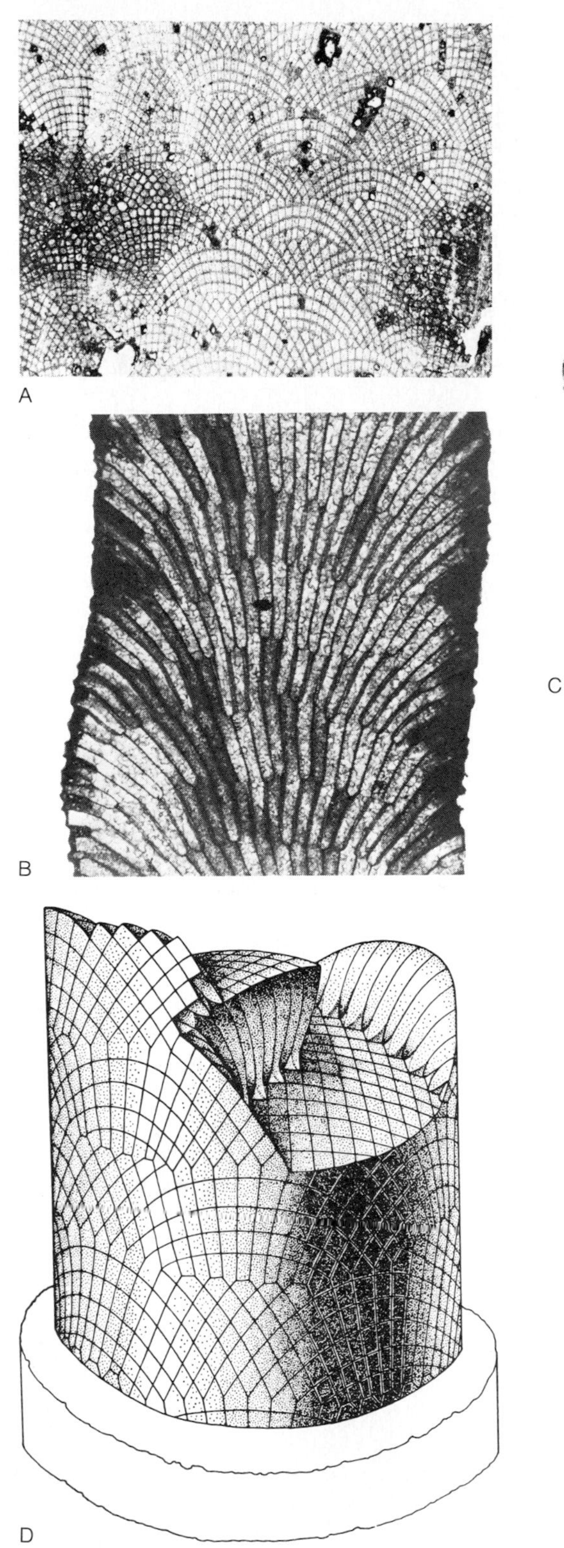
A
B
D

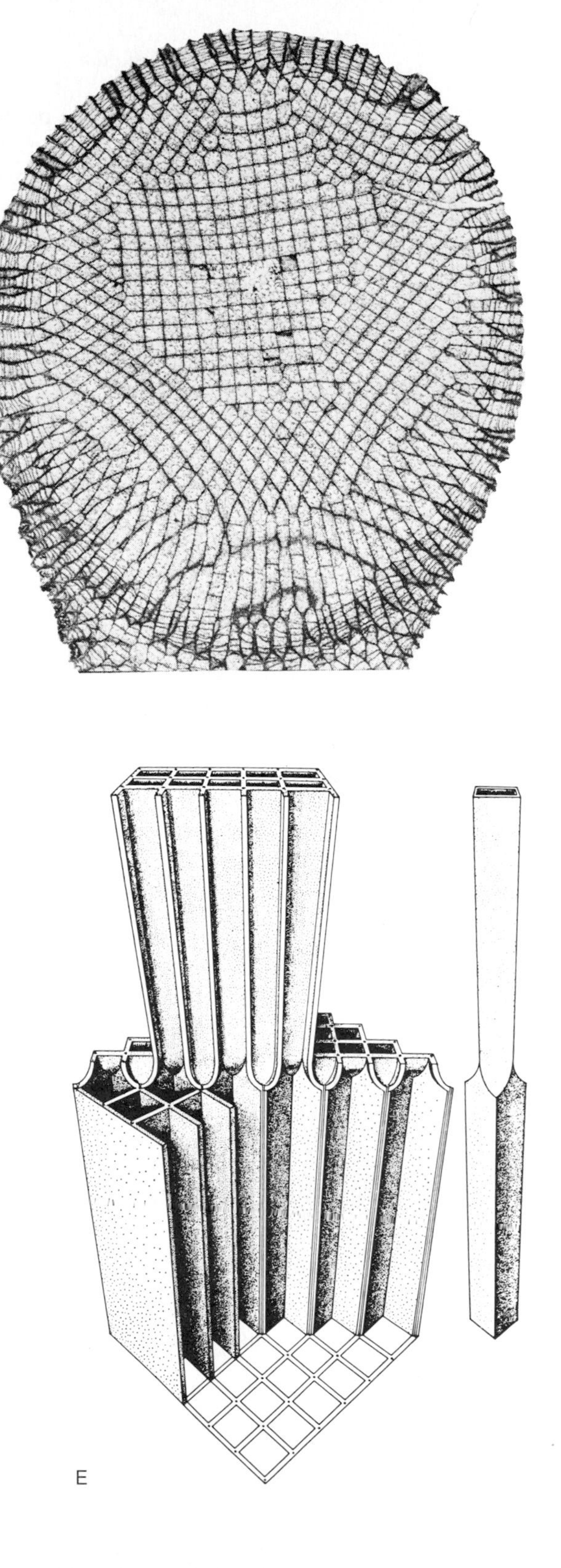
C
E

in colonial growth. The three photographs show different views of a single branch of a Paleozoic trepostome bryozoan. The individual units (zooecia) are almost perfectly quadrate in cross-section (Figure 3-15,C). New zooecia are formed by budding, and groups of zooecia form distinct cycles in colony development. In this species, buds appear at the intersections of the quadrate zooecia and develop with their sides at 45° to the pre-existing zooecia. As is indicated by the schematic drawing in Figure 3-15,E, the new cycle of zooecia is made up of half new individuals and half old ones. Several cycles of this mode of growth coupled with systematic changes in growth direction produce the complex geometry seen in the other schematic drawing (Figure 3-15,D) and in the photographs.

In some colonial organisms (including corals and bryozoans), there is morphologic and physiologic differentiation between individuals in a colony. That is, certain individuals perform a specialized function, such as reproduction or defense, which benefits the colony as a whole. These individuals may or may not be genetically different. The differences between them may be such that we must consider their ontogenies as individuals as well as the growth of the colony as a unit.

Supplementary Reading

Boardman, R. S., Cheetham, A. H., and Oliver, W. A., Jr., eds. (1973) *Animal Colonies: Development and Function through Time.* Stroudsburg, Penns., Dowden, Hutchinson, & Ross, Inc., 603 p. (An up-to-date collection of articles on many aspects of colonial organisms, inlcuding problems of growth.)

Bonner, J. T. (1952) *Morphogenesis.* Princeton, Princeton University Press, 296 p. (An outstanding biological treatment of growth and form.)

Gould, S. J. (1966) Allometry and size in ontogeny and phylogeny. *Biol. Rev.,* **41**:587-640. (An up-to-date and authoritative review of the problem of allometry. The author uses the term "allometry" to include all anisometric growth.)

Huxley, J. S. (1932) *Problems of Relative Growth.* New York, Dial Press, 276 p. (A classic reference on ontogenetic phenomena.)

Le Gros Clark, W. E., ed. (1945) *Essays on Growth and Form Presented to D'Arcy Wentworth Thompson.* Oxford, The Clarendon Press, 408 p. (A collection of essays on various aspects of ontogeny.)

Macurda, D. B., ed. (1968) *Paleobiological Aspects of Growth and Development, a Symposium.* Menlo Park, Calif., Paleontological Society. Memoir 2, 119 p. (A series of articles on skeletal growth.)

Thompson, D'A. W. (1942) *On Growth and Form.* Cambridge University Press, 1116 p. (A classic work on ontogeny. Essential reading for all interested in the interpretation of morphology.)

CHAPTER 4

The Population as a Unit

All plant and animal species exhibit variation; even genetically identical twins exhibit slight differences, due to differences in environment and to chance. Because of variation, description of a single specimen does not suffice to describe a species.

Populations in Biology

Most species do not have continuous or uniform geographic distribution. Individuals tend to be clustered because of chance, uneven distribution of habitats, gregarious behavior, or geographic barriers. Figure 4-1 shows an example of the discontinuous distribution of a terrestrial plant. For our purposes, we may define ***population*** as a group of individuals living close enough together that each individual of a particular sex has an approximately equal chance of mating with a certain individual of the other sex. (The terms ***local breeding population*** and ***deme*** are often given similar definitions.) The population shares a single ***gene pool***. The gene pools of adjacent populations of a species may be partially or completely isolated from each other. When two populations interbreed, there is said to be ***gene flow*** between them.

Subdivision of species into discrete populations is dramatically illustrated in the West Indies and other island provinces. Restriction to islands inevitably segregates terrestrial species into populations that are genetically separated. Similarly, many

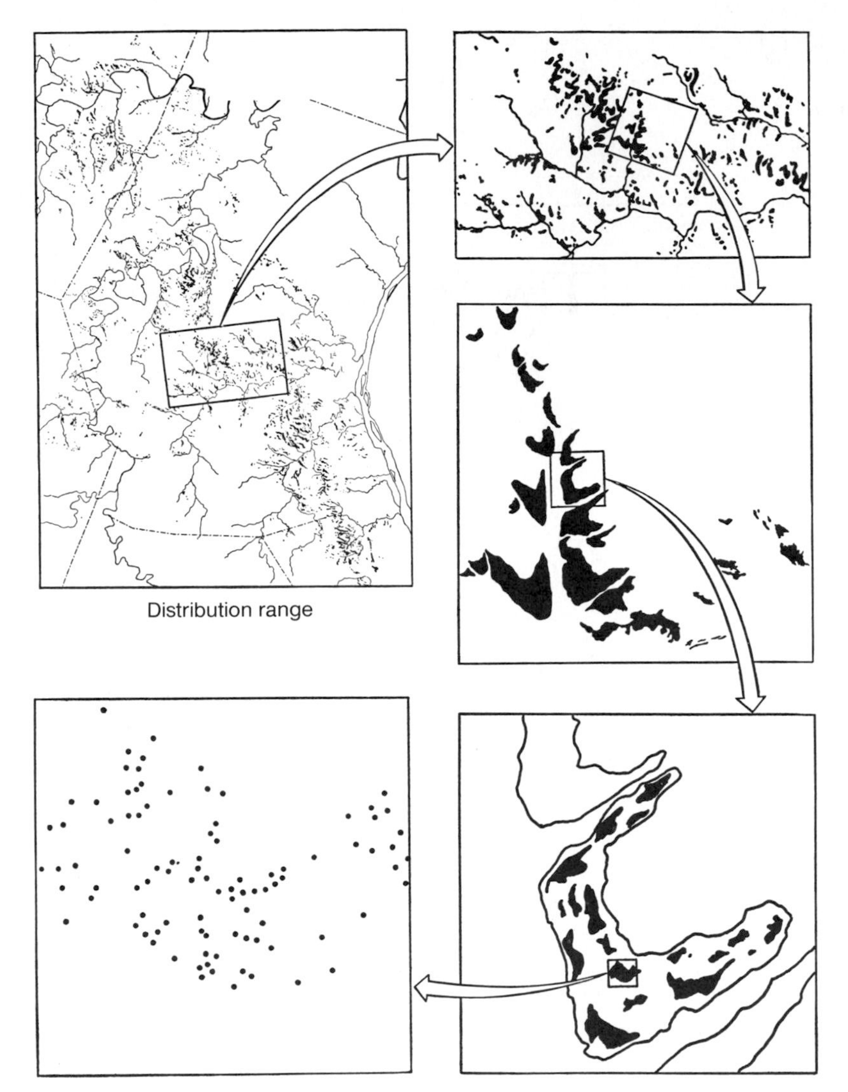

FIGURE 4-1
Discontinuous distribution of the plant *Clematis fremontii.* Dots (lower left) indicate individual plants. Note that the plants are clustered at more than one scale of observation (there are clusters within clusters). (Courtesy of Ralph O. Erickson.)

marine organisms are limited to shallow-water zones and thus are isolated from their counterparts around other islands. These examples are somewhat extreme, but similar distribution patterns are found throughout the whole biosphere. Furthermore, even if there are no geographic barriers, distance alone tends to reduce gene flow between organisms of a species that live in different parts of the geographic range.

Individual Variation within Populations

ONTOGENETIC DIFFERENCES

In the preceding chapter, we discussed changes that take place during the ontogeny of a single individual. Obviously, a living population at any time contains variation due to age differences. At the very least, individuals will differ in size.

In a species whose breeding season is fairly well defined, size variations may depend largely on time of year. Figure 4-2 shows frequency distributions (histograms) of specimen size in a living population of the bivalve *Mytilus edulis* for April and November of the same year. This species spawns during a short period of the year. The histogram for April is based on sampling just after a heavy spatfall and is clearly bimodal: the small, newly settled individuals make up the sharp peak on the left and the rest of the population (consisting of individuals one year old or older) makes up the broad area on the right. The November frequency distribution is also bimodal but both peaks have shifted to the right because most of the individuals in the population have grown. The left-hand peak has become considerably lower because some of the younger individuals have died, and broader because the growth rate varies from individual to individual.

The average age in a population may affect variation in characteristics other than size. In the preceding chapter we discussed a brachiopod species whose shell shape

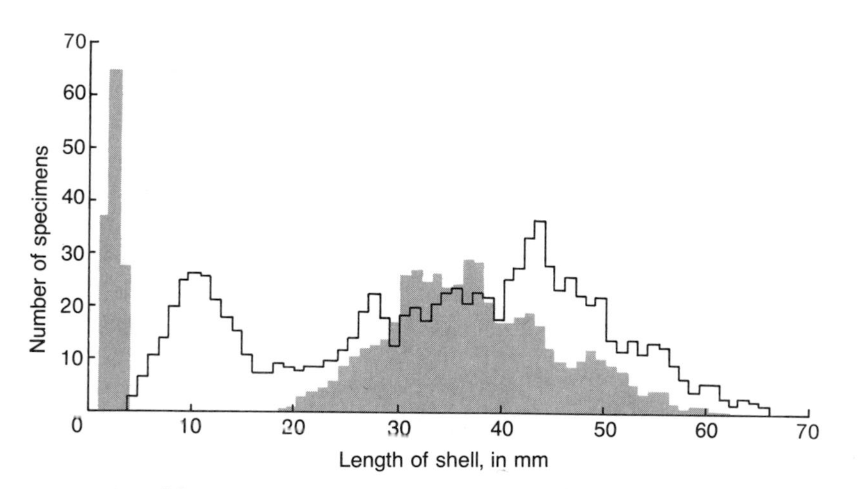

FIGURE 4-2
Relationship between age and size in a certain population. Frequency distributions of shell length in a population of *Mytilus edulis* living in Gosford Bay, Scotland. The population was sampled in April, 1961 (shaded diagram), and again in November of the same year (outlined diagram). The peak on the left in each diagram represents the younger bivalves. (From Craig and Hallam, 1963.)

changed systematically through life (Figure 3-8). Any description of shape in a population of such a species is strongly biased by the age distribution of the specimens studied.

The complex life cycles of certain taxonomic groups produce different types of ontogenetic variation. Most species of Foraminifera, for example, have a life cycle in which a generation produced sexually alternates with a generation produced asexually. Consequently, many species are dimorphic. The dimorph resulting from sexual reproduction is usually larger, but grows from a very small initial chamber. The dimorph produced asexually is usually smaller, but grows from a larger initial chamber. The dimorphs of a common Cenozoic genus *Pyrgo* are shown in Figure 4-3. In the cross-sectional view, some chambers are numbered to facilitate comparison. Dimorphism is especially pronounced in large species of Foraminifera, in which the sexually produced dimorph may be as much as five times as large as the asexually produced dimorph.

GENETIC DIFFERENCES

In most bisexual organisms, the two sexes differ morphologically because of basic genetic differences. Sexual differences may or may not be reflected by fossilized skeletal morphology. In the example of probable sexual dimorphism in fossil

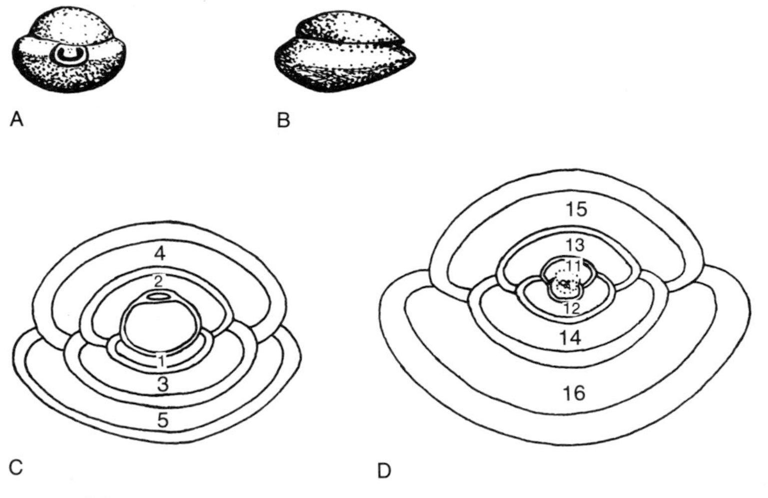

FIGURE 4-3
Pyrgo fischeri (Recent). A: External apertural view. B: External side view. C: Equatorial section of megalospheric (asexually produced) test. D: Equatorial section of microspheric (sexually produced) test. Some chambers in C and D are numbered. (From Easton, 1960.)

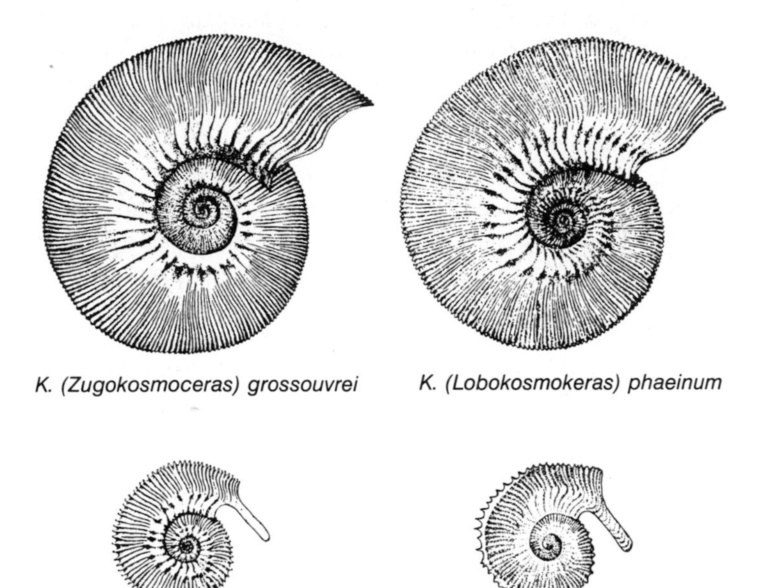

FIGURE 4-4
Sexual dimorphism as an expression of genetic variation. Jurassic ammonoid cephalopods from the Middle Jurassic of England show what has been interpreted as sexual dimorphism. Each larger form (the "macroconch") is considered to be the dimorph of the smaller form immediately below it (the "microconch"). In this case, the names predate the interpretation of dimorphism. All these forms were assigned to the same genus but different subgenera (given in parentheses) and species. (From Callomon, 1963.)

ammonites that is included as Figure 4-4, the two sexes differ in overall size, in several aspects of ornamentation, and in aperture form. The differences are about as great as those observed between many different species and genera, and, indeed, the two sexes have sometimes been classified as separate species. In making a case that particular specimens in the fossil record are sexual dimorphs, the paleontologist must consider such factors as mutual occurrence and relative numbers of the two variants in question.

Even individuals of the same sex and growth stage differ, owing to genetic variation. Each individual carries only a small part of the total gene pool. Some genetically controlled attributes of an organism are more variable than others. In human populations, for example, the number of fingers and toes is virtually constant, but other genetic characteristics, such as hair color, body size, and fingerprints, vary widely.

Genetic variation in a population may be continuous or discontinuous. That is, it may be expressed by discrete and notable differences or by a complete spectrum of intermediate forms that overlap each other. Coiling direction in certain molluscs and foraminiferans is an example of discontinuous variation. Coiling may be to the left (sinistral) or to the right (dextral). Specimens of a species in which both coiling directions are expressed are usually identical except that one type is a mirror image of the other. Both coiling directions may be present in the same breeding population.

It is much more common for genetically determined variation to be continuous than discontinuous. Continuous variation in overall size of individuals of the same growth stage and environment is especially common, but may, however, be produced by chance events during growth or undetected environmental differences as well as by variations in genetic material.

How much genetic variation should be expected within a single population? A simple answer to this question would be useful, especially to the paleontologist, who could then expect to find a certain amount of variation within a fossil population; a larger amount could be used as evidence that what was being studied were the remains of more than one population. Unfortunately, there are inherent differences in size and complexity of gene pools, making rigorous prediction of variation difficult or impossible. The problem is further complicated by the fact that in fossil populations, we can observe a genetic difference only through its morphologic expression.

NONGENETIC DIFFERENCES

In Chapter 3 the "norm of reaction" was discussed. We saw that the genetic control over the form of an organism is not complete and some variation is caused by the environment and some by chance.

Most nongenetic variation is attributable to ecologic differences. An extreme example of discontinuous nongenetic variation is found among social insects, such as honeybees and termites, where the distinction between a worker and a queen may simply result from the fact that one received more or different food than the other during an early stage of development.

The shells of many invertebrate species that nestle in crevices and cavities in hard substrata exhibit continuous nongenetic variation. The shells of such species tend to conform to the shapes of confining cavities.

Some nongenetic effects of the environment are related to function and may even benefit the organism; others seem quite unrelated to function. An example of the former is the development of calluses on the hands and feet of primates. A callus is an acquired characteristic, being the result of the reaction between the animal's tissue and the physical environment. The callus is often an advantage as protection from abrasion and bruising.

Genetic composition of an organism (or a population of organisms) can rarely be observed directly. Separation of genetic and nongenetic factors is especially difficult because the same morphologic variants may sometimes be under genetic control and sometimes under environmental control. Among echinoderms, for example, amounts of the isotopes oxygen-16 and oxygen-18 in the skeleton may differ markedly from species to species as an evolutionary phenomenon (that is, under genetic control), but differences of the same magnitude may also be produced by the effects of water temperature.

Fossil Populations

Fossil preservation commonly affects the variation evident within a population. The principal geologic process tending to increase apparent variation is distortion produced by compaction of sediments or deformation of sedimentary rocks. A particularly striking but by no means rare case is illustrated in Figure 4-5. Bambach (1973) analysed shape variation in a large sample of the infaunal bivalve *Arisaigia postornata* from Silurian rocks of Nova Scotia. For each specimen, the orientation of the rock's cleavage relative to the bivalve's morphology was noted. Because these specimens are actually two-dimensional molds of the original shells, the shapes taken by the fossils leave an accurate account of the deformational history of the rock, with the direction of maximal compression being perpendicular to the direction of cleavage. Figure 4-5 shows drawings of four typical specimens covering the range of geometric relationships of cleavage direction to morphology. The fifth drawing (center) is a reconstruction of an undeformed specimen. As can be seen, the deformation produced marked changes in morphology that are consistent with the

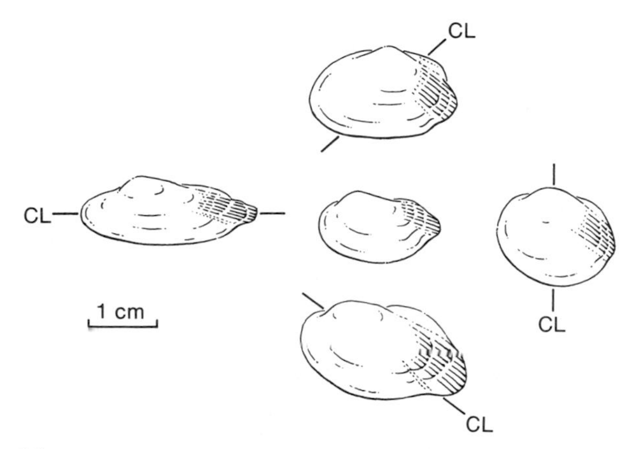

FIGURE 4-5
Effect of rock deformation on fossil morphology. The drawing in the center shows an undeformed specimen of *Arisaigia postornata.* The four drawings surrounding it show different patterns of deformation; in each case, the direction marked "CL" corresponds to rock cleavage and is perpendicular to the direction of maximum shortening. (From Bambach, 1973.)

idea that maximal compression is perpendicular to cleavage. In Bambach's collections, all of the specimens were deformed: that is, the form in the center of Figure 4-5 was not found! When dimensions of the shapes of specimens were graphed, a wide scatter of points was evident, but there was a hole or empty space in the scatter representing the space on the graph which would have been occupied by all the data had the specimens not been deformed. Thus, not only did deformation add to the amount of morphologic variation in the fossil population, it also produced an extraordinarily biased representation of the original morphology.

Reduction of variation as a result of fossilization is also common, owing to information loss stemming from incomplete preservation. Fossilization destroys many features of organisms and obscures differences that may have been very striking in the living population. Thus, fossil populations often exhibit much less variation than comparable living populations.

The paleontologist does not expect to find that an organism's soft tissues have been preserved; the absence of information about them is therefore predictable and must simply be tolerated. More serious is the biasing effect of fossilization. The preservation process serves as something of a filter for variation; that is, readily preserved organisms represent a disproportionately large part of the fossil record. Generally, individual animals with robust and solid skeletons are more likely to be preserved than members of the same species that happen to have lighter, more fragile skeletons, which may lead the unsuspecting paleontologist to characterize a species as having had a heavier skeleton than actually typified it. Selective destruction of ontogenetic stages is also common. For example, because of their fragility, few skulls of children are found in the Pleistocene fossil record (except at burial sites). The phenomenon of selective preservation is exceedingly complex and not yet well understood, but is potentially one of the most fruitful areas of paleontologic research. With the odds against preservation being high, a subtle morphologic difference may determine whether a particular specimen is preserved.

Sorting or selection during post-mortem transport is also important in influencing the composition of a fossil assemblage. A particularly striking example (with considerable paleontologic implication) is illustrated in Figure 4-6, based on a study by P. Martin-Kaye of bivalve shells collected on Trinidad beaches. The species he studied comprise burrowing individuals that live in nearshore environments. He was surprised to find that in a given area of beach, the left and right valves were not present in the equal proportions that would be expected from the fact that each living organism had one of each. Furthermore, he found that the ratio of left to right valves changed quite systematically along a single beach. Each of the species that Martin-Kaye worked with has virtually indistinguishable right and left valves except that one is the mirror image of the other. Representative specimens from one of the species are illustrated in Figure 4-6. The sorting is readily explained by the fact that wave action is not the only agent transporting the shells. Because waves usually approach beaches obliquely and produce a longshore current, most shells are not deposited on the beach directly opposite their life habitat, but are carried along by the current, finally being destroyed by abrasion or thrown up on the beach at some distance from their habitat.

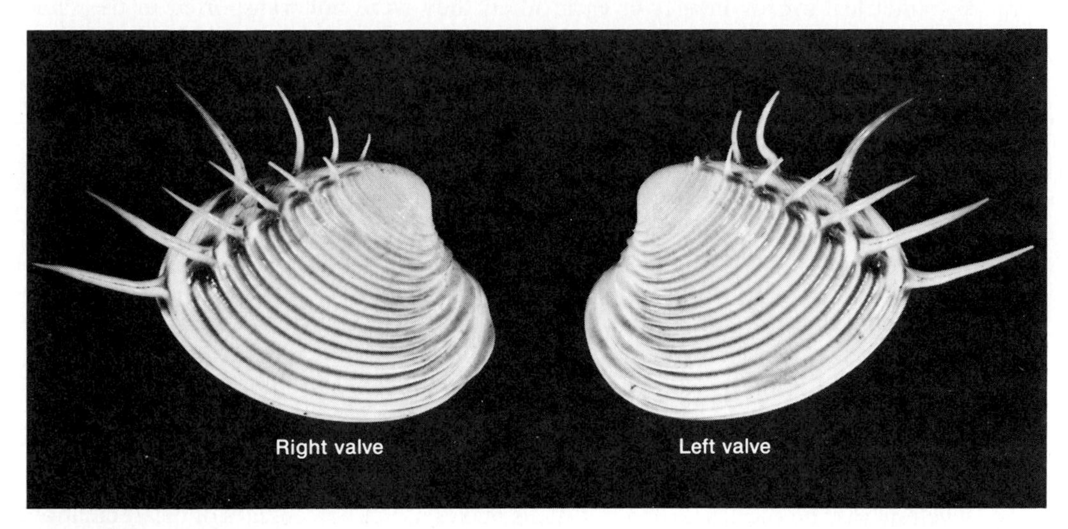

FIGURE 4-6
Effect of selective transport on bivalves. The proportions of right and left valves in beach assemblages collected on the eastern coast of Trinidad varied from place to place, but varied systematically relative to geography. The two valves of the bivalve *Pitar dione* are identical except that one is the mirror image of the other. (From Martin-Kaye, 1951.)

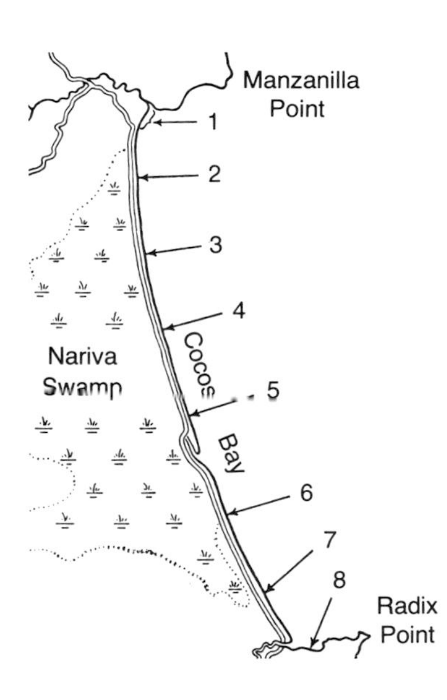

Locality	*Total Number*	*Percent right valves*
1	45	11
2	154	13
3	11	18
4	51	approx. 50
5	73	approx. 50
6	284	46
7	27	approx. 50
8	450	87

It appears that because left and right valves of the species in Figure 4-6 are asymmetrical mirror images of each other, they were not transported in the same manner by the waves. One valve tended to be sheared toward the beach and preferentially cast up on it.

The effect of differential transport of the Trinidad shells is striking, and its interpretation is reasonably obvious because we know that the ratio of left to right valves in the original population must have been exactly one. If we were to compare the assemblage of shells on the beach with original populations we probably would find other relations between morphology and transport.

Biases in the fossil record produced by differential transport are difficult to evaluate. In some groups of organisms, such as buoyant cephalopods, transportation after death is probably the rule rather than the exception. But we do not know how important post-mortem transport is among various types of plants and animals that normally live on a firm substratum. Recent work in oceanography and submarine geology has demonstrated widespread currents that are capable of transporting some bottom-dwelling organisms considerable distances.

A method commonly used for estimating transport distance for those bottom-dwelling marine species having two shells makes use of percentages of disarticulation. Valves of some brachiopods, for example, are more easily separated after death than others. From structural considerations it is often possible to estimate the types and amounts of agitation or transport that would be necessary to produce disarticulation of particular species. Amount of breakage and wear may be interpreted in the same manner.

Boucot, Brace, and Demar (1958) studied the effects of transport on three brachiopod genera in a large block of Lower Devonian sandstone. They found that two genera were each represented by nearly equal numbers of pedicle and brachial valves (*Leptocoelia*—879 pedicle valves, 893 brachial valves; *Platyorthis*—561 pedicle valves, 548 brachial valves). A third genus, *Leptostrophia,* was represented by 378 pedicle valves but only 35 brachial valves. The authors concluded that members of the first two genera had not been transported far from where they had lived and that the third genus had probably lived in a distant environment and had been transported after death by bottom currents. Of the two genera considered to have been preserved near their life site the percentage of whole *Leptocoelia* fossils was much higher than that of *Platyorthis* fossils. The authors concluded that *Leptocoelia* tended to disarticulate less readily than *Platyorthis.*

Population Dynamics

Population dynamics is the study of changes in population size and composition through time. Birth rate, growth rate, and death rate are the kinds of basic information considered. Reproduction in most animal groups is limited to a certain part of ontogeny, beginning at sexual maturation and ending late in ontogeny or at death. Birth rate (rate of offspring production) may remain constant for a single parent

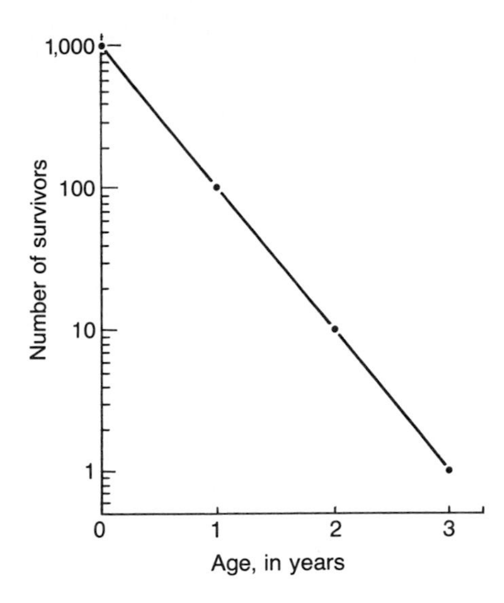

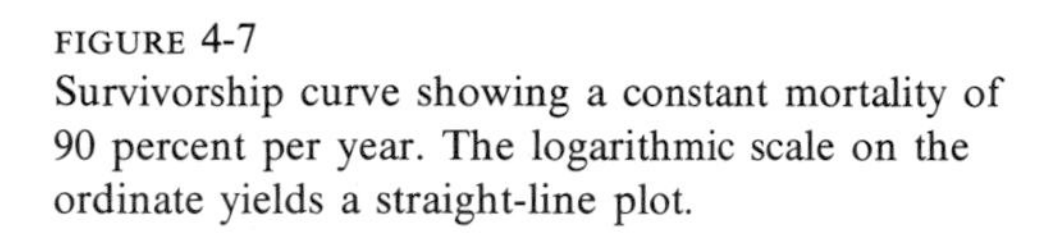

FIGURE 4-7
Survivorship curve showing a constant mortality of 90 percent per year. The logarithmic scale on the ordinate yields a straight-line plot.

during this time or may change. Growth is usually rapid at first and then slower; it either ceases rather abruptly (as in humans) or its rate decreases continuously until death (as in most invertebrates). Death rate is often high early in life (as in humans); it often increases again later, either abruptly (as in humans) or gradually (as in many invertebrates).

Population biologists commonly make use of what is called a survivorship curve—a graph on which the number of survivors is plotted against their age. The data may represent a single population or several populations of a species. Usually the data are plotted on a logarithmic scale because, as illustrated in Figure 4-7, constant mortality then produces a straight line. (In Figure 4-7, 90 percent of existing individuals die each year.)

The paleontologist seeking to study population dynamics first faces the problem of determining whether a fossil assemblage actually represents a former population. As demonstrated in the previous section, this problem may not be easily solved. Not only do destructive and sorting effects produce change, but it is often difficult to determine whether all of the specimens in a fossil assemblage lived at the same time.

It was once assumed by many paleontologists that a simple size-frequency plot could be used as a means of distinguishing between altered and unaltered fossil assemblages of shells or skeletons. It was assumed that high infant mortality is the rule for most animal species and that accumulation of dead shells or skeletons with little or no selective transport or destruction should produce a fossil assemblage in which small individuals greatly outnumber large individuals. The size-frequency distribution for such an assemblage would resemble the one shown in Figure 4-8,A. A normal, bell-shaped size-frequency distribution (Figure 4-8,B) was then taken as evidence of selective sorting during transport, with the mean size determined by factors such as the velocities of transporting currents.

FIGURE 4-8
Left-skewed (A) and normal (B) size-frequency curves for a single-species fossil assemblage, once (but no longer) thought to be evidence of preservation with little transport or sorting (A) and of sorting by currents (B).

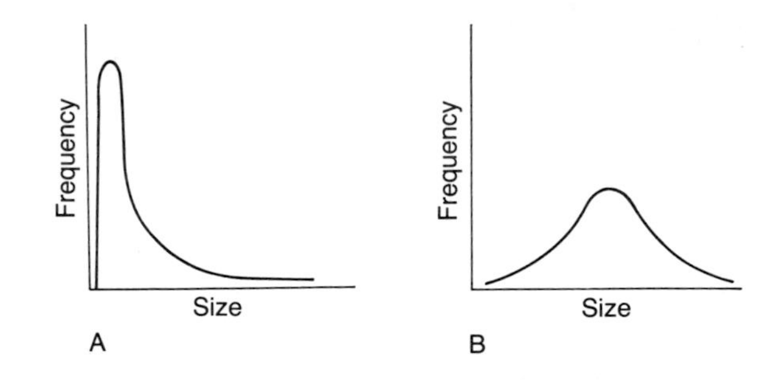

However, Craig and Hallam (1963) pointed out that for most bottom-dwelling marine invertebrates, infant mortality occurs chiefly in planktonic larval stages, which are not preserved with adults. Hence, post-larval fossil assemblages of these organisms do not reflect high infant mortality. In fact, mortality increases during the post-larval life history of most species. Craig and Hallam showed that, by considering measured growth rates for *Cardium edule* (the European cockle) and various possible survivorship curves, the paleontologist can mathematically generate hypothetical death assemblages with a variety of size-frequency distributions. In Figure 4-9, A shows the normal growth curve for *Cardium, B* shows four possible survivorship curves, and *C* and *D* show the size-frequency distributions generated by combining the growth curve with constant mortality (survivorship curve II) and constantly increasing mortality (survivorship curve III).

In other words, a variety of size-frequency distributions can develop depending on growth rate and survivorship in a particular population or species. The distributions represented in *C* and *D* are, in fact, similar to those of Figure 4-8, indicating that a normal, bell-shaped curve does not necessarily give evidence of sorting by currents and that a skewed distribution does not necessarily give evidence of high infant mortality.

It is important to understand that growth and survivorship curves vary from population to population within species and from time to time within populations. The most important factors causing variation in survivorship curves within species are environmental. The many variables in production of a fossil assemblage, however, often make analysis of fossil population dynamics very difficult.

A critical factor in applying population dynamics to fossil taxa is recognition of relative age groups within a species, based on such features as tooth wear (mammals), growth rings (fish scales, echinoderm plates, and mollusc shells), and molt stages (arthropods). Some of these features permit segregation into distinct year classes. Growth rings, for example, commonly represent very slow growth during winter, and molt stages of arthropods are sometimes produced by annual shedding of the exoskeleton. In the absence of distinct growth markers, size-frequency plots may yield distinct age groups (Figure 4-2).

The conventional vehicle for working with the age structure of a population is called a ***life table.*** Three hypothetical examples are shown in Table 4-1. In the first example,

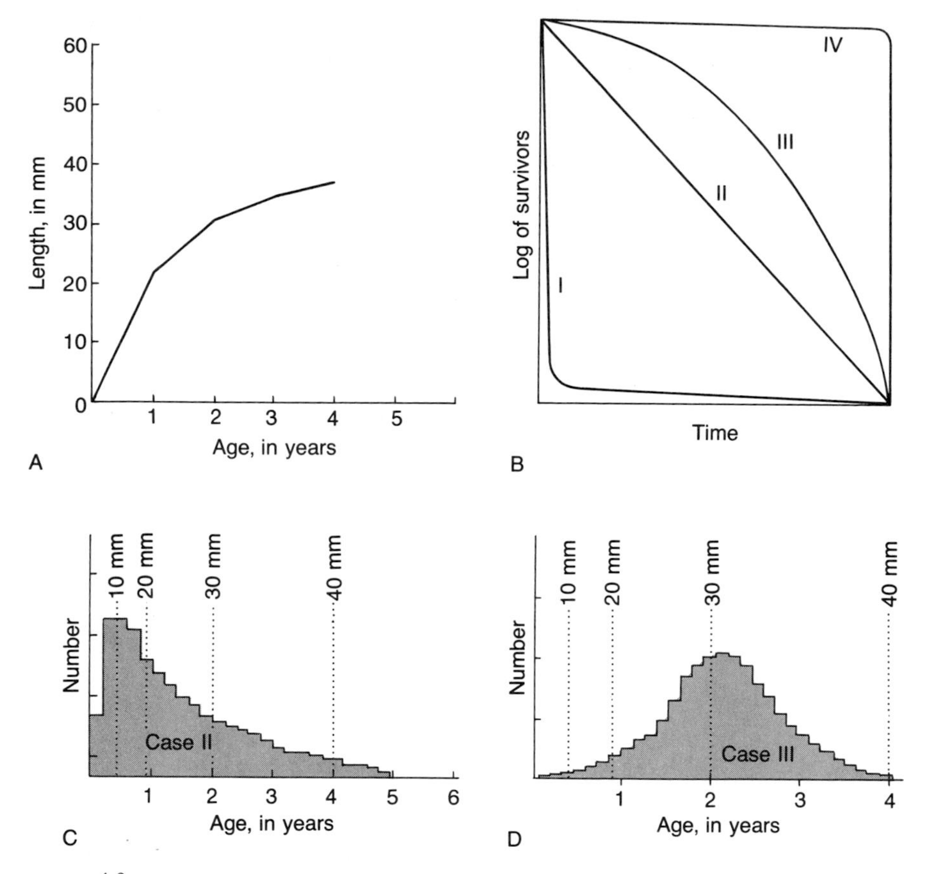

FIGURE 4-9
The production of left-skewed and normal size-frequency curves by combining various death rates with a single growth curve. A: Typical growth curve for *Cardium edule* (European cockle). B: Four possible survivorship curves. C: Size-frequency curve for shells of dead cockles, based on growth curve A and survivorship curve II, which represents a constant death rate. D: Size-frequency curve for shells, based on growth curve A and survivorship curve III, which represents a constantly increasing death rate. (From Craig and Hallam, 1963.)

the probability of death (q_x) is constant throughout ontogeny so that the same proportion of each age group dies during each time interval. In this case, 1,000 individuals are "born" during each time period (one unit of "x") so that there will always be approximately 1,000 in the first age class. Of these, 900 will die sometime during the zero-to-one period and the remaining 100 survive to start the one-to-two period. During the one-to-two period, 90 of these 100 will die. The straight survivorship curve in Figure 4-7 is based on this example: it is a plot of x versus log (l_x) in the life table.

The other two examples in Table 4-1 differ from the first in having increasing and decreasing mortality rates (q_x), respectively. Note that in all three cases, the l_x column

TABLE 4-1

Life Tables for Three Hypothetical Cases. A: Constant Mortality Rate; B: Increasing Mortality Rate; C: Decreasing Mortality Rate.

Age interval (x)	*Number dying in age interval* (d_x)	*Survivors at start of interval* (l_x)	*Mortality rate* (q_x)
	CASE A		
0–1	900	1,000	0.90
1–2	90	100	0.90
2–3	9	10	0.90
>3	1	1	—
	CASE B		
0–1	200	1,000	0.2
1–2	320	800	0.4
2–3	288	480	0.6
>3	192	192	—
	CASE C		
0–1	600	1,000	0.6
1–2	160	400	0.4
2–3	48	240	0.2
>3	192	192	—

is the cumulative form of the d_x column: the first l_x value is the sum of all the d_x values, the second l_x is the sum of the second through last d_x, and so on. Because of this relationship, l_x can always be calculated from d_x, and vice versa. As we shall see, this is important in paleontologic applications.

All the fossils in an assemblage may have died at the same time—as would be the case if a marine benthic population were suddenly killed and buried by an influx of sediment. In such situations, the paleontologist is looking at the age structure of the population at an instant in time and his counts of specimens in age classes appropriately belong in the l_x column of the life table. But fossil assemblages can also be formed as a gradual accumulation of individuals dying naturally. Then, the frequency data are most appropriately considered as d_x and must be put through the cumulation process to get l_x—the data necessary for a survivorship curve. Most fossil assemblages are probably a mixture of l_x- and d_x-type mortality records, but it is often possible to use sedimentologic and other evidence to determine which kind of mortality predominates.

Let us look at an actual example. Figure 4-10,A shows the size frequency distribution of a collection of Silurian ostracodes from the Mulde Marl of Gotland.

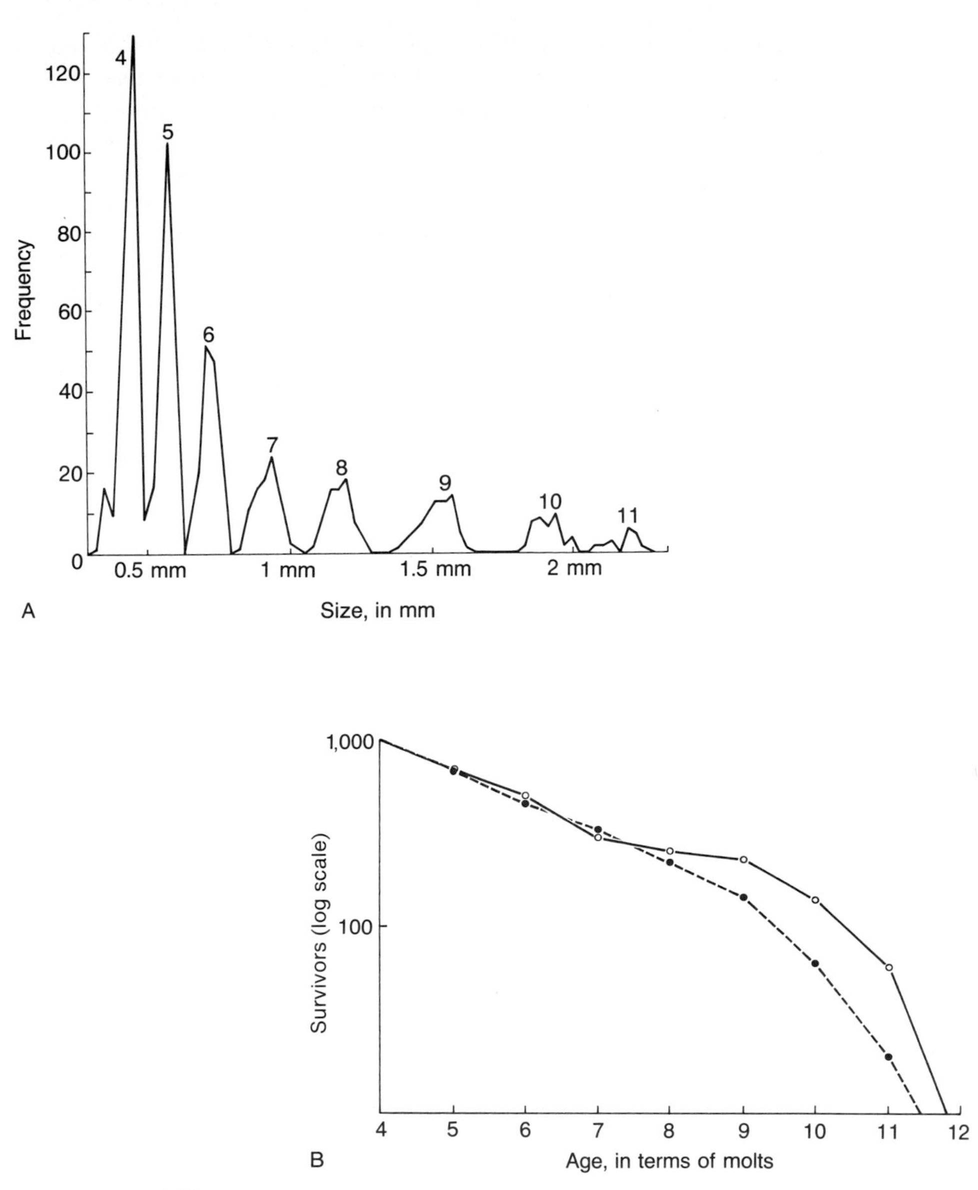

FIGURE 4-10
Survivorship in the Silurian ostracode *Beyrichia jonesi* from the Mulde Marl of Gotland. A: Size-frequency distribution for a collection of 972 valves. Numbered peaks represent successive molt stages. B: Survivorship curves based on A, using data from Table 4-2. The solid curve is based on *time-specific* analysis. The dashed curve is produced by *dynamic* analysis. Both assume that no molts have been preserved. If it is assumed that molts *are* preserved, the survivorship pattern will be identical to the solid curve. (A from Spjeldnaes, 1951; B modified from Kurtén, 1964.)

Because ostracodes molt periodically during growth, sizes tend to cluster into a succession of classes which can be identified readily from the numerical data. In this case, eleven molt stages are recognized but only the last eight are identified in the frequency data because of poor preservation in the smallest stages. A total of 972 *valves* are represented in the graph but because the ostracode is a bivalved organism, the numbers in each size class are halved for life table purposes in order to approximate the number of *individuals.* The halved values are included in the l_x column of the first life table in Table 4-2. By putting the frequency data under l_x, we are *assuming* that all the ostracodes died at the same time—as the consequence of some catastrophe. The d_x values were calculated from l_x and then the mortality rate (q_x) was calculated as a ratio of the two.

The second life table in Table 4-2 makes the alternative assumption—that the ostracode assemblage was formed as an accumulation of natural deaths. The raw frequency data provided the d_x column; the l_x and q_x columns were calculated from the d_x values. The two approaches exemplified by the tables are called ***time-specific***

TABLE 4-2
Life Tables for the Silurian Ostracode *Beyrichia jonesi*

Age at start of interval (x)	*Number dying in age interval* (d_x)	*Survivors at start of interval* (l_x)	*Mortality rate* (q_x)
	TIME-SPECIFIC METHOD[1]		
4	50	155	0.32
5	32	105	0.30
6	27	73	0.37
7	8	46	0.17
8	3	38	0.08
9	14	35	0.40
10	11	21	0.52
11	10	10	—
	DYNAMIC METHOD[2]		
4	155	483	0.32
5	105	328	0.32
6	73	223	0.33
7	46	150	0.31
8	38	104	0.37
9	35	66	0.53
10	21	31	0.68
11	10	10	—

[1] Raw data entered as l_x.
[2] Raw data entered as d_x.

and ***dynamic,*** respectively. They can be compared graphically by looking at the two survivorship curves in Figure 4-10,B. The two methods yield surprisingly similar patterns in this particular case, though this is not usually true. The preservational circumstances of the Mulde Marl assemblage do not suggest mass mortality and thus the dynamic method is probably most appropriate for these ostracodes.

Because of the molting system common to ostracodes and other arthropods, the Mulde Marl case is open to yet another interpretation. The fossils may not represent the death of the organisms that made them. If we assume that *all* the specimens in the assemblage are just molts and therefore, that death is not recorded, we may interpret the numbers in the several molt stages as indicating the proportions of living individuals of various ages in the population. This means that the numbers reflect l_x rather than d_x data even though the molts may have accumulated in the sediment over a long period of time. The result of this reasoning is numerically identical to the first life table in Table 4-2! This may explain why the two survivorship curves of Figure 4-10,B are so similar.

Survivorship analysis of the sort just described is a valuable part of the paleontologist's arsenal of methods. Although rates of growth, and especially rates of mortality, may vary markedly from population to population within species and from time to time within populations, certain fossil species may be found to possess characteristic survivorship curves. It may then be possible to relate such survivorship curves to the basic life history of that species, its mode of life, and its reproductive habits.

Methods of population dynamics have also been applied to the study of evolutionary changes, with extinction substituting for death in the logic of the life table and evolution of new biologic groups substituting for birth. Thus, the life table and its interpretation is not confined to local populations of a single species. Evolutionary applications will be discussed in Chapter 11.

Procedures in Describing Variation

We have discussed many of the causes, both biologic and geologic, of variation within and between fossil assemblages. It is important to express this variation in a clear and meaningful form.

VARIATION WITHIN ASSEMBLAGES

The simplest statistical representation of variation is the frequency distribution, which may be expressed graphically by a histogram or frequency diagram. This representation is an example of ***univariate analysis.*** From it we can learn much about variation of a single attribute—the amount of departure from some average value, whether the variation is symmetrical or skewed, and so on. Many highly sophisticated techniques may be used to analyze univariate distributions. The paleontologist should become familiar with these so that he can express and interpret univariate distributions effectively.

As we saw in discussing ontogenetic variation (Chapter 3), many morphologic relations are better expressed if two attributes are handled at a time. The scatter diagram is the simplest graphic form of such ***bivariate analysis.*** The scatter of points may approximate a straight line passing through the origin (as, for example, in a diagram of perfect isometric growth). The *average* relationship between the two variables is most commonly expressed by the slope of a straight line that represents the line of "best fit" to the data. The slope of such a line can be calculated in several ways. Very often, the scatter of points defines a line that, though straight, does not pass through the origin. To obtain the average relationship between the attributes, not only the slope of a line but also its vertical position must be found.

What does one do when a scatter of points cannot reasonably be approximated by a straight line? We encountered this situation in discussing ontogenetic variation. Allometry is sometimes approached closely enough that a logarithmic plot renders the distribution nearly linear. Where this is not the case, more elaborate methods of "curve fitting" must be used. In a bivariate distribution, the amount and kind of variation can be expressed only after some line (straight or not) has been fitted to the data. Variation from the "norm" or average is expressed by departure from this line.

Describing a group of individuals by citing variation in only one or two attributes greatly oversimplifies the situation. We have seen that some attributes may be nearly constant in a population, while others vary greatly. A fuller picture of variation can be drawn only by including many attributes and applying univariate and bivariate techniques to them sequentially or simultaneously. This is ***multivariate analysis,*** which depends upon a purely numerical analysis to measure the kind and amount of variation in a sample. Multivariate analysis will be discussed more fully later in this chapter. Because accumulating and processing data for multivariate analysis requires a good deal of time, high-speed computers are of great assistance, not only in accumulating data through various digitizing techniques, but also in processing statistical data. Few areas of science possess more raw data than paleontology and it is thus not surprising that the statistical evaluation of variation by computer has become prominent in paleontologic research.

Even with the most ambitious and sophisticated systems of digitizing and computing, the quantitative description of variation is subject to the problems described in Chapter 2. Since we cannot describe all aspects of variation, nor do we even want to, we must make a selection. The selection requires a subjective appraisal, often in advance, of what characteristics in the population are likely to be significant or interesting, or are likely to be directly applicable to some larger problem. Because variation can be described in an almost infinite number of ways, no single way is unique or perfect.

VARIATION BETWEEN ASSEMBLAGES

Many pictorial, graphic, and numerical methods are useful in describing differences between living populations or fossil assemblages. Some are like those used to describe variation within a population or a single fossil assemblage. The primary difference is that *individuals* are compared in describing variation within an assemblage and *groups* of individuals in describing that between assemblages. Sometimes, however, a particular group of individuals (the population or fossil assemblage) may be characterized by a single individual or by the description of a hypothetical average individual. Whether or not to characterize a population by a single individual depends upon the variation shown by the assemblage and upon expected or observed differences between assemblages. If the differences between assemblages are relatively large, a single specimen from each may suffice to describe them. In fact, the differences may be so great that *any* individual, typical or not, drawn from each assemblage may be used to describe differences reasonably.

A problem in every study is deciding how many morphologic characteristics should be judged. Much depends on whether the ultimate purpose is to explore the *differences* or the *similarities* between the assemblages. If we are only concerned with showing that two assemblages differ, a single characteristic may suffice. Figure 4-11 shows frequency distributions of specimen size (the single characteristic measured is maximal width) for two subspecies of brachiopods. The two frequency distributions overlap but clearly differ. In Figure 4-12, the frequency distributions for the width-to-length ratio of the same brachiopods are shown. Even without elaborate statistical treatment, it is evident that the two assemblages do not differ significantly in

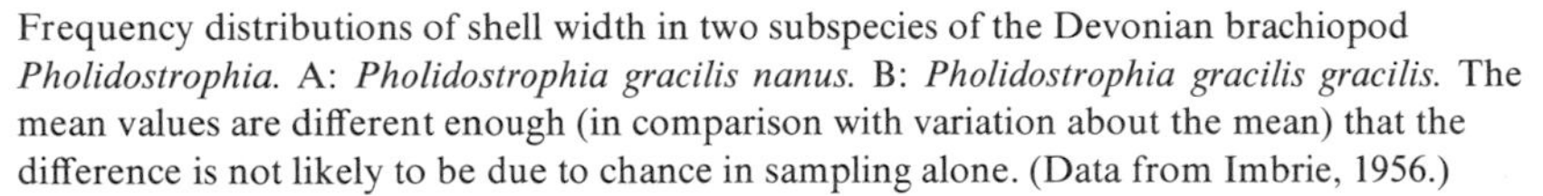

FIGURE 4-11
Frequency distributions of shell width in two subspecies of the Devonian brachiopod *Pholidostrophia*. A: *Pholidostrophia gracilis nanus.* B: *Pholidostrophia gracilis gracilis.* The mean values are different enough (in comparison with variation about the mean) that the difference is not likely to be due to chance in sampling alone. (Data from Imbrie, 1956.)

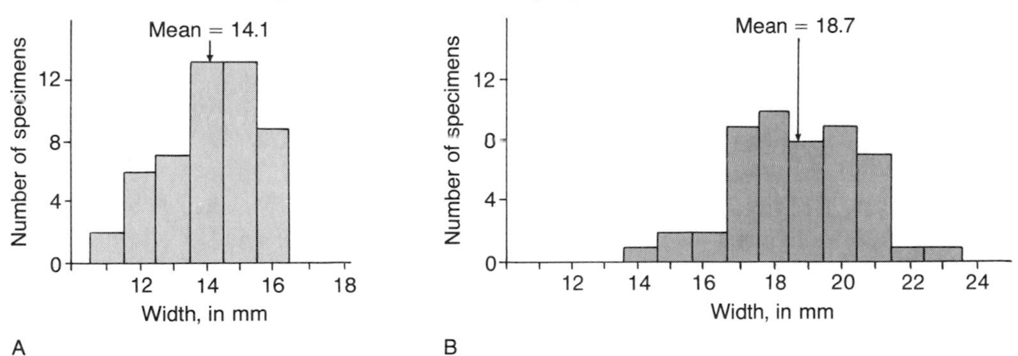

width-to-length ratio. We cannot argue from this, however, that the two assemblages do not differ because we saw in Figure 4-11 that they differ in width! It is common in analyses of this sort for the unsuspecting researcher to find near identity in frequency distributions for some morphologic characteristics and conclude that the samples do not differ, when there may actually be striking differences that are not reflected in the measurements used.

The two examples illustrated in Figures 4-11 and 4-12 are extremes and between them lie an infinite number of intermediates. We must have statistical devices to distinguish the intermediates. Above all, we must be able to determine whether an observed difference between frequency distributions should be looked upon as real (and, therefore, subject to geologic or biologic interpretations) or whether it is simply the result of chance (in sampling). In a univariate analysis, a statistical test may be used to determine whether an observed difference between the mean values of two frequency distributions might have occurred by chance alone. Inevitably, this requires assessment of the difference between the means as well as assessment of the variation within each of the two assemblages. To explain the importance of assessing variation, Figure 4-13 shows two pairs of hypothetical frequency distributions. The differences between the means are identical, but the variation in the individual sample is much greater in B than in A. Obviously, the difference between the means in A is less likely to have occurred by chance than that in B. It is reasonable, therefore, that an expression of variation within the sample enter into the judgment of the significance of the difference between the means. The "answer" given by the statistical test is not a simple "yes" or "no" but an expression of the probability that the observed difference between the means could have occurred by chance.

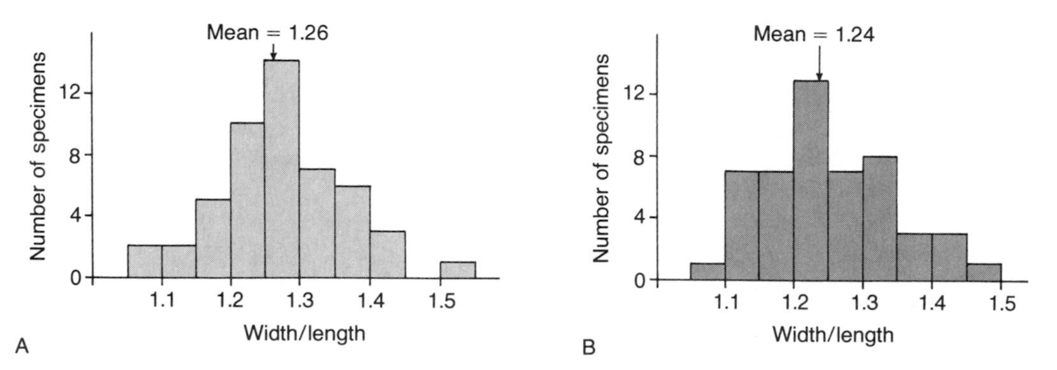

FIGURE 4-12
Frequency distributions of width-to-length ratio in two subspecies of the Devonian brachiopod *Pholidostrophia*. A: *Pholidostrophia gracilis nanus*. B: *Pholidostrophia gracilis gracilis*. The slight difference between the mean values could easily be due to chance errors in sampling. (Data from Imbrie, 1956.)

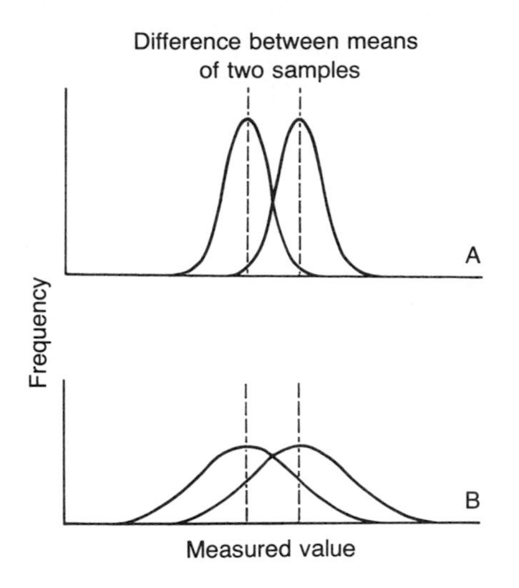

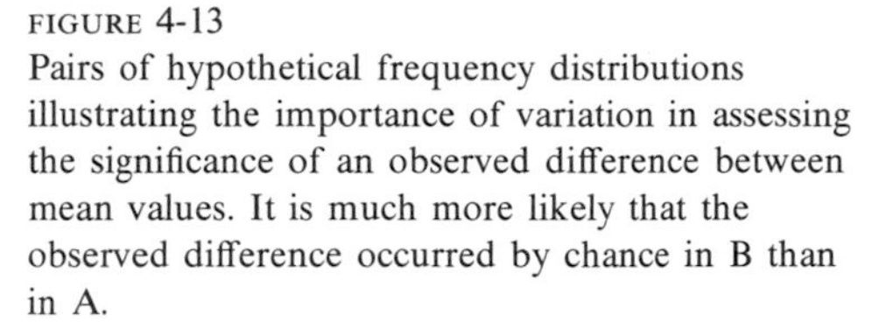

FIGURE 4-13
Pairs of hypothetical frequency distributions illustrating the importance of variation in assessing the significance of an observed difference between mean values. It is much more likely that the observed difference occurred by chance in B than in A.

An ever present problem is that of deciding how low the probability must be before we are safe in concluding that the difference between the means is significant. A value of 0.05 is commonly used as the cut-off point. In other words, we must be 95 percent sure that the difference could not have occurred by chance before we accept the hypothesis that the samples are different. At first glance, this may seem to be a rather stringent requirement. There are several good reasons for it, however. Perhaps the most compelling is the possibility for error that exists even with this cut-off point: if a researcher makes twenty decisions and each is based on a 95 percent chance of being correct, then the odds are that one of the decisions will be wrong! Even more important, if a researcher makes a final decision, which is dependent upon a sequence of several decisions made earlier, the probability that the final decision is correct will be substantially less than 95 percent. If the calculated probability of chance occurrence is less than 0.05, the difference between the means is said to be "statistically significant" at the 5 percent level. The test of statistical significance in itself gives no specific geologic or biologic information, but only tells us whether the observed differences may be due to chance in sampling. If a difference "passes" this test, we are left with the problem of *interpreting* the difference in terms of causal factors.

Many quantitative studies of morphologic differences between fossil assemblages are based on bivariate analysis. Figure 4-14 shows a typical bivariate comparison of data from two assemblages. A straight line has been statistically fitted to the data for each sample. The two lines differ in slope and vertical position. Thus, the two subspecies differ in shape and in ontogenetic development of shape. But are the differences statistically significant? Tests of significance must be applied. If the difference between the lines turns out to be statistically significant, we may proceed to make a biologic or geologic interpretation.

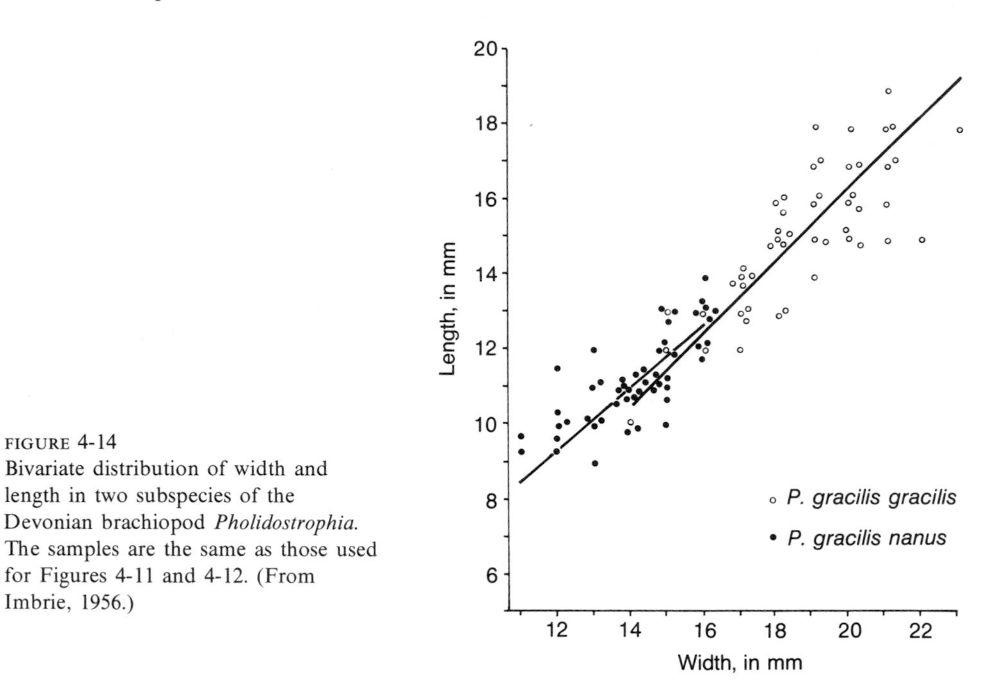

FIGURE 4-14
Bivariate distribution of width and length in two subspecies of the Devonian brachiopod *Pholidostrophia.* The samples are the same as those used for Figures 4-11 and 4-12. (From Imbrie, 1956.)

Note that the data used for Figure 4-14 are the same as those for Figures 4-11 and 4-12. The fact that Figure 4-11 showed a large difference between the samples, and that Figure 4-12 did not, should be understandable from the bivariate relationship shown in Figure 4-14.

MULTIVARIATE ANALYSIS

So far, we have been concerned primarily with treating morphologic characters singly (univariate analysis) or in pairs (bivariate analysis). These approaches are attractive because they are readily amenable to graphical display (histograms and scatter plots, respectively) and because the statistical methodology is well developed and reasonably simple. Also, the computations can often be done without computers. But most morphologic situations are sufficiently complex that simultaneous treatment of more than two variables (*multivariate analysis*) is virtually required. In Figure 4-14, we saw that two variables (length and width) were quite effective in separating brachiopod subspecies. There was some overlapping of the subspecies in the scatter plot of the data, however. Would combinations of other variables have provided as good or better separation, or are length and width the only distinguishing characters in this case? Would the addition of more information make it possible to further subdivide each group? One solution to this sort of problem is to repeat the bivariate analysis on a succession of pairs of variables—to search for differences when variables other than length and width are paired. This is done commonly in paleontologic studies, though it becomes rather cumbersome as the number of characters is increased. Also, it is unlikely to reveal relationships that depend on minor contributions of several characters at the same time.

An informal but often effective method of multivariate analysis is illustrated in Figure 4-15. It is known as the "glyph" system. The basic framework of the analysis is a bivariate plot. In this case, several specimens from each of two populations (two species) were measured for six characters. Measurements of two of the characters were expressed on the vertical and horizontal axes and the other four were added to the diagram by embellishing the symbols (glyphs). By this method, we can gain a subjective but quite valid impression of the patterns of morphologic variation in six characters at once. In Figure 4-15, the striking feature is that the two species are distinct with respect to all six variables. This would have been evident no matter which pair was chosen for the primary axes of the plot.

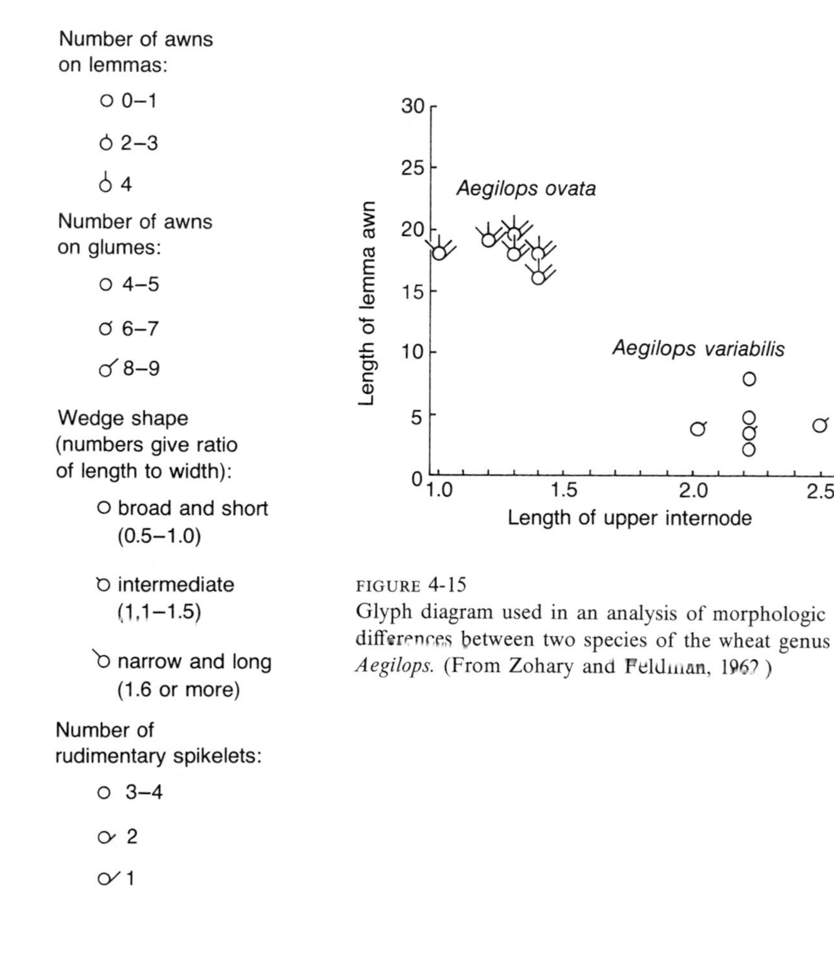

FIGURE 4-15
Glyph diagram used in an analysis of morphologic differences between two species of the wheat genus *Aegilops*. (From Zohary and Feldman, 1962.)

The glyph system cannot handle very many characters at once, especially if the relationships are not as clear-cut as in the example just given. Thus, multivariate analysis usually requires more complex numerical techniques. The end result may be a graphical or pictorial display but the data must pass through a series of transformations before graphical treatment is feasible. Usually, the primary objective of multivariate analysis is to reduce the effective number of variables so that relationships can be visualized and interpreted more readily. The variables that are finally used are often synthetic in the sense that they are combinations of the original variables. In the following paragraphs, some of the salient principles and methods of multivariate analysis will be introduced. Fuller treatments are listed in the bibliography at the end of this chapter.

The foundation of any multivariate analysis is the ***data matrix.*** This is simply a table showing the observations (usually counts or measurements) made on a series of specimens. An example is shown in Table 4-3. In this case, 15 measurements were made on each of 10 specimens, thus yielding 150 pieces of information. Data matrices are often larger than this one, having a larger number of variables or specimens or both. The data in Table 4-3 are somewhat unusual in that all measurements were made

TABLE 4-3
Data Matrix Showing Multivariate Analysis of 15 Measurements and 10 Specimens

Specimens	*Morphologic Characters* 1	2	3	4	5	6	7	8	9	10	11	12	13	14	15
1	74	12	79	27	105	26	7	48	31	39	76	24	2	4	7.3
2	69	12	74	24	120	24	7	44	29	33	75	23	2	4	6.9
3	71	11	72	24	130	24	8	44	30	32	73	22	2	4	7.0
4	82	14	86	30	115	25	8	47	35	40	60	29	3	6	8.0
5	77	13	81	30	110	27	9	50	34	37	70	26	1	3	7.5
6	67	11	72	22	115	24	7	46	28	33	75	23	2	4	6.7
7	58	10	62	20	110	21	6	40	25	29	67	22	2	4	5.6
8	59	9	62	19	100	20	5	36	25	27	62	18	1	2	6.1
9	78	15	80	28	95	27	9	50	33	38	85	30	1	2	7.8
10	65	10	70	22	105	22	5	41	28	32	68	20	2	4	6.3

R-mode (boxed: characters 3 and 4)

Q-mode (boxed: specimens 3 and 4)

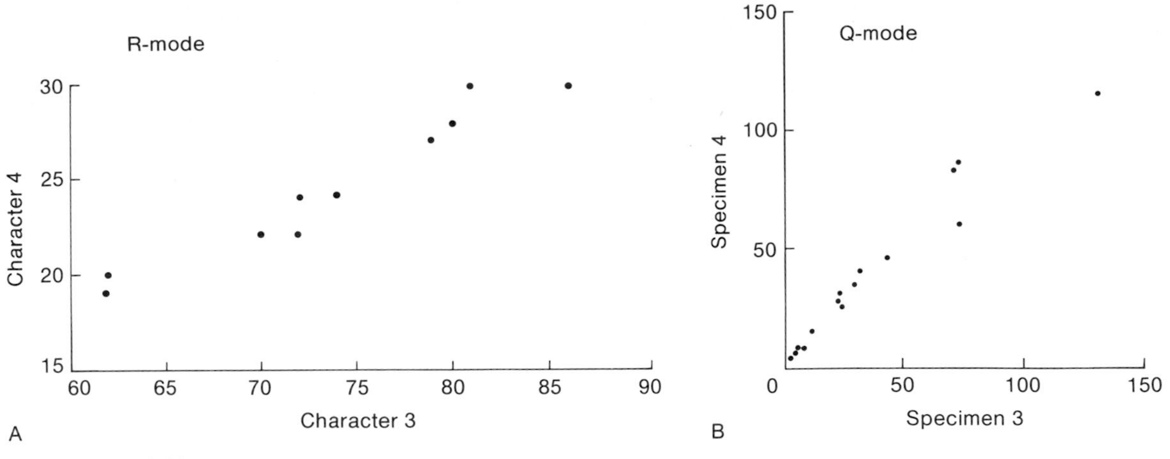

FIGURE 4-16
Bivariate plots of data from Table 4-3.
A: R-mode plot of values for two of the morphologic characters.
B: Q-mode plot of values for two of the specimens.

on all specimens. More commonly, the data matrix contains gaps where specimen breakage or poor preservation has made it impossible to obtain certain measurements on some specimens. Multivariate methods are usually designed to cope with such deficiencies.

The specimens in the data matrix may or may not be divided *a priori* into groups such as species, localities, or lithofacies. In the case of Table 4-3, groups have not been defined in advance: here the job of multivariate analysis might even be to *find* natural groups within the data.

The data matrix can be analysed in two fundamentally different ways: we can look at the behavior of all specimens with respect to a pair of characters or we can concentrate on all characters with respect to a pair of specimens. In the former, the interaction between *columns* is emphasized and in the latter, the *rows* are of most interest. The former is commonly called ***R-mode analysis*** and the latter ***Q-mode analysis.*** The two modes are illustrated by scatter plots in Figure 4-16. The R-mode plot of character 3 versus character 4 tells us whether the specimens fall into natural groupings or clusters and thus yields information either of taxonomic interest or on the presence of growth stages, depending on the nature of the morphology. The same kinds of information can come out of the Q-mode plot (specimen 3 versus specimen 4) but the emphasis is different. The particular Q-mode plot in Figure 4-16 shows that the two specimens are very similar in all characters and that certain characters have nearly the same values. As a very general rule, the taxonomist uses Q-mode analysis and the comparative morphologist uses R-mode analysis. But the two approaches are intimately tied together and it often happens that the taxonomist can use R-mode techniques to lead to conclusions about relationships between species.

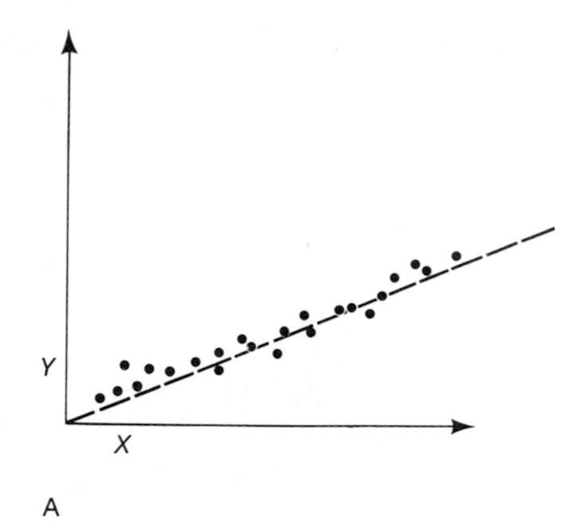

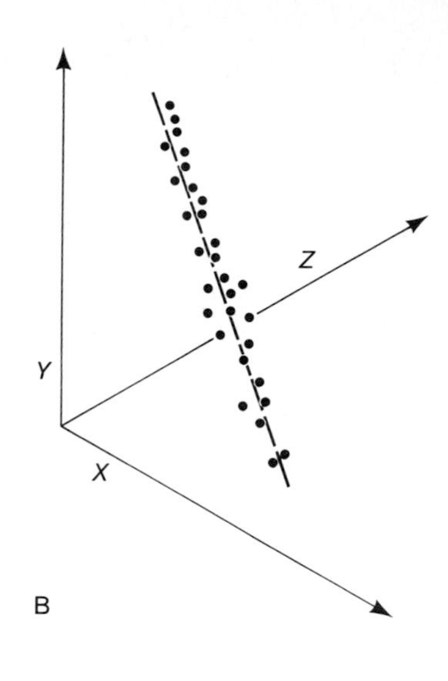

FIGURE 4-17
Hypothetical scatters of points in two-dimensional space (A) and three-dimensional space (B). In each case, most of the variation can be explained by position along a straight line fitted to the data. (Modified from Koch and Link, 1971.)

If we look at the whole data matrix in Table 4-3 and take an R-mode approach, we see that each specimen is defined by 15 characters. This means that each specimen is precisely located as a point in a 15-dimensional coordinate system or a 15-dimensional space (a hyperspace). The full data matrix can be imagined as a cloud or swarm of 10 points in this space. With the Q-mode approach, we have a 10-dimensional space in which 15 points are plotted. The mind can conceptualize neither coordinate system effectively, so the hyperspace must be treated numerically rather than as a visualization. The specific technique that is used depends on the kind of question that is being asked. Most techniques attempt to reduce the number of dimensions in the system while losing as little information as possible.

We may wish to learn about the structure or shape of the cloud of points in multidimensional hyperspace. It often turns out that the cloud of points is much simpler in *its* morphology than would be implied by the number of variables used to define it. Two hypothetical examples are shown in Figure 4-17. In Figure 4-17,A, a familiar bivariate cloud of points is shown. Although some scatter exists, the essential element is the distribution along a straight line. Once the position of this line is known, most of the observed variation is in reality univariate with the primary variable being "size," which may be expressed by either *X* or *Y* or by a combination of the two. In Figure 4-17,B, a similar linear array of points exists in three-dimensional space: most of the variation is again displayed along a line, the position of which may be defined with respect to one of the three axes.

A common multivariate technique known as ***factor analysis*** undertakes the job of defining the shape and structure of a cloud of points so that the distribution of points may be described by specifying a number of "factors" that are fewer in number than the original measured variables. In Figure 4-17,A, nearly all the variation (perhaps 95 percent) can be accounted for by one factor, namely position along the plotted line. If the line can be located relative to one of the original axes (X, for example), the additional measurements (Y) become redundant because they provide little new information. Factor analysis as applied to large data matrices cannot, of course, be visualized in two or three dimensions but the principle illustrated in Figure 4-17 applies to the multidimensional case. When used in complex cases, factor analysis has several advantages. Most important, it brings to light combinations of characters that are highly correlated in the sense of providing the same information (as the two variables in Figure 4-17,A, or the three in Figure 4-17,B). This is often of biologic interest and importance. Groups of characters that are related by mechanisms of growth or by functional relationships will usually be grouped by factor analysis. Or, if Q-mode work is being done, groups of specimens or species which behave similarly in the hyperspace will be isolated as a single factor. In addition to the biologic knowledge gained, the number of variables has been reduced by factor analysis—perhaps to the point where the essential information of the data matrix can be shown in a two- or three-dimensional plot.

An actual example of factor analysis of paleontologic material is shown in Figure 4-18. Kaesler (1970) measured 33 characters on each of 34 specimens of the Permian

FIGURE 4-18
Stereographic view of the results of principal components analysis of 34 specimens of the Permian fusulinid *Pseudoschwagerina.* Five species or subspecies are identified by the letters A–E. When viewed through a stereoscope, the five taxa are seen to form reasonably distinct groups in principal component space. A: *Pseudoschwagerina beedei;* B: *P. uddeni;* C: *P. gerontica;* D: *P. texana texana;* E: *P. texana ultimata.* (From Kaesler, 1970.)

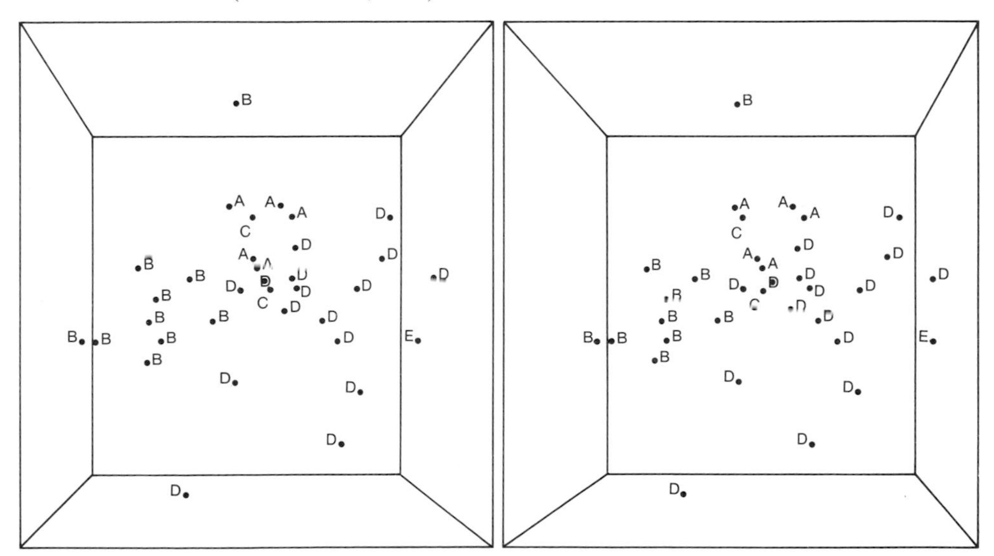

FIGURE 4-19
Phenogram developed by clustering for the same 34 specimens that were used in Figure 4-18. (Data from Kaesler, 1970.)

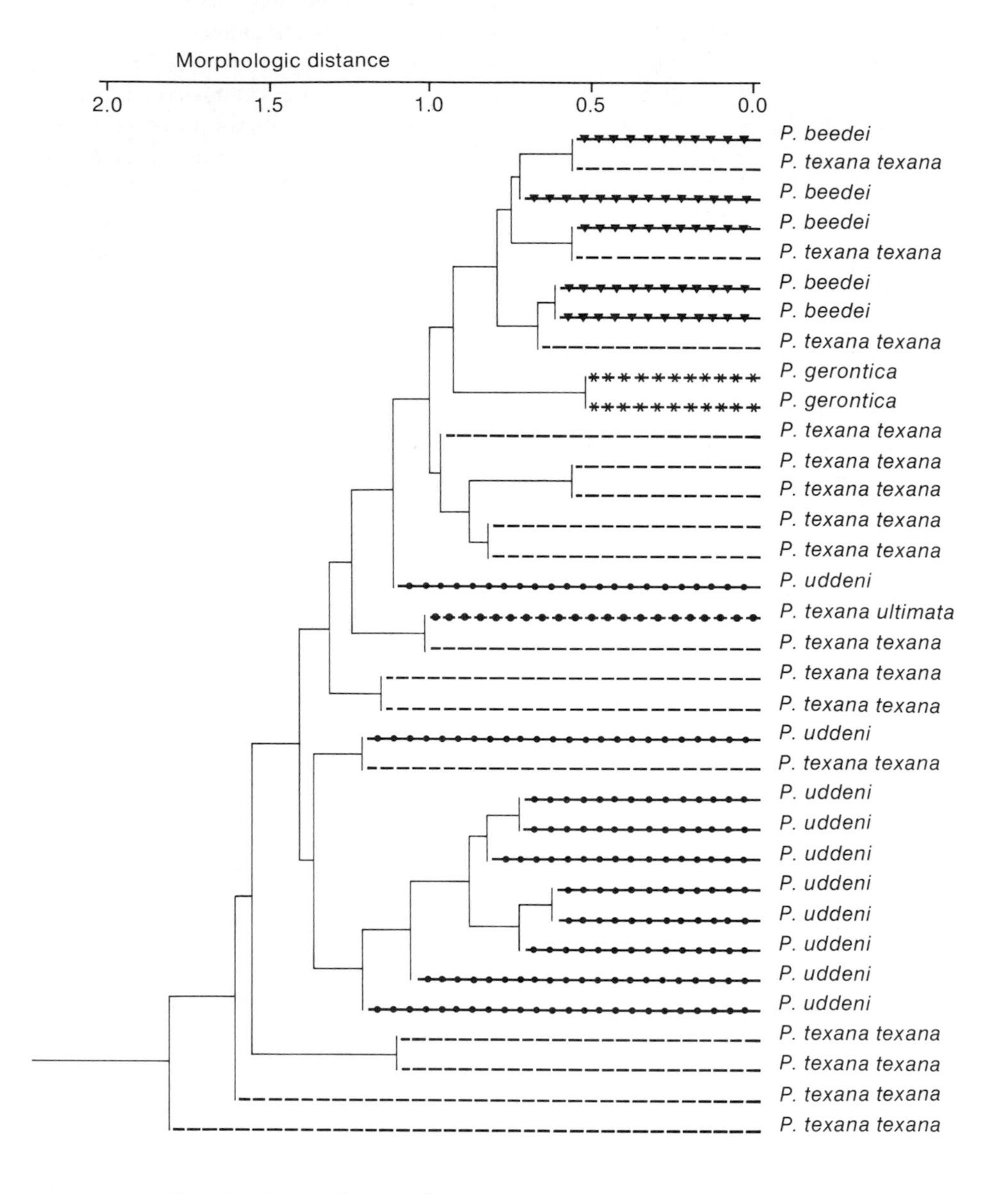

fusulinid *Pseudoschwagerina.* The 34 specimens had been identified as to species or subspecies and Kaesler was interested in finding out whether the whole data matrix supported the species distinctions and whether the taxa were arranged in the 33-dimensional space in a biologically reasonable way. He executed a type of factor analysis called ***principal components analysis,*** which identifies axes comparable to those in the simplified cases discussed above (Figure 4-17). The so-called first principal component is the axis that accounts for the greatest amount of the variation in the data matrix, the second principal component accounts for the second largest amount, and so on. Figure 4-18 is a stereographic plot of the 34 specimens in a coordinate space defined by the first three principal components. The five species and subspecies are identified by letters; their spatial relationships can be seen when the illustration is viewed stereographically. There is relatively little overlap between the taxa, and their arrangement is a reasonable one in terms of what is known about the evolution of this genus. In this case, the first three principal components explain about 55 percent of the variation displayed in the original data matrix, the remaining 45 percent being distributed among the other principal components. Some of the unexplained variation may be biologically real and some of it may result from error in measurement.

The amount of information explained by a few principal components is often higher than in the fusulinid case just described. For example, in a recent R-mode analysis of 27 measurements made on each of 44 specimens of Silurian eurypterids (Andrews et al., 1974), the first principal component explained 95.2 percent of the information and the first three combined accounted for 97.7 percent. The high values in this case resulted from the fact that many features of these arthropods are very tightly integrated, both functionally and in terms of growth mechanics.

In the fusulinid case described above (Figure 4-18), the *a priori* grouping of the 34 specimens into five taxa appears to be "reasonable" when the data are viewed with respect to the first three principal components. But it is unlikely that the same five groups would have been selected from the display in Figure 4-18 if the points were unlabelled. This is another way of saying that the fusulinid points do not fall into obvious, completely nonoverlapping clusters. In other cases where factor analysis has been used, the results do suggest clear-cut clusters: much depends on the nature of the data. Factor analysis is thus not primarily a *group-finding* or *clustering technique,* though it may help the paleontologist to find groups in some cases.

A great variety of clustering techniques are used in biology and paleontology to find and define taxonomic groups—subspecies, species, genera, or yet higher taxa. Kaesler used one such technique on the fusulinid data matrix. The result is shown in Figure 4-19 and is in the form of a ***phenogram.*** A phenogram is a tree-like structure that displays the degree of similarity (or dissimilarity) between specimens or groups of specimens. The scale at the top is a measure of dissimilarity or "morphologic distance" between the specimens which are listed down the side of the phenogram. Thus, the first two specimens in the list differ by slightly more than 0.5 units; these two as a group differ from the third by about 0.7 units, and so on. The clustering method used by Kaesler computes the morphologic distance between specimens in all possible pairs of specimens and uses this information to construct the phenogram. Although the known taxonomy of the 34 specimens was ignored in the clustering process, the original identifications are included in the phenogram so that we can see how successful the clustering has been in segregating the five taxa. You will notice that it has been more successful for some than for others. For example, all but two of the specimens of *Pseudoschwagerina uddeni* are in a single cluster in the phenogram—a cluster that is separated from the other specimens of this species at a relatively high distance level (about 1.3). By contrast, the subspecies *P. texana texana* is scattered throughout the phenogram. It appears that *P. texana texana* is inherently more variable than *P. uddeni.* This was also evident from the results of the factor analysis (Figure 4-18).

Clustering and clustering methods will be discussed more fully in Chapter 6 under the heading of "Numerical Taxonomy."

In the preceding examples, the main objective was to learn something about the structure of a single cloud of points in multidimensional space. If subgroups in the cloud had already been defined (as in the fusulinid case) the classification could be evaluated or, in the absence of defined subgroups, the multivariate methods could be used to find subgroups of clusters. In general, therefore, we have been concerned with a *classification* process. Suppose, on the other hand, that we already have subgroups—or several clouds—and wish to investigate rigorously whether they are distinct or overlapping and, if distinct, how far apart they are. This then becomes a problem in *discrimination.* To be sure, factor analysis makes possible some discrimination, but specific techniques have been developed which do a better job. One such is called ***canonical analysis.*** Figure 4-20 illustrates a simplified example of canonical analysis. Again, we are working in two dimensions, where geometric relationships can be visualized most readily. Three clouds or clusters are shown and each is specified in advance as constituting a separate group. Clouds b and c are distinct in the y direction but largely overlap in the x direction. The reverse is true for a and b. In short, all three groups are obviously distinct but it takes a combination of the two axes to express the separations. It is possible to transform the coordinate system and rotate the axes in such a way as to maximize the separation. With the transformation shown in Figure 4-20 (the new axes x' and y'), the three groups can be clearly distinguished on *either* axis. The job of canonical analysis is to compute the positions of axes (called *canonical axes*) that will maximize the separations between the groups that have been provided. As was the case with factor analysis, the number of canonical axes necessary to do the job (or most of the job) is generally fewer than the original number.

Once the canonical axes have been defined and the data transformed to a new coordinate system based on these axes, the distance between clouds can be measured. This is done by a Pythagorean method in the multidimensional canonical space. If most of the information is embodied by variation in the first two or three canonical axes, the results can be expressed in a two- or three-dimensional plot. An example is shown in Figure 4-21. Here, the problem was to assess the morphologic similarity between a hominid fossil ("Hominid 10") from Olduvai in Tanzania and several other primates, including modern man. The Olduvai fossil is a single bone—a terminal toe

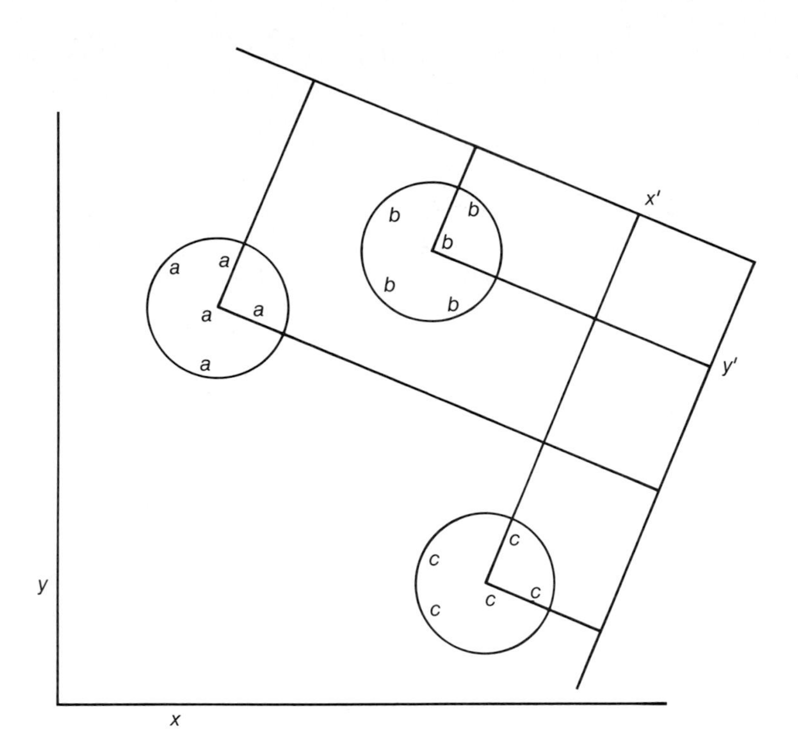

FIGURE 4-20
Hypothetical data to illustrate the method of canonical analysis. The canonical axes (x' and y') are more effective in distinguishing the three clouds of points than were the original axes (x and y). (Adapted from Oxnard, 1973.)

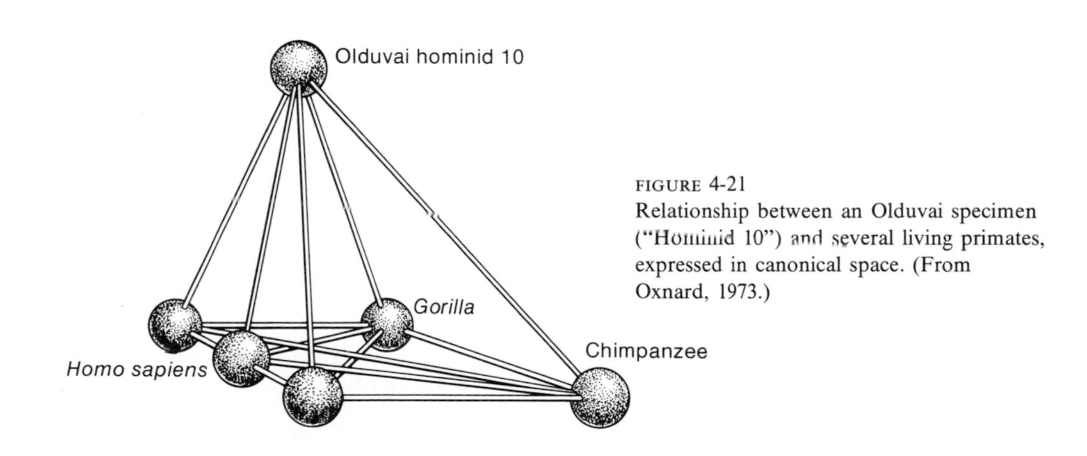

FIGURE 4-21
Relationship between an Olduvai specimen ("Hominid 10") and several living primates, expressed in canonical space. (From Oxnard, 1973.)

phalanx—on which nine dimensions were measured. The same measurements were also made of 122 specimens representing modern man (East African, European, and Bushman), the gorilla, and the chimpanzee. Figure 4-21 shows the positions of the several sets of data in a space defined by distance in canonical space. It is clear from the diagram that the Olduvai specimen is well separated morphologically from modern man as well as from the two African apes. Notice, however, that most of the separation between modern man and Olduvai 10 is in one dimension only: the three points (or "balls") representing modern man and the two representing African apes all fall nearly in one plane, which is approximately the plane of the first two canonical axes. It is the third canonical axis (vertical direction in the diagram) that provides the separation between Hominid 10 and modern man. This illustrates the importance of simultaneous treatment of many variables.

The foregoing pages provide an extremely brief and simplified introduction to multivariate techniques. We have attempted to show a few of the things that can be done with the technique and to give a few glimpses of its mathematical methodology. We will come back to some of these details in later chapters but the student will want to develop a thorough knowledge of the field through the readings listed at the end of this chapter and through other sources.

Supplementary Reading

Dobzhansky, T. (1970) *Genetics of the Evolutionary Process.* New York, Columbia University Press, 505 p.

Imbrie, J. (1956) Biometrical methods in the study of invertebrate fossils. *Bull. Amer. Mus. Nat. History,* **108**:211-252. (A brief but excellent summary of univariate and bivariate statistical techniques applicable to the analysis of variation in fossil assemblages.)

Newell, N. D. (1956) Fossil populations. *In* Sylvester-Bradley, P. C., ed. *The Species Concept in Paleontology.* London, Systematics Assocation. Publication 2, p. 63-82. (A summary and analysis of some of the problems encountered in the biologic treatment of fossil assemblages.)

Oxnard, C. (1973) *Form and Pattern in Human Evolution.* Chicago, University of Chicago Press, 218 p. (Contains an excellent introduction to methods of multivariate analysis and their applications to morphological problems.)

Simpson, G. G., Roe, A., and Lewontin, R. C. (1960) *Quantitative Zoology.* New York, Harcourt, Brace, and Co., 440 p. (A standard reference on quantitative techniques applicable to the study of morphology.)

Sokal, R. R., and Rohlf, F. J. (1969) *Biometry: The Principles and Practice of Statistics in Biological Research.* San Francisco, W. H. Freeman and Co., 776 p.

CHAPTER 5

The Species as a Unit

The tremendous diversity in form, physiology, and behavior of living organisms is not random. If our imaginations were good enough to conceive of the total spectrum of possible biologic forms, we would find that all of the living and extinct organisms, taken together, are only a small part of the imagined whole. Living species are clustered within this spectrum; there are relatively few basic types, with each type capitalizing on the requirements and the advantages of particular roles in particular habitats. In other words, we find that most species are adapted to one or a few environments.

If biologic diversity is to be maintained, *species units must remain distinct.* In biologic terms this means that hybridization between species must be uncommon. Natural hybridization does, of course, occur between species, particularly in plants. Sometimes it produces new structures or new combinations of structures unknown in either parent species. In the long run, successful hybridization blends parental characteristics and results in the formation of a single species where there were originally two. The fact that high species diversity has been maintained over a long period is the best evidence we have for the relative rarity of hybridization.

Diversity is maintained because species are reproductively isolated from each other (even though they may live in the same geographic area and occupy similar habitats.) Therefore, *the evolution of reproductive isolation* is basic to the origin of species. We must understand how reproductive isolation evolves in order to understand the concept of the biologic species. The paleontologist usually deals only with the *results* of species evolution rather than the process itself, but knowledge of the process is as important to the paleontologist as an understanding of hydrodynamics is to the sedimentologist.

The Biologic Species Definition

The most widely accepted biologic definition of the species was formulated by Ernst Mayr: *"A species is an array of populations which are actually or potentially interbreeding and which are reproductively isolated from other such arrays under natural conditions."* As an operational definition, this is not directly applicable to the fossil record because hybridization cannot be observed between fossil species. In fact, most working biologists are rarely able to apply the definition *directly* either. Biologists have therefore developed several indirect methods of identifying the boundaries between species. Many of these methods are also applicable to fossils.

The species definition just given has several important elements. The species is referred to as an array of populations, emphasizing the fact that most species are divided geographically into subunits or breeding populations. It is explicit in the definition that such breeding populations are actually or potentially interbreeding with each other. We can define two types of relationships between populations. ***Allopatric populations*** are those that occupy separate geographic areas; ***sympatric populations*** are those having the same or overlapping geographic ranges. Allopatric populations may be separated by geographic barriers which make gene flow between them impossible. We refer to such populations as being geographically isolated but by convention do *not* refer to them as being reproductively isolated (even though the geographic barriers may in fact prevent interbreeding). Two populations are said to be reproductively isolated *only* if interbreeding would not occur if they both lived in the same area. Thus, "potentially" in the species definition is particularly critical.

The crux of the species definition is that the array of populations constituting one species is reproductively isolated from other such arrays: two species can live sympatrically without interbreeding. Must the reproductive isolation between species be complete? If not, how much gene flow can be tolerated? If two populations are found to hybridize successfully (that is, to produce fertile offspring), are they necessarily one species? The answers to these questions must remain indefinite, but it is possible to point out—since the issue is the maintenance of distinct species—that an amount of gene flow between species that does not lead to the breakdown of the differences between them is compatible with their being properly described as reproductively isolated.

An important part of the species definition is that populations of different species are reproductively isolated from each other under "natural conditions." There are many recorded examples of species hybridizing readily in captivity or under domestication. This stems from the fact that reproductive isolation often depends on rather minor ecologic or behavioral barriers that tend to break down in captivity. The ability of organisms to interbreed in captivity is generally ruled out as a criterion for concluding that they belong to the same species.

We may consider an example that demonstrates how difficult it is to make a practical application of the species definition: assume that we suspect that two allopatric populations belong to the same species. According to the definition, we will never know for certain unless we transplant individuals of one population into the environment of the other. But to do this we would inevitably disturb the natural conditions under which one or both populations lived and, thus, might upset certain behavioral or ecologic causes of reproductive isolation.

The Origin of Species

Speciation is the evolutionary process of species formation. There is some confusion about the use of this term: it is most commonly used as a label for the process by which lineages split, leading to an increase in the total number of species, but paleontologists sometimes use the term in a different way, to describe the appearance of a new species by the gradual evolution of an entire lineage. Because this second process, which does not alter species diversity, is sometimes called phyletic speciation, to avoid confusion we will call it ***phyletic transition:*** phyletic evolution is simply evolution within an established lineage, and the word transition connotes change from one species into another. We will confine our use of the term speciation to the splitting of lineages.

NATURAL SELECTION

Phyletic evolution usually results from the operation of natural selection on the genetic composition of one or more populations. Variation in the gene pool produces variation in the ***phenotype*** (form, structure, physiology, and behavior) of organisms making up the population. This variability arises in various ways. The most fundamental way is by point mutations, which represent changes in single genes. These are usually accidental, resulting, for example, from exposure to certain kinds of radiation or chemical compounds. Thus, there is a large degree of randomness in the origin of genetic variability within a population. Most mutations are reversible, however, and while genetic changes between parent and offspring originate by way of new mutations, most of the total variability within populations arises as the changed genetic material is recombined or reshuffled among all the members of the population in the process of reproduction. Part of this reshuffling involves entire chromosomes. Since any individual of a sexual species receives only one chromosome of any of its pairs of chromosomes from each parent, particular chromosomal arrangements within a population will continually be paired with different ones as time passes. Some pairings will yield more successful phenotypes than others. Also, some arrangements will not be passed on to offspring and will be lost from the population. In addition, segments of chromosomes may be rearranged or lost during the process by which chromosomes are replicated. All of these changes result in phenotypic variability.

The process of natural selection operates upon phenotypic variability. Any change in genetic composition, no matter how slight, constitutes evolution; natural selection really amounts to differential reproductive success. Certain kinds of individuals will tend to leave a larger number of offspring with successful genetic inheritance than others. These kinds of individuals will tend to become more common as generation follows generation. As long as new genotypes continue to arise within the population, change by natural selection can continue until the population of all individuals with unfavored genotypes dies out. There are two ways in which a particular kind of individual can be favored in natural selection. One is by *reproducing* at an unusually high rate. The other is by *surviving* for a long time so as to have an unusually high total reproductive output.

Evolutionary change in the genetic composition of a population may represent either progressive adaptation to *constant* environmental conditions, or adjustment to *changing* environmental conditions. With each succeeding generation, the gap in genetic composition with respect to an arbitrary starting generation is widened. As a population changes, it becomes almost inevitable (statistically) that differences will develop that will inhibit individuals expressing the changed phenotype from breeding with individuals of the original ancestral phenotype.

If we think of a chronologic series of populations and choose one of these as an arbitrary starting point, then, barring termination of the lineage, there will be a point at which the accumulated differences are such that the later populations *would* be reproductively isolated from the initial population *if* they were living at the same time. At this point a new species has been formed by phyletic transition.

GEOGRAPHIC SPECIATION

Let us consider populations of a species living at a given time but not in geographic contact with each other. That is, one or more of the populations is geographically isolated from the other populations. Two or more segments of the species thus evolve and undergo phyletic transition independently. If the geographic barriers between populations remain intact long enough, reproductive isolation may develop. When this happens, geographic speciation has occurred: a single species has become two or more contemporaneous species.

The distinction between phyletic transition and geographic speciation is to some extent artificial in that both processes depend on natural selection. The critical difference is that phyletic transition is accomplished in the absence of geographic isolation and geographic speciation requires geographic isolation.

In addition, a variety of evidence suggests that much geographic speciation occurs by the rapid divergence from parent species of small, local peripheral populations that have become geographically isolated (Mayr, 1963, 1971). It is obviously far more likely that small populations will become geographically separated from the main gene pool of a species than that an entire species will be more or less divided in half. Many small, isolated populations probably exist at the fringe of the geographic range of their parent species, where conditions are not optimal. As a result, a large number probably become extinct rapidly. It is generally believed that only rarely does a geographically peripheral population adapt successfully to its marginal habitat and evolve into a new species. We might suspect that geographic speciation should inevitably lead to a steady, overall increase in number of species but, in fact, extinction counteracts this increase. Lineages are periodically terminated because of changes in environmental conditions which are so rapid that species cannot adjust to them by phyletic evolution. The number of species living on the earth at any given time is therefore the resultant of the positive process of geographic speciation and the negative process of extinction.

THE SUBSPECIES

In the present context, a subspecies may be looked upon as an incipient species. Subspecies is a term usually applied to geographically isolated populations, or groups of populations, that are genetically different but not sufficiently different to be reproductively isolated. From the paleontologic viewpoint the subspecies can also be used to denote the intermediate stage in the replacement of one species by another through phyletic transition.

CLINES AND RING SPECIES

Some allopatric populations are separated by environments not hospitable to the organism, such as, for example, bird populations in an archipelago separated by considerable distances of water, or lowland terrestrial plant populations separated by mountain ranges. Other allopatric populations are not separated by such pronounced barriers.

The gene flow between populations situated at the extremities of the geographic range of a species may be greatly reduced simply because of the distance between them, and a gradient in the genetic composition of the populations may be produced, giving rise to what is termed a ***cline.*** For certain species, it has been shown that the populations living at the extremities of the geographic range are reproductively isolated from each other even though they are connected by a chain of interbreeding populations.

If such reproductively isolated populations from the extremities of a species' range migrate so that their geographic ranges overlap, what is known as a ***ring species*** is formed. The populations originally at the extremities live sympatrically and are thus fully separate species, but are connected through a chain of interbreeding populations.

Fortunately for the biologist, ring species and pronounced clines are relatively rare. It is nearly impossible for the paleontologist to recognize clines and ring species because the fossil record is not complete enough and time determinations are not accurate enough.

ADDITION OF SPECIES WITHOUT GEOGRAPHIC ISOLATION

Although there has been considerable controversy over the question of whether a single species can evolve into two or more contemporaneous species *without* the geographic isolation of two or more of its breeding populations, for our purposes the question may be dismissed once it has been noted that the consensus among biologists and paleontologists is that it is a rare phenomenon.

RATES OF EVOLUTION AND SPECIATION

Because evolution is generally defined as change in gene frequencies from generation to generation, and because there are many genes and many individuals within most species, evolution is virtually ubiquitous. Not all change occurs by natural selection. Directional change in gene frequency that arises by chance but that persists without being favored by natural selection is called ***genetic drift.*** Genetic drift is much less likely in large populations, where chance changes tend to average out, than in populations of only a few individuals. It may play a role in the formation of some new species, as when a new species arises after a few wayward individuals form an isolated breeding population in a local area far from the rest of the populations of the species to which they belong. Also, if many or all of the isolated individuals are somewhat atypical representatives of their species, the population that grows from them may automatically differ significantly from that of the parent species. This second factor is called the ***founder effect.***

It is generally agreed that most evolution is produced by natural selection. Traditionally, three important factors have been thought to govern rates of evolution by natural selection. One is the amount of variability within a population. The Fundamental Theorem of Natural Selection, derived mathematically by R. A. Fisher, shows that, all other things being equal, rate of evolution should be proportional to the degree of variability upon which selection can operate. Variability is, after all, the raw material for directional change. Another factor is generation time. Within a population, selection occurs generation by generation. The faster the turnover of lifetimes, the faster the process will operate. Finally, rate of evolution depends on the intensity of selection. Actually, this last statement, when put into practice, says the same thing twice. The reason the corollary is tautological is that intensity of selection is generally defined and measured in terms of the amount of genetic or morphologic change per generation, and amount of genetic or morphologic change per generation is also a measure of evolutionary rate. This may at first seem puzzling, but a bit of thought should make it clear that there is really no way to measure selection pressure except through its effect on the composition of a population from generation to generation.

How rate of evolution is related to population size has been widely debated. One argument has been that evolution should occur most rapidly within large populations divided into many partially isolated subgroups, or demes. The idea here is that new mutations and combinations of genes will be "tested" in different ways in different demes, each of which will occur in a unique habitat. Then, it is argued, a wide variety of adaptations will tend to arise, and those that turn out to be of general adaptive value to the species will spread by the small amount of gene flow that occurs among the demes. However, Mayr (1963) has argued that evolution actually tends to proceed slowly in situations of the type just described. One of the reasons, he believes, is that the selection pressures tending to move different subpopulations in different evolutionary directions will tend to cancel out because of gene flow among these subpopulations. If this happens, then neither the individual demes nor the species as a whole will evolve very rapidly.

Mayr (1954, 1963) and Grant (1963), among others, believe that most well-established species evolve slowly for another reason as well. This reason relates to what Mayr terms "the unity of the genotype," or the cohesive, well-integrated structure of an adaptively successful arrangement of genes. Mayr (1954) noted that major evolutionary change seems to require the wholesale restructuring of at least part of the genotype. Piecemeal restructuring of the genotype and gradual transition from one adaptive type to another may be difficult in large populations. Rather, the transition is most likely to occur in small, completely isolated populations that evolve rapidly into new species, as described above. Successful speciation is not a commonplace phenomenon. Most isolated populations either re-establish gene flow with the parent population or become extinct without successfully establishing a new adaptive type. A very important consequence of this view is that most evolutionary change should be concentrated in speciation events—relatively swift, dramatic changes within small populations. According to this view, well-established species do evolve, but they tend to evolve slowly. We will examine fossil evidence relating to this matter in Chapter 11.

Biologic Methods of Species Discrimination

The practical problem of distinguishing species usually amounts to deciding whether two populations that differ morphologically are different species or whether they are simply minor variants (perhaps subspecies). The biologist's greatest asset in making such a decision is the generally accepted idea that distinct populations cannot live sympatrically unless they belong to different species. This assumes, of course, that the division of a single species into two or more without reproductive isolation is such a rare event that it may be discarded from general consideration. Following this reasoning, if a biologist finds that the geographic ranges of two morphologically different organisms overlap, he may conclude with considerable confidence that they belong to different species.

Some differences can be observed among nearly all allopatric populations. These may be morphologic, physiologic, behavioral, or ecologic. In assessing the probability that two such populations have become reproductively isolated from each other, the paleontologist works largely from comparisons with difference observed between populations known to be distinct species because they occur sympatrically. Distinguishing species on this basis involves varying degrees of uncertainty. It is widely recognized that some species differ very slightly (the so-called "sibling species"), while others are so different that there could be little question that it is correct to designate them as different species. It is generally true that the amount of difference between species is reasonably constant within a single evolutionary group and therefore the specialist usually possesses a backlog of experience that assists him as he interprets species differences among allopatric populations. It is also true that most living species are quite distinct from one another because they have been reproductively isolated for long enough to have developed many and notable differences.

The fact that species discrimination depends largely on the experience of the person making the discrimination has led to an informal definition of the species that is invoked with surprising frequency: "A species is a species if a competent specialist says it is."

The geographic distribution of allopatric populations can also be used as an aid in making species discriminations. If populations have been widely separated for a long time, the chances for complete speciation are greater than if the populations are close to one another and have not been geographically isolated for very long. For this reason, biologists often use biogeography as a taxonomic aid. It may be reasonably postulated that widely separated populations belong to different species even though the morphologic differences between them are not great. Although this reasoning may be quite useful if used as supporting evidence, it has led to considerable difficulty in studies in which taxonomists have based species discrimination solely on geographic separation.

In summary, the biologist relies primarily on phenotype differences between populations to define species boundaries. Occasionally he can bring biogeography to bear and occasionally he can gain factual knowledge from finding populations living sympatrically. Above all, biologic species discrimination is an *interpretation* of the evolutionary history of the organism being studied.

At various times the suggestion has been made that we should do away with a system based on a species definition that can be applied objectively only rarely and that the definition should be replaced by one based on morphology, in which boundaries between species could be established by arbitrary rules of statistical discrimination. To this end, various "morphospecies" definitions have been proposed. For example, it has been suggested that two populations differing in one or more morphologic characteristics to a statistically significant degree should be called different species. Such a system has obvious appeal in that it would do away with many of the practical problems.

These proposals have never been widely accepted, primarily because the evolutionary process does not operate in a manner amenable to the use of arbitrary boundaries between species. It has been shown repeatedly that the difference between populations necessary to effect reproductive isolation is only crudely correlated with the absolute amount of difference between the populations. There can be no question but that reproductive isolation is the fundamental condition responsible for the segregation of organisms into discrete adaptive units. Therefore the consensus of taxonomists is that it is better to describe the results of evolution by a theoretically valid method, but one difficult to apply, than by a theoretically invalid method that owes its existence to convenience.

The Species Problem in Paleontology

If we could construct a three-dimensional model of the evolution of life, with time represented by the vertical dimension and evolutionary differences between lineages by distance of horizontal separation, it would take the form of a complex tree, branching toward the top. What the biologist sees in his study of living forms is only the horizontal, two-dimensional upper surface of the tree—the tops of the uppermost branches. In other words, he does not deal with evolutionary lineages (except very short ones, in which little evolution occurs). Biologic species tend to be distinct because most belong to lineages that have been reproductively isolated from other lineages for a considerable time. If a lineage is branching at the present time, the biologist may, as we have seen, have difficulty judging whether the incipient species are reproductively isolated or not.

For the paleontologist, the presence of the time dimension makes it impossible to apply the commonly accepted biologic species definition (based on reproductive isolation) and at the same time recognize species as discrete, nonarbitrary entities. This conflict has given rise to what, in paleontology, is referred to as "the species problem." An analogous problem is posed in biology by the existence of clines and ring species, but these are not common enough to cause major difficulties.

A wide variety of opinions are held as to the ways in which fossil species should be defined and recognized. Some workers believe that the biologic species concept should be abandoned, primarily because there is no way of applying the concept to an evolutionary continuum, and because fossils cannot be tested for reproductive isolation. Probably most paleontologists, however, favor applying the biologic species concept, if only in an indirect and imperfect way. A worker can do this by including in a fossil species only specimens that, by his judgment, would have belonged to a single "biologic" species had they lived at the same time. We have emphasized that although the biologist *defines* species on a *genetic* basis, most often he *recognizes* them on the basis of *morphologic* criteria. The paleontologist can do the same, and can use as a reference the degree of morphologic difference between biologic species (preferably species of a living group closely related to the fossil group being studied).

There remains one major problem. Having decided on the amount of morphologic variation within recognized species, where should the boundaries within a lineage be placed? In Figure 5-1, three alternative interpretations of speciation are presented for a simple branching pattern of evolution. Some workers prefer to place species boundaries at branching points and to avoid the situation illustrated in part B of the figure. Because of the added time dimension, some workers use a special term such as "successional species" or "paleospecies" for a species that represents an arbitrarily delimited segment of an evolutionary lineage.

It is rather striking that the concept of reproductive isolation was employed in biologic species definitions long before the idea of evolution was accepted in biology. Furthermore, the modern definition of a biologic species ignores evolution altogether. Consequently, Simpson (1961b) has proposed a special definition for an ***evolutionary species:*** "An evolutionary species is a lineage . . . evolving separately from others and with its own unitary evolutionary role and tendencies." The word "role" here means way of life within a particular habitat. This definition is generally consistent with the genetic definition of a biologic species, but does permit a small amount of interbreeding between species (according to Simpson, "as much as does not cause their roles to merge"). Although the criteria used to delimit species boundaries are not described in his formal definition, Simpson suggests that morphologic differences between nonliving species should be at least as large as those between living species of the same taxonomic group (or a similar group).

There is no way of eliminating the species problem in paleontology; all general methods of defining and recognizing fossil species must be both subjective and

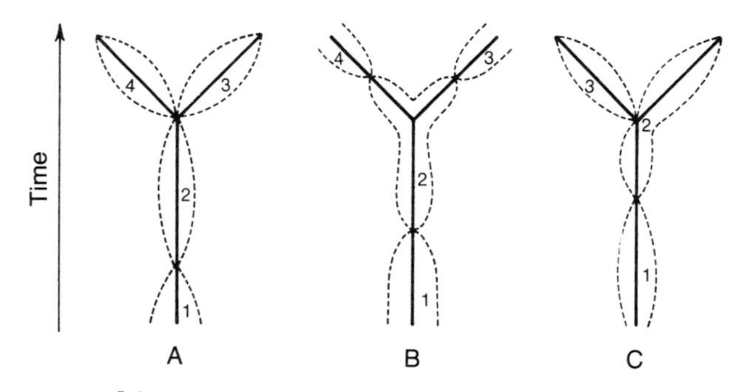

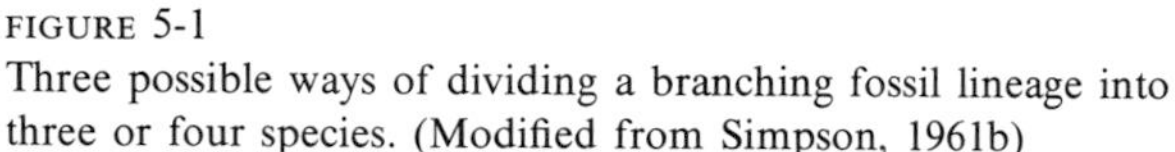
FIGURE 5-1
Three possible ways of dividing a branching fossil lineage into three or four species. (Modified from Simpson, 1961b)

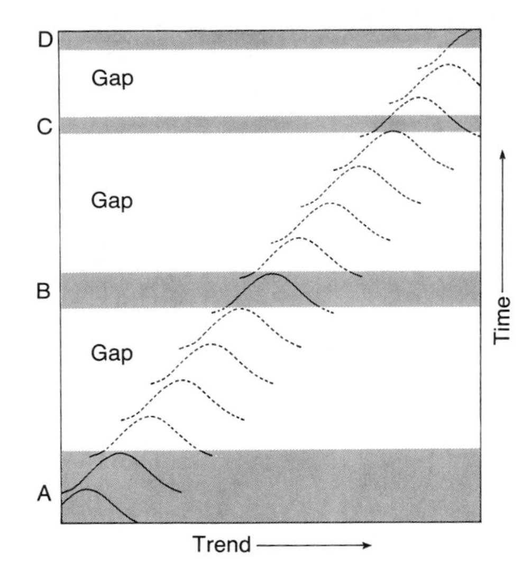

FIGURE 5-2
Diagram showing how gaps in the fossil record can provide arbitrary limits for species discriminations within an evolutionary lineage. The curves are size-frequency plots for an evolving character within a hypothetical lineage. Each curve is for the population that represented the lineage at a particular time. More or fewer curves could have been included. The succession of curves reveals an evolutionary trend. There are major gaps in the fossil record of the lineage, however. Populations are only preserved within four rock units (A, B, C, and D), which together represent less than half the time during which the lineage evolved. Suppose the morphologic differences between the populations in the four rock units are so great we judge that, if the populations in A and those in B had lived at the same time, they would have been unable to interbreed. Similarly, suppose we judge that interbreeding would have been impossible between the populations in B and those in C, and between those in C and those in D. Suppose, on the other hand, we judge that interbreeding would have been possible among the populations within each rock unit. We would conclude that the population group within each rock unit should be assigned to a distinct species. The gaps in the stratigraphic record have made it unneccessary for us to divide continuous lineages subjectively into species. (From Newell, 1956.)

arbitrary. There are, however, two very important factors that eliminate subjectivity from the assignment of many lineage "boundaries." One such factor is the widespread presence of gaps in the fossil record. Our preserved sample of the phylogenetic tree of life is so small and so fragmented that often only short segments showing little evolutionary change are found. Successive fossil populations within a hypothetical lineage are represented in Figure 5-2 to illustrate this point. Some preserved segments of the lineage exhibit a small amount of evolutionary change, and others virtually none. The large gaps between preserved segments of the lineage provide convenient locations for species boundaries. Thus, nature has greatly reduced the species problem in paleontology (but at the expense of much valuable information).

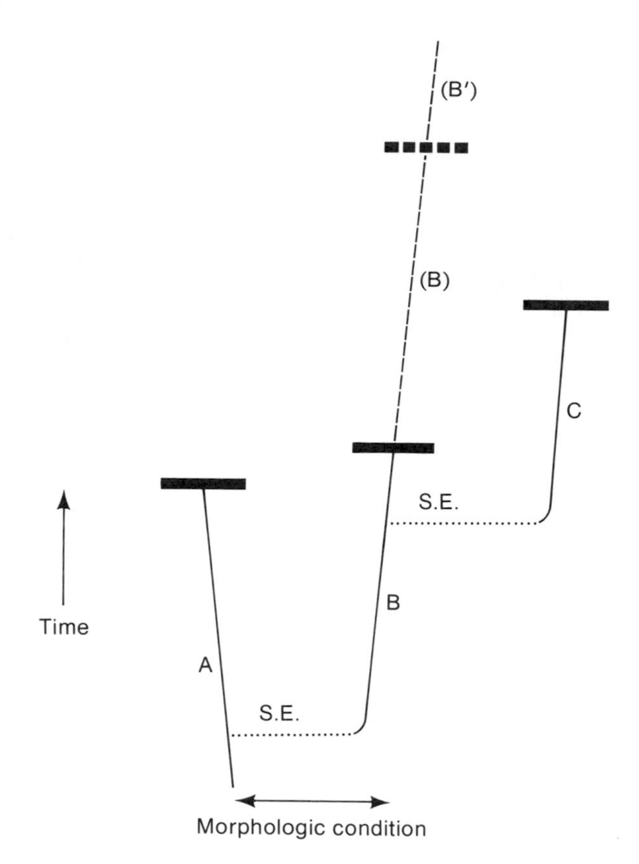

FIGURE 5-3
Diagram showing how the species problem is partially alleviated if it is held that most evolutionary change occurs during or shortly after speciation events (S.E.) and that most new species become extinct before undergoing phyletic transition that would distinguish them indisputably as new species. Solid horizontal bars indicate terminations of lineages. Continuous thin, vertical lines represent three species (A, B, and C). Speciation events giving rise to species B and C occur rapidly in small populations and are not likely to be documented and discovered in the fossil record. The dashed line projected from species B shows that the rate of phyletic evolution of this species would not have produced a new species (B′) until long after the actual termination of the B lineage. The dashed horizontal bar represents the position where the transition would have been placed subjectively by a taxonomist specializing in the group to which the lineages belong. In actuality, the three lineages that are found in the record (A, B, and C) represent quite distinct species.

Another factor that many workers think reduces the species problem is the alleged tendency of evolution to be concentrated in speciation events, as discussed above. If a new species (B) diverges rapidly from its parent species (A) and evolves very slowly thereafter, then the lineage of species B is likely to be terminated before some new species (B′) is formed (see Figure 5-3). If most evolution occurs in association with the splitting off of new species from old ones and if most extinction occurs by termination of lineages, then much of the phylogeny will automatically be divided up into rather discrete species. Furthermore, because small populations from which new species form are not likely to be discovered in the fossil record, most lineages that are preserved will represent distinct species (Ruzhentsev, 1964; Nevesskaya, 1967; Ovcharenko, 1969; Eldredge, 1971; Eldredge and Gould, 1972). There is no question that phyletic evolution occurs and that competent taxonomists consider certain well-documented lineages to be divisible into successional species. The extent to which such transitions constitute a taxonomic problem will depend on how frequently they occur. In other words, the severity of the species problem is in large part a question of what percentage of species arise by phyletic transition and what percentage arise by discrete, divergent speciation events. In other words, is the typical pattern of phylogeny most accurately described by Figure 5-3? We will evaluate this question in Chapter 11.

The incompleteness of the fossil record also contributes to the paleontologist's problem of assigning fossils to species. Not only certain morphologic features, but also many behavioral, ecologic, and geographic traits useful in taxonomy are not preserved. A perusal of several taxonomic articles on local fossil faunas will illustrate the overall problem. Usually a significant percentage of the fossils described in such studies are referred to species equivocally and the species assignments of earlier workers are commonly questioned or rejected. The paleontologist's problem arises not only because he is attempting to divide up continua (lineages), but also because he works with fossil material lacking many features commonly available to biologists. (Many of these features are morphologic, but others, as we have seen, are behavioral, ecologic, and geographic.)

We have already stated that a single specimen does not suffice to describe a species and have provided examples of variation within and between populations. In the past, even after evolution was widely accepted, many taxonomists assumed that an ideal form existed for each species and that a single specimen, the ***type specimen,*** or ***holotype,*** could be chosen to represent this ideal form; the idea of variation within the species was suppressed. With better understanding of the genetics of populations, this "typological approach" has been largely supplanted in the twentieth century. Species morphology is now fully recognized to be best described from a statistical point of view that allows for considerable variation.

Any single worker's *concept* of a living or fossil species is represented by all specimens known to him that he believes belong to the species. For this group of specimens, Simpson (1940b) has proposed the useful word ***hypodigm.*** All specimens of the hypodigm, including any formerly designated as type specimens, have equal weight in the species description. A practical problem is that of communicating one's concept of a species to other workers. Obviously, all specimens of most hypodigms cannot be listed in taxonomic articles. They may number in the thousands, or even millions, and many may lack numbers or other means of identification. The problem is partly solved by providing what is called a ***synonymy*** at the beginning of a species description. The synonymy is a listing of specimens or populations described by earlier workers that are included in the present worker's hypodigm. Many of these specimens or populations may have been assigned other species names by earlier workers, but these names are rejected by the synonymy writer and are listed only for historical reference. Thus the synonymy seldom represents the complete hypodigm but nonetheless provides a shorthand account of the writer's hypodigm that may be used for purposes of comparison.

Formal Naming and Description of Species

A new species may be erected in biology or paleontology either because previously unnamed specimens have become available or because a previously recognized species is judged actually to be two or more species. In paleontology, a combination of the two reasons is common: a worker may erect a new species based on new fossil material but include in it specimens or populations formerly assigned to different species. The individual worker's *concept* of a species and his views on the division of genera into species are subjective matters, as we have seen. In contrast, the *naming* of species is governed by objective rules. Were this not the case, nomenclatural chaos would result.

A wide variety of systems for naming and describing species have been proposed and used for varying lengths of time. Systems have often differed from country to country and from taxon to taxon, leading to considerable confusion. For the last fifty years or so, however, there has been increasing agreement among workers in different countries and among workers studying different plant and animal groups. Most matters concerned with species names are now under the control of international organizations. One of the most important of these is the International Commission on Zoological Nomenclature, which (operating under the continuing International Congress Organization) is responsible for administering and updating the *International Code of Zoological Nomenclature.*

The Code was adopted by the Fifth International Zoological Congress, which met in Berlin in 1901. It is a rather long legal document dealing with procedures to be followed in establishing names of species and other taxonomic groups and with problems posed by names proposed under earlier systems. The Commission acts as a combination court and legislature for treatment of questions of taxonomic procedure.

The International Code applies equally to fossil and living organisms, which is indeed fortunate because relatively few taxonomists actually work with both groups. Without the unifying effect of the International Code, two quite independent and perhaps contradictory systems of nomenclature might develop.

The International Code applies to taxonomic categories from the subspecies to the superfamily. Emphasis in this chapter will be on its application to the species category.

The International Code is almost universally applied to animal taxonomy. A comparable set of procedures for plants is known as the International Rules of

Botanical Nomenclature. In addition, one or two sets of international rules are commonly used for special groups like the bacteria. For our purposes, the differences between these codes are minor; in our examples we will concentrate on the Zoological Code.

Regarding the establishment of a new species name, the most important rules of the Zoological Code cover the following topics: choice of the name, publication of the name, description of the new species, and designation of one or more type specimens.

In order for a species to be officially recognized it must be given a name. The choice is limited by certain conditions. Most important is that the name must be in binomial form; that is, it must consist of two words. The official name for the human species is *Homo sapiens: sapiens* is the specific name and *Homo* is the name of the genus to which the species belongs. According to the International Code, a specific name like *sapiens* (sometimes called the trivial name) is meaningless unless associated with a genus name. In practice, most newly discovered species can be assigned readily to an existing genus and thus the act of describing a new species involves the invention of only one name. If the new species is apparently distinct from all established genera, a new generic name is assigned at the same time.

Except for the generic assignment just mentioned, the International Code does not insist on the complete taxonomic classification of a new species, in recognition of the fact that the complete classification is often difficult or impossible, particularly if the new species is quite distinct from all other known species.

The most important requirement of the new name is that it not be already in use ("occupied"). This restriction against homonyms refers to the combination of generic and specific names. That is, the name used for the second part of the binomial can be one that is used in other genera, but cannot be one already in use with the genus name accompanying it. By convention, repeating species names in closely related genera is also avoided because generic affiliations are often changed as knowledge of the evolutionary relationships between species changes.

The names for species and genera must be Latin words or words that have been latinized. There is considerable latitude in the choice of words to be used as names—latinized place names, names of people, or descriptive words are all used. Certain practical and aesthetic limitations are placed on the selection, however, and full detail on these matters may be found in the International Code.

In order for the name of a new species to be officially recognized it must be published in an approved medium; that is, it must be in print, published in quantity, and circulated through normal bibliographic channels. The precise rules governing the form of publication are complex but their intent is to insure that the announcement of a new name is readily available to taxonomists throughtout the world. A new name is not officially recognized if it has been used only in the labelling of a museum specimen or described orally before a scientific meeting.

Species cannot be described anonymously. Thus, the authorship of a new species is an important part of the official procedure. Although most species are described by a single person, more than one person may participate in the official authorship.

The requirement of publication is obviously necessary. The resulting bibliographic problems are, of course, immense because thousands of species are described each year in a great variety of languages and media, from museum monograph series to large international journals. The International Commission has made some attempt to simplify the bibliographic problems by urging that publication be made in one of a relatively few recommended languages. French, German, English, and Russian are generally recommended although this does not have the status of an absolute rule in the International Code.

The International Code does not require that the publication of a new species name be accompanied by a photograph or other illustration of specimens. Illustrating is, however, strongly recommended in appendices to the Code and a considerable amount of convention has developed; for all practical purposes, illustration of specimens may be considered mandatory.

The International Code says surprisingly little about the manner in which a new species should be described. The principal requirement is that the name be accompanied by a "statement that purports to give characters differentiating the taxon." A set of general recommendations published by the International Commission and appended to the International Code expands on this statement and a large body of convention has developed.

The International Code specifies that each newly described species be accompanied by the designation of a type specimen or set of type specimens. These are the only specimens that officially bear the name of the new species. For practical reasons, the Code requires that type specimens be clearly labelled and that suitable measures be taken for their preservation and accessibility to interested scientists, which means that type specimens are usually deposited in a major museum where curatorial facilities are available.

As well as the specifications for naming species just mentioned, the International Code contains a vast array of rules covering various contingencies brought about by historical changes in procedure and by the need to evaluate species that have not been described in adherence to the rules. It often happens that a new name is proposed for a species that has, in fact, already been described and named by someone else. In such a case we apply what is known as the ***law of priority.*** Except in special circumstances, the name proposed first has precedence over all subsequently proposed names. The International Code contains a set of procedures to be followed for the clarification of such duplications and for the arbitration of any disputes that may arise.

The complications that can and do arise in species nomenclature are many and varied. This discussion presents only an introduction to the general problem of assigning names to species.

THE FORMAT OF A DESCRIPTION

The following format for proposing a new species has been suggested by Mayr, Linsley, and Usinger (1953).

Scientific Name
Taxonomic references and synonymy (if any)
Type specimen (including information about where it was found and its present repository)
Diagnosis
Description
Measurements and other numerical data
Discussion
Range (geographic)
Habitat (ecologic notes) and Horizon (for fossils)
List of material examined

The scientific name is in binomial form, following the rules of the International Commission. "Diagnosis" refers to a listing of characteristics by which the new species can be distinguished from other species. "Description," in this context, means a full assessment of the characteristics without particular reference to similarities and differences with respect to other recognized species. The number of measurements and other numerical data included in a description varies from author to author and from one biologic group to another. At a minimum, the major dimensions of the type specimen or specimens should be included. The "discussion" section may include information about nongenetic variation, ontogenetic stages, the derivation of the name, evolutionary affinities with other species, and, for fossils, state of preservation.

The "geographic range" is usually a list of places at which the new species has been found (in addition to the type locality). With regard to habitat, the paleontologist is most concerned with the geologic setting (rock type, for example). The "list of material examined" usually includes reference to museum repositories of specimens other than those formally designated as types.

To illustrate further the format of species description, three actual examples are given on the following pages, in Boxes 5-A, 5-B, and 5-C, with photographs of type specimens. They illustrate both the consistency of species descriptions and some of the variation among them.

DIAGNOSIS AND DESCRIPTION

The form and content of diagnosis and description has varied considerably from time to time and from place to place. The increased emphasis on intraspecific variation has had substantial and significant effects on the manner in which species are described and diagnosed. Because one of the prime objectives is communication of information, there is an understandable premium attached to consistency. This means, for example, that standardized morphologic terminology is used wherever possible. Chaos would result if each taxonomist invented new terms for describing fossils.

BOX 5-A *Diplocaulus parvus* Olson, n. sp.

HOLOTYPE. UCLA VP 3015, partial skull and skeleton including vertebrae, ribs, shoulder girdle, humerus, radius, and ulna.

HORIZON AND LOCALITY. Chickasha Formation (Permian: Guadalupian, equivalent to the middle level of the Flowerpot Formation) about 2 miles east of Hitchcock, Blaine County, Oklahoma. Site BC-1 (Olson, 1965)[1] SW 1/4 SW 1/4, sec.6, T.17N., R.10W., Blaine County, Oklahoma.

[1]Literature mentioned in the description is cited in Olson, 1972.

DIAGNOSIS. A small species of *Diplocaulus* in which the adult ratio of skull length to skull width is attained when the skull length is approximately 60 mm, as contrasted to *D. magnicornis* and *D. recurvatus* in which the adult ratio is reached at skull lengths of between 80 and 110 mm. Otherwise similar in all features to *D. recurvatus.*

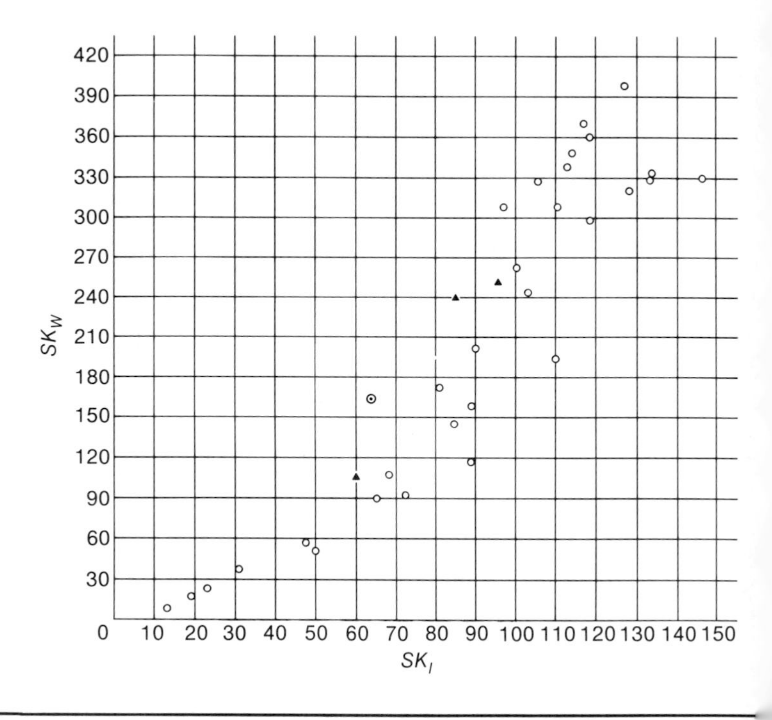

The relationships of skull length and skull width in *Diplocaulus magnicornis* (open circles), *D. parvus* (circle with center dot), and *D. recurvatus* (triangles). (From Olson, 1972.)

A truly complete diagnosis or description is impossible. As we saw in Chapter 2, an infinite number of attributes can be used to describe a fossil. It is neither possible nor advisable to describe exhaustively either a single specimen or an array of specimens. For diagnosis, enough attributes should be included to distinguish specimens of the new species from those of closely related species. Emphasis is usually placed on

Measurements of skull dimensions based on UCLA VP 3015. Measurements as described and figured in Olson (1953).

	mm
Skull length	63.0
Skull width	172.0
Pineal-frontal length	5.0
Interparietal length	14.1
Parietal length	19.6
Frontal length	25.2
Orbito-snout length	14.5
Interorbital width	10.8
Orbital width	10.2
Orbital length	9.9
Parietal width[1]	84.0
Interparietal width	94.0

[1]Based upon measurement of right-hand element and multiplied by 2 to give full width as used in various other papers.

Permian amphibian *Diplocaulus parvus* Olson. The original species description is printed in this box. The drawing is a dorsal view of the holotype. (From Olson, 1972.)

attributes that are most noticeable in all states of preservation. If too many are included, the system becomes cumbersome and obscures the most significant attributes. Therefore, if no reference to a particular character is found in a diagnosis, it cannot be assumed that the character is not diagnostic. This is where the actual specimens become most important in serving as a "backup" for the diagnosis.

BOX 5-B *Hattonia spinosa* Pickett and Jell, n. sp.

HOLOTYPE. UQF50894 from the Emsian (Jell, 1968; Strusz, 1972)[1] Martins Well Limestone Member, Broken River Formation, at Martins Well, 8 km northeast of Pandanus Creek homestead, north Queensland (GR611847 Clarke River 1:250,000 sheet).

OTHER MATERIAL. Three specimens UQF63992, UQF63993, and UQF63994, from the same locality.

DIAGNOSIS. *Hattonia* with numerous short horizontal septal spines; mural pores oval, in a single vertical series in the center of each face; tabulae tending to be grouped in pairs with incomplete or sometimes complete tabulae between pairs.

DESCRIPTION. The holotype is an abraded part (10 × 10 × 15 cm) of a tall corallum with corallites radiating upwards and outwards from its base. The distal surface of the colony was probably domed. The other specimens are fragments of probably similar coralla. The corallites are polygonal, four- to six-sided, the sides in some being a little curved between the angles, and are 0.9 to 1.4 (mean 1.02) mm in diameter. The walls vary in thickness from 0.1 to 0.15 mm in the center of the faces and thicken towards the angles, which in some become rounded. In transverse section the wall consists of a thin median dark line (as seen in transmitted light) which is the same width throughout but discontinuous and may be represented by a line of closely spaced dark spots, and a wider zone of lighter-colored material on either side which thickens towards the angles. In most places the microstructure of the wall is not obvious; however, in parts where the central dark line is represented by dark points, lighter tissue on either side seems to consist of fibers radiating outwards from these points. In longitudinal section, tangential sections of the wall show a fibrous structure with the fibers directed upwards and inwards from the angles towards the center of the faces. Whether the fibers are grouped into trabecular bundles or are all parallel is not known. This structure resembles that seen in the Australian species of *Squameofavosites*. Juvenile corallites develop by the insertion of a new partition across the angle of an adult corallite. Septa are

Devonian tabulate coral *Hattonia spinosa* Pickett and Jell. The original species description is printed in this box. The photographs are of microscope thin-sections of the holotype. (From Pickett and Jell, 1974.)

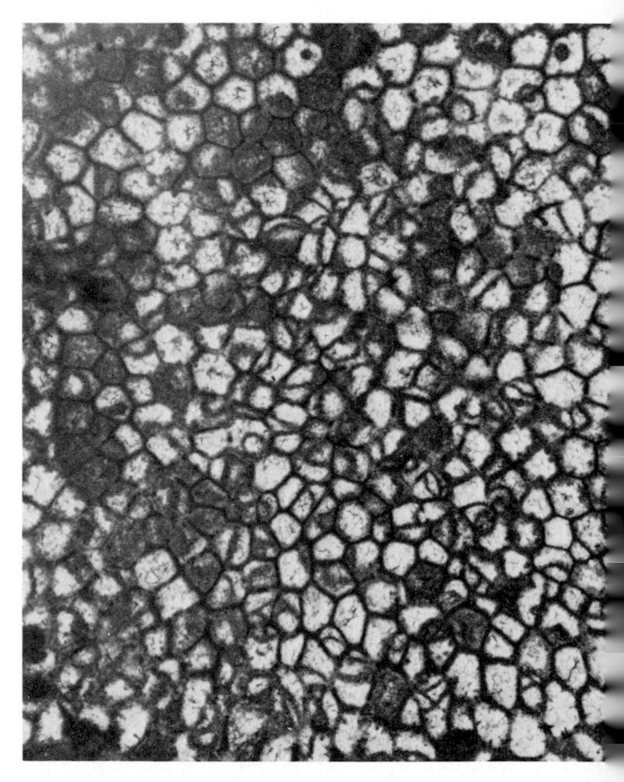

[1]Literature mentioned in the description is cited in Pickett and Jell, 1974.

represented by numerous short, blunt, nearly horizontal septal spines arranged in two or three subopposite to alternating vertical series on each face. Each spine seems to have a median dark line and is possibly trabeculate. The mural pores are commonly oval, 0.25 to 0.3 mm in width, and 0.3 to 0.35 mm in height, or less commonly circular, 0.3 mm in diameter. They are arranged up to 2.5 mm apart in a single vertical series in the center of each face and seem to be developed at the same level as the tabulae. The tabulae are grouped in pairs and tend to be at or nearly at the same level in adjacent corallites. However, there is only a general regularity in their arrangement and local disparities are common. The lower tabula of a pair is usually saucered and the upper horizontal or arched; they are usually 0.4 to 0.5 mm apart. Only occasionally are there any tabulae developed between those of a pair. The pairs are 1.5 to 2.5 mm apart and sometimes incomplete tabulae are developed between them, which are commonly convex in longitudinal section and are usually based above and below on the one wall.

REMARKS. Commensal worm tubes similar to those in *Favosites duni* (Etheridge Jr.) are seen occasionally in the wall at or near the angles of the corallites.

This species differs from the type species in its slightly larger corallites, more numerous septal spines (though their absence in the holotype of the type species may be due to the recrystallized nature of the specimen), less regularity in the arrangement of the tabulae, and the more common incomplete tabulae between tabular pairs.

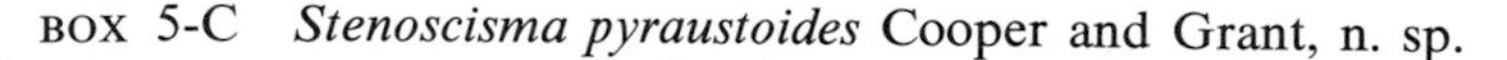

BOX 5-C *Stenoscisma pyraustoides* Cooper and Grant, n. sp.

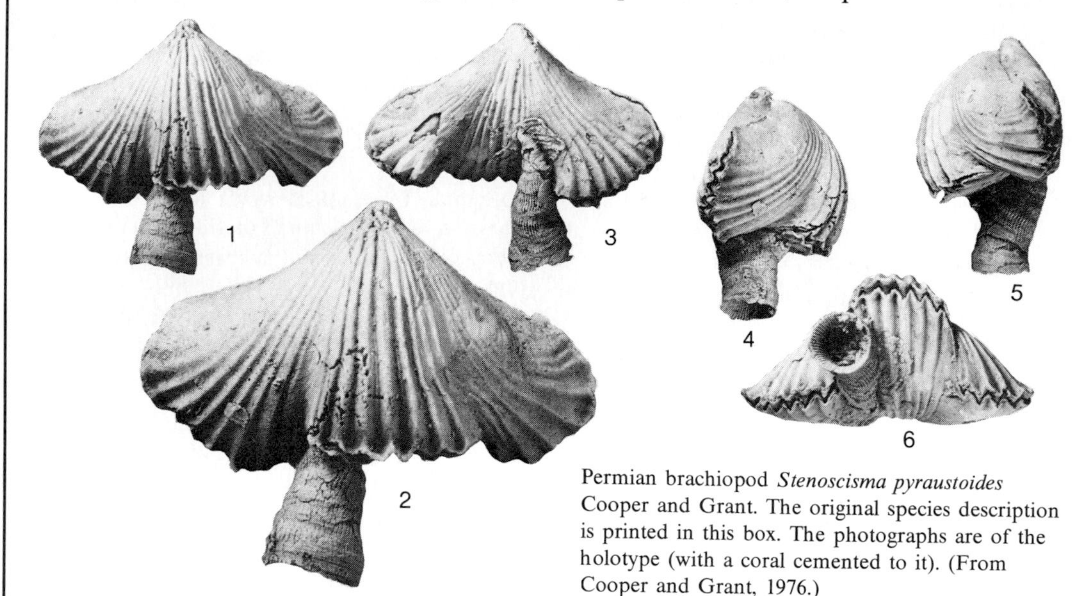

Permian brachiopod *Stenoscisma pyraustoides* Cooper and Grant. The original species description is printed in this box. The photographs are of the holotype (with a coral cemented to it). (From Cooper and Grant, 1976.)

Large for genus; outline broadly subelliptical to subtrigonal, sides diverging between 80° and 125°, normally over 100° in adults, maximum width near midlength, normally slightly farther toward the anterior; profile strongly biconvex to subtrigonal; commissure uniplicate, fold moderately high, standing increasingly high anteriorly, beginning 1–5 mm anterior to brachial beak; sulcus rather shallow, but dipping steeply at anterior, extending forward as broad tongue, producing emargination of anterior. Costae strong and sharp crested on fold and in sulcus, lower, broader, and rounder on flanks, beginning at beaks, frequently bifurcated, especially on fold and sulcus, numbering 6–10 on fold (normally 9), one less in sulcus, 4–9 on each flank, number not necessarily equal on both sides; stolidium better developed on brachial valve, varying from broad and fan-like to nearly absent.

Pedicle valve flatly convex transversely and from beak to flanks, strongly convex longitudinally through sulcus; beak short, only moderately thick, suberect to erect but not hooked; beak ridges gently curved, ill-defined; lateral pseudointerareas elongate, narrow, normally covered by edge of brachial valve; delthyrium moderately large, sides only slightly constricted by small, normally widely disjunct deltidial plates; foramen large for genus, nevertheless small, opening ventrally.

Brachial valve strongly convex transversely, only moderately convex along crest of fold owing to anterior increase in height of fold, convexity uniform without swelling in umbonal region; beak bluntly pointed, apex only slightly inside pedicle valve.

Pedicle valve interior with small teeth, continuous with dental plates that form short, boat-shaped spondylium just above floor of valve; median septum low, extending slightly forward of spondylium; troughs of vascula media diverging from midline of valve just anterior to median septum, extending directly across floor of valve; muscle marks in spondylium faint and undifferentiated.

Brachial valve interior with short, broad hinge plate, semicircular to crescentic; cardinal process at apex of hinge plate, located just beneath apex of valve, low or rather high, knob-like, normally not polylobate, shallowly striate for muscle attachment; hinge sockets short, narrow, at lateral extremes of hinge plate, finely corrugated; crural bases slightly diverging anterior to cardinal process, space between filled by narrow crural plates dipping along center line attaching crural bases to top of intercamarophorial plate; brachial processes not observed, presumed to be normal for genus; median septum high, thin, exceptionally short, length increasing greatly with height; camarophorium narrow, relatively short, anteriorly widening; intercamarophorial plate low, thick, relatively long; muscle marks not observed.

STRATIGRAPHIC OCCURRENCE. Skinner Ranch Formation (base); Hess Formation (Taylor Ranch Member); Cibolo Formation.

LOCALITIES. Skinner Ranch: USNM 705a, 705b, ?709a, 711o, 711z, 715c, 716p, 720e, 726j, 729j. Taylor Ranch: USNM 716o. Cibolo: USNM 739-1.

DIAGNOSIS. Exceptionally large and wide *Stenoscisma* with numerous bifurcations of costae on posterior of fold and flanks.

TYPES. Holotype: USNM 152220i. Figured paratypes: USNM 152219a–d; 152220b,c,k; 152221a,b; 152225. Measured paratypes: USNM 152220a–h,j; 152225. Unfigured paratypes: USNM 152220a,d–h,j.

COMPARISONS. *Stenocisma pyraustoides* is characterized by its exceptional width, large maximum size, numerous and frequently bifurcating costae on flanks, short beak with small disjunct deltidial plates, relatively short spondylium and camarophorium. The only known species that is closely related to *S. pyraustoides* is *S. multicostum* Stehli (1954)[1] from the Sierra Diablo. *Stenoscisma pyraustoides* is larger, wider, and less strongly costate, especially on the flanks where the costae are lower, broader, and fewer. The species bears superficial resemblence to *S. trabeatum*, new species, which is smaller, more triangular in outline, less strongly convex, has a longer beak, and a stolidium that is continuous from flanks to fold.

[1]See Cooper and Grant, 1976.

Measurements (in mm); measurements exclude stolidium

Localities and types	*Length*	*Brachial valve length*	*Width*	*Thickness*	*Apical angle (°)*
USNM 705a					
152220a	13.0	10.7	14.5	circa 6.0	95
152220b	15.0?	13.0	16.7	10.3	89
152220c	13.5	12.8	18.4	11.0	104
152220d	18.2	16.2	23.5	14.0	103
152220e	19.0	16.8	26.0+	14.0	107
152220f	23.7	22.4	28.0	16.0	93
152220g	26.0	25.2	35.9	21.3	116
152220h	28.3	26.6	45.1	22.7	104
152220i (holotype)	32.5	30.5	50.0	26.6	114
152220j	34.7	32.5	56.0?	21.0?	118
USNM 716o					
152225	35.5	33.5	50.5	23.2	109

The description, as distinct from the diagnosis, serves several purposes, not the least important of which is to provide an assessment of attributes that may at some future time be critical in diagnosis. If the species is part of a well-known group and is similar in most regards to other species, much of the description may be neglected in deference to the existing descriptions of closely related species. The weight may then be carried by a simple diagnosis. For a species belonging to a relatively unknown group, the description must be more comprehensive in order that relevant comparisons may be made if related species are discovered subsequently.

If possible, a description should include discussion of ontogenetic development, particularly if the organism's ontogeny is accompanied by a change in form. Also important is an assessment of variation encountered within and between populations of the species. When a new species is recognized on the basis of few specimens or fragments of specimens, this is not possible.

There has been considerable controversy over how much specimen material is necessary to establish a new species. It has been argued that because recognizing the variation within a species is very important, a single specimen or a fragment should never be used to establish a new species. No unequivocal answers can be given to this question because what is necessary in a particular description depends on the amount of difference between related species. Often a single specimen demonstrates that the organism is different from all other known organisms, making it folly to wait for the accumulation of large numbers of specimens. Many significant discoveries in paleontology are based on single specimens or on fragmentary material. On the other hand, if a new species belongs to a well-known group in which differences between species tend to be rather subtle, a large amount of material must be accumulated to make the description complete and effective.

Many conventions concerning the amount and quality of description are dictated by practical considerations. A description should not be so long that it cannot be read and absorbed efficiently. The combined lengths of diagnosis, description, and discussion of a new species usually do not exceed two thousand words. Because automatic methods of storage, processing, and retrieval of data are developing rapidly, many taxonomists are looking to these new facilities to extend the practical limits of description. It would be a relatively easy matter in many instances to codify the description of a new species so that it could be stored in compact form and compared with others at will. Several proposals have been made recently and pilot studies have been carrried out. It is tempting indeed to contemplate that the information contained in a two-thousand-word description of a fossil species might be reduced to an assessment of 30–40 morphologic characters and might then occupy only a few inches on a magnetic tape. It might even be possible to do away with the diagnosis as we know it because descriptions of attributes of several species could be compared automatically and the differences noted by the machine would represent the diagnosis.

An attribute of an organism used in diagnosis is referred to as a ***taxonomic character.*** How does a taxonomic character differ from other characters? The answer is inevitably circular because a taxonomic character is one shown to be useful in taxonomic discrimination. It is a character useful in showing *differences* between taxonomic groups.

To be useful in discriminating between species, a character must satisfy several conditions. Because species discrimination is ultimately a question of assessing genetic differences between populations, the taxonomic character must be one that shows a minimum of nongenetic variation. To be effective, it must also be a fairly obvious attribute, especially in fossil material in which shortcomings of preservation often yield an incomplete picture of the total organism. Ideally, a taxonomic character should be present and recognizable throughout an organism's ontogeny, and not just during certain growth stages.

Many morphologic characters satisfy the conditions just mentioned but do not constitute effective taxonomic characters. These are most usually characters that, although genetically controlled, are subject to great genetic variation within populations. Hair color in the human species is an example.

How effective a particular taxonomic character is may depend on which taxonomic category is under consideration. Many characters are too variable to be useful in distinguishing species but may be extremely important in describing subspecies. Other characters show no variation from species to species but are valuable in distinguishing genera and families.

The biologist often chooses taxonomic characters based on soft-part anatomy or behavioral traits that rarely leave clues in the fossil record. Problems thus arise in the taxonomy of groups of organisms with both fossil and living representatives. Very often, the paleontologist restudies the living species and develops a set of secondary taxonomic characters based on readily preservable parts of the organism. Certain living species, however, cannot be distinguished by preservable characters. The usual result is that several fossil species are combined under a single name although they would be separated into distinct species were complete knowledge of their morphology available.

In summary, taxonomic characters are chosen primarily for their usefulness in taxonomic discrimination. Their choice is based in large measure on hindsight. Taxonomic characters as a group are probably the most significant characters biologically, but the system contains no assurance of this.

TYPE SPECIMENS

The selection of type specimens presents several problems. If a single specimen is designated, it is called the ***holotype.*** If several specimens serve this purpose they are called ***syntypes.*** Both alternatives are officially acceptable although the International Commission strongly urges the use of a holotype rather than a series of syntypes because it is always possible that the series will be judged by later workers to contain representatives of more than one species.

Type specimens were once widely used to define species and as special standards for comparison. We now know, owing in part to the adoption of the population approach in taxonomy, that many of these type specimens are not the most legitimate representatives of their species. Even in a large sample a single specimen may not exist that is average for all observable morphologic characters. Furthermore, the mean value for any variable feature tells nothing of the range of variation.

In modern biology and paleontology the type specimen no longer serves to define a species (unless no other specimens are available). It is relegated, instead, to the position of "name-bearer." When a species is named, the name is formally attached only to the one or more specimens that are designated as type specimens. In practice, fossil type specimens actually tend to be somewhat unusual representatives of the hypodigm. The most common bias is toward large size and good preservation.

Several other kinds of type specimens figure prominently in various phases of taxonomic work. A ***paratype*** is a specimen, other than the holotype, which is formally designated by the author of a species as having been used in the description of the species. The designation of a single holotype and a series of paratypes thus contains some of the advantages of both the holotype system and the syntype system. The holotype remains as the name bearer but the paratypes, which may be numerous, serve to express more fully the author's concept of the species.

A ***lectotype*** is a specimen originally designated as a syntype but subsequently singled out as the definitive type specimen for a species. This duplicates what the author would have accomplished if a holotype and a set of paratypes had been designated originally.

It often happens that type specimens are destroyed or lost. The International Commission has established procedures for designating new type specimens to replace those that are lost. The ***neotype*** is such a replacement.

For species described many years ago, it is often important to redescribe and re-illustrate specimens, especially if the original description was written under a different set of rules. ***Plesiotypes*** are type specimens used for redescriptions of existing species.

A great many other kinds of types have been proposed and are occasionally used. For example, a ***topotype*** is a specimen that is not part of the original type material but that has been collected at the type locality.

Changing Species Names

A worker may change the name of a taxon either because he has found that its use violates a rule of nomenclature or because he judges that the taxon has been improperly classified. We will restrict our discussion of name changes to species names.

Homonyms are identical species names that denote different species groups. There are two varieties. ***Primary homonyms*** are identical names that were erected for different taxa (with different holotypes) belonging to the same genus. The author of the later-named homonym was in error, not knowing that the species name had been occupied when he used it. Once such an error is discovered, only the first published, or ***senior homonym,*** can be retained. The Code states that the ***junior homonym*** must be permanently rejected. The difference between primary and ***secondary homonyms*** is that the latter originate by transfer of one species to a new genus that contains a species with the same specific name. The author of neither species is in error according to the Code, because the same specific name can be used for species belonging to different genera. The problem arises when a later worker has decided that the two species belong to the same genus. The worker who discovers secondary homonyms should formally reject the junior name. A rejected primary or secondary homonym must be replaced by the oldest available name, or if no previously published name is available, by a new name.

Synonyms are two different names applied to the same taxon. There are two varieties. ***Objective synonyms*** are different names that are based on the same type specimen or specimens. Here there is no question of taxonomic opinion; the senior (first published) synonym must be retained, and the junior synonym must be permanently rejected. ***Subjective synonyms*** are names that were established for different type specimens that are later judged by a worker to belong to one species. This worker must then apply the senior synonym to both of the type specimens and all other specimens belonging to the hypodigm in which he places them. Another worker, however, may judge that the type specimens belong to separate species; he will not consider the names to be synonyms and will retain both. In other words, while a junior objective synonym is eliminated automatically by the Code, a junior subjective synonym remains available as a name, its use depending entirely upon taxonomic opinion.

Rejection of names on the basis of priority is sometimes unfortunate because it eliminates familiar names. The formal change in generic name of the familiar Eocene "dawn horse" from *Eohippus* to *Hyracotherium* was unpleasant to many workers. The Commission on Zoological Nomenclature is empowered by the Code to exercise its plenary powers in order to suspend the rule of priority at special request. Many familiar names found to be junior homonyms or synonyms have been retained by this procedure.

Perspective

Taxonomic procedures form an extremely complex and important subject for the working paleontologist. In actual fact, procedural problems are considerably more complex than outlined here. The student is often impressed and discouraged by this complexity and by the many seemingly archaic and inefficient practices. The possibility of working with the biology and paleontology of organisms from outer space has given impetus to many reform movements. The contention has become common that the exploration of space should provide an opportunity to correct the errors of the past and to start with a new, modern system.

It cannot be denied that the current procedures could be improved upon. To date, the greatest deterrent to sweeping reform has been the obvious cost. With 1,500,000 living species and 250,000 fossil species the problem of effecting change has been overwhelming. The mere size of the problem has been a deterrent to progress but at the same time has been a stabilizing influence and has effectively prevented the confusion that would undoubtedly result from partial reforms. In the next generation, we can probably look forward to gradual improvement and streamlining, but it is unlikely that the system will be altered significantly.

Supplementary Reading

Blackwelder, R. E. (1967) *Taxonomy: A Text and Reference Book.* New York, John Wiley & Sons, 698 p. (A comprehensive sourcebook of taxonomic theory and practice.)

Eldredge, N., and Gould, S. J. (1972) Punctuated equilibria: an alternative to phyletic gradualism. *In* Schopf, T. J. M., ed. *Models in Paleobiology.* San Francisco, Freeman, Cooper and Company, p. 82–115. (A provocative discussion of the relative roles of speciation and phyletic transition in evolution.)

Mayr, E. (1963) *Animal Species and Evolution.* Cambridge, Mass., Harvard University Press, 797 p.

Mayr, E. (1969) *Principles of Systematic Zoology.* New York, McGraw-Hill, 428 p. (An excellent and well-organized treatment of taxonomy.)

Savory, T. (1962) *Naming the Living World.* London, English University Press, Ltd., 128 p. (A brief summary of taxonomic nomenclature.)

Schenk, E. T., and McMasters. J. H. (1956) *Procedure in Taxonomy,* 3rd ed. Revised by A. M. Keen and S. W. Muller. Stanford, Calif., Stanford University Press, 119 p. (A detailed discussion, with examples, of the rules of taxonomic nomenclature.)

Simpson, G. G. (1961) *Principles of Animal Taxonomy.* New York, Columbia University Press, 247 p. (Emphasizes taxonomic theory.)

Stoll, N. R., et al., eds. (1961) *International Code of Zoological Nomenclature.* London, International Trust for Zoological Nomenclature, 176 p. (The official version of the International Code; contains a glossary and extensive appendices.)

CHAPTER 6

Grouping of Species into Higher Categories

The naming of species has usually gone hand in hand with arranging them in some kind of classification. Species having characteristics in common have been grouped together and thus distinguished from other such groups, which assists us in communicating information about them. One might say, for example, "I saw a bird eating a worm." There are about 8,600 species of birds living today and about 25,000 species of worms and the statement just given would be meaningless to all but the most specialized audience if species names were substituted for "bird" and "worm." By using the word "bird" we describe the predator as being one of a clearly defined group of similar organisms and by so doing exclude all but about $\frac{1}{10}$ of one percent of the animals living today. Furthermore, the purpose in making the statement might have been such that specific identification would be quite irrelevant or the species might have been very difficult to identify.

Perhaps the most practical reason for grouping species into ***higher categories*** is that each person cannot learn and remember the names and distinguishing characteristics of a million and a half different organisms. In the example of the bird and the worm, a good ornithologist might be able to recognize the species of the bird (particularly if he were familiar with the fauna of the region), but it is unlikely that he could identify the worm. In this sense, the grouping of species compensates for the lack of storage space in the human memory. The usual procedure for identifying a specimen is to determine its phylum first, and then gradually proceed from the general to the specific.

How is the classification of plants and animals best accomplished? A single group of species can be subdivided in many different ways. The species could, for example, be arranged alphabetically in 26 categories, each determined by the initial letter of the species name. Somewhat more logically, they might be grouped according to year of discovery. Neither of these systems, however, would be of any assistance to the person who observes the bird eating the worm, because the organisms have no visible characteristics that serve as an aid in relating them to these systems.

Most classifications are based upon readily observable characteristics, either of appearance, habitat, behavior, or geographic or stratigraphic occurrence. Many of the earlier classifications of vertebrates were based on habitat, with aquatic vertebrates (fish) being distinguished from land-dwelling vertebrates (reptiles, birds, and mammals) and so on. The fact that there were exceptions to these classifications (such as aquatic mammals) did not negate the usefulness of the classifications for most purposes.

Most successful classifications have been based on morphology. We use morphology here in a rather broad sense to include not only external form, but also internal anatomy and even some details of physiology, biochemistry, and behavior.

Most classifications used today had their beginnings long before organic evolution was widely accepted. Pre-Darwinian classifiers assumed for the most part that the species they were dealing with were created spontaneously and independently. Nevertheless, some early classifications, particularly at the higher levels, have changed little as the understanding of evolution has become widespread.

Classification versus Evolution

If classification is to serve primarily for communication and identification, utility is the principal criterion for choosing one system over another. With the rise of interest in evolution there has inevitably been a move toward using classification to express evolutionary relations also.

In most plant and animal groups, increased emphasis on evolution has not brought about very dramatic changes in the actual form and content of classification, primarily because morphologic similarity is highly correlated with evolutionary proximity. Thus a classification intended to reflect evolution is largely dependent upon the same kinds of similarities and differences as one designed simply for identification.

Classifications are constantly undergoing change, partly as a result of our increase in descriptive knowledge. New higher categories are often needed to express the distinction between newly discovered species and ones that have been known for a long time. Classifications also change from time to time because of increase in theoretical knowledge of evolutionary mechanisms. Considering this we may note that classifications at a given time are an index of evolutionary thought. Furthermore, classifications by different but equally well-qualified scientists commonly differ at any one time simply because interpretations of evolution differ.

Taxonomic Categories

Linnaeus, in 1759, used only six taxonomic categories: kingdom, class, order, genus, species, and variety. Of these, kingdom, class, order, and genus qualify as higher categories in the present context. Since the ***kingdom*** as a taxonomic category was used by Linnaeus only to separate plants from animals, in his work as a zoologist he used only three higher categories: class, order, and genus.

Since the number of species known in the time of Linnaeus was relatively small, his categories were quite adequate to divide the total array into manageable groups. The tremendous increase in the number of known species has demanded the addition of categories intermediate between kingdom and species. The principal additions are ***phylum*** and ***family,*** with the most commonly used hierarchy being: kingdom, phylum, class, order, family, and genus. For some groups, further categories such as subphylum, subclass, superorder, suborder, superfamily, subfamily, and subgenus have been added.

How is each of these categories defined? In simplest terms each category may be defined as a subdivision of the next higher category.

The International Code of Zoological Nomenclature covers many procedures used in dealing with higher taxonomic categories. The Code itself includes rules for formation of categories up to the superfamily level. The recommendations of the International Commission attached to the Code, however, cover the entire range of higher categories (by implication if not by explicit rules). The rules are approximately parallel to those for species. We have already seen, for example, that a new species must be assigned at least to the next higher category (the genus). It is generally assumed that all genera can be assigned to families, orders, and so on, although allowance is given for the possibility that the evolutionary affinities of a genus may be unknown. In such a case, the genus may be defined in isolation ("incertae sedis"). Several rules governing the formation of names of higher categories make it possible to recognize the taxonomic rank on sight from the spelling of the ending of the name.

The type concept also extends to the definition of higher categories. When a new genus is proposed, a ***type species*** designation must accompany the original description. The type species thus becomes the name bearer for the genus. Similarly, a family must have a ***type genus,*** and so on.

When a newly discovered species is named and described, every attempt is made to find a plausible assignment in existing higher categories. If this proves to be impossible, new categories are sometimes proposed; not only might a phylum name be proposed, but also new names for a class, order, family, and so on. This illustrates an important feature of the conventional system of classification: a higher category may be ***monotypic.*** A phylum need have only one class; a class might have no more than one order, and so on.

Most classifications assume that evolution follows some well-defined model. The most common graphic method of portrayal of the model is a tree-like structure called the ***phylogenetic tree.*** The main topologic characteristic of the tree is that it is continuously branching; that is, a branch does not rejoin the ancestral branch or some other branch. This means, in effect, that hybridization between lineages is assumed not to occur.

Quite different phylogenetic trees are shown in Figure 6-1 and 6-2. In Figure 6-1, a few closely related species evolved over a short span of time: each segment of the "tree" was part of an evolving lineage. Figure 6-2 shows the evolution of a large group of species over a long time. The smallest taxonomic unit is the order. Thus, a single line in Figure 6-2 represents a complex of many evolving lineages, each of which contains evolving populations.

FIGURE 6-1
Cenozoic phylogeny of *Venericardia (Claibornicardia)* in the Gulf and East Coasts of North America. (From Heaslip, 1968.)

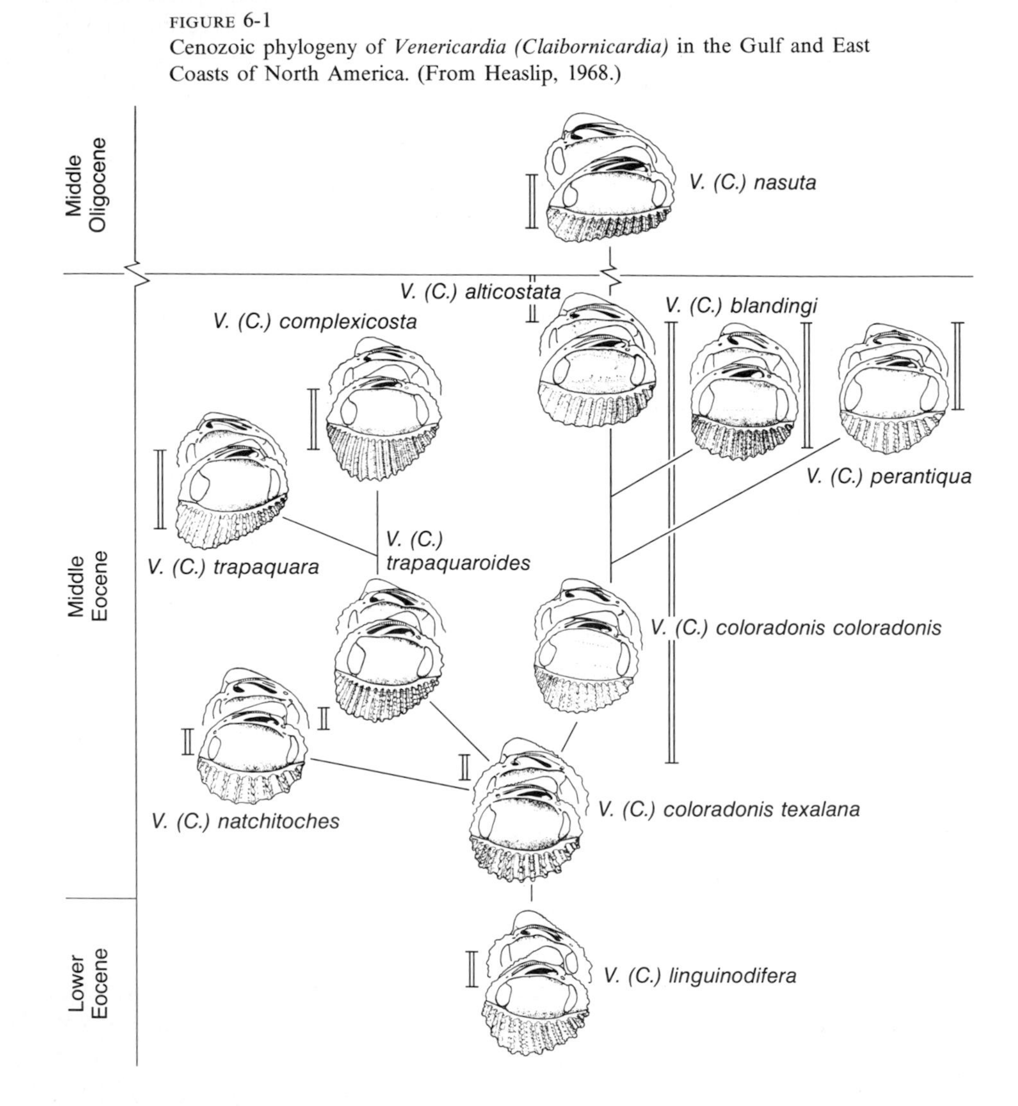

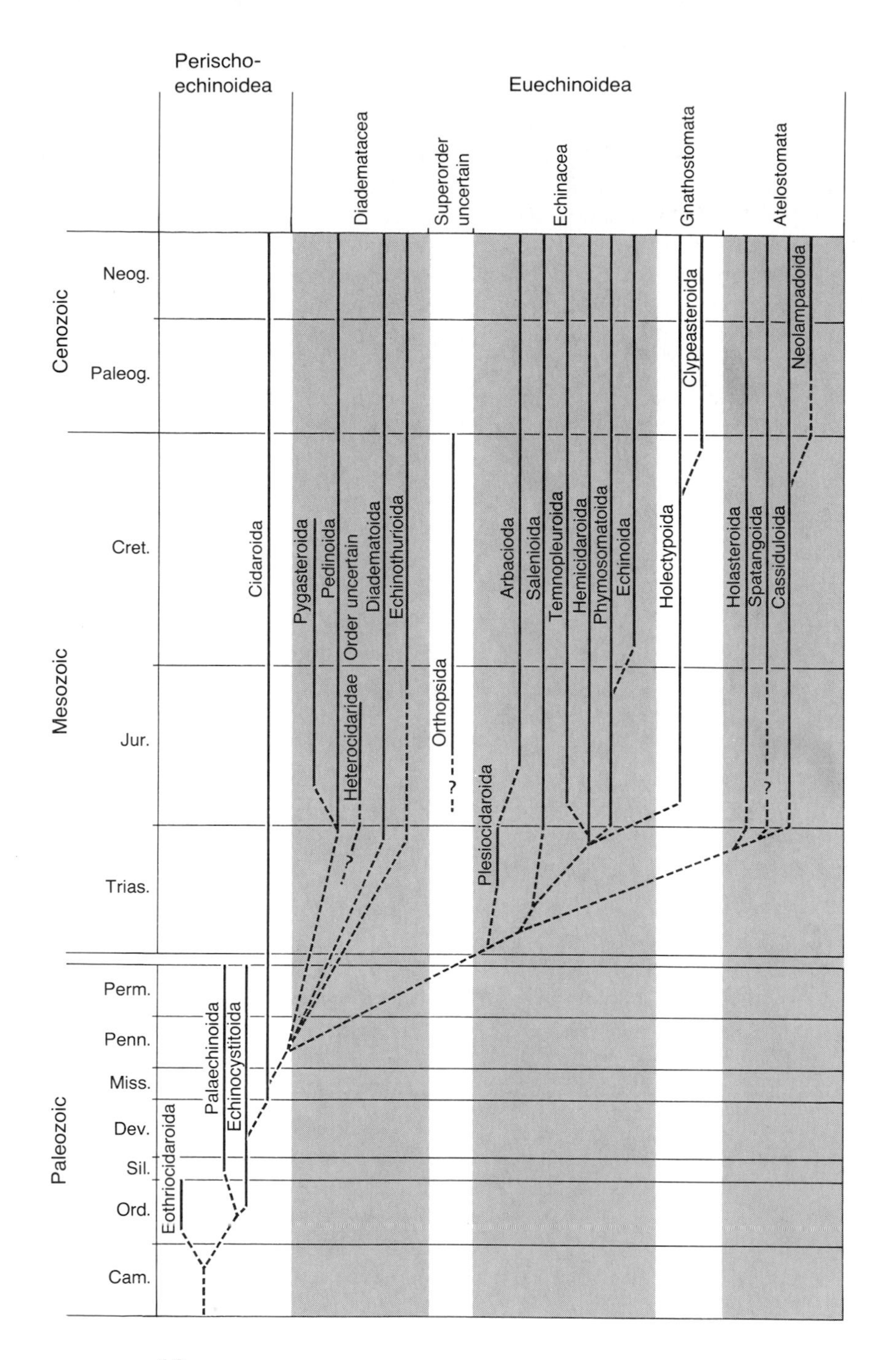

FIGURE 6-2
Phylogeny of the class Echinoidea, based on ranges in the stratigraphic record of specimens of the various orders and on inferred evolutionary relationships. Gap above Permian indicates a change in the vertical scale. (From Durham, 1966.)

At best, the phylogenetic tree is a simplification of nature. At any point in time, a species is usually represented by many geographically distinct populations. Elements of a lineage may diverge with the onset of geographic isolation and later converge if the geographic barriers are removed. The "internal" character of an evolving lineage is shown diagrammatically in Figure 6-3. The internal structure of the lineage shows an anastomosing pattern (like a braided stream), rather than a continuously branching pattern such as is shown in the phylogenetic trees of Figures 6-1 and 6-2. The anastomosing pattern indicates that, within a lineage, gene flow among populations is common.

What coordinate system is used in a phylogenetic tree? Many different systems are used and it is important to understand the differences. In the examples shown in Figures 6-1 and 6-2, the vertical dimension is time; the horizontal dimension shows the relationships between lineages. In phylogenetic trees such as these, the top of a line indicates the extinction of the lineage unless it reaches to the top of the diagram.

Figure 6-4 shows a phylogenetic tree that is essentially dimensionless. The end of a line does not imply the extinction of a lineage, though it may coincide with it, nor does the distance between branches indicate geographic or evolutionary distance. Time plays a role in the diagram only in that the chronologic sequence of evolutionary events is established.

In Figure 6-5, the vertical and horizontal dimensions are time and geographic area. It was possible to construct the phylogeny in this way because the evolving lineages rarely overlap geographically.

Figure 6-6 shows the classification of thirty-one fossil brachiopod genera into eleven subfamilies (names ending in "-inae") and four families (names ending in "-idae"). If we assume that this classification is a valid interpretation of evolution, we may look

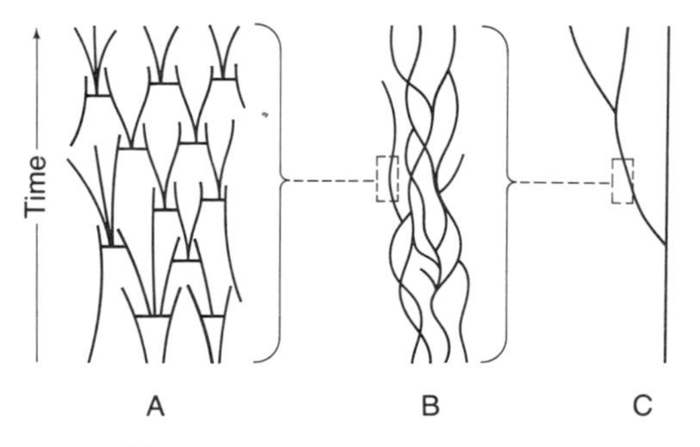

FIGURE 6-3
The anatomy of an evolutionary lineage.
A: Ontogenies of individuals in a population. Horizontal lines represent mating. B: Evolving array of populations making up the lineage. C: Relationship between several lineages. (From Simpson, 1953.)

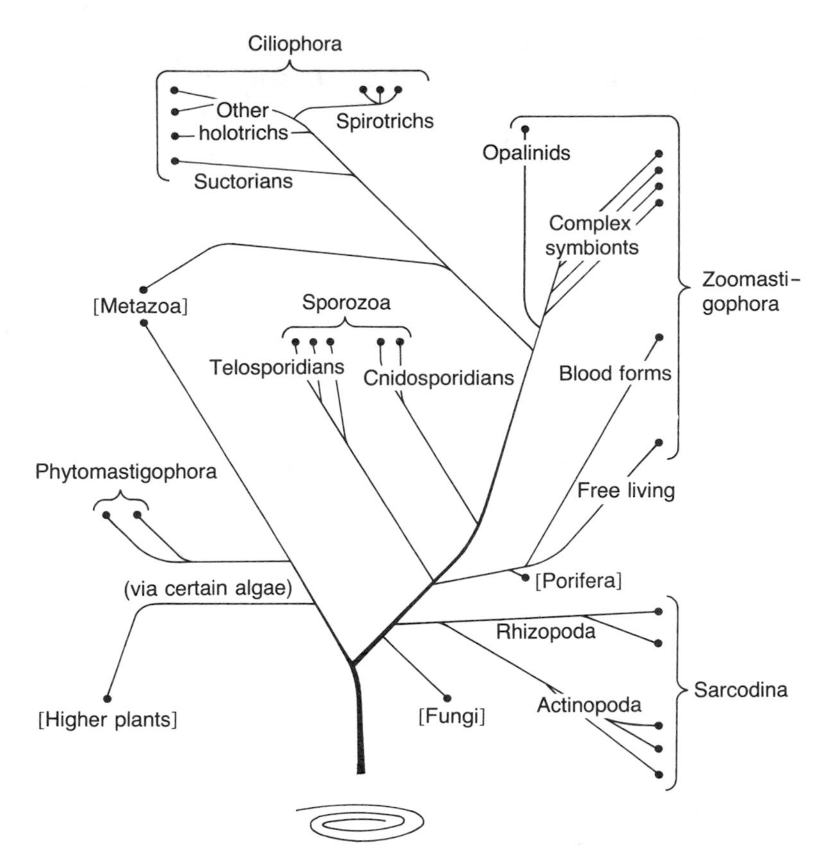

FIGURE 6-4
An inferred phylogeny for the Protozoa. Presumed evolutionary descendants of the Protozoa are indicated in brackets. (From Corliss, 1959.)

upon it as a phylogenetic tree (without geographic or time dimensions). Genera considered to be most closely related are linked in the diagram as members of a subfamily; closely related subfamilies are linked as members of a family. The thirty-one genera, as a group, constitute a superfamily.

At this point, we might wonder whether higher categories have an objective basis. In a sense they do and in a sense they don't. We define the biologic species by using the concept of reproductive isolation and look upon reproductive isolation as a basic biologic phenomenon that makes it possible for evolving lineages to remain genetically distinct. In this sense, Simpson's definition of the ***evolutionary species*** has some objective basis even though division of lineages into species is arbitrary. The family, on the other hand, has no comparable objective basis. Higher categories are used to express evolutionary relationships but we do not have, in the present state of

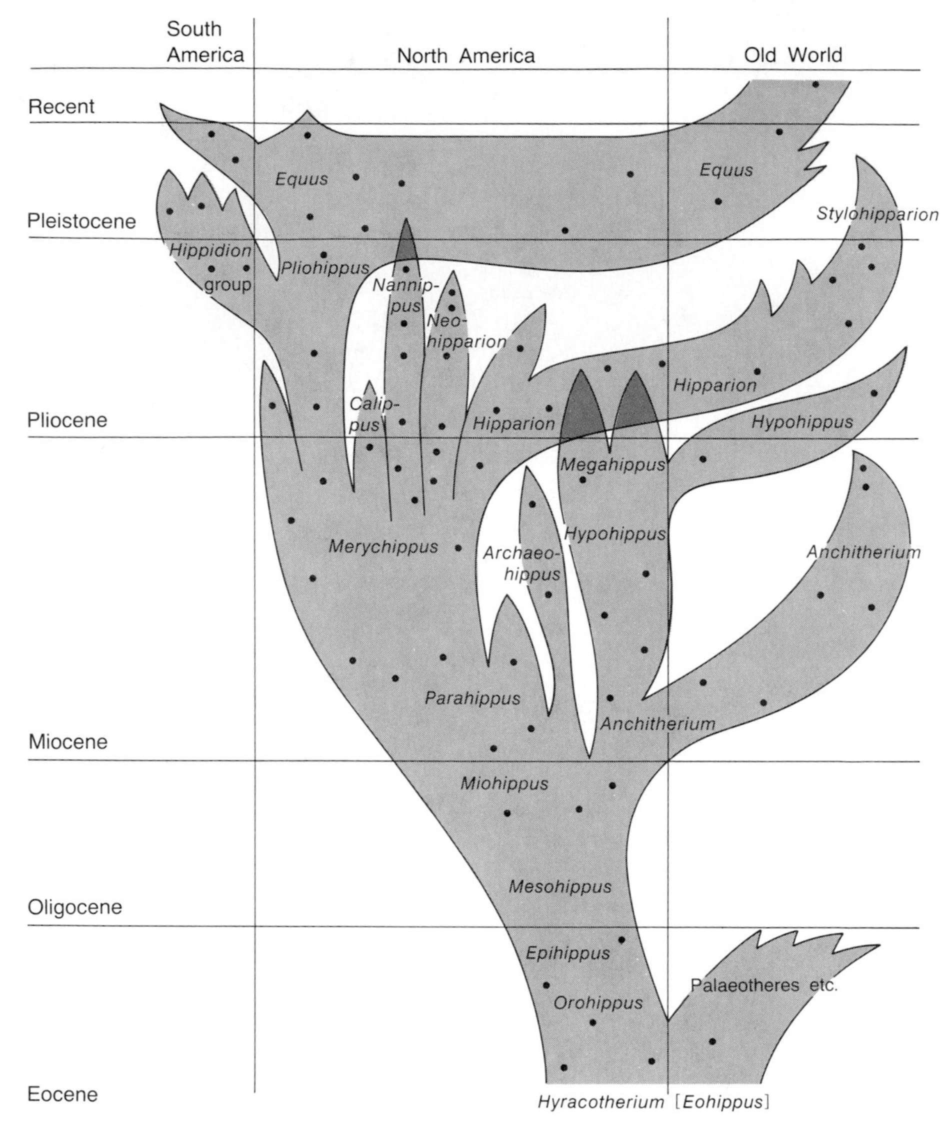

FIGURE 6-5
Inferred phylogeny of the horses. (From Simpson, 1951. Copyright © 1951 by Oxford University Press. Reprinted by permission.)

our understanding, specific rules that can be applied. In other words, there are no thresholds in the transition from genus to family comparable to the threshold between the subspecies and the species (at a single moment in time).

Some objectivity does enter into the development of higher-category classification. It is implicit, for example, in the general classification model, that all species of a genus have evolved from one ancestral lineage. This follows from the fact that at the

Family	Subfamily	Genus
Chonetidae	Chonetinae	*Chonetes*
	Strophochonetinae	*Strophochonetes*
	Devonochonetinae	*Devonochonetes*
		Longispina
		Notiochonetes
		Protochonetes
	Anopliinae	*Anoplia*
		Anopliopsis
		Chonetina
		Notanoplia
		Tornquistia
	Retichonetinae	*Retichonetes*
	Rugosochonetinae	*Rugosochonetes*
		Dyoros
		Eolissochonetes
		Lissochonetes
		Mesolobus
		Pliocochonetes
		Quadrochonetes
	Chonetinellinae	*Chonetinella*
		Neochonetes
		Waagenites
	Semenewiinae	*Semenewia*
Eodevonariidae		*Eodevonaria*
Chonostrophiidae		*Chonostrophia*
		Chonostrophiella
		Tulcumbella
Daviesiellidae	Daviesiellinae	*Daviesiella*
	Delepineinae	*Delepinea*
		Megachonetes
	Airtoniinae	*Airtonia*

FIGURE 6-6
Classification of the fossil brachiopod superfamily Chonetacea, arranged in the form of a dendrogram. (From Rowell, 1967; after Muir-Wood, 1965.)

supraspecific level we assume that phylogeny is a process that expresses itself by continuously branching. If two species can be shown to have a common ancestor that is not also the ancestor of a third species, then the first two species can objectively be placed in a higher category different from that of a third. This does not lead, however, to an objective basis for deciding where the family level leaves off and the order level begins because the difference between the levels is one only of degree. It is not surprising that classifications developed by different taxonomists often differ in assessment of higher categories. If classifications are based on the same understanding of phylogeny, however, the classifications should be correlatable. Rarely does the biologist or paleontologist have a truly accurate knowledge of phylogeny; classification is thus an expression of knowledge at a given time.

The approaches of paleontologists and biologists toward questions of higher categories differ somewhat because the paleontologist is able to observe the time dimension directly whereas the biologist must always infer the chronologic development of lineages. The paleontologist is sometimes able to say that one species evolved from another because he finds the two in stratigraphic succession.

Criteria for Definition of Higher Categories

If classification is to reflect phylogeny, we must assess the "phylogenetic distance," or amount of genetic difference, between species. Genetic difference cannot be observed directly in fossils any more than it can in most living forms. We can, however, use morphologic difference as a rough measure of genetic difference (just as we did when dealing with species). The two species in a group that have the largest number of morphologic characters in common are most likely to be descended from a common ancestor and thus are most likely to qualify as members of a single higher category.

A large number of species, each defined by many taxonomic characters, may be divided up into many quite different systems of groups. How do we choose between several alternative classifications? Which alternative is most likely to express phylogeny? It sometimes seems that there are just as many different answers to such questions as there are taxonomists. It has even been stated that taxonomy at this level is an art rather than a science and that the methods defy a clear and logical explanation. Let us explore some commonly used methods.

Very often, a few characters are singled out (on *a priori* grounds) as being of greater evolutionary significance than the rest and division is based on these characters in preference to others. This ***weighting of characters*** takes several forms. In the example of the Solnhofen bird discussed in Chapter 1, the presence of feathers was used as a criterion for assignment to the class Aves. This was based partly on the knowledge that feathers were of vital importance in enabling the organism to occupy a different set of habitats from those of its ancestors and therefore, to undergo tremendous evolutionary expansion.

In other cases, choice of characters to be weighted is based on a more nearly objective judgment of the effect of weighting on the resulting classification. Table 6-1

TABLE 6-1
Coded Morphology of Ten Hypothetical Species

Species	*Characters* 1	2	3	4	5	6	7	8	9	10
A	−	+	−	+	−	+	+	−	+	−
B	−	−	−	−	+	+	+	−	−	+
C	+	+	+	+	−	−	−	+	−	+
D	+	+	−	−	−	−	−	+	+	+
E	+	−	+	−	+	+	−	+	−	−
F	+	−	−	+	+	−	−	+	−	+
G	−	−	+	−	+	+	+	−	+	−
H	−	+	−	+	−	+	+	−	+	+
I	+	−	+	−	+	−	−	+	+	+
J	−	+	−	−	−	+	+	−	−	+

shows purely hypothetical data that we will use to illustrate this type of weighting: ten imaginary species are defined by ten morphologic characters; each character is expressed in a given species as one or the other of two possible states, indicated by plus and minus. The two states might be "large" and "small," or "red" and "white," or "having teeth" and "toothless."

Let us pose the problem of dividing the ten species into two groups (without specifying the relative sizes of the groups). Let us assume that the two groups occupy different branches of a phylogenetic tree and that we wish to use the morphologic information to determine them. There are 637 possible divisions of ten species into two groups. Our job is to select one having a high probability of being correct.

Some of the possible divisions are more plausible (morphologically) than others. Table 6-2 shows three of the alternatives. The first subdivision produces two groups that are each heterogeneous in all morphologic characters. The second produces groups that are homogeneous for one of the characters (4) but heterogeneous for all others. The third produces groups homogeneous for three of the ten characters (1, 7, and 8). The last alternative may be said to be "supported" by 30 percent of the characters. Note also that an additional character (6) almost supports the classification: one group is completely homogeneous for character 6 (all plusses) and the other group is nearly homogeneous (all but one species are minus). The three characters infallibly supporting the grouping might be called "excellent" taxonomic characters because they each suggest the same classification. Character 6 might be called a "usable" taxonomic character because it suggests a classification very similar to that supported by the other three.

If we were to expand this hypothetical taxonomic problem to include many more species (perhaps several hundred) we would increase the complexity of the problem to a point at which there would be an almost infinite number of possible classifications. We could not begin to consider all of them. On the basis of the preliminary study of

TABLE 6-2
Three Possible Classifications of Ten Species into Two Genera

Classification 1

Species	Characters 1	2	3	4	5	6	7	8	9	10	
A	−	+	−	+	−	+	+	−	+	−	
B	−	−	−	−	+	+	+	−	−	+	Genus 1
C	+	+	+	+	−	−	−	+	−	+	
D	+	+	−	−	−	−	−	+	+	+	
E	+	−	+	−	+	+	−	+	−	−	
F	+	−	−	+	+	−	−	+	−	+	
G	−	−	+	−	+	+	+	−	+	−	Genus 2
H	−	+	−	+	−	+	+	−	+	+	
I	+	−	+	−	+	−	−	+	+	+	
J	−	+	−	−	−	+	+	−	−	+	

Classification 2

Species	Characters 1	2	3	4	5	6	7	8	9	10	
A	−	+	−	+	−	+	+	−	+	−	
C	+	+	+	+	−	−	−	+	−	+	Genus 1
F	+	−	−	+	+	−	−	+	−	+	
H	−	+	−	+	−	+	+	−	+	+	
B	−	−	−	−	+	+	+	−	−	+	
D	+	+	−	−	−	−	−	+	+	+	
E	+	−	+	−	+	+	−	+	−	−	Genus 2
G	−	−	+	−	+	+	+	−	+	−	
I	+	−	−	−	+	−	−	+	+	+	
J	−	+	+	−	−	+	+	−	−	+	

Classification 3

Species	Characters 1	2	3	4	5	6	7	8	9	10	
A	−	+	−	+	−	+	+	−	+	−	
B	−	−	−	−	+	+	+	−	−	+	
G	−	−	+	−	+	+	+	−	+	−	Genus 1
H	−	+	−	+	−	+	+	−	+	+	
J	−	+	−	−	−	+	+	−	−	+	
C	+	+	+	+	−	−	−	+	−	+	
D	+	+	−	−	−	−	−	+	+	+	
E	+	−	+	−	+	+	−	+	−	−	Genus 2
F	+	−	−	+	+	−	−	+	−	+	
I	+	−	+	−	+	−	−	+	+	+	

the ten species we might postulate, however, that characters 1, 7, and 8 would be significant in dividing the larger group of species. Character 6 could be added tentatively because of the 90 percent agreement between its distribution and that of the selected three. Characters 1, 7, 8, and 6 could then be tested with the larger group of species. If their separate use produced the same (or nearly the same) divisions of the larger group and if the divisions were geographically, ecologically, or stratigraphically reasonable, we might conclude that the four characters are critical in determining a major evolutionary division. We might forego further testing and agree to give disproportionate importance to these characters in all subsequent taxonomic decisions. By so doing, we would have *weighted* the taxonomic characters.

In the hypothetical example, the classification would be completed by further subdivision of each group. We might call the original group of ten species a family. The inital division would provide two genera, a subdivision would produce subgenera, and so on. The characters used to determine genera could not, of course, be used in the determination of subgenera because the genera are homogeneous for those characters. Thus, the process of weighting characters excludes them from use at lower taxonomic levels.

Most weighting of characters is probably done in a manner similar to that used in the hypothetical example. The presence or absence of feathers, for example, is weighted heavily in vertebrate classifications because feathers are always present in a group (the birds) that is united by other characters as well; feathers are absent in other vertebrates. The presence or absence of feathers is thus analogous to characters 1, 7, and 8 in our example. The absence of teeth is almost universal in birds, though *Archaeopteryx* had teeth. The presence or absence of teeth, as a character, is thus analogous to 6 in our example. It is a usable taxonomic character.

The taxonomist rarely follows a clear routine and he may not be aware of the logical steps he follows. It is not surprising that the classification process has been called intuitive or an "art".

Probably the greatest pitfall in the history of higher-category taxonomy has been the natural tendency to search for a single definitive character for grouping species. Many taxonomic problems with invertebrates have arisen in attempts to establish orders or subclasses. Attempts to group brachiopod families into orders solely on the basis of shell microstructure or morphology of the lophophore support have failed. Likewise, attempts to classify trilobites into orders strictly on the basis of facial suture configuration have proved unsuccessful. In the Bivalvia, two schemes, each based on a single morphologic character, were proposed. Many biologists favored use of gill type in classification of families into orders whereas other biologists and most paleontologists favored use of dentition (configuration of the hinge teeth of shells). Two contradictory classifications developed. It is now well known that the eulamellibranch type of gill evolved both in burrowing clam groups with heterodont dentition and in oysters, which lack hinge teeth and were derived from scallops. Certain types of dentition also arose independently in two or more taxonomic groups. Modern workers no longer attempt to use a single character in grouping bivalve families into orders.

Most use several characters, including dentition and gill type, which they weight in a variety of ways.

Few modern workers employ the single-character approach in higher-category taxonomy. Still, for a few groups it has proven to be generally adequate. As we have seen, feathers in birds represent a definitive taxonomic character. Similarly, pelvic structure alone is sufficient to divide dinosaur families into two major orders (Figure 6-7). In a sense, however, the union of the two dinosaur orders in a single subclass is artificial; they are no more closely related to each other than either is to crocodiles, which constitute a distinct order that arose from the same ancestral group.

Numerical Taxonomy

It is tempting to wonder whether classification could be accomplished by machine. Given the information in Table 6-1, could a computer be programmed to investigate all 637 alternative divisions and choose the one supported by the largest number of characters? A modern digital computer could probably do this job in less than a second! Furthermore, it could be programmed to apply a variety of criteria in selection of the "best" classification. Selecting the classification supported by the largest number of characters is only one of many possible criteria.

Numerical taxonomy is the science of classifying organisms by purely mechanical or mathematical means. If we were to devise a numerical method for assessing the 637 possible classifications of the ten hypothetical species in Table 6-1, we would be doing numerical taxonomy. Numerical taxonomy is almost as old as taxonomy itself but nearly all major developments have come since the middle 1950's because of advances in computer technology.

So far, classifications developed by numerical methods have been based entirely on observable characteristics of the organisms. Such factors as biogeography, stratigraphic distribution, and ecology have not been used. The resulting classifications are thus called ***phenetic classifications*** because they are based on the phenotype (defined broadly to include physiology, biochemistry, behavior, as well as what we conventionally mean by morphology). There has been considerable controversy over whether such classifications are "natural" in the sense of reflecting phylogeny. The answer to this depends partly upon the importance one attaches to nonphenetic information in classification. The fact is that most conventional taxonomies are based largely or completely on morphology and thus numerical taxonomy does not necessarily represent a substantial departure.

Most methods of numerical taxonomy are designed to operate at any taxonomic level: the basic units to be classified may be individual organisms, species, or even genera or higher groupings. The unit to be classified is called the ***operational taxonomic unit,*** or OTU.

The usual procedure in numerical taxonomy includes the following steps:

1. Selection of the OTU's to be classified.

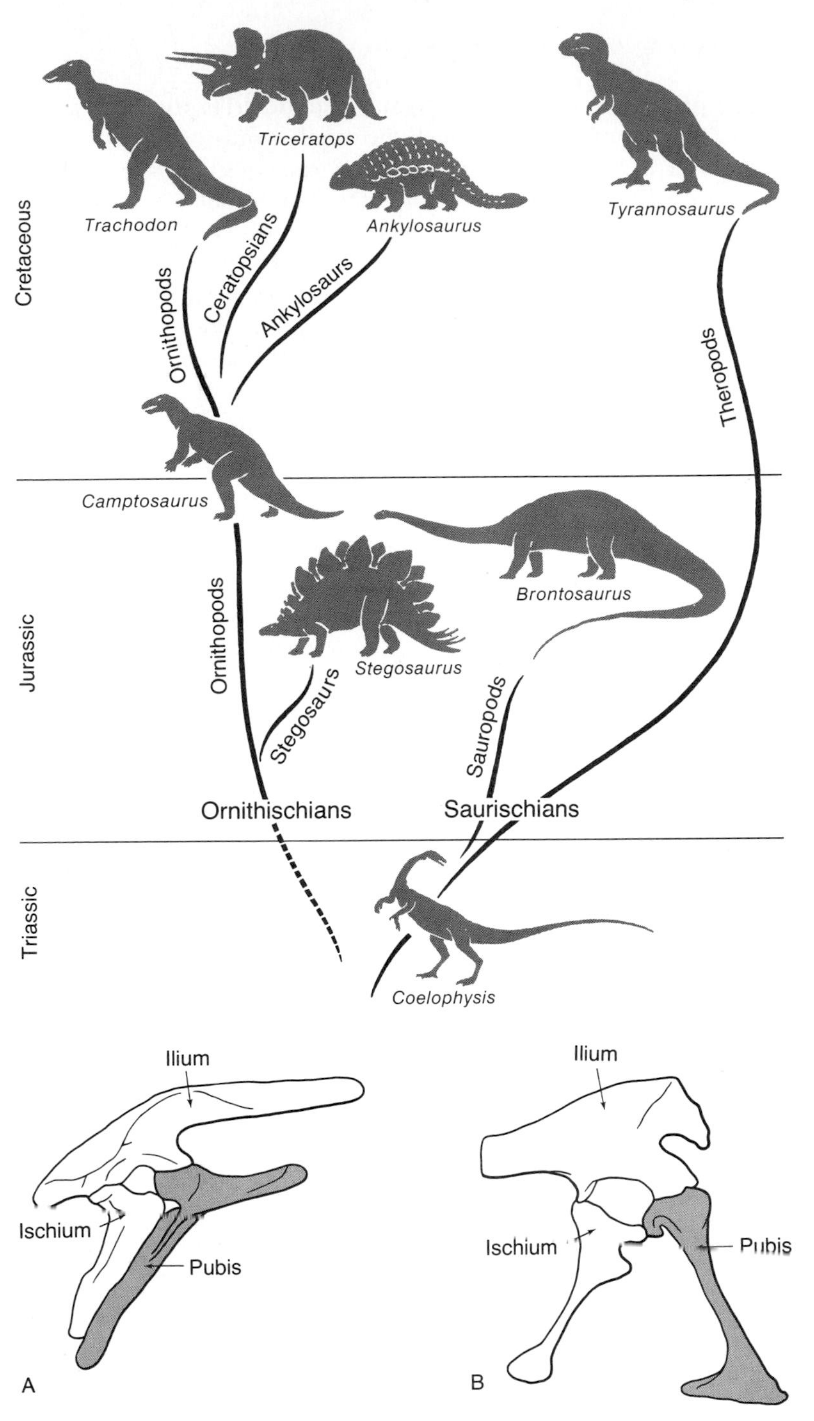

FIGURE 6-7
The two dinosaur orders, which share a common ancestry but are characterized by distinct pelvic structures. A: Ornithischian ("bird-hipped") structure. B: Saurischian ("lizard-hipped") structure. (From Colbert, 1955.)

2. Selection of a group of phenetic characters to describe the OTU's. Between 50 and 100 characters are often used though the number may be larger or smaller. The characters themselves can take several different forms. In some cases, a character may have only two expressions or states, such as "large" or "small," "present" or "absent," and so on. The characters used in Tables 6-1 and 6-2 are all of this type and are what is called ***two-state characters.*** More commonly, characters have several states. They may be counts, as in the number of vertebrae of a skeleton, or measurements, such as length or width. Occasionally, a ***multi-state character*** is such that the various states cannot be ranked (as counts or measurements can). For example, one character might be color, with the several states reflecting the presence or absence of several discrete pigments. The several types of phenetic characters call for somewhat different numerical treatment.

3. Comparison of each OTU with every other OTU. This involves using some measure of phenetic similarity or dissimilarity which makes possible the construction of a *similarity matrix*. Table 6-3 is a similarity matrix for the data already presented in Table 6-1. In this example, we have used the simplest possible measure of similarity applicable to two-state characters: the percentage of characters for which two OTU's coincide, or match. This measure thus varies from 0 (no matches) to 100 (perfect correspondence in all characters). A wide variety of other more complex techniques are used by numerical taxonomists—each being designed for different sorts of data matrices and taxonomic objectives.

4. Determination of groups or clusters of OTU's on the basis of the computed similarities. Again, many methods for clustering are available and the results take a variety of forms. In simple cases, the clustering can be done almost by eye. Table 6-4 shows the same data as Table 6-3 but with the OTU's rearranged to show two natural clusters. OTU's A, B, G, H, and J are more similar to each other than to OTU's C, D, E,

TABLE 6-3
Similarity Matrix for Ten Hypothetical Species (Based on Data from Table 6-1)

	A	B	C	D	E	F	G	H	I	J
A		50	30	40	20	20	60	90	10	70
B			20	30	50	50	70	60	40	80
C				70	50	70	10	40	60	30
D					40	60	20	50	70	50
E						60	60	10	70	30
F							20	30	70	30
G								50	50	50
H									20	80
I										20
J										

TABLE 6-4
Similarity Matrix from Table 6-3 Rearranged to Show Clusters of Species

	A	B	G	H	J	C	D	E	F	I
A		50	60	90	70	30	40	20	20	10
B			70	60	80	20	30	50	50	40
G				50	50	10	20	60	20	50
H					80	40	50	10	30	20
J						30	50	30	30	20
C							70	50	70	60
D								40	60	70
E									60	70
F										70
I										

F, and I. Similarity values are generally high when members of a cluster are compared but low when members of different clusters are compared.

5. Graphic display of the results of clustering. A common method is the construction of a ***phenogram*** or ***dendrogram.*** We saw a phenogram of fusulinid data in Chapter 4 (Figure 4-19). Figure 6-8 shows a phenogram based on the similarity matrix already developed (Tables 6-3 and 6-4). The scale on the left indicates the similarity between OTU's or groups of OTU's that are connected by horizontal lines. The actual value of the similarity between groups of OTU's is usually calculated by some method based on averaging the individual similarity values. Note that the two groups or clusters listed as the third alternative in Table 6-2 (and segregated in Table 6-4) are seen in the phenogram as being separated at the lowest point of the similarity scale. Also note that smaller groups or subclusters are evident in the phenogram (A–H, B–J, and so on).

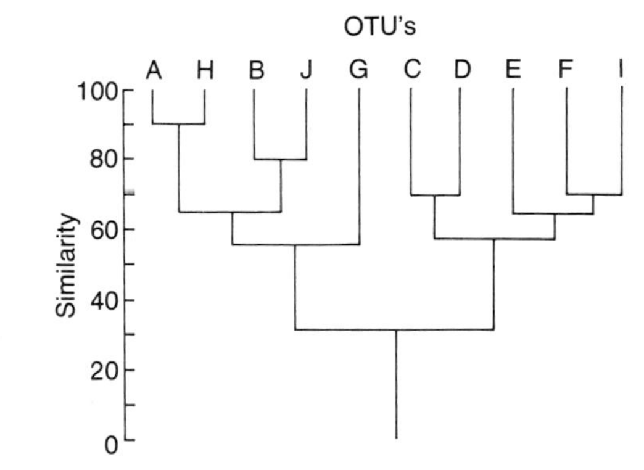

FIGURE 6-8
Dendrogram based on numerical taxonomic data given in Table 6-4.

What relationship, if any, is there between a dendrogram and a phylogenetic tree, or between a phenetic classification and one developed by standard taxonomic means? Many proponents of numerical taxonomy claim that the phenetic classification is as real as a conventional one. They look upon the clusters in a phenogram as denoting higher taxa of rank proportional to their separation on the similarity scale. In Figure 6-8, separations falling between 60 and 70 might denote *genus rank* differences and those less than 60 might be *subfamily rank,* and so on. The boundaries between ranks are arbitrary—just as they are in conventional taxonomy.

An effective way to test numerical taxonomy is to compare the results with conventional taxonomy for groups of organisms that are well known. Where such tests have been run, the results are encouraging for numerical taxonomy. In a wide variety of instances, the two approaches are in substantial agreement. Many conventional taxonomists who have claimed that classification is an art rather than a science have been forced to examine and explore their methodology. Weighting of characters, in particular, is at issue. Most of the techniques of numerical taxonomy provide no weighting of characters, yet the results are often comparable to classifications in which weighting is an important methodological element. This may mean that weighting serves only as a useful time-saving device and is unnecessary where the work is being done by computer. On the other hand, it may be that weighting is biologically important and that numerical taxonomists have been successful in spite of not weighting because the clusters they normally deal with are sufficiently well defined that a weak method still comes fairly close to the truth.

Figure 6-9 shows a dendrogram, which was produced by numerical taxonomic methods, of nearly the same set of brachiopod genera that were classified by conventional means in Figure 6-6. Although there are obvious differences between the two classifications, there is surprisingly good agreement in essential features. For example, the genus *Eodevonaria* in Figure 6-6 is separated from all other genera at the family level; in Figure 6-9, it is distinct from the other genera at nearly the lowest similarity (phenetic) level. The genera *Dyoros* and *Eolissochonetes* are the most closely linked in Figure 6-9 and are members of the same subfamily in Figure 6-6. The same sort of agreement is found when the genera *Chonostrophia, Chonostrophiella,* and *Tulcumbella* are considered.

Cladistic Taxonomy

In the past few years, an approach to classification quite different from that of numerical taxonomy has gained support from many taxonomists. This approach, which we will call ***cladistic taxonomy,*** has been discussed in detail by Hennig (1966) and more briefly by Schaeffer et al. (1972). In part, the approach of cladistic taxonomy resembles that of numerical taxonomy, in that morphologic similarity is stressed while little attention is paid to geologic evidence on the relative times of origin of the

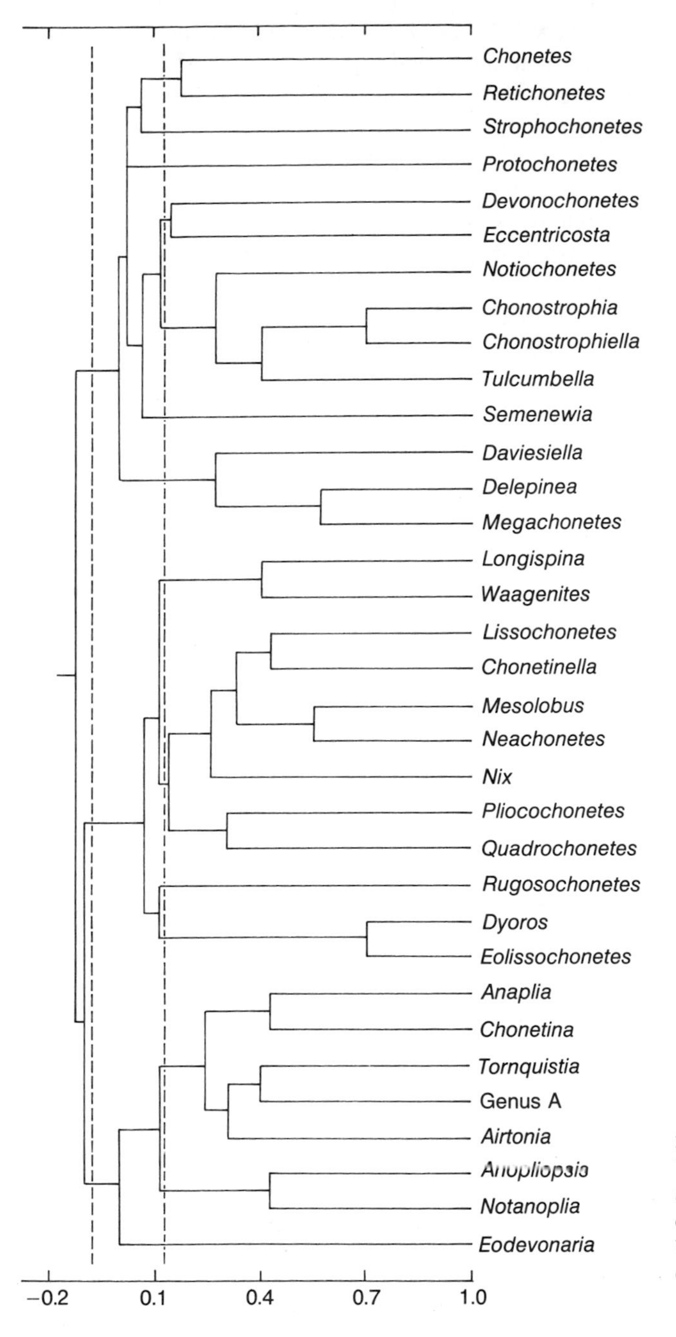

FIGURE 6-9
Dendrogram produced by numerical taxonomic analysis of chonetacean brachiopods (for comparison with the classification developed by nonnumerical methods shown in Figure 6-6). (From Rowell, 1967.)

subtaxa being classified. Here, however, the similarity ends. In cladistic taxonomy the notion of most numerical taxonomists that characters are to be chosen objectively and weighted equally is rejected. Instead, before classification, characters of the subtaxa to be classified are subjectively judged as to degree of primitiveness, or relative time of origin in phylogeny. Various criteria are used. Occurrence of a character within a variety of subgroups is generally regarded as an indication of primitiveness, and characters that arise in the ontogeny of many taxa during similar developmental sequences are also commonly regarded as primitive. Another approach is to arrange character states among subtaxa in a step-like or intergrading series (a ***morphocline***) without regard to geologic age; in this manner an evolutionary sequence can often be established, but it must then be decided which end of the morphocline is primitive. This can sometimes be accomplished by noting that the character state at one end of the morphocline is shared by other subtaxa not belonging to the morphocline. This shared character state is regarded as primitive because it is not likely to have evolved separately in several lineages.

In cladistic taxonomy the clustering of subgroups into higher categories is accomplished by assessing the number of nonprimitive characters shared by subgroups. The larger the number of nonprimitive ("derived") characters found to be shared by two subgroups, the more closely related the subgroups are considered to be. Because they are shared by many subtaxa, primitive characters are not considered to be useful in the clustering of subgroups into higher taxa.

Figure 6-10,A is a ***cladogram,*** a representation of phylogeny. It depicts the relative recentness of divergence of subgroups, based on number of derived characters

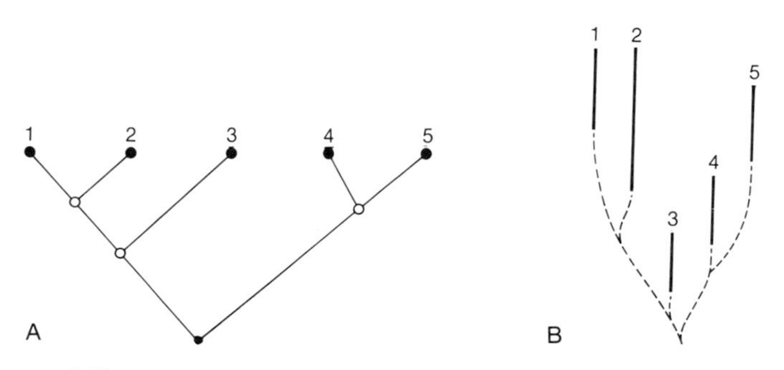

FIGURE 6-10
Graphic representation of cladistic taxonomy. A: A cladogram depicting degrees of similarity among 5 subtaxa (labelled 1 through 5). The vertical scale depicts time in a nonlinear fashion. The order in which pairs of subtaxa are shown to have diverged reflects the number of derived characters that each pair possesses in common. B: A phylogeny inferred from the cladogram. (From Schaeffer et al., 1972.)

possessed in common. Figure 6-10,A indicates that the subtaxa labelled 1 and 2 are very similar to each other and that subtaxa 4 and 5 are equally similar (share with each other the same number of derived characters). On the other hand, subtaxon 3 is more similar to subtaxa 1 and 2 than to subtaxa 4 and 5. A phylogeny inferred from the cladogram would look something like Figure 6-10,B.

Cladistic taxonomy takes its name from the word ***clade,*** which refers to a phylogenetic cluster of taxa: a set of taxa that are more closely related to each other than to other taxa. Phylogeny consists of a hierarchy of clades, just as higher taxa have a hierarchic arrangement, and it is the aim of cladistic taxonomy to produce a classification that reflects the natural hierarchy of clades.

Stratigraphic data on times of appearance of subtaxa are generally ignored in cladistic taxonomy because of a belief that the fossil record is not complete enough to be a reliable guide. After a cladogram has been constructed, however, it can be checked against the fossil record to see if relative positions of subtaxa in the cladogram correspond to the order of their appearance in the record. Nevertheless, orthodox "cladists" attribute any disparity found in such a comparison to the imperfection of the record. This viewpoint is not shared by some paleontologists, who believe that evolution has followed many pathways more than once. Opponents of cladistic taxonomy also believe that the direction of evolution has been reversed commonly enough that many characters appearing to be primitive have evolved secondarily. The future of cladistic taxonomy is uncertain, but this interesting methodology promises to have a profound impact on the philosophy of classification of higher categories.

Supplementary Reading

Hennig, W. (1966) *Phylogenetic Systematics.* Urbana, Ill., University of Illinois Press, 263 p. (A major pioneering contribution to cladistic analysis.)

Heywood, V. H., and McNeil, J., eds. (1964) *Phenetic and Phylogenetic Classification.* London, Systematics Association. Publication 6, 164 p. (A collection of articles on the relative merits of orthodox taxonomy and numerical taxonomy.)

Kaesler, R. L. (1967) Numerical taxonomy in invertebrate paleontology. *In* Teichert, C., and Yochelson, E. L., eds. *Essays in Paleontology and Stratigraphy.* Lawrence, Kans., University Press of Kansas, p. 63–81.

Mayr, E. (1969) *Principles of Systematic Zoology.* New York, McGraw-Hill, 428 p. (A classic, especially with respect to taxonomy at the species level.)

Ross, H. H. (1974) *Biological Systematics.* Reading, Mass., Addison-Wesley Pub. Co., 345 p. (An up-to-date textbook on taxonomic methods and principles.)

Simpson, G. G. (1953) *The Major Features of Evolution.* New York, Columbia University Press, 434 p. (Chapter XI is a discussion of problems of the evolution of higher categories.)

Simpson, G. G. (1961) *Principles of Animal Taxonomy.* New York, Columbia University Press, 247 p.

Sneath, P. H. A., and Sokal, R. R. (1973) *Numerical Taxonomy.* San Francisco, W. H. Freeman and Company, 574 p. (A general textbook on numerical methods in taxonomy.)

CHAPTER 7

Identification of Fossils

To identify a fossil is to assign it to a taxon of some pre-existing classification. Actually, no worker who specializes in a particular taxonomic group is likely to do this without formulating his own concept of the selected taxon, which may differ from the concepts of all other workers. Thus, for the specialist, identification and classification are not clearly separable. In this book it is necessary to put that consideration aside: this chapter describes the approach of the nonspecialist who is willing to accept the classifications of earlier workers.

Clearly, some classifications are more widely accepted than others. Because of the species problem in paleontology, identification of a species by a nonspecialist can seldom be done with certainty. The precision of fossil identification required varies, however, and commonly a species designation is unnecessary. Paleoecologic analysis, given a well-understood stratigraphic framework, seldom requires species identification. It may, for example, be possible to establish a marine origin for sediments simply by noting the presence of fossil remains of Cephalopoda (a class) or Echinodermata (a phylum). Similarly, in stratigraphy or geologic mapping it may be possible to distinguish between nearly identical local rock units of differing age by fossil identification of phylum, class, or order.

Normal Procedures

In establishing a classification, lower taxonomic categories are generally delineated first and then assigned to higher categories. The procedure is usually reversed in identification. By first recognizing a phylum or class we quickly eliminate most of the vast number of recognized species.

Thus, the initial step is phylum identification, and it is important to have a good working knowledge of the taxonomic characters most useful in distinguishing phyla. The necessary information is little enough to be easily grasped. There are approximately twelve important phyla in the fossil record and each is defined by an average of six to ten characters.

Identification of the class of a fossil specimen is considerably more difficult. About thirty-two classes are generally important paleontologically. Few people can correctly identify representatives of all classes at a glance but the distinguishing characteristics of most classes are well enough summarized in recent monographs to permit identification by nonspecialists.

With each successively lower taxonomic category, it is necessary to dig more deeply into the specialized literature. The nonspecialist may therefore find it expedient to send his fossils to a recognized specialist for identification.

We illustrate the general procedures by tracing the steps that might be followed in identifying the fossil shown in Figure 7-1. The specimen was collected from Lower Carboniferous rocks near Moscow, U.S.S.R. It is somewhat broken and therefore photographs of a similar living organism are also shown (Figure 7-2).

The fossil is covered with polygonal, plate-like elements arranged radially. Columns of plates extend from one "pole" of the roughly spherical skeleton to the other. Most plates have one or more prominent knob-like structures called tubercles, which, by analogy to the living relatives, are points of attachment for movable spines (see Figure 7-2). This combination of characters suggests that the specimen belongs in the phylum Echinodermata and within it, in the class Echinoidea. No other classes of this phylum or any other phyla share these characters. We thus eliminate at sight all but one class, thereby reducing the identification problem to a choice among about 8,000 species (assuming that the specimen does not represent a previously undiscovered species). More than 240,000 fossil species and more than a million living species have thus been eliminated from consideration.

The characters used to identify the *class* are not the only ones that could have been used: they are simply the most obvious and best preserved in this specimen.

About seventy-five *families* are generally recognized in the class Echinoidea. Approximately two-thirds show a distinct bilateral symmetry superimposed on the radial symmetry. This has led to a somewhat informal differentiation of echinoids into two groups: the regulars (radial) and the irregulars (bilateral). The fossil illustrated lacks bilateral symmetry so the identification is narrowed to about twenty-five families.

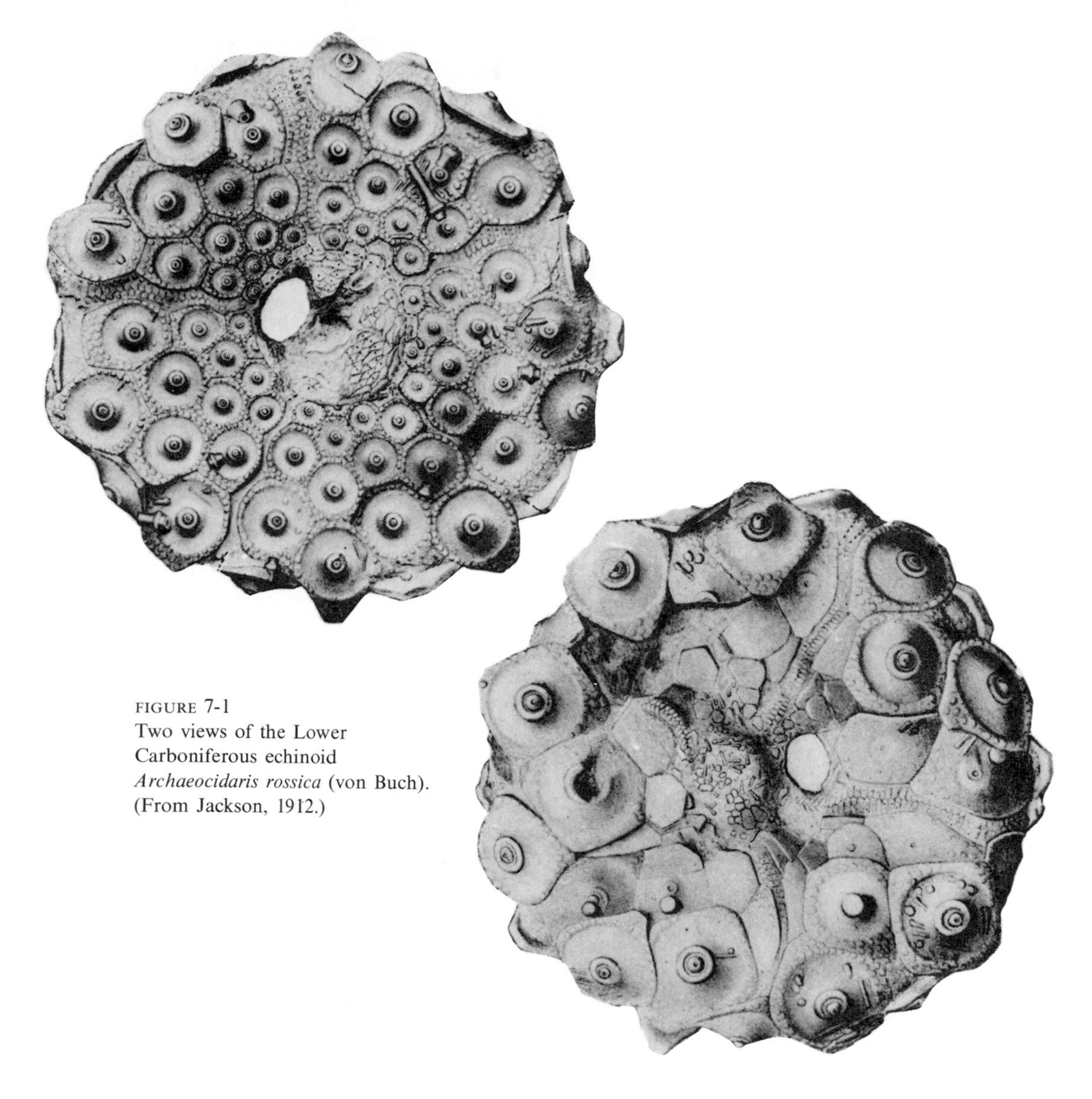

FIGURE 7-1
Two views of the Lower Carboniferous echinoid *Archaeocidaris rossica* (von Buch). (From Jackson, 1912.)

A more detailed look shows that the plates making up the prominent columns are of two distinctive types. One type (ambulacral) has small holes or pores (two per plate), the other (interambulacral) does not. Furthermore, the ambulacral plates are much smaller than the interambulacral plates. (Because of post-mortem sliding of plates, some of the ambulacral columns in the fossil specimen cannot be seen.) In the living species there are five double columns of interambulacral plates. In the fossil specimen, however, there are more than ten columns of interambulacrals. Only six echinoid families contain species with more than ten interambulacral plate columns. All are found in Paleozoic rocks. Of the six, only one (Archaeocidaridae) has the type of tubercles seen on the interambulacral plates in Figure 7-1.

It is important to note that the tuberculation of the living species shown in Figure 7-2 does not differ (at the level of the present discussion) from that of the fossil. But the living species does not belong to the Archaeocidaridae. We eliminated its family (Cidaridae) because of the number of interambulacral plate columns. Tuberculation made it possible to narrow the choice to a single family *only* after all but six families had been eliminated on other grounds.

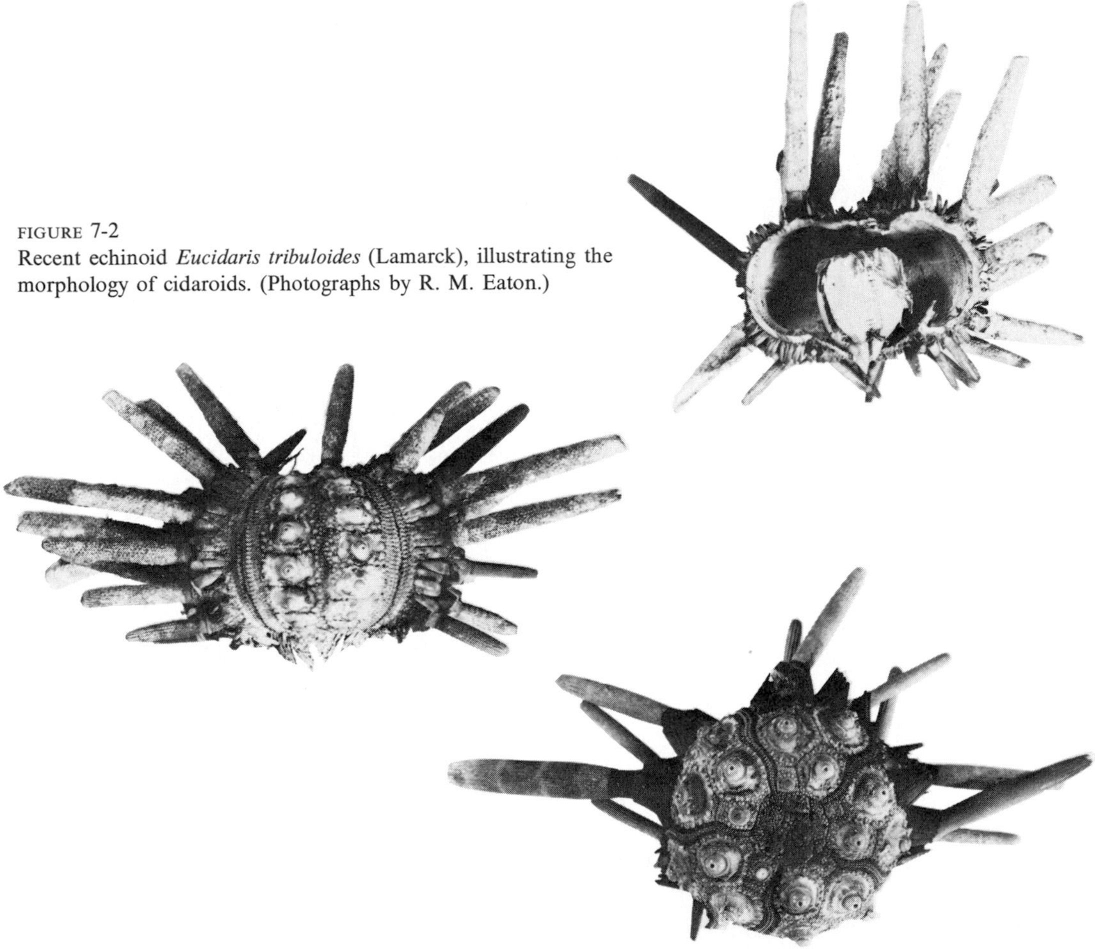

FIGURE 7-2
Recent echinoid *Eucidaris tribuloides* (Lamarck), illustrating the morphology of cidaroids. (Photographs by R. M. Eaton.)

It should be re-emphasized that the characters used here are not the only ones that could have been used. Our analysis has been limited to as few characters as possible and to those most easily diagnosed. It is normally prudent to use more characters so that the results from using one can be checked by the others.

To summarize, identification of the family to which the fossil specimen belongs was accomplished by the following steps: (1) choice of phylum and class, (2) recognition of the specimen as being a regular echinoid, (3) elimination of all but six families of regular echinoids, and (4) elimination of all but one of the six families. Quite different routes could have been followed. We ignored the subclass and order level and used instead the informal distinction between regular and irregular. This distinction is not an official taxonomic one because the two groups are not considered to be real evolutionary groupings. Nevertheless, it is a convenient and practical aid to identification.

We shall not carry the echinoid identification beyond the family level. Identification of genus and species is done by the same procedures although more characters are used. As the species identification is approached, it becomes more critical to be able to compare the specimen being identified with photographs or with actual specimens known to be members of the various species to which the specimen being identified might possibly belong.

Keys

Many attempts have been made to make the process of identification more systematic. Foremost among the tools devised for this purpose is the ***key.*** On page 156 is reproduced a key commonly used in echinoid identification. A key consists of a series of paired statements concerning particular morphologic attributes. From each pair the investigator chooses the one that most closely expresses the morphology of the specimen. His choice then leads him to another pair of statements, and so on. The path leads ultimately to the name of a taxonomic group. This may be a species (as in the key on page 156), or a family, depending upon the taxonomic level of the key. Each key assumes that the investigator starts with the knowledge that his specimen belongs to a certain higher category. The key helps him to identify lower taxa. Thus, each key has an upper and a lower taxonomic limit.

The individual statements in the key may describe single characters or combinations. The characters chosen must be selected with care. Obviously the investigator can be "thrown off the track" or stopped at any point if he finds it impossible to choose between the alternatives. For a key to be workable and broadly applicable, the characters must be ones that are commonly preserved and can be diagnosed unequivocally. The choice depends largely on the material that the key is intended for. Botanists, for example, often establish quite different keys for the same group of plants depending on whether the key is to be used with or without foliage or whether flowers

Key to the Species of the Genus *Histocidaris*

1. Primary spines perfectly smooth, at most with some longitudinal ridges without serrations. Ambital primaries downwards-curved	*H. magnifica.*
Primary spines with more or less fine serrations or thorns; ambital spines not downwards-curved	2.
2. Primary spines uniformly serrate	3.
Primary spines of aboral side with scattered larger thorns, mainly in the basal part	11.
3. Serrations of primary spines very fine, microscopical	4.
Serrations of primary spines coarser, distinctly visible to the naked eye	9.
4. Marginal series of ambulacral tubercles very irregular	*H. cobosi.*
Marginal series of ambulacral tubercles regular or, at most, slightly irregular in larger specimens	5.
5. Interporiferous zone with a well-marked, deeply sunk middle line	*H. sharreri.*
Interporiferous zone without a well-marked, deeply sunk middle line	6.
6. Valves of large tridentate pedicellariae spoon-shaped, the whole inside concave	7.
Valves of large tridentate pedicellariae more or less slender, not spoon-shaped, usually with a small concavity above the apophysis	8.
7. Primary spines cylindrical	*H. variabilis.*
Primary spines fusiform	*H. misakiensis.*
8. Primary spines slender, cylindrical	*H. elegans.*
Primary spines rather thick, fusiform	*H. crassispina.*
9. Ambital primaries upwards-curved, somewhat flattened towards the end, the serrations mainly arranged so as to form a pair of lateral keels	*H. formosa.*
Ambital primaries not upwards-curved, not flattened in the outer part	10.
10. Apical system approximately 50 percent horizontal diameter	*H. acutispina.*
Apical system approximately 36 percent horizontal diameter	*H. denticulata.*
11. Ambital spines somewhat flattened and upwards-curved in the outer part	*H. recurvata.*
Ambital spines not flattened or upwards-curved in the outer part	12.
12. Valves of large tridentate pedicellariae very broad	*H. carinata.*
Valves of large tridentate pedicellariae narrow, slender	13.
13. Primary spines scarcely exceeding horizontal diameter	*H. australiae.*
Primary spines approximately $2\frac{1}{2}$ times horizontal diameter	*H. nuttingi.*

are to be used. Living forest trees can usually be identified by what is known as a twig key, which may be based entirely upon such characters as bark, general form of the tree, and morphology of branches and twigs. Such a key can be applied at any season of the year in the study of deciduous trees. Completely parallel keys have been developed that depend only on leaf form and the characteristics of flowers and fruit. Similarly, the key developed by a biologist may be quite different from one made by a paleontologist. The differences emphasize the fact that taxonomic characters commonly used in species discrimination are chosen from a larger group of characters that could have been used. It is important that the student not assume that the characters commonly used in species discrimination are the only characters distinguishing species. All too often a biologist or paleontologist finds it impossible to distinguish between two species by using a certain key and concludes that there are no differences between the two species although they are actually distinguishable by many characteristics in the formal descriptions of the species or the type specimens.

Figure 7-3 shows the structure of the echinoid key in the form of a dendrogram. Each bifurcation corresponds to a numbered pair of statements. What relation is there between this type of dendrogram and one designed to reflect phylogeny? In other words, what relation is there between identification and phylogeny? It is possible to construct a key based literally on phylogeny, but because the primary purpose of a key is to aid identification, many of the most efficient keys do not follow phylogeny rigorously. For example, as we proceeded in the identification of a specimen earlier in this chapter we eliminated about two-thirds of the echinoid families because the specimens lacked bilateral symmetry. This symmetry in echinoids has evolved independently several times and a phylogentic tree thus does not contain a single bifurcation separating radial and bilateral echinoids. Nevertheless, the distinction is relevant to almost every echinoid species and is extremely useful in a key.

Keys are used much more widely for some taxonomic groups than for others, partly because of tradition but also because of differences in identification problems in different organisms. To construct a workable key, the taxonomy of a group must be reasonably well known. The group must lend itself to key construction; that is, its classification must be based on relatively discrete characters, preferably ones that can be expressed as the presence or absence of a morphologic feature.

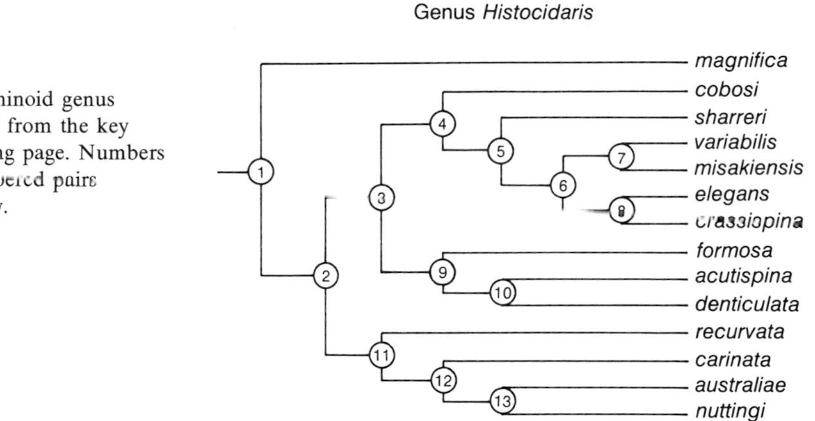

FIGURE 7-3
Dendrogram for the echinoid genus *Histocidaris* constructed from the key reproduced on the facing page. Numbers correspond to the numbered pairs of statements in the key.

Automatic Methods of Identification

We begin the identification procedure with the highest taxa for convenience and practicality. The alternative approach requires an awesome amount of work because it entails systematically comparing the unknown specimen with specimens of each species in turn until a strong similarity is found. If time were not a factor we might choose a quite different system, because the normal procedure is not without pitfalls. Consider, for example, the problems that arise if an error is made at the higher taxonomic levels. If in the echinoid identification we had chosen the incorrect family, it could lead only to the erroneous establishment of a new species.

The computer has freed scientists to deal with things other than procedures originally selected because they made it possible to process data in a minimal amount of time. With a computer, millions of logical decisions can be made in the time required for the human mind to make a single decision. It is technologically possible to design an analysis that will determine the lowest taxon of a totally unclassified specimen—that is, a program to compare the morphologic characteristics of an unknown specimen with the characteristics of all recognized species within a very large taxonomic group.

We saw in Chapter 6 some of the basic methods developed for codifying the morphologic description of a species: a definitive species description may be recorded on one or two punched cards or on a short segment of magnetic tape. Once the distinguishing characteristics of many species are thus recorded it is simple to program a computer to compare those of an unknown specimen with them.

To date, computerized identification of fossils has been infrequently attempted because the technology and methodology needed are very new (little more than a decade old). It is clear however, that these methods will be used more and more. Another important factor delaying computer identification is the difficulty of putting information about morphology into machine-recognizable form. As long as this difficulty exists, the "human element" will remain important in taxonomic identification. A major problem in this context is to select those features most amenable to computerization without jeopardizing the scientific validity of the result.

Presentation of Results

Whether a species identification is for publication or only a museum label, it must be presented in usable form. Certain conventional formats have been developed.

By convention, the generic and specific names are italicized when printed or underlined when written or typed. The generic name is capitalized and the specific name is not. Immediately following the species name, the name of the author of the species is given. These conventions are arbitrary and are used to avoid confusion in communication.

A list of fossils found in a particular Tertiary assemblage follows this paragraph. Notice that the names of some authors of species are enclosed within parentheses, which indicates that the genus has been changed since the species was erected. Notice also that a generic name may be abbreviated to its initial letter in a list of two or more congeneric species. This is acceptable as long as it is unambiguous. In the text of a paleontologic paper generic names are often abbreviated and the names of authors of species deleted.

Typical List of Species Found in a Fossil Assemblage
Rhabdocidaris sp. cf. R. *zitteli* de Loriol
Porocidaris schmidelii (Münster)
Pedinopsis ? *melo,* n. sp.
Porosoma lamberti Checchia-Rispoli
Ambipleurus rotundatus, n. sp.
A. douvillei (Lambert)
Echinolampas fraasi de Loriol
Plesiolampas curriae, n. sp.
P. auraduensis, n. sp.
Conoclypus delanouei de Loriol
Echinocyamus polymorpha (Duncan and Sladen)
Brightonia macfadyeni, n. sp.
Leviechinus gregoryi (Currie)
Pharaonaster sp. cf. *P. ammon* (Desor)
Opissaster farquharsoni Currie
O. auraduensis, n. sp.
O. somaliensis Currie
O. derasmoi Checchia-Rispoli
O. derasmoi var. *angulatus,* n. var.
Hemiaster (*Trachyaster*) sp.
Schizaster africanus de Loriol
S. (*Paraster*) *hunti,* n. sp.
S. (*Paraster*) sp. cf. *S.* (*P.*) *meslei* Peron and Gauthier
S. (*Paraster*) *karkarensis,* n. sp.
S. (*Paraster*) *beloutchistanensis* (D'Archiac)
S. (*Paraster*) *duroensis,* n. sp.
Linthia somaliensis Currie
L. cavernosa de Loriol
Lutetiaster maccagnoi Checchia-Rispoli
Arcaechinus auraduensis, n. sp.
Migliorinia migiurtina Checchia-Rispoli
Eupatagus cairensis de Loriol
E. dainelli (Checchia-Rispoli)
E. fecundus (Checchia-Rispoli)
E. sp. cf. *E. cordiformis* Duncan and Sladen
Brissopsis sp. cf. *B. raulini* Cotteau

In one entry in the list, the subgeneric taxonomic rank is given (*Paraster*). By convention, names of subgenera are enclosed in parentheses and follow the generic name. An "sp.," rather than a species name, following a generic or subgeneric name indicates that the species could not be identified with confidence. In a few cases, "sp." is followed by "cf." and a species name, indicating a questionable or doubtful species identification. When "n. sp." follows a species name, this means that the author of the list is naming the species for the first time.

In dealing with groups of organisms whose species and generic affiliations are well known and for which the classification and nomenclature are relatively stable, the simple combination of genus and species names and species authorship is an unambigous identification. For many organisms, however, identifications like those shown in the list could lead to considerable confusion. To exemplify this, let us consider the echinoid shown in Figure 7-1.

This echinoid has been identified as *Archaeocidaris rossica* (von Buch). Following this paragraph is a list of the sort known as a ***synonymy*** of the species. A synonymy is a brief history of the taxonomic treatment of a species, with bibliographic citations to important works.

Archaeocidaris rossica (Buch)

(?) *Cidaris deucalionis* Eichwald, 1841, p. 88. [Description is unrecognizable so the name cannot hold.]
Cidaris rossicus Buch, 1842, p. 323.
Cidarites rossicus Murchison, Verneuil, and Keyserling, 1845, p. 17, Plate 1, figs. 2a–2e.
Palaeocidaris rossica L. Agassiz and Desor, 1846–'47, p. 367.
Echinocrinus rossica d'Orbigny, 1850, p. 154.
Palaeocidaris (*Echinocrinus*) *rossica* Vogt, 1854, p. 314.
Eocidaris rossica Desor, 1858, p. 156, Plate 21, figs. 3–6.
Echinocrinus deucalionis Eichwald, 1860, p. 652.
Eocidaris rossicus Geinitz, 1866, p. 61.
Archaeocidaris rossicus Trautschold, 1868, Plate 9, figs. 1–10b; 1879, p. 6, Plate 2, figs. 1a–1f, 1h, 1i, 1k, 1l; Quenstedt, 1875, p. 373, Plate 75, fig. 12; Klem, 1904, p. 55.
Archaeocidaris rossica Lovén, 1874, p. 43; Tornquist, 1896, text-fig. p. 27, Plate 4, figs. 1–5, 7, 8.
Archaeocidaris rossica var. *schellwieni* Tornquist, 1897, p. 781, Plate 22, fig. 12.
Cidarotropus rossica Lambert and Thiéry, 1910, p. 125

This particular species was apparently first described by Eichwald in 1841 under the name of *Cidaris deucalionis* but the name is disallowed by the author of the synonymy (Jackson, 1912) because Eichwald's description was too vague. The next entry is to von Buch's description of the species as *Cidaris rossicus.* As the first valid description of the species, the name *rossicus* has priority over all names subsequently applied to the species (although the spelling has been altered to conform grammatically to a change in generic affiliation).

The third entry in the synonymy records the assignment of the species to the genus *Cidarites* (meaning "fossil *Cidaris*") by Murchison, Verneuil, and Keyserling. Several subsequent entries record similar shifts in generic affiliation, most reflecting changes or differences of opinion regarding the taxonomic relationships of the species. One entry in the synonymy stands out from the others: *Echinocrinus deucalionis.* This is credited to Eichwald (1860) who evidently recognized as valid his 1841 publication of the name *Cidaris deucalionis.* The use of the genus name *Echinocrinus* raises another nomenclatural problem. This name was proposed (quite validly) in 1841 by Agassiz. *Archaeocidaris* was proposed independently for the same group of echinoids three years later by McCoy (1844). Technically, the name *Echinocrinus* is the correct name because it was proposed first. A special exception was made in 1955, however, by the International Commission on Zoological Nomenclature partly because *Echinocrinus* had rarely been used by echinoid specialists and partly because it was misleading in being very similar to generic names common in nonechinoid echinoderms (particularly crinoids).

Synonymies often contain names of species that are completely unrelated nomenclaturally to the species in question. This occurs when the synonymy writer feels that two or more species previously considered distinct are actually one. The name proposed first is then used for the species, unless an official exception is made. Because a synonymy is in part an historical record and in part an interpretation of a taxonomic situation, it is common that synonymies written by different specialists for a single species name do not agree.

Bibliographic Sources

Paleontologic information (particularly taxonomic information) has been published in a vast literature extending back well into the eighteenth century. It is published in all major languages and a wide variety of publication media, from regularly scheduled periodicals to occasional monographs of museums, governments, and even private individuals. The paleontologist is thus more dependent on bibliographic aids than is, for example, the nuclear physicist or the electrical engineer.

Identification is greatly aided by definitive monographs on either a specific taxonomic group or fossils found in a particular part of the geologic column. If the monograph has been well prepared, it includes reference to all important literature. The reader need then consult other bibliographic sources only for articles that have been published since the publication of the monograph. In using a given monograph, the reader must understand to what taxonomic categories the writer's definitive summary reaches. The writers of the *Treatise on Invertebrate Paleontology,* for example, attempt to be comprehensive in listing genera but do not try to include all known species.

When no up-to-date summary treatment is available, the paleontologist must turn to published bibliographies, such as the *Zoological Record* published by the Zoological Society of London. Each volume is a reasonably comprehensive survey of the zoological and paleozoological literature published during the preceding year. The *Zoological Record* is divided taxonomically into eighteen sections and within each is a list of titles, authors, and bibliographic references and a comprehensive subject matter index. The index includes a comprehensive list of all taxonomic names used in the papers cited. The *Zoological Record* is a valuable aid in the development of a synonymy because it permits tracing the bibliographic citations to a genus or species year by year.

The *Zoological Record* is not complete, however. No such bibliography could be and still be issued within a reasonable time after the publication of the literature on which it is based. Therefore the *Zoological Record* must usually be supplemented by other bibliographies such as *Biological Abstracts* and *Bibliography of North American Geology.*

PART II

THE USES OF PALEONTOLOGIC DATA

CHAPTER 8

Adaptation and Functional Morphology

We have discussed evolution from several viewpoints and in several contexts. A central idea has been the assumption that organisms evolve by natural selection and thereby adapt. Adaptation makes it possible for organisms to cope with changing environmental conditions, invade new environments, and function more efficiently in a given environment.

As a result of adaptation having taken place in countless independent evolutionary lineages (over hundreds of millions of years), we now have a truly staggering array of forms. In part, this array must reflect diversity in possible ways of life. We cannot argue, however, that there are as many ways of life as species. Many organisms that have evolved independently now live in very similar ways. Some such organisms display ***convergence:*** the independent evolution of similar morphologies that function similarly. Other organisms adapt to the same environment in quite different ways. Thus, the number of morphologic types produced by adaptive evolution is considerably larger than the number of general ways of life.

In this chapter we will take a more detailed look at the results of adaptation and attempt to measure the success of organisms in coping with and exploiting their environments. We will assume that morphology is primarily adaptive; that is, that the observed morphologic features are functional in terms of life activities. It is theoretically possible for morphologic features to evolve which are functionally neutral—features that do not benefit the organism. The present consensus among evolutionary biologists and paleontologists is that it is extremely rare for such "neutral" features to evolve in the form of important skeletal parts—rare if it occurs at all. The argument is a difficult one to make, however, because though it is often easy to prove that a structure has a function, it is virtually impossible to prove the negative proposition, that a structure has no function.

The Nonrandomness of Adaptation

If we were able to describe the spectrum of biologically feasible morphologies and compare it with the forms that have *actually* evolved, we would find that all possible forms have not been used. Those that have been used have evolved in markedly unequal numbers. In a sense, this is analogous to the array of forms *actually* produced by the automotive engineer as compared with the spectrum of forms that *could* be produced. If we could somehow express all possible automobile designs and then plot the frequency of actual types, we would find that the distribution of actual cars (both living and extinct) would be an extremely spotty one and would concentrate on a relatively few designs found to be efficient.

The nonrandomness of adaptation has attracted the attention of many evolutionary biologists. The problem has been stated by Dobzhansky (1951), as follows:

> Every organism may be conceived as possessing a certain combination of organs or traits, and of genes which condition the development of these traits. Different organisms possess some genes in combination with others and some genes which are different. The number of conceivable combinations of genes present in different organisms is, of course, immense. The actually existing combinations amount to only an infinitesimal fraction of the potentially possible, or at least conceivable ones. All these combinations may be thought of as forming a multidimensional space within which every existing or possible organism may be said to have its place.
>
> The existing and the possible combinations may now be graded with respect to their fitness to survive in the environments that exist in the world. Some of the conceivable combinations, indeed a vast majority of them, are discordant and unfit for survival in any environment. Others are suitable for occupation of certain habitats and ecological niches.

Another evolutionary biologist, Sewall Wright, suggested a method of expressing these principles graphically. Figure 8-1 is a hypothetical example based on Wright's method. The vertical and horizontal axes represent genetically controlled characters of an organism so that the area bounded by the axes includes all possible combinations of values of the two characters. The characters might be length and width, number of ribs and spacing of ribs, or any pair of characters or character combinations. The contours in Figure 8-1 indicate varying "fitness" or adaptiveness. In any situation, some morphologic combinations will be better suited to the organism than others. The contoured surface reflecting these differences is called an ***adaptive landscape:*** the topographic highs are ***adaptive peaks*** and the lows are ***adaptive valleys.*** The reasons for differences in fitness may have to do with the mechanics of the functioning of a

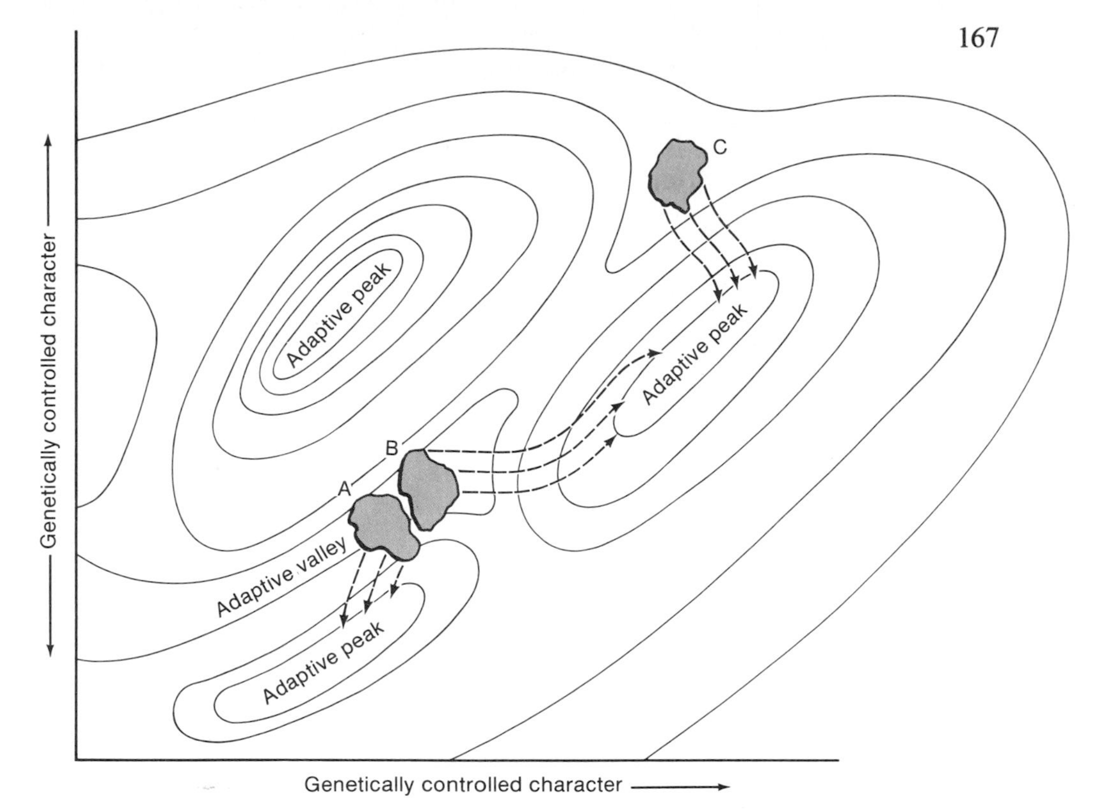

FIGURE 8-1
Hypothetical adaptive landscape showing three populations (A, B, and C) and possible evolutionary paths followed by each (dashed lines). The sizes of the shaded areas indicate the amount of genetic variability in the three original populations. Populations B and C climb the same adaptive peak and thus illustrate adaptive convergence. Populations A and B, on the other hand, by climbing different adaptive peaks, undergo divergence.

structure or may relate to the environment of the organism. Environmental factors may be physical (such as water density, sediment characteristics, and so on) or biological (prey–predator relations, food supply, and so on).

Figure 8-2 shows a paleontologic example which may be viewed from the standpoint of adaptive landscapes. The two axes are attributes of the coiled shells of cephalopods (one attribute, "W," is the square of the ratio of radii used in Figure 2-9, page 40). In Figure 8-2,A, computer-generated drawings show the morphologic effect of varying the two attributes. The curved line passing through the lower-right portion of the diagram has the equation $W=1/D$ and separates shell forms in which successive whorls are not in contact from those where whorls overlap. In Figure 8-2,B, the frequency of actual occurrences of Paleozoic and Mesozoic ammonoid cephalopods is contoured. The greatest concentration is near the point $W = 2$, $D = 0.35$, and

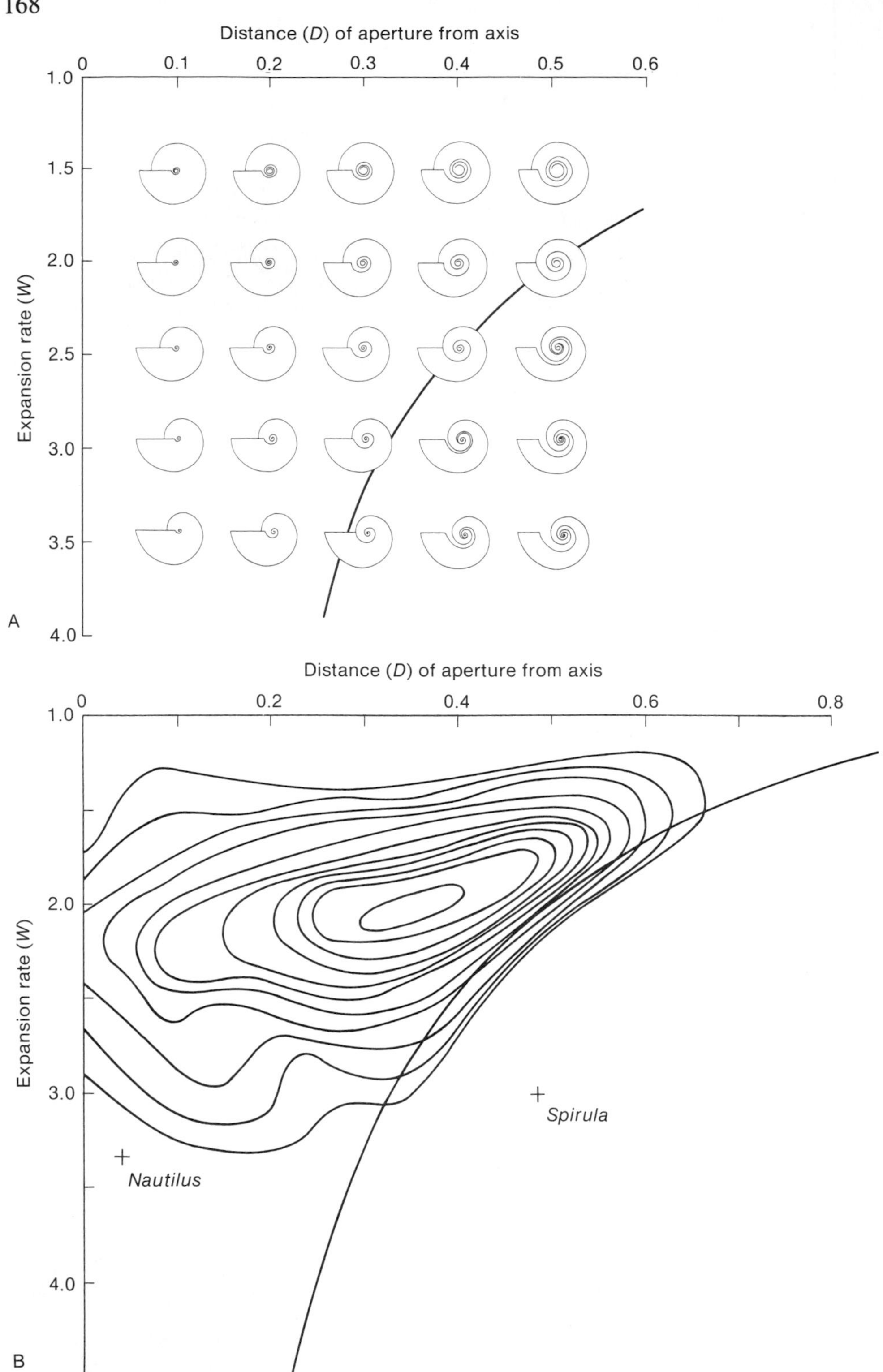
Distance (D) of aperture from axis
0
0.1
0.2
0.3
0.4
0.5
0.6
1.0
1.5
2.0
2.5
3.0
3.5
4.0
Expansion rate (W)
A
Distance (D) of aperture from axis
0
0.2
0.4
0.6
0.8
1.0
2.0
3.0
4.0
Expansion rate (W)
Spirula
Nautilus
B

FIGURE 8-2
Variation in the coiled form of coiled cephalopods. A: Computer simulations of 25 combinations of two important variables that contribute to shape. The curved line (W = 1/D) separates forms with overlapping whorls from those forms having whorls that do not touch. B: Contoured frequency distribution of 405 genera of Paleozoic and Mesozoic ammonoids. Ninety percent of the sample lies within the outermost contour. Crosses indicate the form of two living nonammonoid cephalopods. (Modified from Raup, 1967.)

abundance falls off away from this point. The $W = 2$, $D = 0.35$ shape was, therefore, the most "popular" shape for this evolutionary group. If we assume that evolution was directed toward adaptation, this shape represents an adaptive peak on a rather simple adaptive landscape. The steepness of the contoured surface on the side toward the curved line ($W = 1/D$) suggests that the open spiral found below and to the right of this line was particularly unsuited to ammonoids—though the existence of a few ammonoids in this region indicates that under some circumstances, at least, open spirals were possible. Most ammonoids were swimming organisms and had extremely thin shells. A shell with whorl overlap would be stronger and also hydrodynamically better than the open spiral. This probably explains why most ammonoids stayed to the left of the $W = 1/D$ line.

Figure 8-2,B also shows single points for two living nonammonoid cephalopods: *Nautilus* and *Spirula.* Both fall outside the area of ammonoid concentration. Does this mean that *Nautilus* and *Spirula* are poorly adapted forms, or do they occupy different adaptive peaks on a more complex adaptive landscape? The second alternative is probably the correct one. *Spirula* has an internal shell (that is, it lies within the animal's body rather than around it) which serves only to provide the organism with buoyancy. Strength is not as critical a factor and, inasmuch as it is internal, streamlining is not a consideration. Thus *Spirula* can exist—and perhaps excel—with an open spiral. The explanation for the position of *Nautilus* is not as clear because we do not fully understand the differences in shell function and mode of life between *Nautilus* and the ammonoids. Because nautiloids as a group cluster around the *Nautilus* position in Figure 8-2,B, we can presume that nautiloids have occupied an adaptive peak separate from the ammonoid peak.

Let us return to Figure 8-1 and to a theoretical consideration of the evolution of species across an adaptive landscape. We have seen that each organism occupies a *point* on the adaptive landscape. It follows that a breeding population occupies an *area* on the surface with the size of the area reflecting the genetic variability within the population. Three hypothetical populations are labeled A, B, and C in Figure 8-1. Inevitably some individuals are, by chance, higher on the adaptive landscape than others and this will control the course of natural selection. If natural selection is operating, populations will move upward on the surface because the better-adapted (most fit) variants will be favored. It is important to remember that natural selection cannot cause movement *down* a slope on the adaptive landscape. Even with this constraint, the course of evolution is not necessarily fixed. Chance may play a large role in determining which adaptive peak is climbed if a population lies between two peaks. Populations A and B in Figure 8-1 started at about the same level on the landscape but moved up different adaptive peaks. In a very diagrammatic way, this illustrates ***evolutionary divergence.*** By contrast, populations B and C started at very different morphologies but came together on one of the adaptive peaks—through ***evolutionary convergence.***

The adaptive landscape model has broad implications bearing on a variety of aspects of evolutionary paleontology. For example, what happens when a population (or species) reaches an adaptive peak? Theoretically, evolution should stop because an optimum has been reached. Notice that this does not mean that *the* optimum has been attained: in the hypothetical case illustrated in Figure 8-1, the highest adaptive peak (upper left) was left unoccupied. This can happen in nature because a population on one adaptive peak cannot get to higher peaks without going down into an intervening valley—unless, of course, the population has enough variability to encompass organisms on higher slopes of adjacent peaks. In any event, it is clear from the fossil record that evolution has not stopped! That is, we have not reached the condition where all species are "stuck" on adaptive peaks. Several explanations have been suggested for this. A few of them are:

1. Not enough time has been available for all species to reach adaptive peaks.
2. Nonadaptive evolution (meaning movement downslope) may be more common than generally realized.
3. The adaptive landscape is constantly changing—peaks are replaced by valleys, and so on—so that species must change to keep up with the changes.

In working with the adaptive landscape model, it is important to remember that the two-character case (Figures 8-1 and 8-2) is a gross simplification when compared with most real world situations. In fact, evolving organisms are operating in a multidimensional character space and the adaptive landscape is extremely complex. Even in the apparently simple ammonoid case illustrated in Figure 8-2,B, the adaptive landscape is more complicated than we have indicated. A glimpse of this can be seen in Figure 8-3, where a subsample of the ammonoid data is shown. This is the order Goniatitina, which includes most of the Paleozoic ammonoids. The maximum abundance of goniatites is far to the left of that for the ammonoids as a whole. Evidently, there was something different about shell functions or habitats of goniatites which made them "climb" another adaptive peak within the general range of ammonoids. Thus, the single peak shown in Figure 8-2,B is probably a generalized or average adaptive peak on which lies a more complex terrain.

Theoretical Morphology

The analysis of morphology in terms of an adaptive landscape requires that one be able to define the *spectrum of theoretically possible forms* that can be expressed by an organism or a structure of an organism. It is not enough to know what the species that

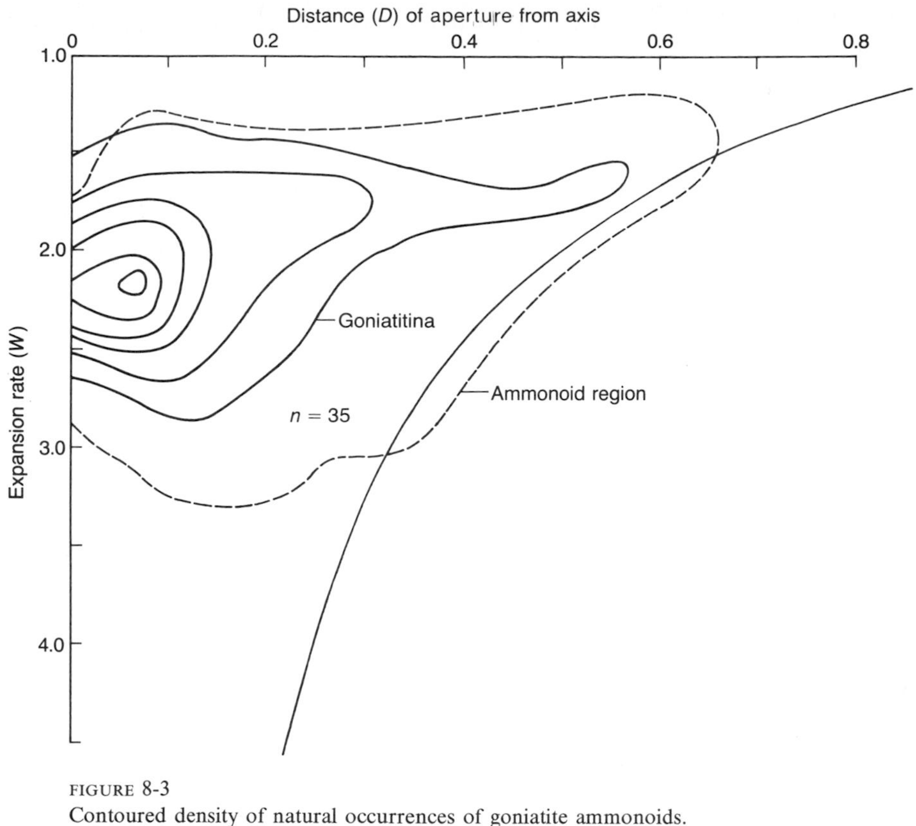

FIGURE 8-3
Contoured density of natural occurrences of goniatite ammonoids. "Ammonoid region" refers to the outermost contour on Figure 8-2,B. (From Raup, 1967.)

have evolved look like; we must also know about the morphologies of those that *have not* evolved and we must know the geometric or topologic interrelations between them. Analysis of this type is referred to as ***theoretical morphology.*** It requires a rather full understanding of the morphology of the biologic group so that variation in form can be expressed by a single, integrated set of characters or attributes.

Among invertebrate animals, those with a coiled shell—exemplified by the ammonoids—are especially well suited to analysis of theoretical morphology. This is because several features of the spiral are remarkably constant during growth and closely approximate a simple equation—that of a logarithmic or equiangular spiral. This consistency is found in many diverse groups, from single-celled foraminiferans to highly complex organisms such as molluscs and brachiopods. Spiral coiling has developed in several distinct evolutionary lines, and its functional significance varies.

The basic geometry of the coiled shell may be illustrated by the common gastropod. The shell is a hollow, tapered cone, or tube, open at its larger end (the aperture). New

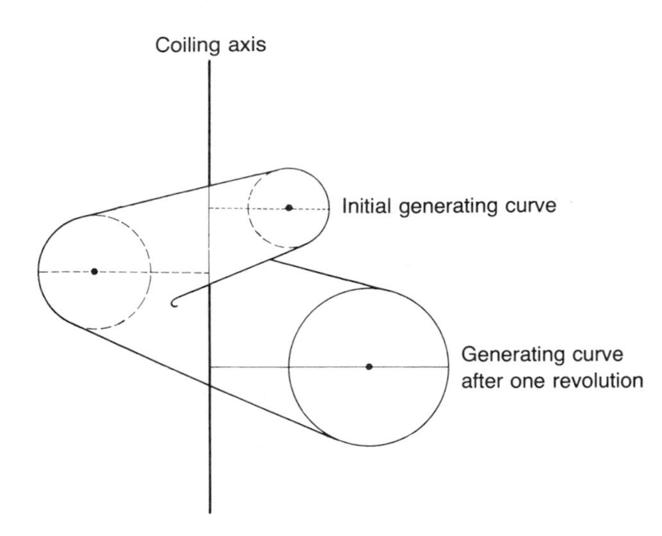

FIGURE 8-4
Schematic diagram of part of a gastropod shell. (From Raup, 1966.)

shell material is added at the aperture so that the cone becomes longer as the animal grows. During growth, more material is added on one side of the aperture than on the other. The effect is to produce a spiral form such that the cone appears to revolve about a fixed axis, which is known as the coiling axis.

Only four parameters are necessary to describe the overall form of most coiled shells. Two of the parameters were used in the ammonoid example (Figures 8-2 and 8-3). The four parameters are as follows (see also the diagram in Figure 8-4):

1. The shape of the cone in cross-section, which is usually referred to as the *shape of the generating curve.* In Figure 8-4 the generating curve is circular. (Technically, the generating curve is defined as the shape of the intersection of the expanding tube with a plane that contains the coiling axis. In most gastropods this is coincident with the shape of the aperture.)
2. The rate of expansion of the generating curve with respect to revolution about the axis. This is often referred to as the *rate of whorl expansion,* and is the ratio between the same linear dimension (such as the diameter) on two generating curves separated by a full revolution. In Figure 8-4 the whorl expansion rate has a value of 2, meaning that any linear dimension of the generating curve is doubled for each revolution about the axis.
3. The *position and orientation of the generating curve* with respect to the axis. In Figure 8-4 the circular generating curve is separated from the axis by a distance equal to half its own diameter. If the generating curve were noncircular, its orientation with respect to the coiling axis would also be critical.
4. The movement of the generating curve along the axis, which is known as *whorl translation.* Translation is most conveniently expressed by the ratio of movement along the axis to movement away from the axis during any interval of revolution

about the axis. The reference point for determining this ratio is the geometric center of the generating curve. In some forms the rate of translation is zero and the tube revolves about the axis in a single plane and produces what is known as a planispiral shell.

The shape of the generating curve, the whorl expansion rate, the position of the generating curve relative to the axis, and the rate of translation are generally (though not always) constant during growth. When different species are compared, however, marked differences become evident. Some have nearly circular generating curves and others have extremely complicated shapes (defying simple mathematical description). The variation in generating-curve shape in gastropods is by no means random, however. If we were to survey a large number of gastropod forms, fossil and living, we would find that certain generating-curve shapes appear over and over again while others are only rarely found and still others seem never to have developed.

Gastropods also show considerable variation in whorl expansion rate: from only a little more than 1.0 (the theoretical minimum) to 4 or 5. Similarly, the generating curve varies from being in contact with the axis to being separated from it by a so-called *umbilicus,* that is, a roughly cone-shaped depression in the base of the shell.

One of the most striking variations is in translation, which may vary from zero (planispiral) to extremely high values (in the high-spired snails). Among living gastropods, there are no truly planispiral forms but among early Paleozoic gastropods, the planispiral form is fairly common.

The four geometric parameters of gastropod form involve much, but not all, of coiled shell morphology. Many parameters could be added to the list. An obvious one, for example, is the direction of coiling. Some gastropods coil in the left-handed fashion (sinistral) and some in the right-handed fashion (dextral). Curiously, the vast majority of gastropods are dextral.

In order to describe gastropod form fully we also require some means of describing departures from the model just presented. An example is illustrated in Figure 8-5 in which the rate of translation decreases during ontogeny so that the resulting shell has a concave lateral profile (instead of the normal straight-sided spire). Such ontogenetic departures from normal geometry are usually consistent within species and are genetically controlled; presumably they are adaptive.

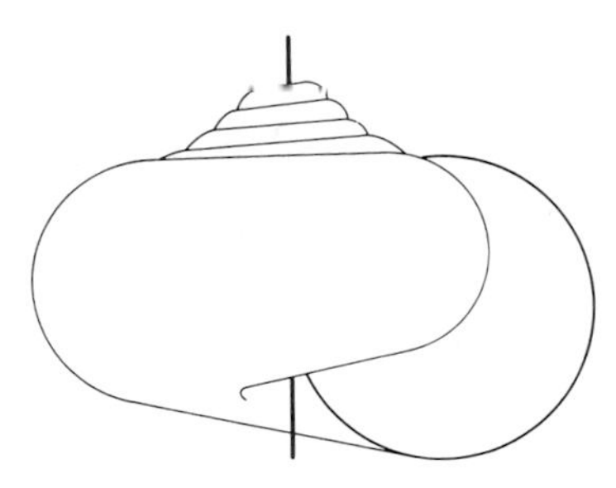

FIGURE 8-5
A gastropod in which the rate of whorl translation decreases during ontogeny. (From Raup, 1966.)

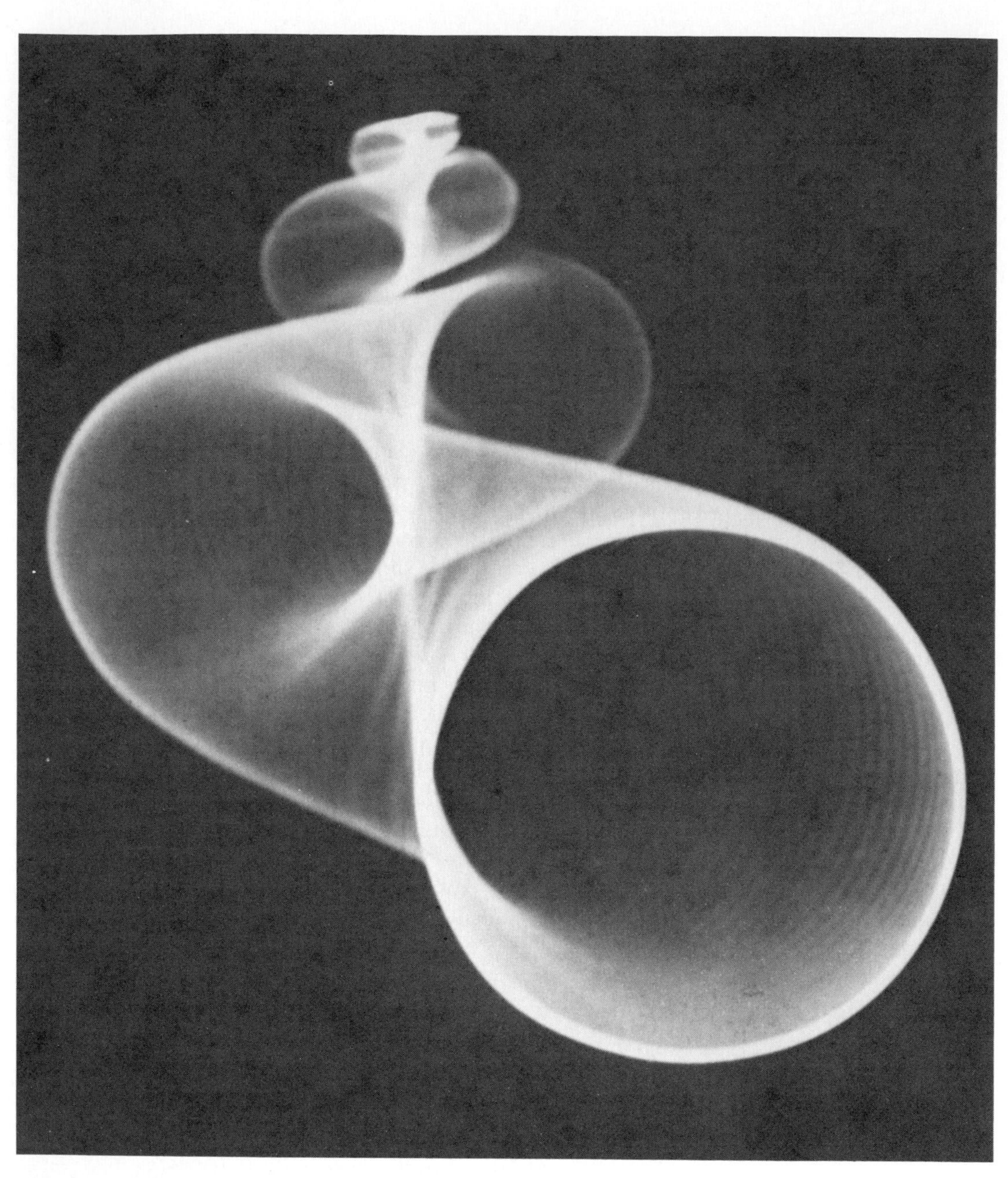

FIGURE 8-6
Analog computer simulation of a common gastropod shell form. (From Raup and Michelson, 1965. Copyright © 1965 by the American Association for the Advancement of Science.)

In addition to the morphologic attributes just mentioned a wide variety of minor features are not included in the general coiling model; for example, impressions of muscles on the interior of the shell, color patterns, operculum shape, the many nonspiral details of ornamentation, and so on.

A useful test of the model is to use it to replicate the form of a shell by computer. If we specify the shape of the generating curve, its position, its rates of expansion and translation, we should be able to calculate the shape of the surface that would be produced by a prescribed amount of growth. A computer simulation of a common gastropod type is shown in Figure 8-6. The generating curve is circular and in contact with the axis. It has an expansion rate of 2.0. The translation rate is fairly high. These characteristics were converted into an electrical circuit for an analog computer. The generating curve was allowed to "grow" on an oscilloscope screen. The simulation in Figure 8-6 is a photograph of the trace of the oscilloscope output and is a generalized replica of the shell surface. As a test of the model it convinces us that the four basic parameters do not overlook or misrepresent important aspects of shell growth. Much more important, the computer simulation provides the opportunity of constructing pictures of gastropod forms that are possible but that have not been produced in nature.

The coiling model can be extended to include coiled forms found in many animal groups. As we have seen, the coiled cephalopods differ little in basic form from many gastropods. Most have planispiral shells (zero translation). The generating curves are elliptical or circular and are symmetrical about the plane of coiling.

Cephalopods and gastropods are, of course, quite different animals and their shells have different functions. The cephalopod shell is partitioned by internal septa into a series of chambers used to hold gas for buoyancy. These chambers do not, however, affect the external form or the overall geometry of the outer shell. It is thus reasonable from a purely descriptive viewpoint to consider cephalopods as variants on the basic model used for gastropod description.

The rate of whorl expansion is generally much higher in bivalves than in gastropods and coiled cephalopods. Whorl expansion rates in gastropods rarely exceed 5, whereas in bivalves they are often as high as one million. The high whorl-expansion rate in bivalves is accompanied by some whorl translation, which is minimal among forms such as the scallops but always exists to some degree and produces an asymmetrical shell form.

The comparison just made between bivalves and other coiled organisms is geometric rather than physiologic. The total skeleton of a bivalve of course comprises two articulated coiled shells (one dextral, one sinistral).

Let us consider yet another coiled group: the brachiopods. The individual brachiopod shell is planispiral. Like that of the bivalve, the total brachiopod skeleton is made up of two articulated, coiled shells. Consistently high whorl-expansion rates mean that it is virtually impossible to confuse the geometry of the brachiopod shell with that of planispiral cephalopods. The principal source of variation among brachiopods is the shape of the generating curve. In the other groups that we have considered, a reasonable simulation of the shell can be constructed by assuming that the leading, or growing, edge of the shell is equivalent to the shape of a cross-section of the expanding shell (the cross-section in a plane that includes the coiling axis). In brachiopods the leading edge of the shell is often nonplanar, and thus is not coincident with the theoretical generating curve. This is most evident in forms whose shells have a strongly developed fold and sulcus on the anterior margin. A nonplanar margin is produced when some parts of the shell are "ahead" of others during growth. This does not mean that the shell lacks a generating curve in a geometric sense but only that the growth at the leading edge departs in time from the simple geometric model. To handle the array of brachiopod forms we must therefore introduce an additional parameter reflecting nonuniform growth rates.

Many foraminiferans have a spiral form that appears to conform to the model established for brachiopods and molluscs. The shell has chambers that have a much greater effect on external morphology than do those of the cephalopods and therefore the geometry does not appear as regular. Where measurements have been made, however, it appears that the basic geometry is the same. A great deal of work remains to be done in this area to confirm and develop a model applicable to foraminiferans.

In Figure 8-7 the general relationships between coiled shells are formalized for forms with a circular generating curve by expressing variation in a three-dimensional block diagram. Translation increases from zero on the right to 4 on the left. Whorl expansion rate increases from 1 (the theoretical minimum) at the top to 1,000,000 at the bottom. The distance between the generating curve and the axis (relative to its own diameter) increases from front to back. We may look upon this block, therefore, as containing *all possible geometric forms within the restrictions of the model.*

Shaded areas in the block (Figure 8-7) indicate regions occupied by most species in four taxonomic groups. The gastropods typically occupy the region of low whorl-expansion rate but show quite a range in translation rate and in distance between the generating curve and the axis. Bivalves generally have low translation and high expansion rate. Brachiopods occupy much the same range in expansion rate but are geometrically separated from the bivalves by their lack of translation. The coiled cephalopods are mostly planispiral, but their whorl expansion rates are consistently lower than those of the brachiopods and thus no overlap between the two regions occurs.

The three-dimensional block in Figure 8-7 indicates that the four evolutionary groups occupy virtually nonoverlapping regions of the geometric space available to them. This is not surprising because the four groups have quite different functional and environmental requirements and inevitably approach different adaptive peaks.

Perhaps the most striking element in Figure 8-7 is the fact that large regions in the block are virtually empty. There are several possible explanations. The empty spaces may represent adaptive valleys. Alternatively, they may represent forms that are biologically impossible. A third possibility is that there has been insufficient time for the evolutionary development of various forms that would exploit the entire geometric range. Of the three alternatives, the first, that empty spaces represent adaptive valleys, seems the most reasonable.

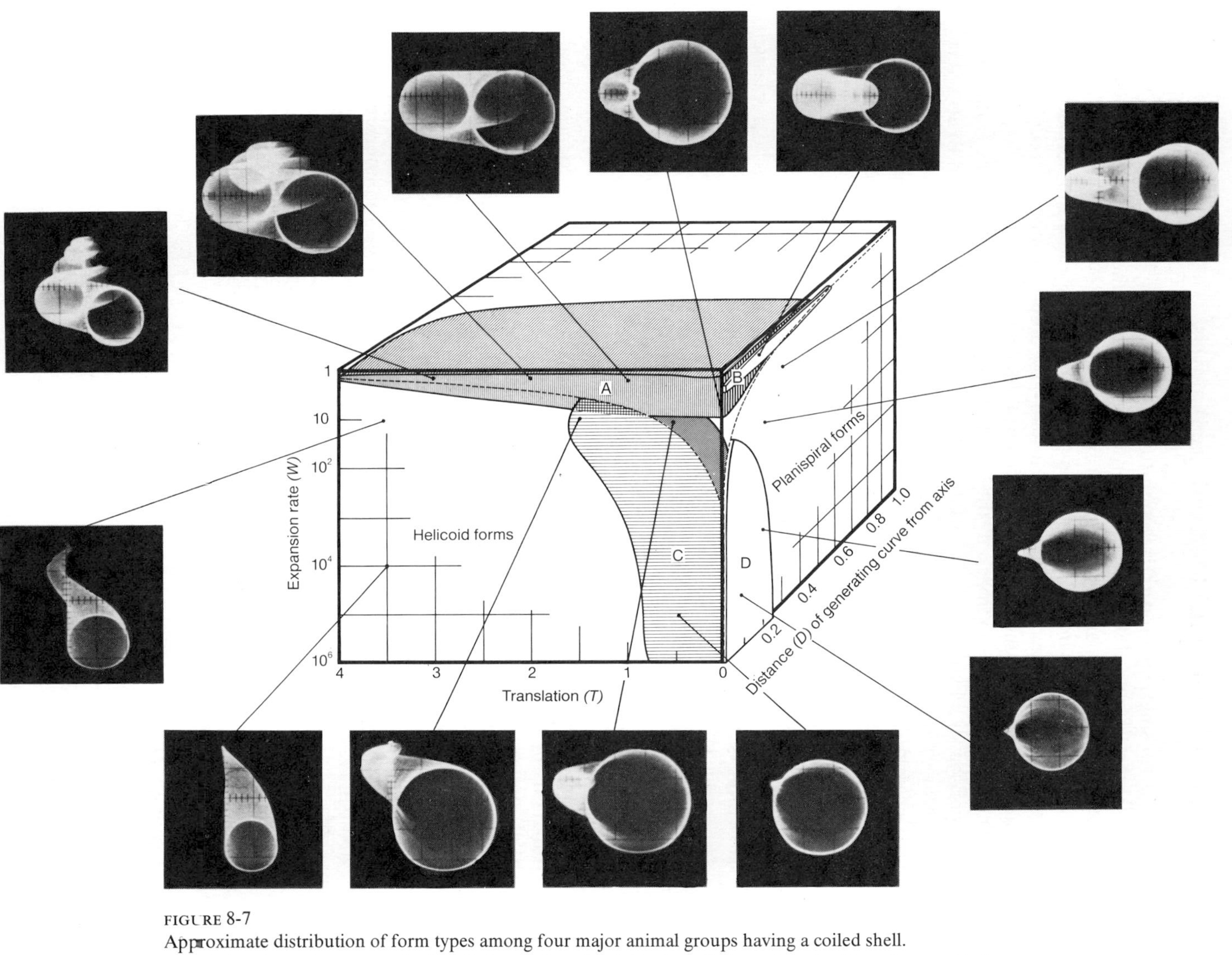

FIGURE 8-7
Approximate distribution of form types among four major animal groups having a coiled shell. Labelled regions: A, gastropods; B, coiled cephalopods; C, bivalves; D, brachiopods. (From Raup, 1966.)

Methods of Functional Morphologic Analysis

There is no single method of inferring function from morphologic features. Many approaches are possible. In part, the approach depends on the information available. For a fossil organism, partial preservation and only spotty knowledge of life habits and habitats impose limits on the method of analysis. For a living organism, soft anatomy, life habits, and habitat preferences can usually be observed more directly and in greater detail, permitting more accurate interpretation.

The most common approach to the interpretation of fossil structures is through comparison with living species. Consider first the case in which fossil and Recent taxa that bear similar structures are closely related and the structures are judged to have had a common origin. The fossil structure may then, by ***homology*** (by virtue of having the same origin), be judged to serve the same function served by the Recent structure. In other instances, the similarity of structures may be more superficial, having arisen independently in separate taxonomic groups. The fossil structure may still be interpreted as having served the same function the Recent structure serves, but by ***homoplasy*** (by its having the same form). Once it is established that homoplastic structures serve the same, or similar, functions, they are known as ***analogous*** structures. Homology and homoplasy represent pathways of comparison, rather than methods of inference of function.

In recent years, it has been recognized that most interpretations of function are ultimately based on mechanical analysis, even when applied to fossil taxa by homology or homoplasy. Because most fossils represent skeletal remains and most skeletons are rigid supportive structures, fossil structures generally served mechanical functions in life. Often we can observe the mechanical function of a structure in Recent taxa and, by homology, apply our observations to closely related fossil taxa. For example, we can observe antler fighting among male members of the deer family in autumn. Inasmuch as antlers are restricted to males, are grown and shed annually, and reach their full development for the autumn mating season, we can conclude that their primary function is to serve as weapons in intraspecies combat for females. (We cannot, however, rule out secondary functions.) By homology we can interpret fossil deer antlers of similar construction as having served the same primary function.

The function of fossil structures is seldom as obvious as that of deer antlers. Even if Recent analogues or homologues of a structure are available it is often necessary to deduce their function or functions before tackling the fossil structure. A common procedure is as follows: first, postulate hypothetical functions for the similar structure in the Recent group. Next, test each hypothetical function in terms of whether the function would be useful to the living organism in light of known life habits and habitats and whether the function is mechanically *feasible* in terms of the morphology, life habits, and habitat preferences. The most reasonable function is chosen on this basis. The conclusion can then be applied to the problematic fossil structure by homology or homoplasy.

Many variations are possible within this basic framework. If the fossil group is closely related to the living group and the structures are considered to be homologous, data from both the living and fossil groups may be used together to test the hypothesis. If the fossil and living structures are not homologous, it may be necessary to justify application of conclusions derived from the Recent group. This can be done, for example, by considering evidence from fossil or lithologic associations, to determine the likelihood that the fossil taxa did, indeed, have life habits and habitat preferences similar to those of the Recent taxa.

When information from living organisms is inadequate or absent, a variant on the procedure described above, called the ***paradigm*** approach, is particularly useful. The use of paradigms was formalized by Rudwick (1964). In Rudwick's scheme, one or more functions for a given structure are postulated, but they are used to define abstract mechanical models called paradigms—one paradigm for each possible function. Each paradigm represents what the stucture *should* look like in order to perform the function best. The result is a set of purely hypothetical structures. (It is sometimes helpful to construct three-dimensional replicas or computer simulations of these hypothetical structures.) The paradigm that most closely fits the actual structure is the one whose associated function is chosen as the most probable for the real structure.

An important requirement that must be observed in using the paradigm approach is that the hypothetical ideal structures must be so formulated as to be consistent with the genetics and physiology of the organism. For example, a cephalopod may be noticed to function like a submarine, but shipbuilding and cephalopod growth are not analogous processes and the differences between them rule out the submarine's being a paradigm. Furthermore, the cephalopod's skeletal composition, basic organ system, and other characters limit—even exclude—the possibility of making a comparison with the submarine.

In any approach to functional morphology, it is important to consider what might be called ***multiple-effect factors,*** of which there are three basic types: (1) a structure may perform more than one function, (2) a structure may be affected by more than one gene, and (3) a gene may affect more than one structure. The third factor is what geneticists call "pleiotropy."

Because of multiple-effect factors, many structures are subjected to selection pressure in more than one direction. The situation is comparable to a mechanical system in which two or more forces pull on an object in different directions. The direction and magnitude of each force may be represented by a vector and the direction and magnitude of the resultant force (which, if strong enough, will move the object) are, in effect, a compromise of the component forces.

Likewise, the direction of evolution of most morphologic features is a compromise. For example, most external mollusc shells serve a protective function in addition to supporting muscular systems. Extremely thick shells would be useful in making molluscs invulnerable to many types of predation. But high mobility—most common among molluscan species having thin, light-weight shells—permits an organism to escape from its predators. Shell thickness in most molluscs thus represents a compromise between opposing selection pressures. We must also bear in mind that a fossil species may represent a stage of evolution in which the final compromise has not yet been attained.

Examples of Functional Analysis

The examples of functional analysis that follow illustrate the variety of existing methods and approaches. Some rely heavily on neontologic information; others are relatively independent of living organisms.

VISION IN TRILOBITES

The eyes of trilobites are similar in many regards to those of living arthropods. Therefore, much can be learned about the functional morphology of the trilobite eye by analogy with the Recent. But there are important structural differences that suggest that the optical systems used by trilobites were significantly different. This is not surprising in view of the fact that trilobites represent many millions of years of evolution independent from the lineages that led to the forms taken by present-day arthropods. Trilobites evidently climbed different adaptive peaks.

We have seen that the evolutionary process of adaptation is basically a process of optimization. This is nowhere better illustrated than by an analysis of the trilobite lens system. Knowledge in this field derives from an unusual collaboration between a paleontologist (Euan N. K. Clarkson) at the University of Edinburgh and a physicist (Riccardo Levi-Setti) at the University of Chicago. When the two met at a conference in Oslo in 1973, both had for several years been active students of trilobite morphology; Clarkson had done considerable work on trilobite vision and Levi-Setti had a physicist's knowledge and understanding of optical systems.

Trilobites possessed a compound eye, consisting of numerous lenses arranged in rows. The lenses were usually deployed in a geometric configuration that provided the closest packing possible. The lenses themselves were made of calcium carbonate in the form of the mineral calcite and are sometimes preserved. In the process of vision, light was transmitted through the calcite lenses to photoreceptive cells within the eyes. It is not known whether the animals could perceive a clear image; certainly they could recognize movements of an object and estimate its size.

Lens morphology varies considerably from one group of trilobites to another. A particularly interesting lens-shape is illustrated in Figure 8-8. It is a doublet consisting of an upper unit which is convex on its upper surface but has a more complex shape on its lower surface. Two variants of the shape of the lower surface are shown at the center in the illustration; in the lens at center-right, a central downward-projecting knob is surrounded by a circular depression. In both lenses the lower part of the doublet has an upper surface which fits the shape of the upper lens and a lower surface which is simply convex. The two lenses together thus make a biconvex compound lens.

Upon examination of Clarkson's reconstruction of trilobite eyes, Levi-Setti noticed that the upper lenses just described are very close approximations of lens designs published by Descartes and Huygens in the seventeenth century. The Descartes and

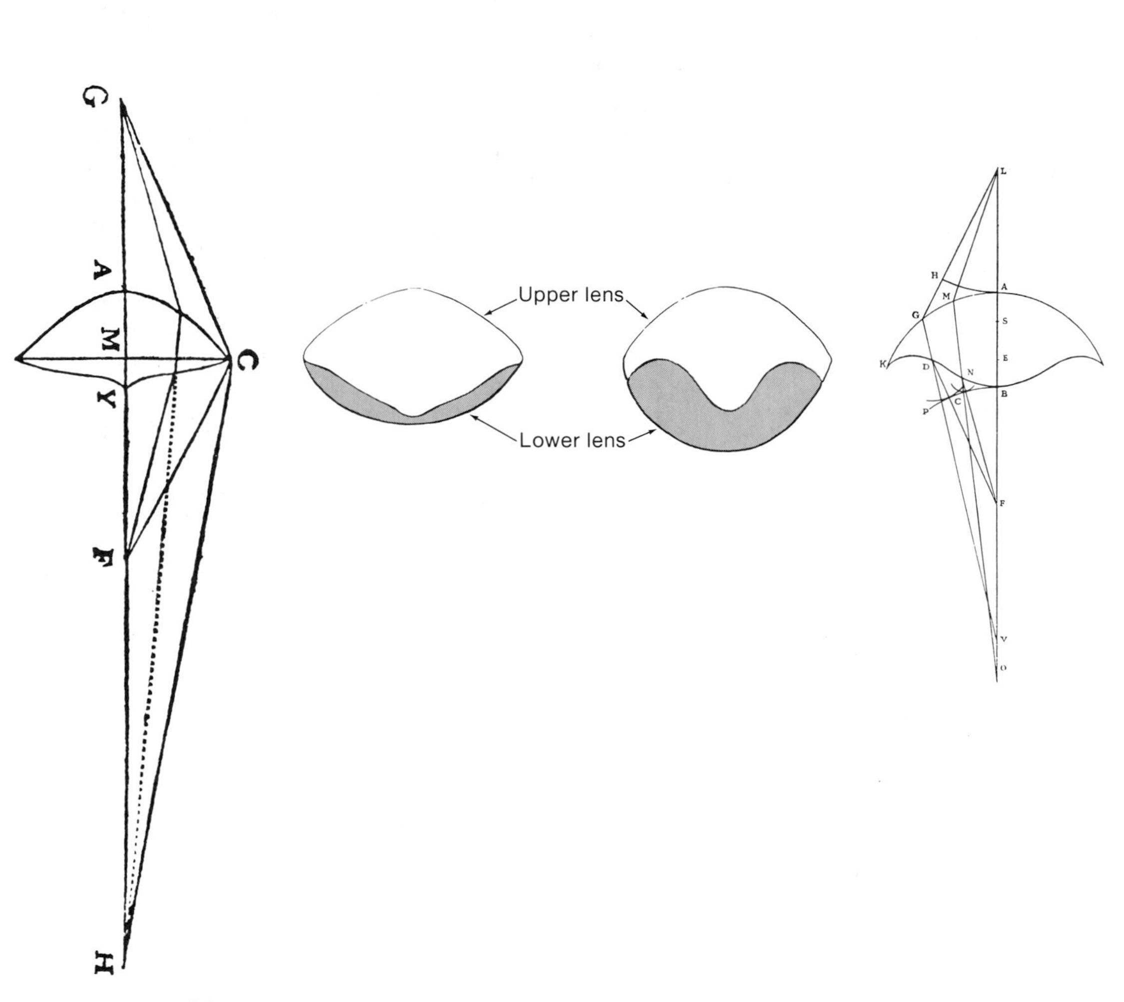

FIGURE 8-8
Lens morphology of two trilobites (center) compared with the original drawings for aplanatic lenses published by Descartes (left) and Huygens (right). (Based on Clarkson and Levi-Setti, 1975.)

Huygens drawings are reproduced for comparison in Figure 8-8, left and right. The purpose of both designs was to produce what is known as an "aplanatic" lens—one that avoids spherical aberrations. The similarity between the shapes of the upper trilobite lens (see Figure 8-9) and the lenses designed by Descartes and Huygens is remarkable; indeed, the lenses differ little, other than in the presence of the lower lens in the trilobite, an element that does not appear in the designs of either Descartes or Huygens. But this is understandable when it is noted that the aplanatic lens was designed to operate in air. Calculations have shown that in the trilobite's aqueous environment the lower lens would be necessary to compensate for the relatively high refractive index of seawater. Thus, the trilobite lens doublet appears to be an optimal modification of basic designs that became a part of human technology only as recently as the time of Descartes and Huygens.

The trilobite lens is optimal in yet another way. The material of the upper trilobite lenses was calcite. This is a good material for the lens of an organism in an aqueous environment because it has a relatively high index of refraction and thus maximizes an organism's light-gathering ability. But calcite has the disadvantage of being highly birefringent: that is, the refractive index of calcite varies greatly depending on its orientation relative to incident light rays. The birefringence can be avoided, however, if the crystal is oriented so that the light is moving parallel to its principal optic axis. And this is precisely the orientation observed in trilobite lenses.

To summarize, it appears from the work of Clarkson and Levi-Setti (1975) that—through natural selection operating on chance variations—trilobites evolved a remarkably sophisticated optical system. For an optical engineer to develop such a system would require considerable knowledge of such things as Fermat's principle, Abbe's sine law, Snell's laws of refraction, the optics of birefringent crystals, and quite a bit of ingenuity. As an application of Rudwick's paradigm to problems of functional morphology, the example provided by the trilobite lens is unsurpassed!

In studies carried out independently by Clarkson (1966, 1973a, 1973b), orientations of the many lenses of a single eye were determined for several well-preserved trilobite species. He plotted stereographic projections of the axial bearings of individual lenses relative to the animal's plane of bilateral symmetry. Together, all the lens bearings for a single eye represent its *visual field* (Figure 8-10,A, B). Knowing the visual fields available to several species has made possible a variety of interpretations of life orientation, mode of life, and other aspects of functional morphology in these trilobites.

BIOMECHANICS OF PTEROSAURS

Pterosaurs are extinct flying reptiles found in rocks of Jurassic and Cretaceous age. Some species, such as the sparrow-size *Pterodactylus* of the Solnhofen Limestone, were quite small, but the order also includes the largest flying organisms known from any age: the genus *Pteranodon* of the Cretaceous (Figure 8-11) attained a wingspan of

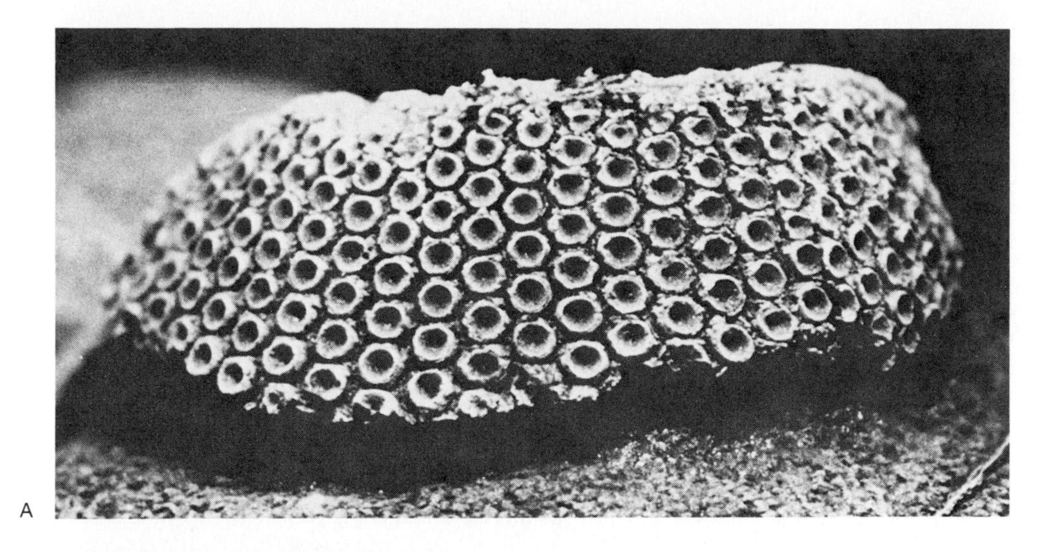

A

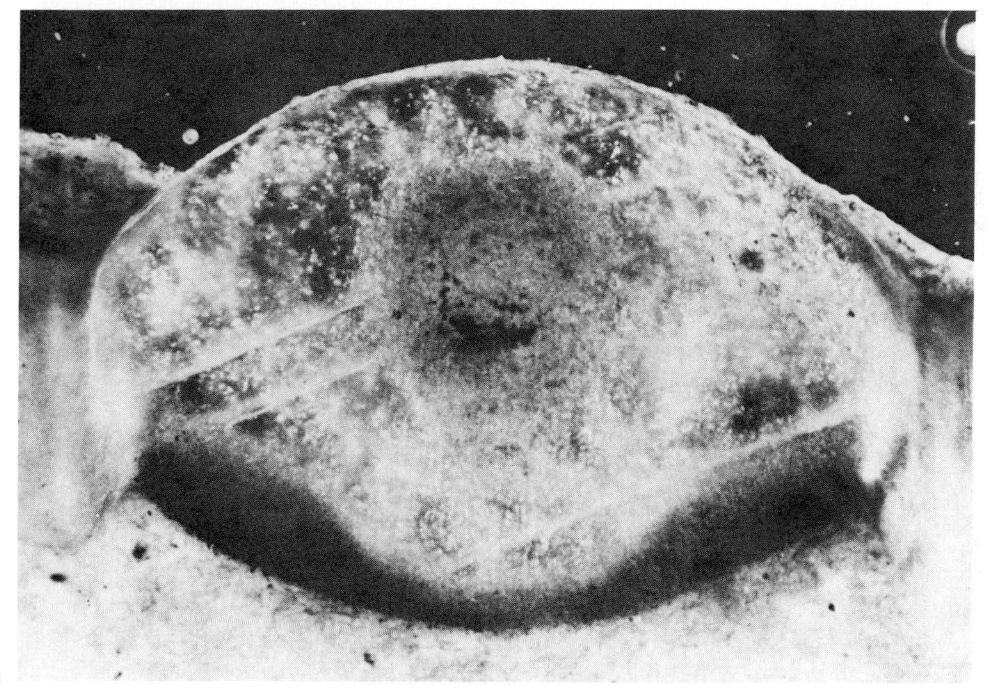

B

FIGURE 8-9
Eye of the trilobite *Dalmanites.* A: Internal mold showing depressions in the center of each lower lens unit. B: Photomicrograph of a cross-section of one lens doublet. Upper lens is calcite showing cleavage planes in the expected orientation; lower lens is calcite mixed with organic material. The geometry of the doublet is intermediate between the constructions of Descartes and Huygens (see Figure 8-8). (From Clarkson and Levi-Setti, 1975.)

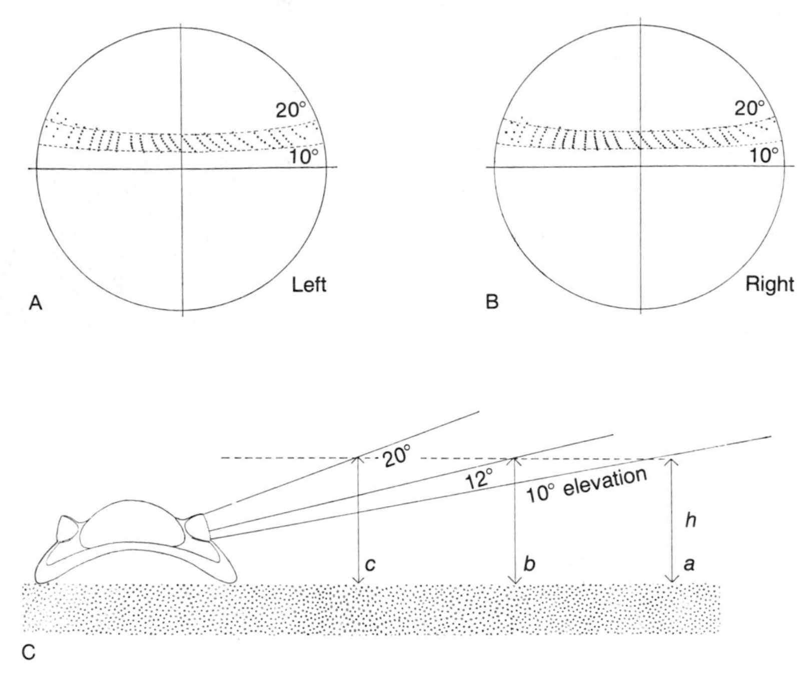

FIGURE 8-10
Vision of *Acaste downingiae.* A and B: Visual fields of left and right eyes shown by stereographic projection. C: Front view of head in the inferred life position, showing "latitudinal" limits of vision as an object of height h approaches from a to c. (From Clarkson, 1966. Copyright © 1966 by the Palaeontological Association.)

at least seven meters, and a recent pterosaur find in Texas (Lawson, 1975) suggests a pterosaur wingspan of as much as fifteen meters. Even the seven-meter wingspan is considerably larger than the largest present-day bird and exceeds the wingspan of some small airplanes. Because of the large size, the mechanics of flight of pterosaurs has attracted much attention from functional morphologists as well as the interest of some airplane designers.

The large size of *Pteranodon* presents several aerodynamic problems. By the principle of similitude (discussed in Chapter 3), a flying organism which increases its size (ontogenetically or phylogenetically) without changing proportions has certain scaling problems. Wing area does not increase as rapidly as the mass to be borne, wing strength lessens, relative to mass, and musculature weakens, relatively. Birds compensate for these factors in a variety of ways: bones evolve that are light in weight because they are essentially hollow structures with thin walls; strength increases through the evolution of optimally placed internal struts; the largest birds often have internal air sacs that displace weight. Moreover, the largest birds tend to be gliding or soaring creatures, making use of a flight mode wherein added weight can actually be

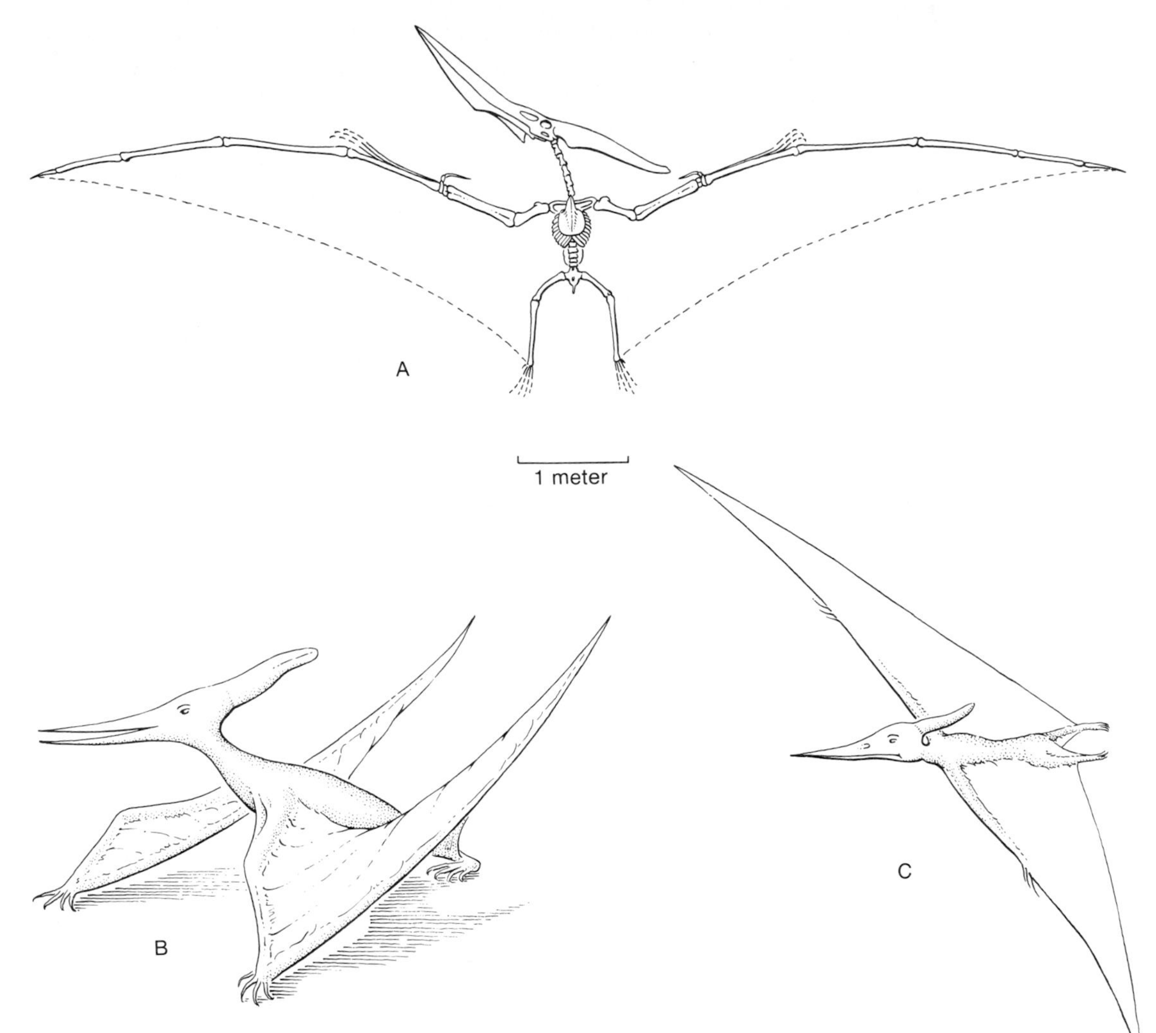

FIGURE 8-11
Reconstructions of the large flying reptile *Pteranodon.* A: Skeleton of *Pteranodon* with dotted line indicating probable placement and extent of the large membranes that formed wings. B and C: Possible appearance of *Pteranodon* while standing and soaring. (A based on Stein, 1975; B, C based on Bramwell and Whitfield, 1974.)

an advantage (up to a point) because it increases flight speed. The largest living bird has a wingspan of about three meters and weighs about ten kilograms. It is doubtful whether the avian anatomy could be modified enough to attain *Pteranodon* size and still be able to fly. It has been estimated that a bird having a wingspan of seven meters would weigh 100 kilograms, whereas the pteranodons of this size probably weighed only about 15 kg.

There has, in fact, been considerable controversy over whether *Pteranodon* truly *did* fly and in particular whether it was capable of self-powered takeoff and flight. The contention that pterosaurs did fly is encouraged by the obvious similarity between their form and that of many living organisms that are capable of flight. Living free-tailed bats have nearly identical wing shapes (Figure 8-12), though the overall size of these bats is but a small fraction of that of *Pteranodon.* It is assumed that such close evolutionary convergence must have resulted from a convergence of function. The main problem therefore is *how* did the pterosaurs fly?

The *Pteranodon* morphology has been subjected to exhaustive theoretical and experimental analysis—including wind tunnel experiments with scale models. As a result, sophisticated estimates of such characteristics as lift and drag have been made. It is clear that these flying reptiles combatted the problem of weight by elegant optimization of the structure and arrangement of skeletal parts: the walls of the bones were extremely thin yet were designed for great strength; the arrangement of bones and their muscle attachment surfaces served to maximize the ratio between performance and weight.

Figure 8-13 shows some of the results of the analyses just described. Here, estimated flying speed is plotted against sinking rate for a seven-meter *Pteranodon,* a living bird,

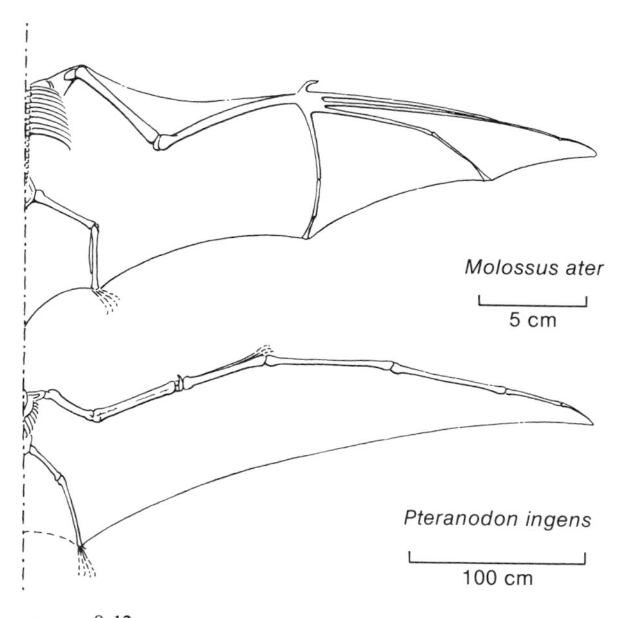

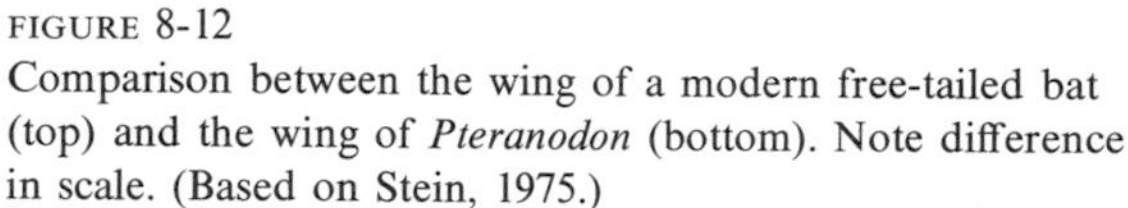
FIGURE 8-12
Comparison between the wing of a modern free-tailed bat (top) and the wing of *Pteranodon* (bottom). Note difference in scale. (Based on Stein, 1975.)

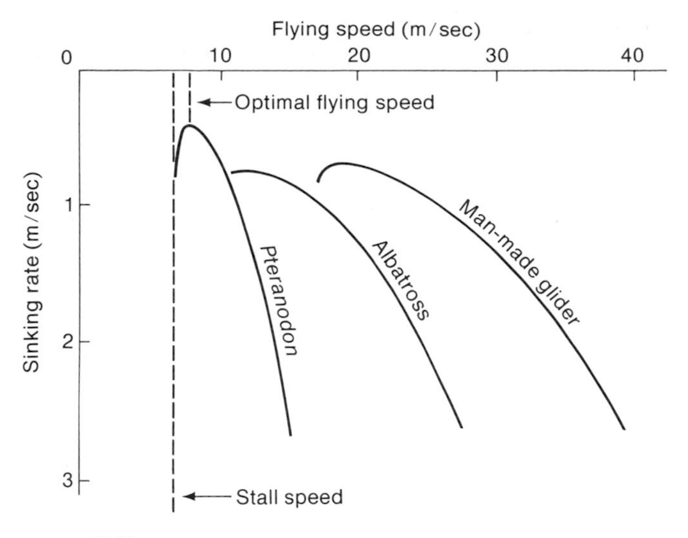

FIGURE 8-13
Relationship between flying speed and sinking rate in *Pteranodon,* a modern albatross, and a man-made glider. *Pteranodon* excels in having a lower stall speed and a lower optimal flying speed. (Modified from Bramwell and Whitfield, 1974.)

and a man-made glider. *Pteranodon* comes off well in the comparison. In any flying structure, there is an optimal relationship between flying speed and sinking rate. For *Pteranodon,* the lowest sinking rate is at a flying speed of about eight meters per second. This flying speed is thus the optimal one. At speeds higher than the optimum, flight would be possible but less efficient. At much lower flying speeds, sinking rate becomes too great and stalling occurs. Bramwell and Whitfield (1974), who did this analysis, estimate that the stall speed for *Pteranodon* would have been about 6.7 m/s. The other curves in Figure 8-13 indicate less favorable relationships. Both show greater sinking rates at optimal flying speeds and the man-made glider has a substantially higher stall speed. Among other things, this means that the glider has to achieve a higher flying speed before gliding and soaring are possible.

A somewhat more elaborate analysis of the same problem by Stein (1975) has indicated that the stall speed of the seven meter *Pteranodon* may have been as low as 4.9 m/s. This would not only make gliding and soaring easier but, more important, it would make self-powered takeoff and actual flight more credible. The matter of powered flight remains something of a problem. Bramwell and Whitfield concluded that *Pteranodon* was barely capable of powered flight. They postulated that it was primarily a glider and depended for takeoff on favorable winds and the ability to climb to and jump from high places, such as coastal cliffs. Stein's analysis, on the other hand, using the lower estimate of stall speed, suggests that *Pteranodon* may have had some power to spare. He calculated that a large individual flying near its stall speed would have been able to generate about 0.10 horsepower but that only about 0.07 hp was needed.

Some attempts have been made to relate knowledge of the aerodynamics of *Pteranodon* to the question of the extinction of the pterosaurs near the end of the Cretaceous. For example, Bramwell and Whitfield concluded that the large forms depended on rather gentle winds for successful gliding and soaring. A cooling trend affecting world climates at the close of the Cretaceous might have meant increased wind velocities and these would have worked to the detriment of the large flying reptiles. This idea is far from proven, however, and, as in so many instances, the rather sudden extinction of an apparently successful group is not easily explained.

The foregoing discussion merely touches the surface of an important and fascinating problem in functional morphology. References to the major recent works (Stein, 1975; Bramwell and Whitfield, 1974) are given at the end of this chapter.

EVOLUTION OF JAW MECHANICS

Many functional morphologic studies of the vertebrate skeleton focus upon the mechanics of jaw movement. The basic jaw mechanism of mammals has proved far more versatile in evolution than the simpler mechanism of reptiles. Crompton (1963) provides a detailed analysis of mechanical changes in the evolutionary transition from the jaw of primitive reptiles to the jaw of mammal-like reptiles, and Alexander (1968) has provided a useful summary of this analysis (Figure 8-14), though to understand it one must grasp a few basic mechanical principles.

For example, for simplicity we can view the muscles and jaw articulation on one side of an animal's skull as lying in a single plane; when an animal bites down on a piece of

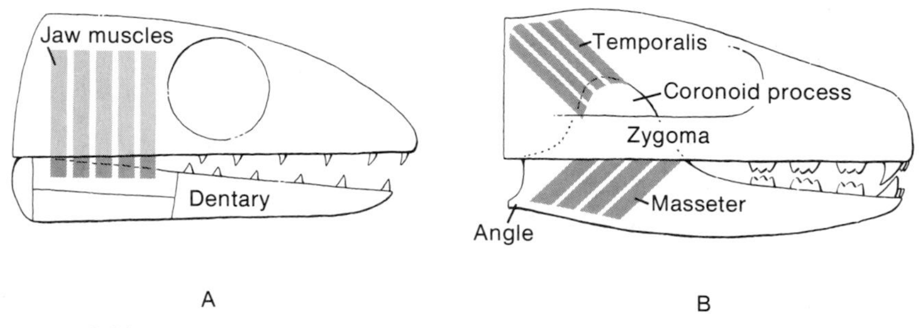

FIGURE 8-14
Diagrammatic views of the arrangement of the major jaw muscles of a reptile (A) and a mammal (B). (After Alexander, 1968.)

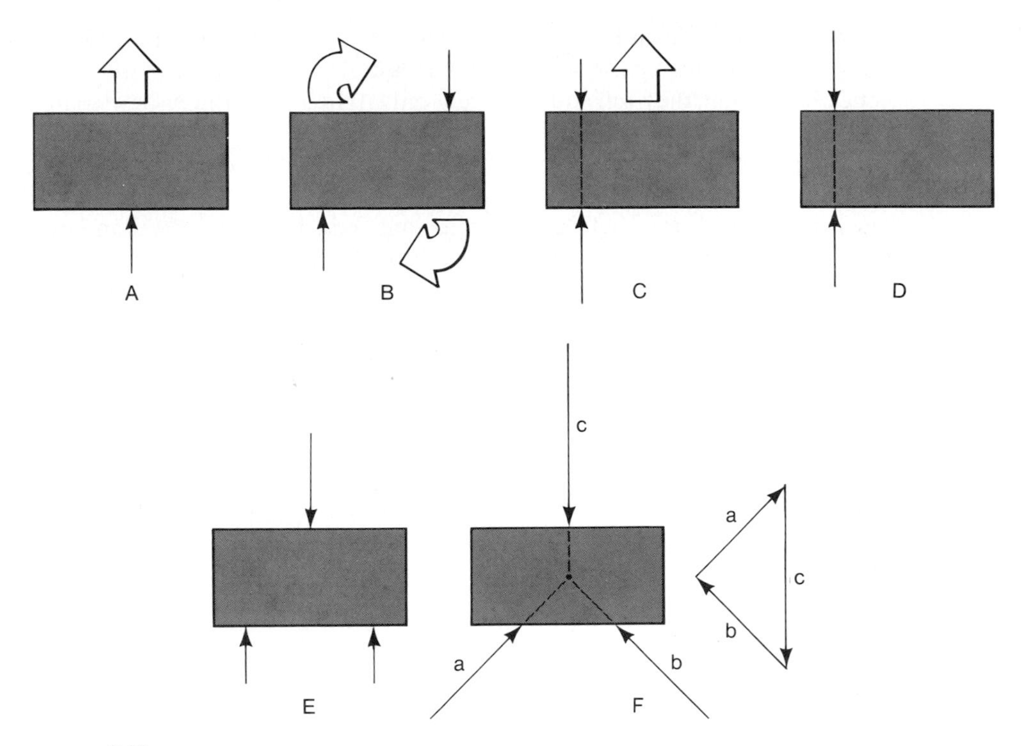

FIGURE 8-15
Diagrams showing different ways in which forces can affect an object. The rectangles represent the object. Forces are shown as vectors (thin arrows), whose magnitude is represented by their length. Thick arrows show direction of movement of the object in A–C. The object remains unmoved in D–F. In F, lines of action of the forces pass through a point so that they can be added together. When added, they form a closed triangle, which means that any two cancel the third, and the resultant force is zero.

food, the animal's jaw remains motionless until the food is crushed or torn. There are only a few sets of conditions under which a structure like a jaw can be motionless when acted upon by forces. Forces are, of course, vectors having both magnitude and direction. If only one force acts upon an object, the object cannot remain motionless. It will move in the direction in which the force acts (Figure 8-15,A). If two forces act in opposite directions, they will cause the object to rotate (Figure 8-15,B) so long as they are not aligned (Figure 8-15,C); if they are aligned, they will cause it to undergo translational (nonrotating) movement unless they are of equal magnitude (Figure 8-15,D). The net force, or resultant force, of two or more vectors can be determined graphically by adding vectors that represent the magnitude and direction of the forces. Three or more forces acting upon a body will cause it to move unless the vectors add up to zero. They can add up to zero only if they are parallel (Figure 8-15,E) or if their lines of action meet at a single point (Figure 8-15,F). Even if one of the latter conditions is met, they will add up to zero only if their magnitudes have relative values that add up to zero.

Diagrams of the jaws and muscles of a typical reptile and a typical mammal are shown in Figure 8-14. The reptilian jaw has several bones; the muscles that close the jaw act more or less at right angles to its long axis. In the mammalian jaw, only the dentary, which bears the teeth, remains little changed from its form in the reptilian jaw; there is an upward projection, the coronoid process, at the rear of the mammalian jaw, and the musculature has become more complex than in the reptilian jaw. The large temporalis muscle pulls diagonally upward and backward on the coronoid process, and the masseter, which attaches to the lower rear part of the jaw, pulls diagonally upward and forward.

How did the simple reptilian mechanism evolve into the mammalian mechanism? The evolutionary pathway is illustrated by focusing on the stages of evolution in the origin of mammal-like reptiles (Figure 8-16). The orientation and approximate sizes of muscles in extinct forms can be reconstructed from the configuration of hollow spaces and attachment surfaces on the bones. Comparisons with living reptiles and mammals are, of course, essential to this sort of reconstruction.

In primitive reptiles, the jaw would have been stationary when clamped against hard food only because the force of the rather simple jaw muscle (*CM*) was balanced by one or more other forces. One such force would have been that of the food (*F*). This would have acted in a direction opposite to the force of the jaw muscle and in a position farther forward. The jaw would have rotated, except that the cranium pressed against it at the point of articulation with force *R*. Until the food yielded, forces *R* and *F* would have exactly balanced force *CM* and the jaw would have remained motionless.

For simplicity in analyzing the jaw mechanics of advanced mammal-like reptiles such as *Diarthrognathus* (Figure 8-16,F), discussion can be restricted to the largest muscle, the temporalis (*T*), and the superficial masseter (*SM*), and related muscles that probably operated approximately parallel to the masseter (Figure 8-16). When *Diarthrognathus* bit into food with its rear teeth, the force (*F*) transmitted by the food may have cancelled the muscle forces (*SM* and *T*), approximately, so that no force at all would have been transmitted from the cranium to the rear point of articulation of the jaw. The reason for this approximate cancellation is that all forces seem to have met at or very near a single point (refer back to Figure 8-15,F). There is no way of knowing whether the relative magnitudes of the muscular forces were such that this was in fact the case. It is clear, however, that less force would have developed at the jaw articulation of the mammal-like *Diarthrognathus* than at the jaw articulation of a primitive reptile (Figure 8-16,A).

It is interesting that the posterior bones of the jaw were progressively reduced and then lost in the evolution of advanced mammal-like reptiles. This coincided with the

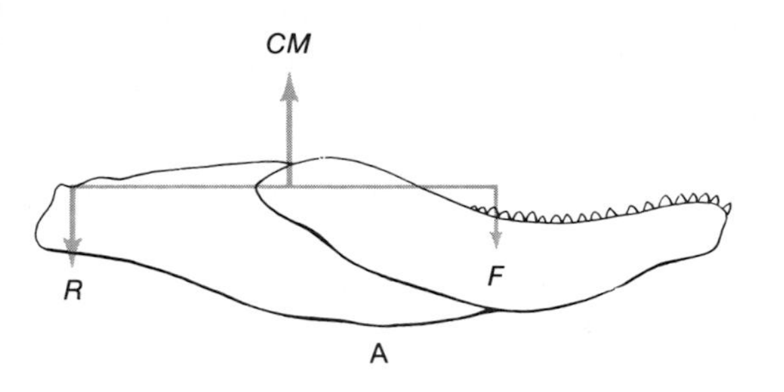

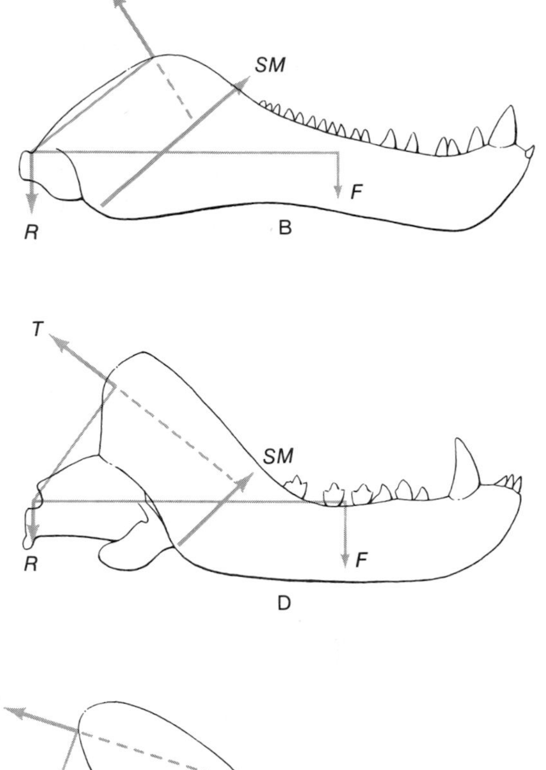

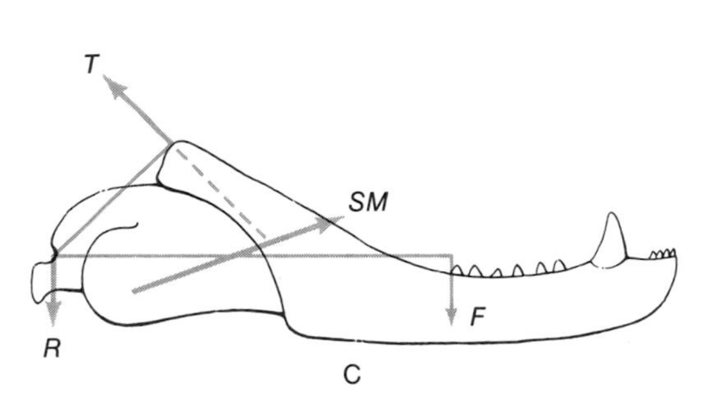

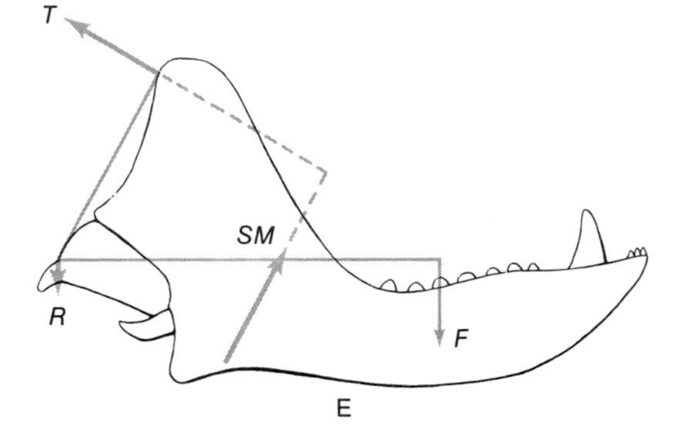

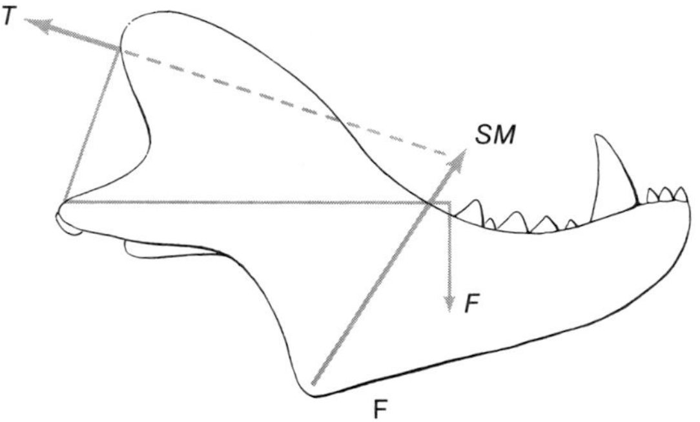

FIGURE 8-16
Lower jaws of a series of reptiles ranging from a primitive form (A) to an advanced mammal-like reptile (F). Arrows represent inferred lines of action of the major forces: *F,* force of food; *R,* force at jaw articulation; *CM,* force of major jaw muscles of a primitive reptile; *SM,* force of superficial masseter; and *T,* force of temporalis muscle.

origin and enlargement of the coronoid process and reduction of force at the articulation, which was originally supported by the posterior bones (Figure 8-16,A). Eventually, in *Diarthrognathus* the dentary came to form the articulation with the cranium. The general mechanism of this genus was passed on to the mammals, allowing for a variety of biting and chewing mechanisms to evolve. More than one mechanism is usually employed even within one mammal species. It is therefore no surprise that the mammals have evolved a remarkable degree of tooth differentiation. The kind of jaw movement and the quality and location of the food influence the kinds of teeth and the forces of the jaw muscles that are most suitable. Tearing, gnawing, and grinding mechanisms have become highly evolved in various taxa.

LIFE-HABIT TRANSITIONS IN MUSSELS

As another example of the importance of observations of living species to functional morphology, we can consider the evolution of marine mussels of the bivalve mollusc family Mytilidae (Stanley, 1972). These animals attach to the substratum by a byssus (a group of horny threads secreted by the foot). The most conspicuous mytilids are the ***epibyssate*** species, which attach to the surface of the substratum (as shown in the diagrams on the left in Figure 8-17). Most of these species attach to hard substrata, but some form clumps or banks on soft substrata by attachment to shell debris and by mutual attachment of individuals. A large percentage of living mytilid species, however, are ***endobyssate,*** living partly or entirely buried in soft substrata and attaching to large sedimentary particles or buried plant structures (diagrams on the right in Figure 8-17). As might be expected, the two life habit groups have distinctive morphologies. Endobyssate species have a cross-sectional shape somewhat like that of a burrowing clam. The ventral region is shaped like the hull of a boat. Epibyssate species, in contrast, have a flattened ventral region that provides stability for resting on hard surfaces. The lateral-view profiles of the two types of mussels also differ. The posterior byssal retractor muscles of endobyssate species are positioned so as to pull obliquely downward, anchoring the upright shell in the substratum. The flattened ventral region of epibyssate species in part reflects the absence of a ventral lobe (Figure 8-18,C). In addition, the posterior byssal retractor muscles are positioned to pull at nearly right angles to the ventral margin of the shell, holding it firmly against the surface to which the byssus attaches. The muscles are relatively large, reflecting the species' need for firm attachment in an exposed, epifaunal position (in which there is no surrounding substratum to contribute to stability). From anatomical studies, it has long been believed that the most primitive bivalves were infaunal, burrowing clams and that epifaunal bivalves were derived from them. The endobyssate condition might therefore be expected to be the primitive condition in the Mytilidae, representing an intermediate stage between burrowing ancestors and epibyssate forms. The fossil record bears out this prediction. Newell (1942) found that all middle Paleozoic species of the Mytilidae were shaped like *Modiolus* (Figure 8-17, right).

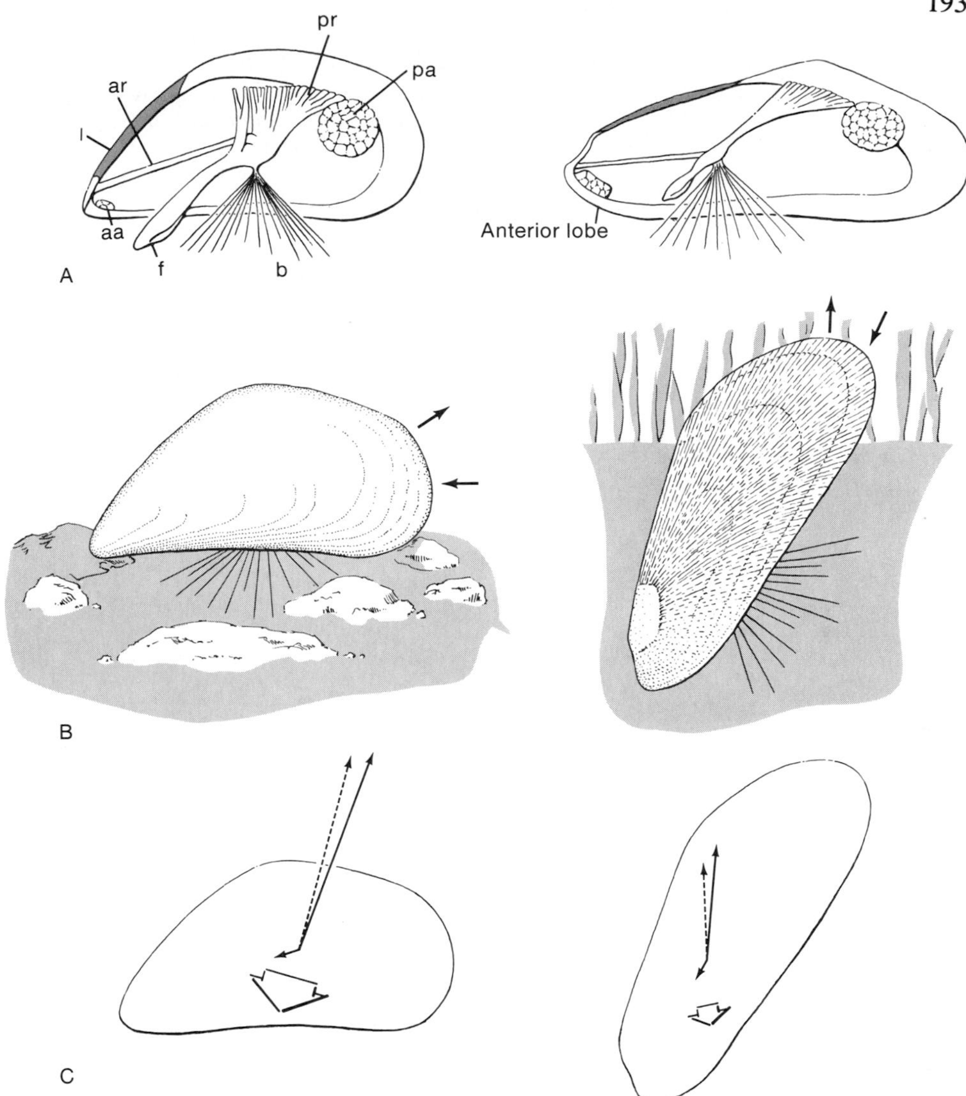

FIGURE 8-17
Comparative adaptive morphology of the living mussels *Mytilus edulis* (left) and *Modiolus demissus* (right). A: anterior and posterior adductor muscles, aa and pa, close the two valves of the shell; l, the ligament, opens the shell; f, foot; b, byssus; ar and pr, anterior and posterior retractor muscles, pull on the byssus to anchor the shell in place. *Mytilus* lacks an anterior lobe. B: The two species in life position. *Mytilus* attaches epifaunally to rocks. *Modiolus* attaches to plant debris in marsh peat, with the shell mostly buried. The arrows at the posterior of each animal depict inhalent currents, from which food and oxygen are obtained, and exhalent currents. C: Relative forces anchoring the shells in place. Solid thin arrows represent force vectors for the retractor muscles; dashed thin arrows represent resultant forces; stubby hollow arrow shows the direction in which the shell is pulled. (From Stanley, 1972.)

FIGURE 8-18
Contrasting morphologies of living endobyssate and epibyssate mussels of the family Mytilidae. A: Cross-sectional shapes of eight living species. The portion of the shell that is ventral to the position of maximum shell width has been shaded. This region is flattened in epibyssate species. B: Coordinate transformation illustrating how *Modiolus* can be deformed to produce the shape of *Mytilus.* C: Four living species representing conditions in a spectrum from fully infaunal habits to epifaunal habits. The line S–S′ shows the surface of the substratum. (From Stanley, 1972.)

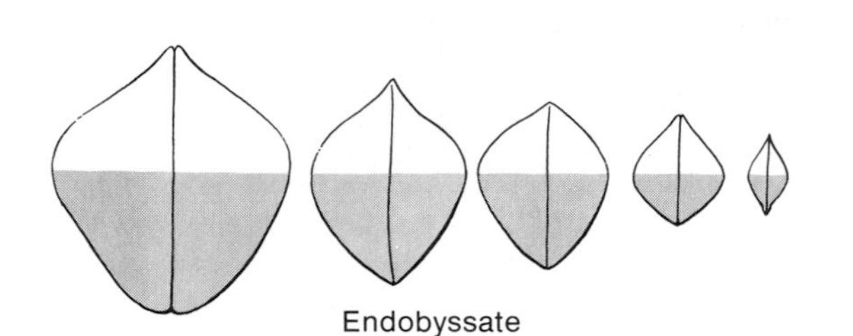

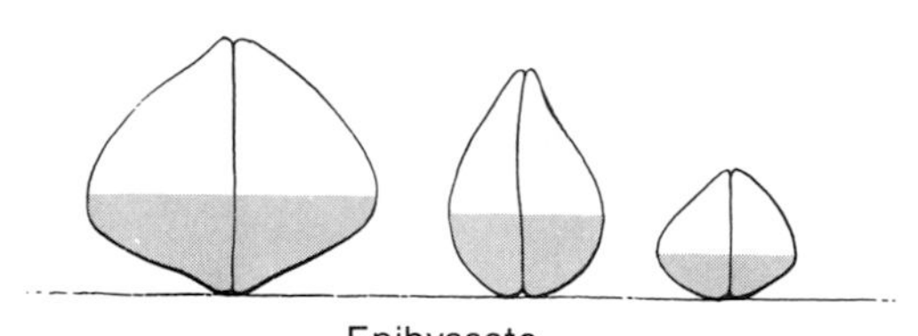

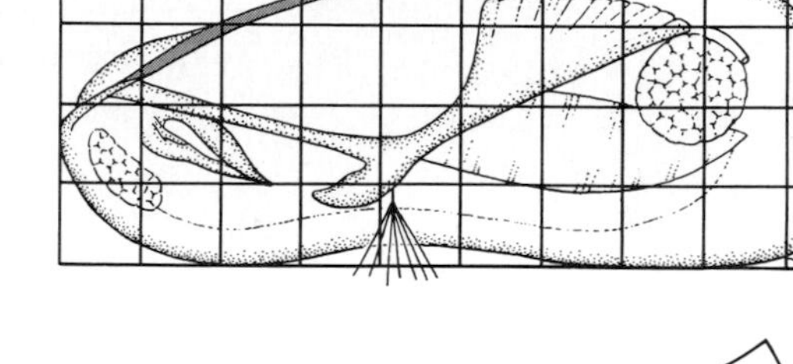

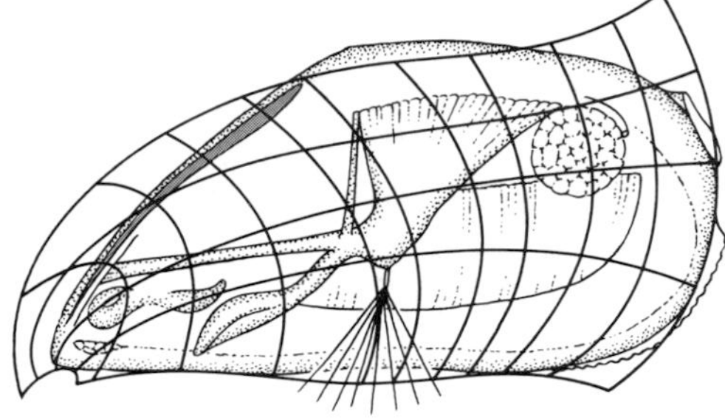

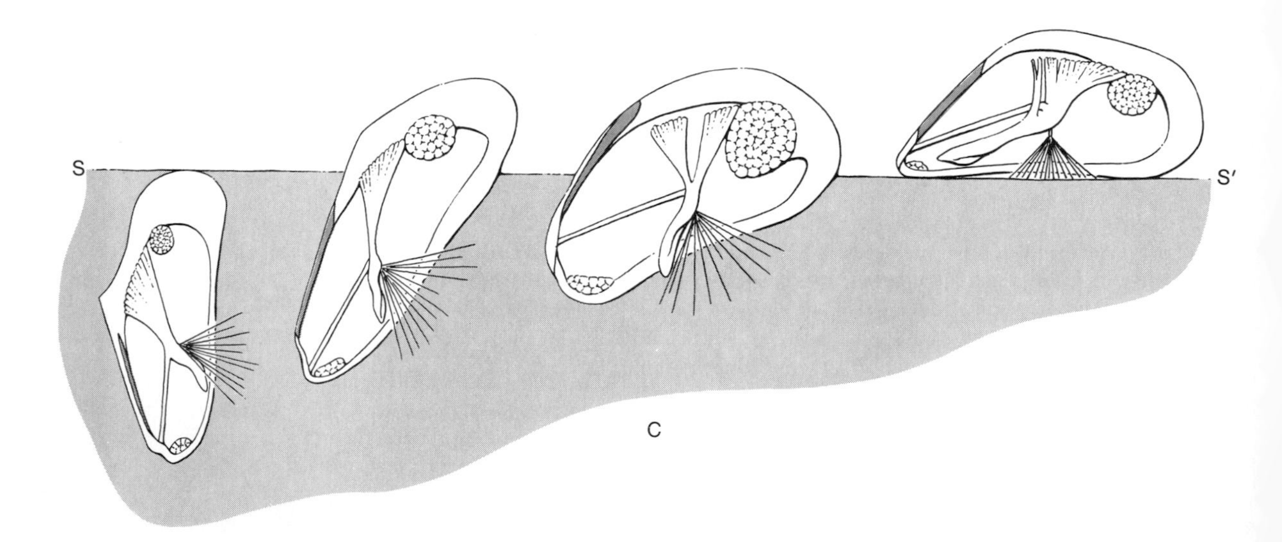

Species shaped like *Mytilus* (Figure 8-17, left) did not appear until the Mesozoic. Species with intermediate shapes are found in the late Paleozoic, and Newell placed these in the genus *Promytilus*. The use of transformed coordinates (see page 58) illustrates the change in shape that occurred with the origin of epifaunal habits (Figure 8-18,B). It is almost certain that the transition occurred polyphyletically. Furthermore, the entire spectrum of shapes and life habits is represented by modern species of the family. In other words, lineages of the ancestral type have survived along with their descendents. (It should be noted that not all of the living *Modiolus*-like species live in soft sediment; some nestle in crevices in hard substrata, and it seems reasonable to suppose that some Paleozoic representatives lived in this way as well.) It is quite interesting that similar trends from endobyssate to epibyssate habits are evident in the Paleozoic records of several other bivalve families.

Conclusion

Although the examples we have considered are but a small sample of many interpretations of function that have been made for fossil structures, they illustrate the great diversity of problems that invite study and the variety of analytic approaches that can be employed. Functional morphology is an extremely important area of research for two reasons:

When applied to an individual fossil species, it can greatly facilitate paleoecologic analysis—the study of the species' interaction with its environment (for example, its life habits, habitat preferences, and relationships to other species).

When applied to taxonomic groups of species, it can reveal how the morphologic trends in evolution that have been observed in the fossil record are adaptive.

Supplementary Reading

Alexander, R. M. (1968) *Animal Mechanics*. Seattle, University of Washington Press, 346 p. (A thorough introduction to biomechanics.)

Bonner, J. T. (1952) *Morphogenesis*. Princeton, Princeton University Press, 296 p. (An outstanding biologic treatment of growth and form.)

Bramwell, C. D., and Whitfield, G. R. (1974) Biomechanics of *Pteranodon*. *Philos. Trans. Roy. Soc. London*, **B.267**:503–581.

Dacqué, E. (1921) *Vergleichende biologische Formenkunde der Fossilen niederen Tiere*. Berlin, Gebrüger Bornträger, 777 p. (The first comprehensive analysis of functional morphology of fossils.)

Gould, S. J. (1970) Evolutionary paleontology and the science of form. *Earth-Science Reviews,* **6**:77–119.

Hertel, H. (1966) *Structure-Form-Movement.* New York, Reinhold Pub. Co., 251 p. (A fascinating and informative analysis of the engineering aspects of form with particular reference to mobile forms.)

Rudwick, M. J. S. (1964) The inference of function from structure in fossils. *Brit. Jour. Philos. Sci.,* **15**:27–40. (A provocative discussion of the logical problems encountered in deducing function from morphology.)

Stein, R. S. (1975) Dynamic analysis of *Pteranodon ingens:* a reptilian adaptation to flight. *Jour. Paleont.,* **49**:534–548.

Thompson, D'A. W. (1942) *On Growth and Form.* New York, Cambridge University Press, 1116 p. (The classic reference on problems of form and function.)

Wainwright, S. A., Biggs, B. D., Currey, J. D., and Gosline, J. M. (1976) *Mechanical Design in Organisms.* New York, Halsted Press, 423 p. (Application of engineering theory to the interpretation of animal skeletons.)

CHAPTER 9

Biostratigraphy

Stratigraphy is the study of the geometry, composition, and time relations of stratified rocks. Its special concern is with the history of these rocks, including their fossil components. Biostratigraphy is simply that branch of stratigraphy that is primarily concerned with fossils. In its broadest sense, biostratigraphy can be considered to include most of the topics discussed in this volume. In its narrower sense, as used here, it concerns the spatial distribution and temporal relations of fossils and fossil-bearing rocks.

In 1961, the American Commission on Stratigraphic Nomenclature published its "Code on Stratigraphic Nomenclature" in the Bulletin of the American Association of Petroleum Geologists. To a much greater extent than the "International Code of Zoological Nomenclature" (page 114), this stratigraphic code embraces scientific concepts rather than simply rules of nomenclature. It therefore contains a larger number of controversial statements than the zoological code. Still, it is a valuable formulation of prevailing stratigraphic opinions and has tended to promote uniformity in classification and naming of stratigraphic units. It will certainly be revised to accommodate changes in prevailing ideas and techniques in the years to come. In general, stratigraphy is a highly subjective field of study for which few unique principles serve as an underpinning. Perhaps the only basic stratigraphic principle is the "law" of superposition, which states that in any undisturbed sedimentary sequence the oldest bed is at the base and the youngest bed is at the top. Many principles of paleoecology, evolution, sedimentology, and geomorphology are commonly applied to stratigraphic problems. For example, the so-called "law of biotic succession," often cited in elementary or historical discussions of biostratigraphy, states that biotas have followed one another in an orderly succession through geologic time. This succession is simply the product of organic evolution and might properly be considered to be a part of evolutionary paleontology borrowed by biostratigraphy.

Rock-stratigraphic Units

Stratigraphy deals with units, primarily units of rock and of time that are used to divide stratigraphic sequences and geologic history. The stratigraphic record is divided into three-dimensional rock bodies on the basis of lithology, the physical character of rock. Rock units, often called ***rock-stratigraphic units,*** are recognized without regard to time relations or fossil content, except insofar as fossils contribute to their physical and chemical composition. Units are recognized on the basis of lithologic uniformity, which is inevitably judged with a great deal of subjectivity. The basic rock-stratigraphic unit is the ***formation.*** Formations may be lumped into higher units called ***groups,*** or divided into smaller units called ***members.***

It is commonly stipulated that to be designated a formation a unit should be "mappable." Because of the great variety of scales used for geologic maps, this stipulation is of questionable value. Much like an arbitrary taxonomic subdivision, such as a "genus" or "family," a formation is a category that defies quantitative description. As with higher taxonomic units, most workers develop a general concept of the traditional magnitude of rock units. Still, in both taxonomy and stratigraphy, opinions differ among various workers as to the amount of subdivision desirable in particular instances. In general, increased detail of knowledge tends to promote increased subdivision.

One of the first facts learned by the introductory geology student is that the stratigraphic record is far from complete. In all regions major unconformities in rocks and erosional surfaces at the earth-air interface reveal the existance of large gaps in the rock record.

The student also learns that most rock units seeming to show continuous deposition were actually formed by discontinuous deposition. Chemical precipitation in certain restricted environments, such as closed evaporite basins, may produce nearly continuous deposition, but in most other settings sediment deposition is sporadic. Nearly all marine detrital sediments undergo some transportation on the sea floor before reaching the site of their final deposition. Many, of course, are carried into marine depositional basins from the land. While sediment may be supplied to a depositional basin nearly continuously (usually with fluctuations in *rate* of supply), at any given time *final* deposition of sediment layers will occur only in certain areas.

Nondeposition, or erosion, will occur simultaneously in others. Periodic "instantaneous" agents such as storms may produce waves and currents that erode thick sediment layers that took many years to accumulate.

A single bed, or lamina, of sediment represents a single depositional event. Surfaces between beds or laminae then indicate gaps in the rock record, just as unconformities do on a larger scale. Early in the twentieth century, Barrell recognized the discontinuousness of deposition and for gaps due to short time periods of nondeposition or erosion he introduced the term ***diastems.***

One of the most striking illustrations of the discontinuousness of sediment deposition comes from study of preserved animal burrows. Figure 9-1, from a report by Rhoads (1966), shows a slab of Silurian rock sectioned perpendicular to bedding. The rock contains three distinct types of burrows, which tended to destroy some of the depositional bedding. Burrowing extended downward from the sediment-water interface to a depth of several centimeters. If deposition had been continuous and burrowing activity had kept pace with it, we would expect the burrowing animals to have maintained their position in the upper few centimeters during sediment accumulation and to have largely destroyed primary bedding throughout.

In fact, the rock section shows four layers (A–D) in which bedding was not entirely destroyed. The upper beds in these layers (especially in the second layer from the bottom) are partly obliterated by burrows, producing irregular upper surfaces. Rhoads has interpreted the rock slab as representing four distinct depositional events. Careful observation reveals that the four distinct bedded layers, each several centimeters thick, were deposited in succession. Each layer was suddenly "dumped" on the one below. The sediment-water interface on which each was deposited is shown in Figure 9-1. Some of the lowermost beds of each layer remained undisturbed because burrowing organisms established themselves only in its upper part, near the sediment-water interface.

Vertical "escape" burrows are represented by heavy arrows in Figure 9-1. They cut across the normal burrows, most of which are oriented horizontally. The escape burrows strongly support the idea of sudden deposition, for apparently they were upward migration routes of animals to a new sediment-water interface. Some animals may not have been so fortunate, perhaps being unable to make their way upward through a new blanket of sediment. The idea of sudden deposition also gains favor from the fact that in layer B, the thickest of the four, the lower beds have been least disturbed by burrowing.

Similar analysis by Seilacher (1962) has demonstrated that many of the famous deep-water flysch deposits of Europe were formed by successive events of deposition, in which thick beds were laid down suddenly from time to time. Most burrows and trails in flysch deposits are preserved on the bottoms of sandstone beds underlain by shale. For years there was disagreement as to whether burrowing took place only at the interface between sand and shale or occurred throughout homogeneous sandstone units but was only well preserved at the lower surface (which individual burrows disrupted). Using various lines of evidence, Seilacher concluded that some animals

10 cm

FIGURE 9-1
Section of rock from the Silurian Stonehouse Formation, Nova Scotia, with preserved animal burrows. (From Rhoads, 1966; sample, courtesy of R. K. Bambach.)

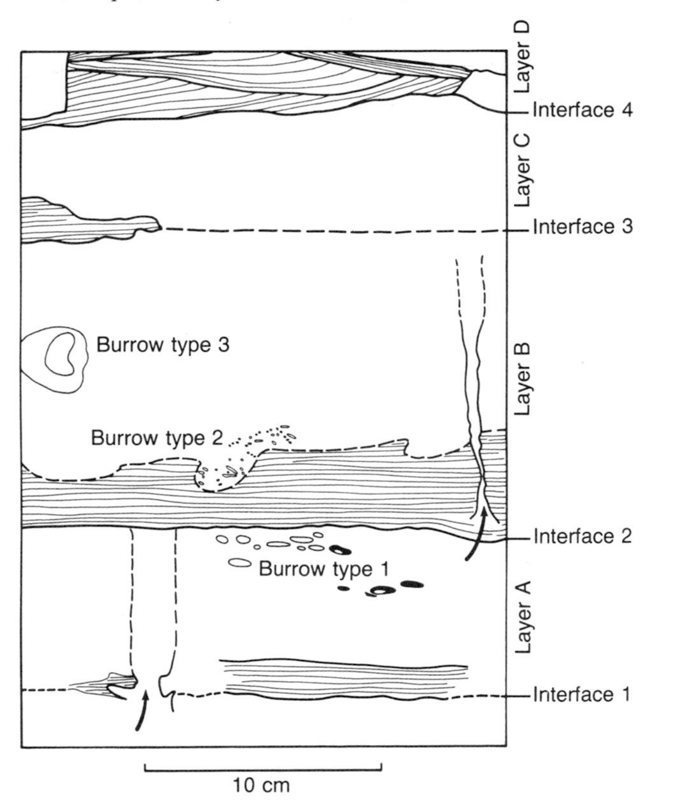

were surface burrowers and others penetrated the sediment to considerable depth. He next restricted his interpretation to the latter group, and reasoned that each species must have burrowed to a characteristic sediment depth. Therefore, if deposition of each bed was sudden, there should be a maximum thickness for all sandstone units on whose undersurfaces each burrow type is found. Figure 9-2 shows that data compiled for various burrow types bear out this prediction and support the idea of sudden deposition.

Even though these examples offer unusually favorable opportunities for demonstration of discontinuous deposition, there is reason to believe that most sedimentary rocks have had similar histories of discontinuous deposition, whether deposition of the included beds was separated by gaps of days, years, or thousands of years.

Thickness of beds, in cm	Total number of beds in section A B C	Postdepositional burrows: *Granularia*	*Fucusopsis*	*Phycosiphon*	*Scolicia*	*Neonereites*	Predepositional burrows: *Lorenzinia*		*Palaeodictyon* Regular	*Palaeodictyon* Irregular	*Scolicia*	*Ceratophycus*
0–1	21 – 3	6 – –	– – –	4 – 1	10 – 2	3 – –	3 – –	2 – –	4 – –	– – –	2 – –	1 – –
1–2	41 12 16	19 3 5	– – –	13 2 8	12 7 6	5 1 2	9 – –	3 – –	4 1 5	3 1 –	1 – –	12 1 –
2–3	33 14 18	18 7 8	– – –	12 1 7	4 5 1	4 2 3	5 – 1	1 – –	7 2 6	1 2 –	5 2 –	8 – –
3–4	19 17 9	15 9 7	1 – –	9 4 1	– 4 –	2 1 2	5 2 –	2 – –	2 3 7	2 – –	2 1 –	4 3 –
–5	18 14 7	15 7 4	1 – –	9 3 4	1 3 1	1 – 1	3 – –	1 – –	4 1 6	3 3 –	1 3 –	1 2 –
–6	17 15 8	10 14 4	– – –	4 4 6	1 – 1	1 – –	4 4 1	1 – –	1 2 3	2 2 1	4 1 –	2 5 –
–7	12 11 10	11 11 9	1 – –	5 1 6	– – –		– – –	– – –	4 1 5	1 1 –	5 2 –	3 – –
–8	10 11 8	7 11 4	– – –	6 – 2	1 – 1		– – –	– 1 –	2 – 6	1 – –	1 2 –	2 3 –
–9	12 14 6	10 13 4	– – –	2 – 2			1 – 2	– – –	1 – 3	1 1 –	4 1 –	3 2 –
–10	12 5 2	12 4 2	1 – –	6 – –			2 – –	1 – –	– – –	2 1 –	2 1 –	– 1 –
–12	13 17 5	11 16 3	– – –	3 1 3			5 1 –	– – –	6 7 3	2 1 –	4 4 –	1 1 –
–14	7 16 4	7 16 4	1 – –	1 – –			– – –	1 – –	2 7 3	– 3 –	1 3 –	– 3 –
–16	5 5 5	6 5 5	– – –				2 – –	1 – –	2 – –	2 – –	3 1 2	1 – –
–18	7 6 1	7 6 1	– 1 –				2 1 –	1 – –	– – –	– 2 –	3 1 –	2 – –
–20	8 8 –	6 8 –					2 – –	1 – –	4 1 –	– – –	2 – –	– 1 –
–25	14 10 ––	12 10 –					5 – –	1 – –	– – –	2 – –	5 1 –	3 – –
–30	9 10 1	7 9 1					3 1 –	– 1 –	– – –	– 1 –	3 1 –	1 1 –
–35	10 6 –	10 6 –					– 1 –	1 – –	– 1 –	– – –	2 1 –	– 1 –
–40	4 3 –	4 3 –					1 – –	– 1 –	1 1 –	1 1 –	2 – –	– 1 –
–45	7 7 ––	7 7 –					– – –	– 1 –	– – –	– 1 –	– – –	– 1 –
–50	4 7 –	3 7 –					1 1 –	– 1 –	– – –	– – –	1 1 –	– – –
–60	8 3 –	8 3 –					1 – –	2 1 –	1 – –	2 – –	3 – –	3 1 –
–80	7 5 –	7 5 –					– – –		– – –	– – –	1 – –	– – –
–100	2 3 –	2 3 –					– – –		– – –	– – –	– – –	– – –
–150	12 8 –	12 7 –					2 – –		– 1 –	– 1 –	2 – –	– – –
–200	4 7 –	3 7 –					1 – –		1 – –	1 1 –	– 2 –	1 – –
Sum	655	493	6	130	60	28	72	24	121	49	89	75

FIGURE 9-2
Diagram comparing burrows and trails preserved on the undersurfaces of beds in three stratigraphic sections (A, B, and C) in the Flysch of northern Spain. Postdepositional burrow types tend to be restricted to thinner beds; predepositional burrows show no such relationship to thickness of bed. (From Seilacher, 1962. Copyright © 1962 by The University of Chicago.)

Just as environments change laterally over the earth's surface, sediments being deposited simultaneously in different places differ in chemical and physical characteristics. The general aspect of a body of rock deposited over a wide area during a certain time interval may, therefore, show considerable variation. Portions of a body of rock showing lateral changes in aspect are known as ***facies.*** In recognizing facies, emphasis may be placed on ***lithofacies,*** based on prominent lithologic features, or ***biofacies,*** based on fossil features. Facies may be found within rock units that are formally classed as groups, formations, or members. Facies may be used to define lateral boundaries between formal rock units, though the boundaries are then usually gradational rather than abrupt.

Biostratigraphic Units—The Biozone

Superimposed on the rock-unit subdivision of the stratigraphic record is a system of subdivision based on fossil occurrences. The units of this subdivision, called ***biostratigraphic units,*** are also tangible rock bodies, but their boundaries are defined by various paleontologic criteria, such as the appearance, maximum abundance, and disappearance of fossil species or genera in various local rock sequences.

The fundamental biostratigraphic unit is the zone. Many sorts of zones have been recognized. We can envisage an abstract kind of zone known as a ***biozone,*** which represents all rocks throughout the world that were deposited during the time interval in which a species lived. A biozone is an abstraction because we can never delineate it physically. No species is present in all rocks of its biozone.

It is instructive to examine the reasons that species have not filled their biozones. We can assume that many species have originated through evolutionary divergence of geographically isolated populations of a preexisting species. The extent to which a new species arising in this manner has come to fill its biozone has depended partly on how rapidly it spread throughout areas of the earth's surface that it could potentially inhabit. In a sense, we are comparing rates of evolution and rates of geographic dispersal. We will discuss geographic dispersal in some detail in the final chapter of the book. A brief introduction will suffice for our present discussion.

If there were no ***effective barriers,*** most species would have spread over large geographic areas quite rapidly relative to rates of species evolution. Many examples of the rapid dispersal of species given access to new regions are presented in the excellent book by Elton (1958) on the ecology of animal invasions. In fact, however, geographic restrictions apply to all species. Biogeographic ranges, particularly their latitudinal extents, are limited by environmental conditions, especially temperature. In addition, oceans are barriers to terrestrial animals, and land masses are barriers to marine animals. Large, abrupt changes in altitude on land or of depth in the ocean may also form effective barriers.

It is difficult to formulate general rules to describe the effectiveness of barriers in preventing species dispersal. In part, dispersal across barriers is a matter of chance. Such agents as floating logs, bird droppings, and strong winds have undoubtedly transported many species across major barriers. Simpson (1940a) has referred to chance mechanisms of dispersal as "sweepstakes routes." Chains of islands often form stepping stones for "sweepstakes" migration of both terrestrial species and shallow-water marine species from one continent, or continental margin, to another.

More continuous, but narrow, dispersal routes are sometimes referred to as "corridors" of migration. Because of their narrowness, corridors of migration are subject to opening and closing, perhaps repeatedly, in the course of geologic history. The Isthmus of Panama was a land corridor to South America for North American placental mammals that enabled them to displace some previously isolated South American marsupial groups near the close of the Pliocene Epoch.

Many plant and animal groups have developed special mechanisms for migration across barriers. For example, planktonic larvae may be viewed as specially adapted dispersal stages in the life histories of certain benthonic marine invertebrates (page 274). It has been found that most planktonic larvae can temporarily delay metamorphosis if they fail to locate a suitable substratum. For many years it was believed that transport of larvae of shallow-water invertebrates across the Atlantic Ocean was virtually impossible. However, Scheltema (1968) has found that larvae of shallow-water benthonic species found on both sides of the Atlantic are commonly transported across the Atlantic from west to east and east to west by major ocean currents (Figure 9-3). Seeds and spores are important wind-transported dispersal agents of many terrestrial plant groups. Under unfavorable environmental conditions, many aquatic and terrestrial unicellular organisms alter to inactive, cyst-like resting stages that, like spores, may be blown by the wind to transport species across otherwise effective barriers.

It is not uncommon for a species to disappear from most parts of its range but persist in a limited region as one or more relict populations (page 431).

There are additional reasons that no fossil evidence of a species is found in many parts of its biozone. Even within their geographic ranges, species can generally inhabit only certain types of local environments. Consequently, many fossil species are noted for their restriction to certain rock types. Such fossils may be known as "facies fossils." In most instances, their local distribution patterns reflect primary ecologic distribution. Furthermore, no species is preserved in all rock units deposited in environments in which it lived. Finally, a species may actually occur in a rock unit but may be misidentified because of inadequate preservation or confusion with another species.

Prevailing winds and ocean currents disrupt latitudinal temperature gradients in many parts of the world. Sharp temperature changes and abrupt physiographic

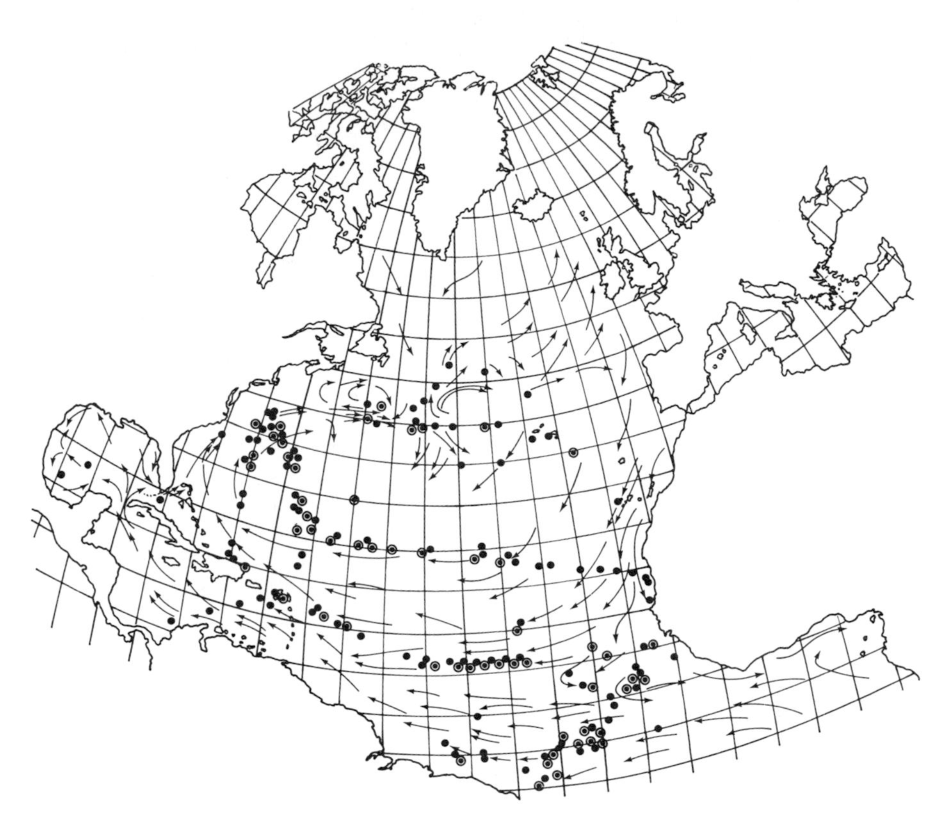

FIGURE 9-3
Distribution of planktonic larvae of the gastropod family Architectonicidae. Circled dots indicate stations where larvae of *Philippia krebsii,* a member of the family, have been found. Adults belonging to the Architectonicidae occur in shallow-water areas throughout the tropics. Arrows show major water currents at the ocean surface. (From Scheltema, 1968.)

barriers tend to confine *groups* of species to certain regions, which are called ***biogeographic provinces.*** Species limited to any given region are termed ***endemic.*** Most recognized biogeographic provinces contain some endemic species, but also share species with other provinces. Hence, provincial boundaries are not always clear-cut; differences of opinion as to precise boundary designations and desirable degrees of subdivision have led to a variety of provincial classifications in certain parts of the world.

Several provincial classifications for the coastal waters of eastern North America are shown in Figure 9-4. Eastern and western boundaries are the continental margin and shoreline. The Gulf Stream swings away from the coast in the vicinity of Cape Hatteras, forming a sharp temperature discontinuity. Likewise, the Labrador Current, sweeping south along the northeastern coast, is diverted seaward at Cape Cod. Northern and southern boundaries in other parts of the world do not necessarily

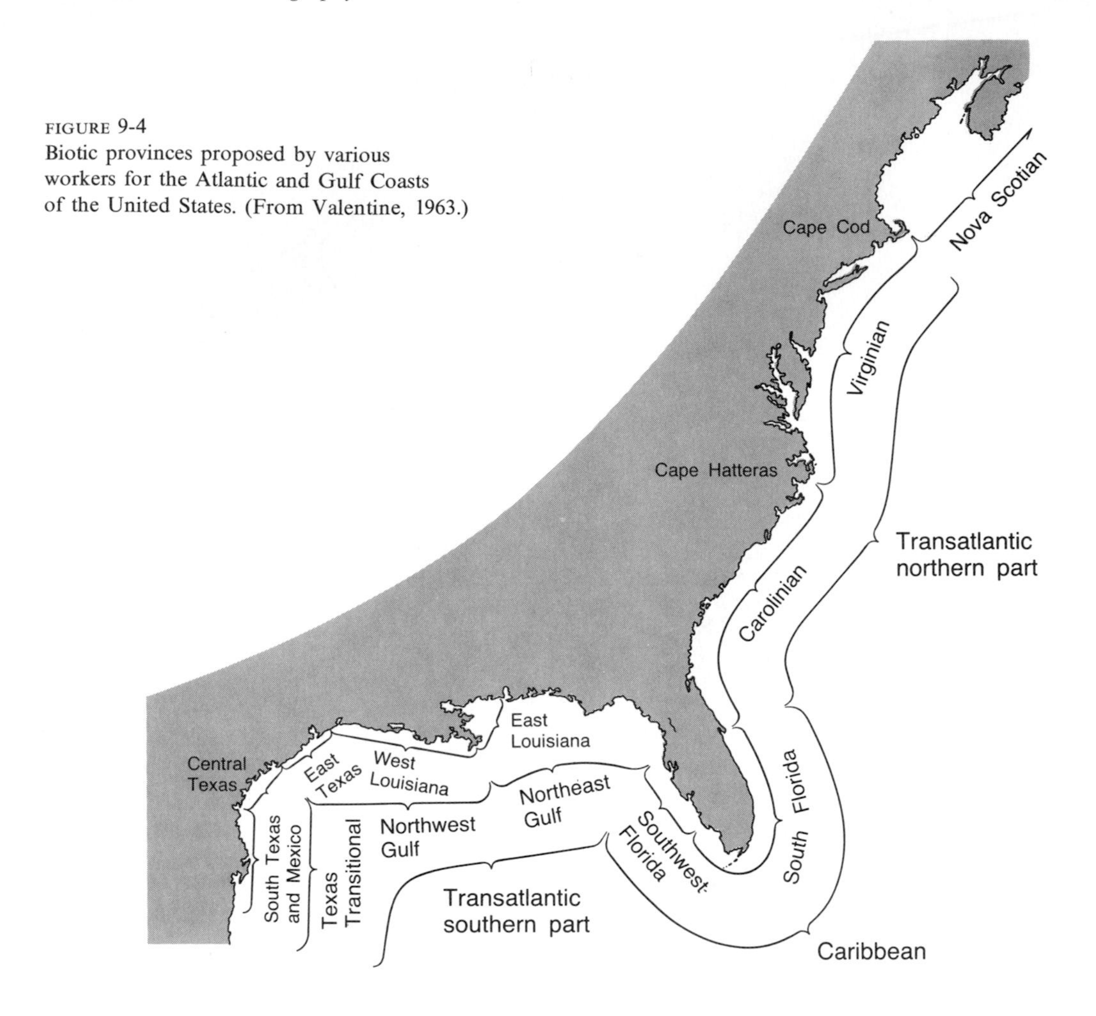

FIGURE 9-4
Biotic provinces proposed by various workers for the Atlantic and Gulf Coasts of the United States. (From Valentine, 1963.)

coincide with those of Figure 9-4 (in either latitude or mean annual temperature).

In the final chapter of the book we will discuss some relationships between temperature gradients and biogeographic distributions. For our present purposes, it is apt to note that we would expect widespread warm climates of the past to have produced broad biogeographic provinces and steep temperature gradients to have produced narrower ones. One of the most interesting and best-studied geologic Periods from this standpoint is the Jurassic. Arkell (1956) and other authors have noted that many Lower Jurassic taxa (especially of cephalopods) were cosmopolitan; for every recognized North American ammonite species there is a similar European counterpart, for example. In contrast, Middle and Upper Jurassic faunas are markedly provincial. Arkell recognized three major Upper Jurassic faunal provinces, which he termed "realms": the Boreal Realm (of northern latitudes), the Pacific Realm (of areas bordering the Pacific Ocean), and the Tethyan Realm (including the rest of the world, but especially well-represented by fossil faunas of the Mediterranean region). Origin of the realms was by no means instantaneous, and their faunas shifted and overlapped from time to time. But the realms tended to persist and are recognized in the Cretaceous as well (see Figure 12-43). Most workers have considered their distinctness to have been caused by the development of steep temperature gradients and physiographic barriers to dispersal. Whatever the cause of Jurassic faunal differentiation, it is easy to see why intercontinental correlation is generally simpler and more accurate for Lower Jurassic sequences than for Upper Jurassic sequences.

Valentine (1963) has pointed out that the geologic record of a biogeographic province represents a biostratigraphic unit according to the Code of Stratigraphic Nomenclature. He has recommended the adoption of a formal system of biostratigraphic nomenclature for provincial units. Such a system might be very useful.

Correlation with Fossils

Given the partial representation of fossil species in their biozones, to what extent can the stratigraphic distributions of fossils be used profitably to determine relative geologic time? From our previous discussions, it should be evident that biostratigraphic units can only be used in an approximate way to determine time relationships. Establishing the time equivalence of two spatially separate stratigraphic units is known as ***correlation.*** Unfortunately, the term "correlation" is also used by some workers to imply equivalence of rock type or fossil content without regard to time. It is less confusing to use terms like "lithologic equivalence" and "biostratigraphic equivalence" for the latter sorts of relationships.

Time correlation, though usually only approximate, can be undertaken by various methods, only some of which make use of fossils. Nevertheless, fossils represent by far the most important tools for time correlation. Establishment of biostratigraphic equivalence is the first step. This equivalence may then be interpreted as demonstrating approximate time equivalence, or correlation. We will review several types of biostratigraphic equivalence in the following sections.

STRATIGRAPHIC RANGES AND ZONES

Most stratigraphic work is based on studies of vertical fossil distribution in local stratigraphic sections (rock sequences exposed for study). A stratigraphic section is really a three-dimensional exposure of rock strata, often on the side of a hill or along a creek bed, where there is considerable relief. For correlation purposes, however, it is commonly treated as if it were one-dimensional. The two-dimensional columns used to represent stratigraphic sections in geologic literature merely serve to permit diagrammatic portrayal of rock types at various levels (Figure 9-5). Techniques used to study faunal distribution in a stratigraphic section vary with structural patterns of the rock units, terrain, lithology, and fossil content. In common stratigraphic practice,

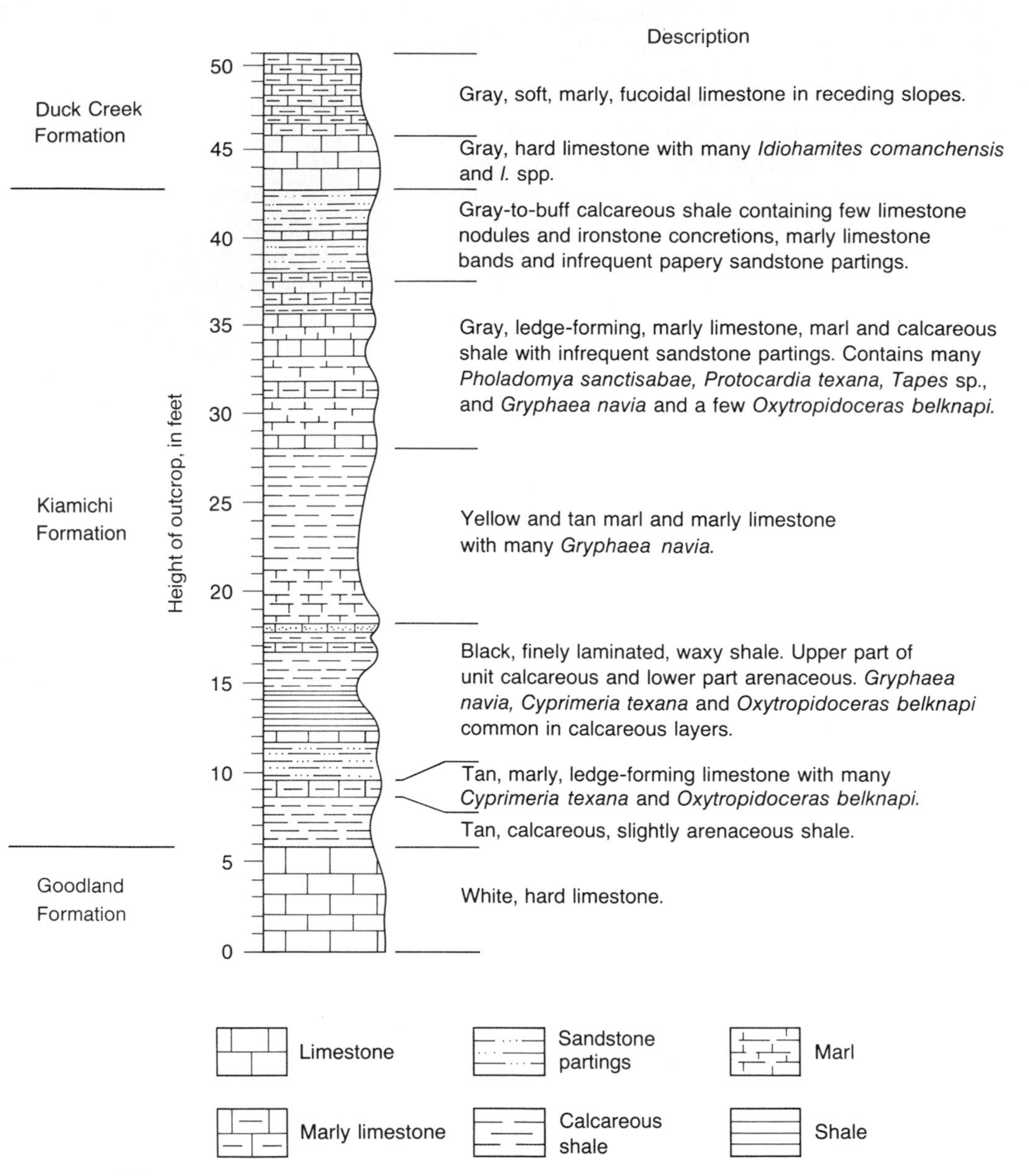

FIGURE 9-5
Stratigraphic column for the Cretaceous Kiamichi Formation, at a bluff southwest of Meacham Field, Tarrant County, Texas. The scale is in feet above the base of the outcrop, which is located within the Goodland Formation. (From Perkins, 1960.)

a section is measured in feet or meters, and significant changes in lithology are then plotted on the vertical scale. Fossil species are carefully collected throughout the section and their vertical positions are recorded. It is then possible to plot the ***stratigraphic range*** of each fossil species on the scale. This is simply the stratigraphic interval bounded by its lowermost and uppermost occurrence. Species ranges of a typical stratigraphic column are shown in Figure 9-6.

A common method of correlation uses upper and lower boundaries of the stratigraphic range of a single species in two or more stratigraphic sections. Under certain circumstances, such "earliest" or "latest" appearances indicate approximate time equivalence. They may, however, represent different times because of species migration or differing time of extinction from place to place. Palmer (1965), for example, has demonstrated large-scale migration of Late Cambrian trilobite faunas of the Great Basin region of the United States. Each of the three faunas shown in Figure 9-7 apparently arose from an offshore stock. Detailed correlation reveals that, after its appearance, the second fauna migrated from deep water toward the shoreline, apparently displacing the preceding fauna as it moved eastward. After undergoing considerable evolution, it was displaced by eastward migration of the third fauna. Thus, the biostratigraphic unit defined by each of the three faunas has non-synchronous boundaries. The probable reason that migration was not an "instantaneous" event in geologic time was that displacement took place gradually and as the result of ecologic competition. There is no evidence that faunas changed in response to environmental changes, for major faunal transitions do not coincide with major changes in sediment character.

A correlation method that has historical importance is based on the maximum abundance, or "acme," of a species in its stratigraphic range. The importance of local environment in determining abundance in life and extent of preservation after death invalidate this approach for correlation over large geographic distances. In a particular place, however, where uniform environmental conditions may have prevailed, correlations judiciously based on maximum abundance may be reasonably accurate.

Obviously, reliance on more than one species would tend to improve the accuracy of correlations. Most recognized biostratigraphic zones are based on the occurrence together of two or more species. Such a zone, known as an ***assemblage zone,*** is customarily named for a single characteristic genus or species, (which need not necessarily be present throughout the entire horizontal or vertical extent of the zone).

PERCENTAGE OF COMMON TAXA

Some correlations have been based on comparison of a local fossil assemblage with "standard" assemblages from regions in which stratigraphic relationships are better understood. Similarities are usually evaluated in terms of numbers of taxa (usually species) found in both the assemblage in question and a standard assemblage. For example, the percentage of taxa of either assemblage also found in the other as-

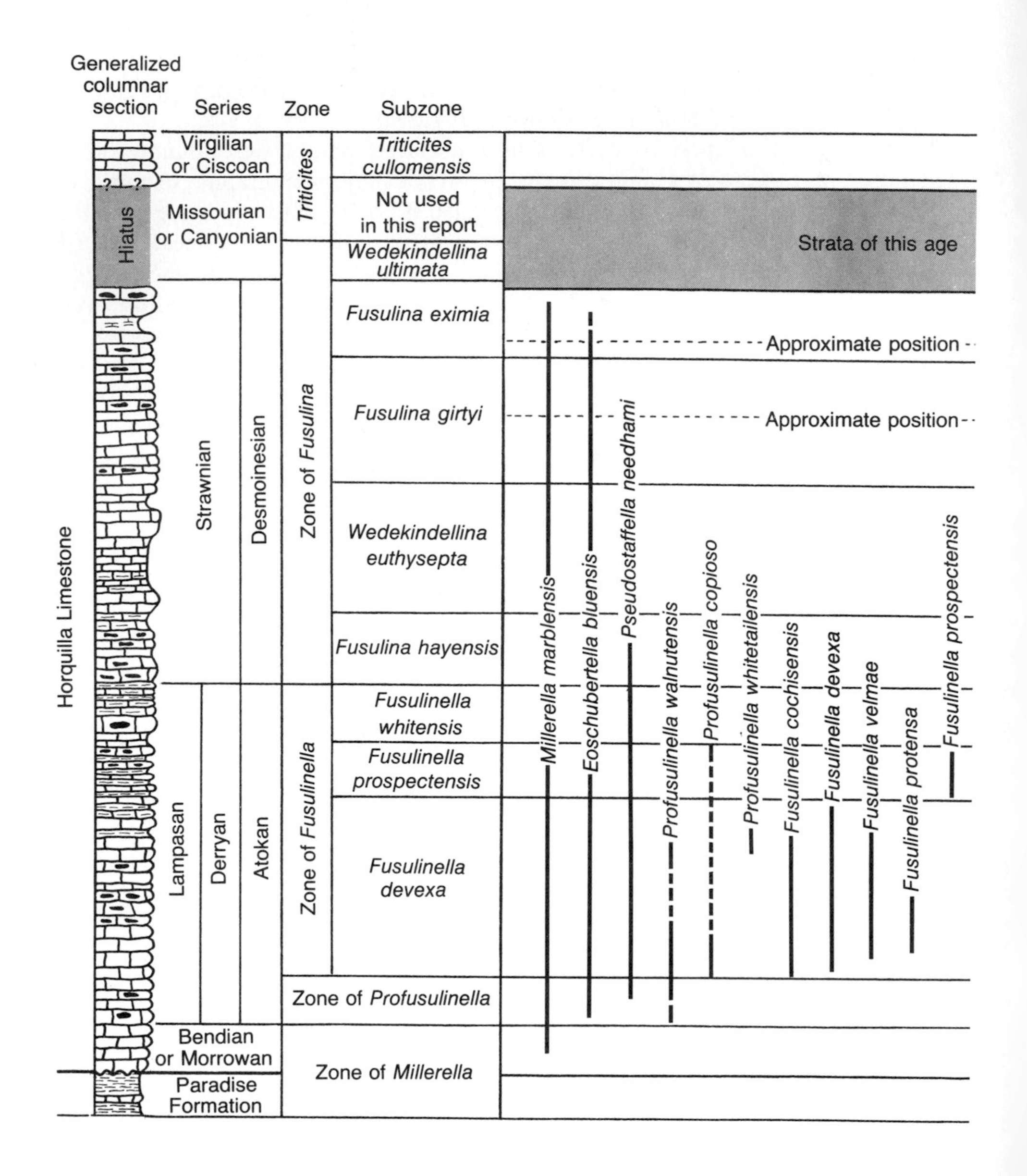

FIGURE 9-6
Generalized section of the lower and middle parts of the Pennsylvanian Horquilla Limestone of Arizona. (From Ross and Sabins, 1965.)

Stratigraphic range of Fusulinid species

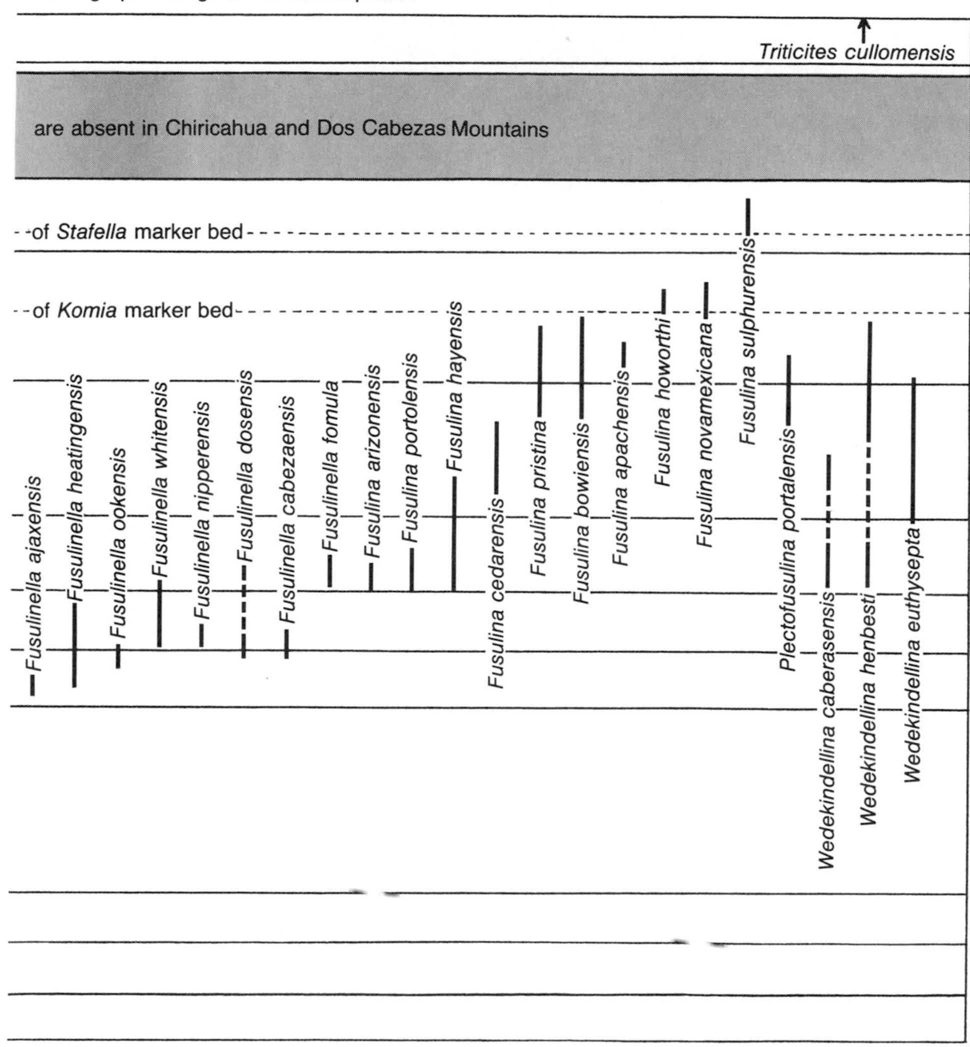

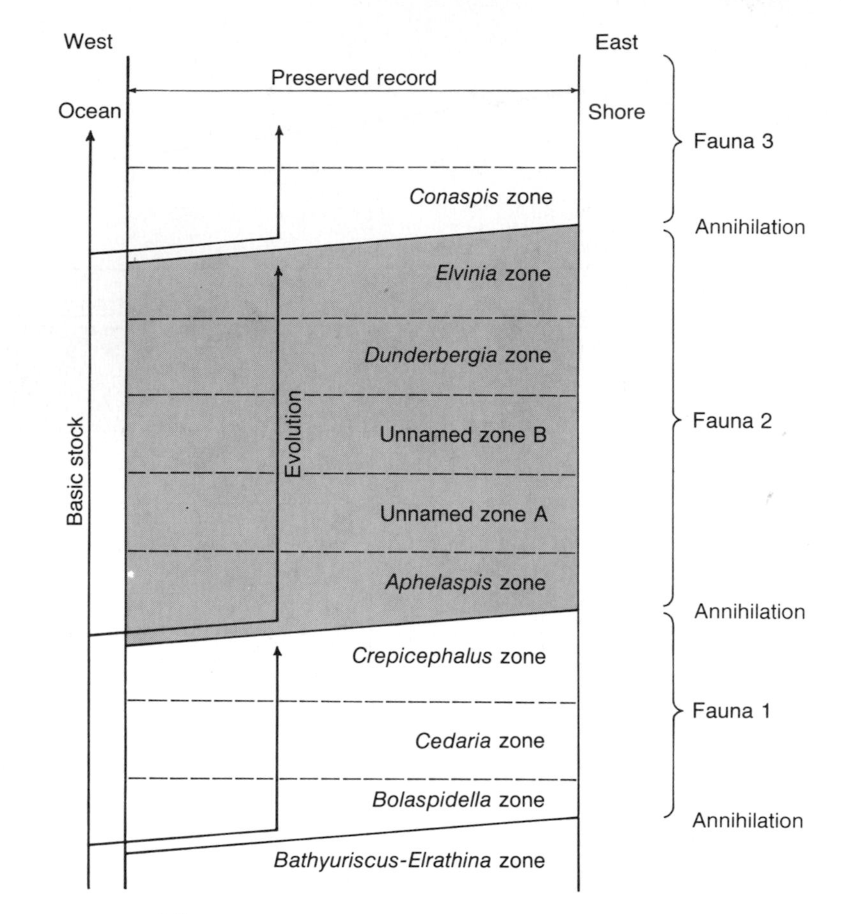

FIGURE 9-7
Schematic diagram showing successive invasions of shallow Late Cambrian seas of the western United States by trilobite faunas evolving from a basic parent stock in deeper water to the west. Horizontal dashed lines represent time. (From Palmer, 1965.)

semblage may be calculated and used as an index of comparison. An alternative is to use the ratio between number of common taxa and total number of unique taxa in the two assemblages.

Simpson (1960a) has advocated use of the following index for comparison of two fossil faunas:

$$\frac{\text{Number of common taxa}}{\text{Number of taxa in the smaller fauna}} \times 100.$$

This index is especially meaningful when one fauna contains many more taxa than the other. (The sparser fauna usually represents less complete preservation of a community.) Simpson has also discussed methods for inclusion of relative abundance data, which are especially important in comparison of closely similar faunas.

Correlations based on percentages of common species can be made without concern for stratigraphic ranges of taxa, but accuracy can be greatly reduced by ecologic biases. It is therefore usually necessary to qualify purely statistical conclusions, depending upon knowledge of biozone ranges and species ecology.

INDEX FOSSILS

Taxa found to be especially useful in correlation are commonly referred to as ***index fossils,*** or ***guide fossils.*** The attributes of an ideal index fossil are: wide geographic distribution, ecologic tolerance, abundance, rapid evolutionary rate, and distinct morphologic features. Index fossils are especially useful for interregional and intercontinental correlation.

Floating and swimming species are generally the best marine index fossils because their occurrence in life is usually widespread and independent of local benthonic conditions. Good examples are provided by many Ordovician and Silurian graptolite taxa and Mesozoic ammonite taxa. Benthonic groups, such as Cambrian trilobites and late Paleozoic fusulinid foraminiferans, have also produced important index fossils.

Unfortunately, a second, spurious usage of the term "index fossil" has become common in some geologic circles. In this usage, which has value in geologic mapping, a taxon is described as an "index fossil" if it is especially common in a local rock-stratigraphic unit and can therefore be used in identifying isolated outcrops of the rock unit when lithology alone is inadequate. This use of the term is confusing and should be avoided.

MORPHOLOGIC FEATURES

Sudden morphologic changes within species and genera can be used profitably for correlation, especially of rocks and sediments of Cenozoic age. Among the most spectacular examples are those involving shell-coiling direction (dextral versus sinistral) in planktonic foraminiferans. Coiling reversal in *Globorotalia menardii* has been widely used for recognition of the Pliocene-Pleistocene boundary in deep-sea-sediment cores (Figure 9-8). Shell-coiling direction of planktonic foraminiferans has also been used for correlation of older Cenozoic deposits. Coiling direction appears to be affected by temperature, and widespread reversal of coiling direction is apparently a response to a major change in climate. It seems likely that coiling direction itself has no adaptive significance. It is probably a nonadaptive manifestation of a single gene or group of genes that also controls temperature adaptation (Mayr, 1963, p. 236).

The great utility of coiling-direction reversals for correlation stems from the fact that reversal is apparently triggered suddenly and occurs over broad geographic areas.

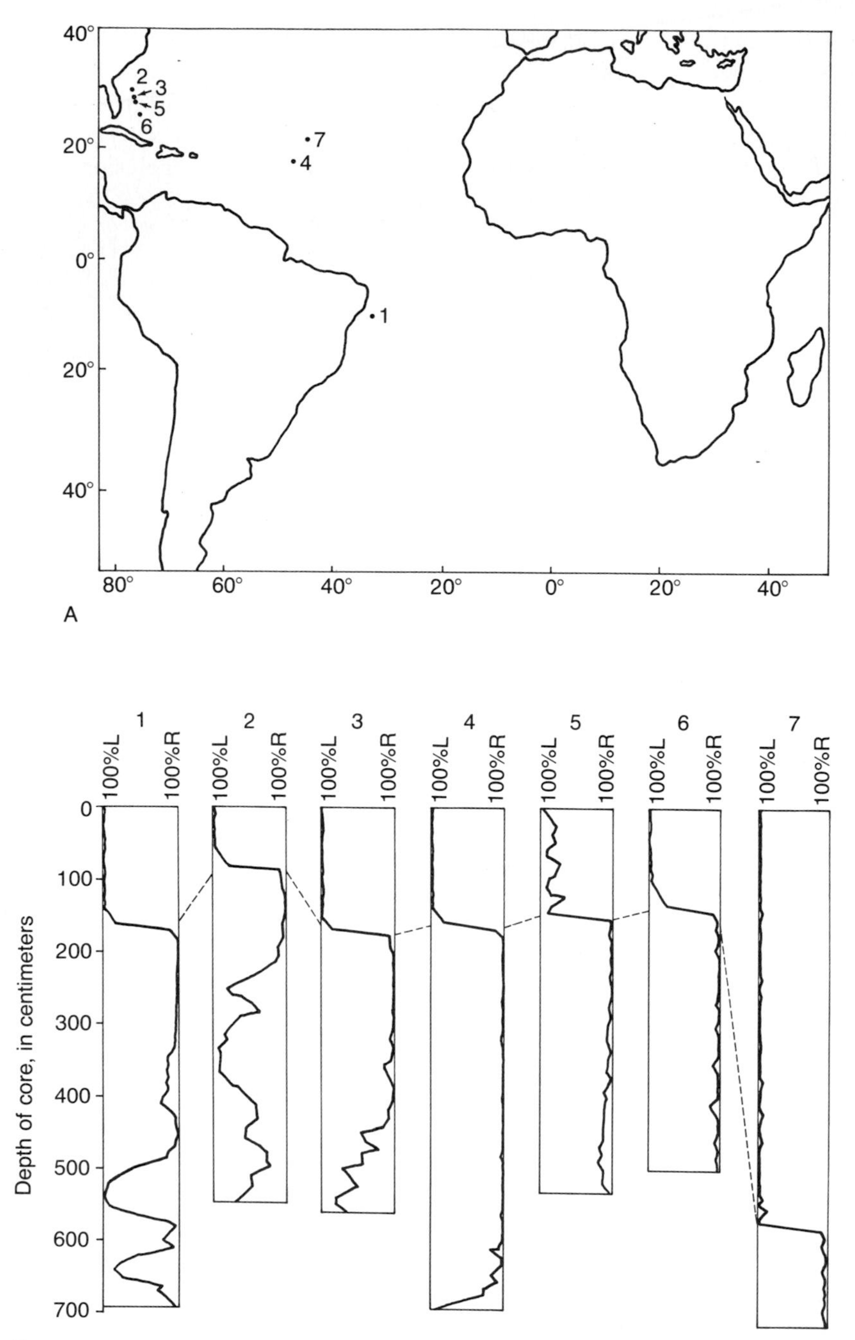
40°
20°
0°
20°
40°
80°
60°
40°
20°
0°
20°
40°
1
2
3
4
5
6
7
A
100%L
100%R
Depth of core, in centimeters
0
100
200
300
400
500
600
700
B

FIGURE 9-8
Changes in direction of shell coiling among fossil foraminiferans which are assigned to *Globoratalia menardii,* but which may actually belong to more than one species. A: Places at which sediment cores were taken. B: Coiling direction of fossils in the sediment cores. L is coiling to the left; R, to the right. The Pliocene-Pleistocene boundary, as delineated by the reversal in coiling direction, is indicated by the dashed lines between cores. (From Ericson, Ewing, and Wollin, 1963. Copyright © 1963 by the American Association for the Advancement of Science.)

ECOLOGIC PATTERNS

Under certain circumstances, facies and their characteristic fossils exhibit stratigraphic patterns that permit correlation of two or more sections. A simple example is illustrated in Figure 9-9, in which a hypothetical stratigraphic cross-section reveals a period of ***transgression*** followed by a period of ***regression.*** During deposition of the hypothetical rock units, land existed in the west and the ocean lay to the east. The shoreline shifted first to the west, as the sea encroached on the land, and later to the east, as the sea receded from the land. Each of the marine rock units represents a distinct facies, with its own characteristic lithology and fossil content. The six marine

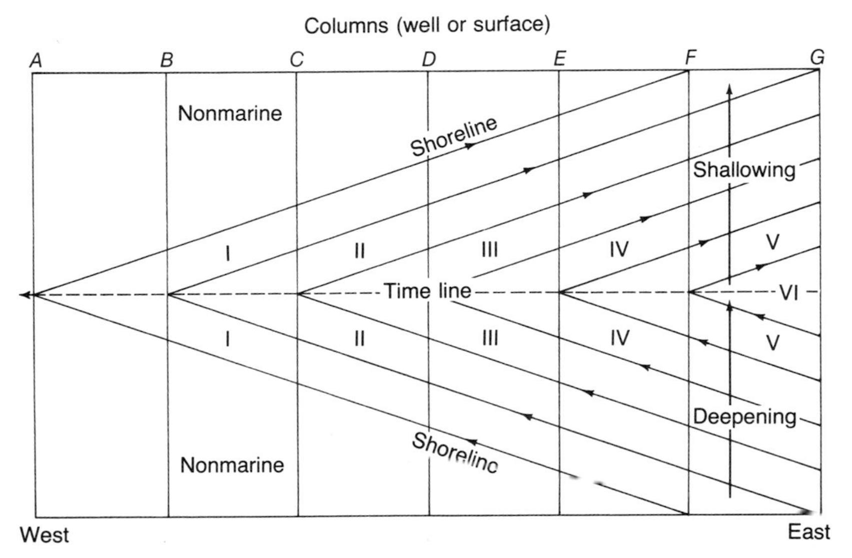

FIGURE 9-9
Cross-section of a hypothetical depositional sequence produced by transgression and regression. The time line (line connecting points of deepest deposition for each column, *A* through *G*) represents the time of maximum westward transgression. (From Israelsky, 1949.)

facies represent adjacent seafloor environments that trended parallel to the shoreline. The point marking the westernmost extent of each facies represents the time of maximum westward transgression of the sea. A line drawn through these points for all facies is a correlation line, representing a single time in the history of development of the sequence. In most geologic settings only isolated outcrops or well cuttings are available for study. Still, the single-time surface can be approximately determined by correlating all sections at the facies deposited in the deepest water.

QUANTITATIVE CORRELATION METHODS

Shaw (1964) has presented a statistical method for correlation based on detailed knowledge of groups of local sections containing numerous fossil species in common.

Each of two local sections to be correlated is first plotted on a different axis of a graph. The units of the graph axes represent distances, usually measured in feet, from the bases of the sections to their tops. Each species occurring in the two sections can then be represented on the graph by points representing the base of its range, the top of its range, or both, in the two sections (Figure 9-10). If species ranges in the two sections (X and Y) are identical, the line connecting all points will be straight and will form an angle of 45° with each axis.

Let us postulate that the *times* of appearance and disappearance of each of the nine species common to both sections were identical at the two locations. We will retain this assumption throughout our discussion. The line connecting points will then be an approximate correlation line. The reason that it is not necessarily an exact correlation line, even though the sections are identical, is that our hypothetical situation indicates only that each of the plotted *points* is a point of correlation. If the rates of rock accumulation differ during a time interval between two points, then the real correlation curve, if drawn, will not follow a straight line between the two points. The limiting cases in an analysis like this would be those in which all rock accumulation of the interval between two points occurred at one section before there was any accumulation in the other section. These cases, which are shown in Figure 9-11, would enable the investigator to determine the *maximum* error that could exist by simply connecting the two points with a straight line. Evaluation of the *actual* amount of error is a matter of judgment, based on geologic information. Shaw believes that use of a straight line is reasonable for most sections containing marine sediments deposited in what are now the central regions of continents.

The phrase "rock accumulation" is used instead of "sediment accumulation" because different types of sediment are compacted to different extents after deposition. Rock thickness in the stratigraphic record is therefore not proportional to original sediment thickness in a constant way.

Consider the case in which rates of accumulation are approximately constant for each of two sections but differ between the sections. We are still assuming that the

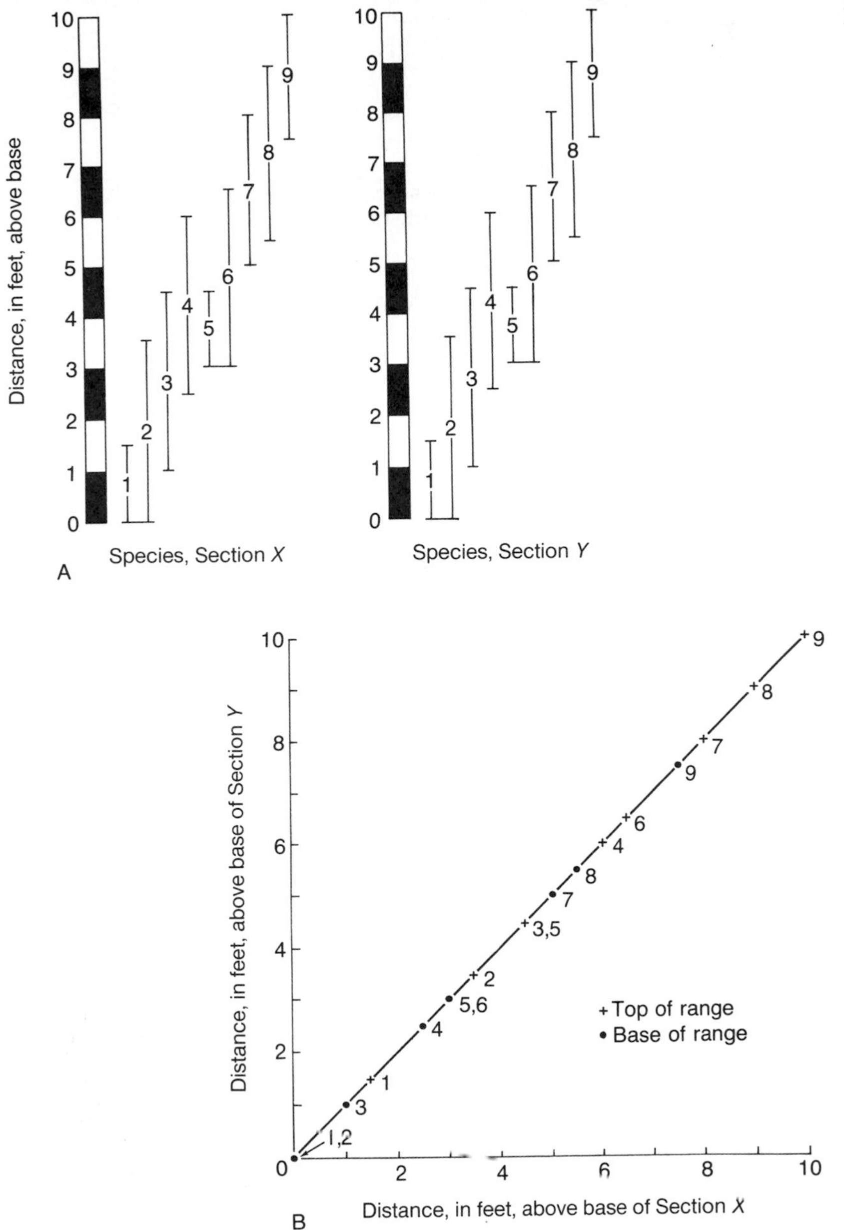

FIGURE 9-10
Confirmation, through use of a graph, that ranges of nine fossil species are identical in two hypothetical sections. A: Diagrams showing vertical ranges of the species in sections. B: Graph on which has been plotted the top and base of the range of each species in the two sections. Connecting all the points plotted gives a straight line at 45°. (From Shaw, 1964. Copyright © 1964 by McGraw-Hill Book Company.)

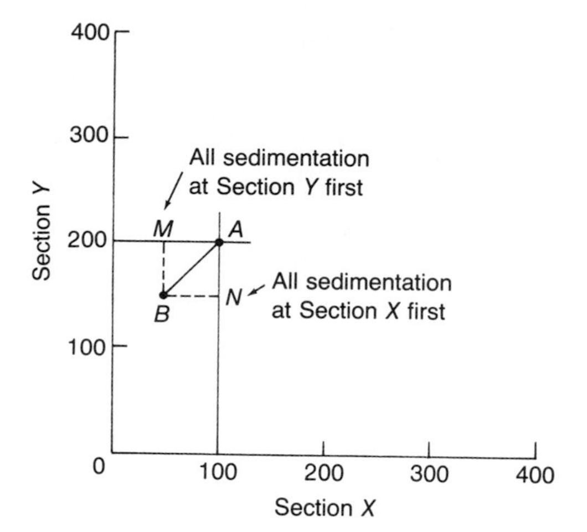

FIGURE 9-11
Portion of a graph like that in Figure 9-10, showing limiting cases of completely noncontemporaneous deposition in two sections during a certain time interval. *A* and *B* are two recognizable events, such as the appearance or disappearance of a species, that mark the beginning and end of a time interval in Sections *X* and *Y*. Interpolation of a straight line between *A* and *B* is perfectly accurate only if rates of deposition were identical in the two sections throughout the interval. The distances *BM* and *BN* show the maximum possible error that could be introduced if the assumption of equal rates of deposition were wrong. This maximum possible error would exist if all sediment that accumulated during the interval in either Section *X* or Section *Y* was deposited before deposition began in the other section. In either case the correlation error would be 50 feet. (Modified from Shaw, 1964. Copyright © 1964 by McGraw-Hill Book Company.)

ranges of a species in the two sections represent a single time interval. The species ranges will be spatially longer in one section than in the other and the line connecting points on the graph will no longer form an angle of 45° with the axes (Figure 9-12).

For any given time there is a ratio between the rates of rock accumulation in two sections being studied. Often this ratio may change. Shaw has found that a "dog leg" pattern (Figure 9-13) is typical of graphs used to compare ranges of species common to two sections. In other words, changes in relative rates of accumulation between sections tend to be abrupt. Otherwise, curved graphic patterns would be more common.

Statistical theory can be applied to the raw stratigraphic data to compute "best-fit" correlation lines for graphic representation. The statistical significance of a correlation line (an estimate of the probability of error) can also be calculated.

Having made correlation-line comparisons of several local sections, it is useful to construct what Shaw calls a "composite standard." Usually, the section that is most complete (having no major depositional breaks) and contains the largest biota is used

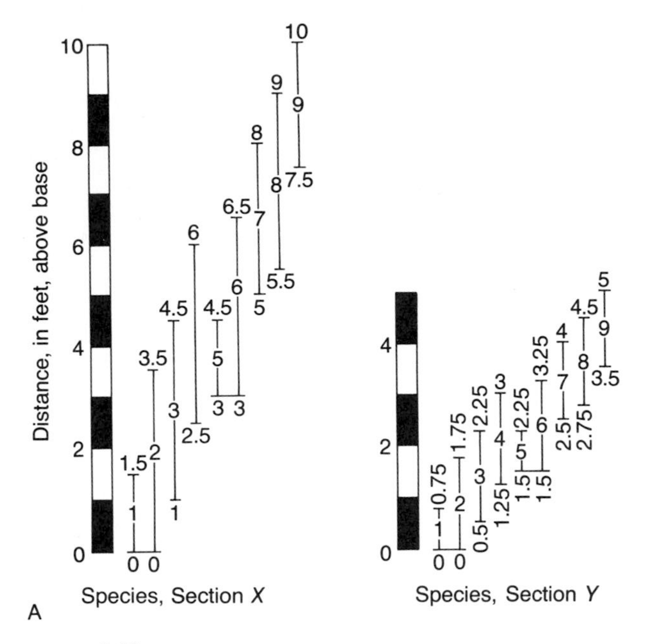

FIGURE 9-12
Diagrams and graphs like those in Figure 9-10, of two sections representing the same time interval and having identical fossil species, but different rates of rock accumulation. (From Shaw, 1964. Copyright © 1964 by McGraw-Hill Book Company.)

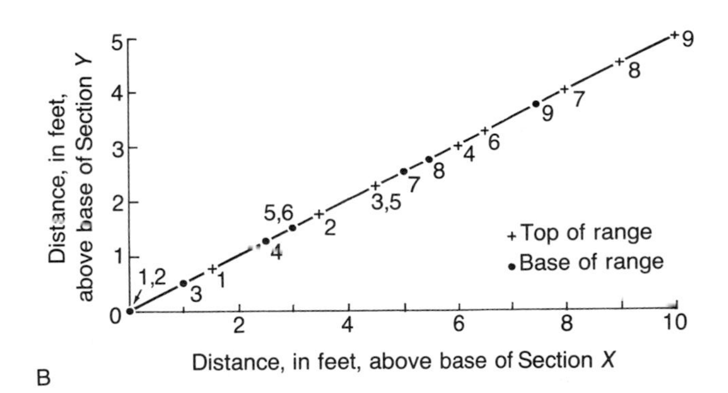

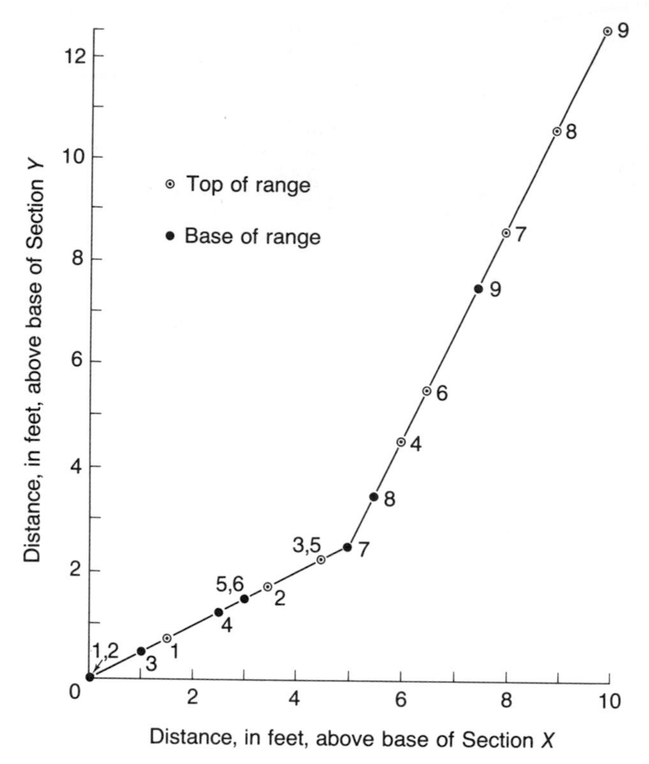

FIGURE 9-13
Graph of species ranges in two sections illustrating a sudden change in relative rate of rock accumulation. (From Shaw, 1964. Copyright © 1964 by McGraw-Hill Book Company.)

as a base. Fossil ranges of all other local sections can be projected onto the base section. This is accomplished for each of the other sections by applying the equation that represents its line of correlation with the base section. In effect, differences in rates of rock accumulation among all the sections and differences in time of local appearances and disappearance of species are cancelled out when the data are thus combined. In a composite standard, such as the one in Figure 9-14, the relative ranges of various species are estimated for the entire local area. In fact, relative ranges shown for species in a composite standard will nearly always be incomplete. However, most will approach the actual relative ranges more closely than will the stratigraphic ranges of any single section. The composite standard is an abstraction based on rates of rock accumulation in the base section. Its vertical scale is, therefore, not in units of feet or time. Its units are undefined.

If constructed from adequate data, the composite standard can be extremely important to stratigraphic, paleoecologic, and evolutionary studies. It can be correlated with a single section or distant composite standard by the same technique used for comparing local sections. The composite standard makes quantitative correlation between widely separated regions possible. Presumably, it can also be used for intercontinental correlation, using ranges of genera rather than species. By comparison of a series of sections with the composite standard, geographic migration of a species could potentially be demonstrated. In a similar manner, relict populations might be recognized in certain regions. In addition, a composite standard nearly always pro-

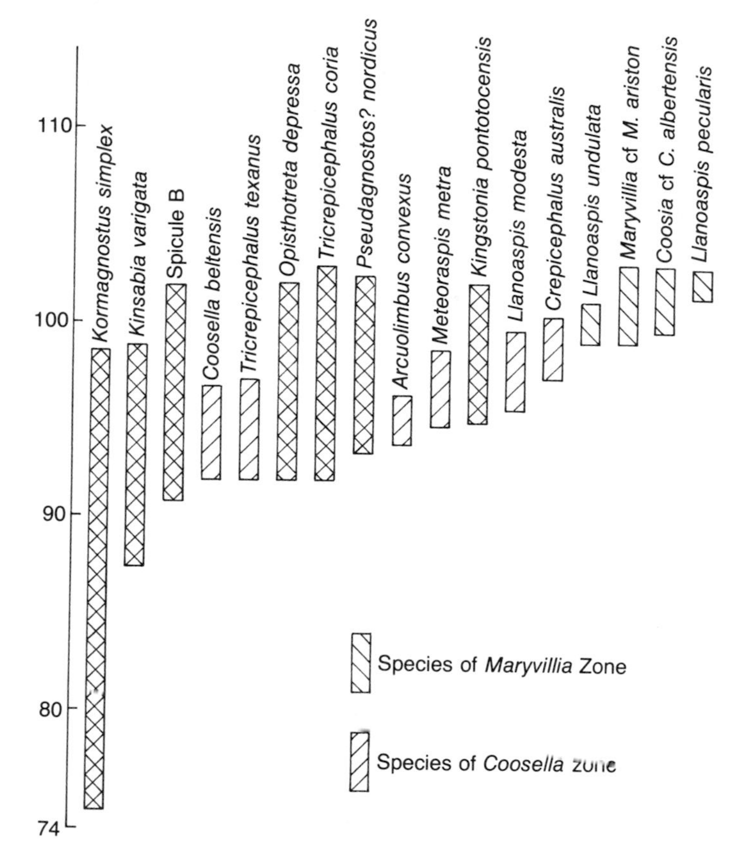

FIGURE 9-14
Graphic representation of composite standard ranges for the seventeen most common species of the *Coosella* and *Maryvillia* (trilobite) zones of the Cambrian Riley Formation, based on eight local sections. (From Shaw, 1964, copyright © 1964 by McGraw-Hill Book Company; data for local sections from Palmer, 1955.)

vides a much clearer picture of relative times of origin and extinction of species than can be provided by any single section. Its use, therefore, makes recognition of evolutionary lineages and phylogenetic relationships less speculative than they would otherwise be.

Time and Time-rock Units

Very early in the history of modern geology it became convenient to divide regional stratigraphic sequences into large units that could be distinguished from one another on the basis of fossil content. Thus, the stratigraphic ***systems,*** from Cambrian through Quaternary, came to be recognized over the course of several decades during the nineteenth century. Delineation of these did not proceed according to any well-planned scheme, nor were they established in order of their temporal relationships. There was a natural tendency for early workers to establish system boundaries at major faunal and floral breaks. Many of the boundaries seem especially well placed, even today, for some breaks that were chosen represent distinct evolutionary changes or extinctions that occurred during short time intervals on a worldwide scale. Nevertheless, there is an unfortunate tendency among students to accept boundaries between geologic systems as being natural breaks carefully chosen at the most appropriate possible stratigraphic levels. In fact, other boundaries of equal or even greater utility could have been chosen. There is a large measure of arbitrariness in the stratigraphic procedures by which our geologic time scale has been established.

The geologic time scale, of course, remained a *relative* time scale for many years. Not until the advent of radiometric age dating (just after the turn of the century) could the scale be related reasonably accurately to *absolute* time, measured in years.

Geologic systems are still defined on the basis of ***type sections,*** which are local stratigraphic sections in the ***type areas*** in which the systems were first recognized. Most type areas are in Europe. One problem with this traditional practice is that not all type sections represent the most useful rock sequences for definition of their respective systems, as now recognized throughout the world. Many modern workers support the idea of redefining the systems by defining their *boundaries* in selected areas in which the boundaries can be placed within fossiliferous rock sequences formed by nearly continuous deposition. Why such an idea is attractive will be shown in the following paragraphs.

In order to understand the problems of extending a system from its type section to other biotic provinces or continents, a diagram like the one in Figure 9-15 is useful. As long as one system is defined in one area, and an adjacent system in another area, the systems may in fact contain rocks that overlap in age. It is just as likely that an age gap may exist between two systems. In other words, even the type sections of geologic systems, as now defined, do not neatly divide the stratigraphic column. The use of fossils to correlate rocks in a type section with rocks of other regions is not perfectly

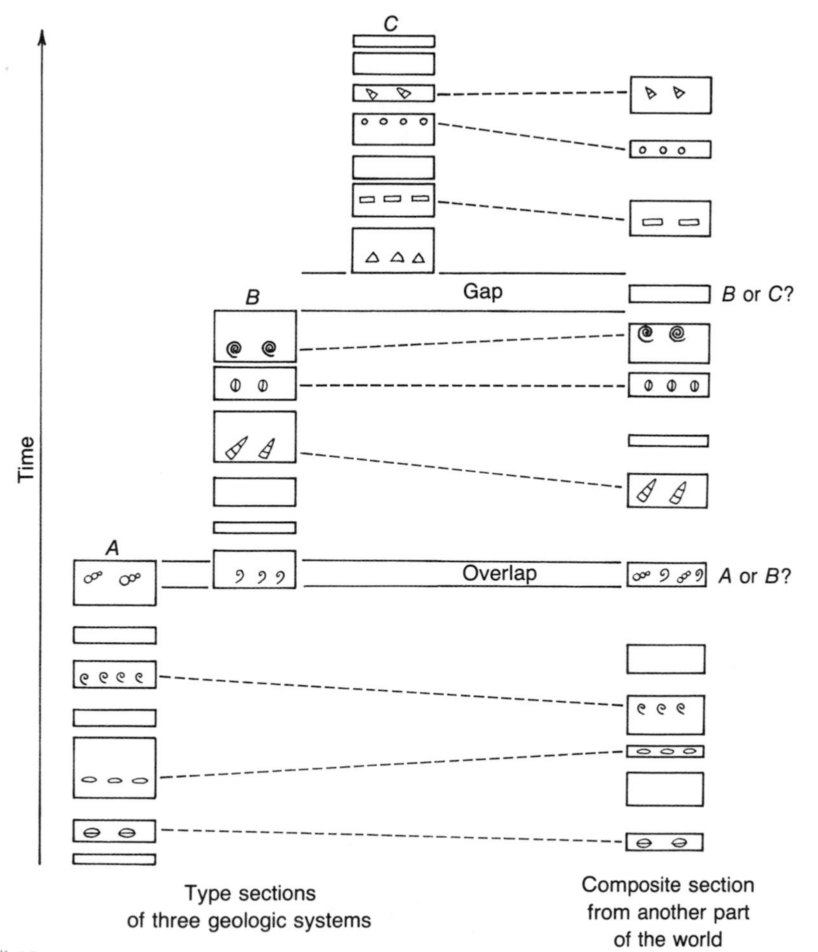

FIGURE 9-15
Problems of correlation with type sections. *A, B,* and *C* are type sections of three geologic systems in different areas. The vertical scale is based on time, rather than on sediment thickness. On the right is a composite section showing the systems in another area. Gaps, due to nondeposition or erosion, are found in all these sections, and are diagrammed as empty boxes. Lines of fossil correlation (dashed) show varying degrees of accuracy, time lines being horizontal. Systems *A* and *B* overlap in time; there is a gap in time between systems *B* and *C*. Even if perfect correlation were possible, parts of the composite section on the right could not be assigned to a system.

accurate, as we have seen, and the inherent inaccuracies further serve to reduce the degree of synchroneity of recognized system boundaries throughout the world.

Systems have been subdivided into ***series,*** and series into ***stages.*** All these subdivisions are most commonly recognized by using fossils. All three of these kinds of units—systems, series, and stages—have traditionally been called ***time-rock units,*** implying that they are ***isochronous units,*** that is, units bounded by surfaces that are everywhere the same age. This view is taken in the Code of Stratigraphic Nomenclature mentioned at the beginning of this chapter. "Time-rock unit" used in this way is a somewhat misleading phrase. Because systems, series, and stages are all recognized primarily on the basis of biostratigraphic equivalence (except in their type areas) the recognized subdivision boundaries are no more synchronous than are the biostratigraphic boundaries used to define them.

This problem is recognized by most modern workers. But the question is really one of accuracy. Some workers, such as Hedberg (1951), argue that because systems and series represent larger time intervals than zones, the percentage of error in their recognized boundaries is smaller. They believe that most system boundaries throughout the world are so nearly synchronous that the phrase "time-rock unit" is appropriate. Weller (1960) advocates retention of the phrase "time-rock unit" to describe systems and series, but only as long as it is understood that most units to which the phrase is applied are only approximately isochronous. However, Dunbar and Rodgers (1957, page 293) reject the attempts of many workers to use the phrase "time-rock unit" in this way: "Historically, series and systems, as they have always been used by stratigraphers, are as dependent on fossils as zone, and if zone is a biostratigraphic term, so are they." The question of whether systems and series should be regarded as time-rock units is still widely debated.

The geologic time scale itself is an abstraction. The time units of the scale are defined by the boundaries between systems in their type areas. As we saw by looking at Figure 9-15, these time units may overlap or be separated by gaps. Thus, though time is a continuum, our geologic time scale, as now recognized, does not neatly subdivide this continuum.

It is easy to understand why many workers now desire that geologic systems be redefined in places other than the original type areas using *boundaries* within fossiliferous rocks, rather than the rock units themselves. Thus, a system could have its lower boundary defined in one place and its upper boundary defined elsewhere. The type section for each boundary could be so selected that the boundary would be defined by good index fossils contained within rocks formed by nearly continuous deposition. The new systemic boundaries could be so chosen as to approximate the traditional ones. Some of the old type sections might be retained but used to define a boundary between two systems rather than a system itself. Series, which also have type areas, might be subjected to similar revision, though many are presently defined in areas in which they are represented by well-exposed, fossiliferous strata.

An alternative proposal has been presented by Bell and colleagues (1961) in a report dissenting from some of the ideas and rules set forth in the Code of Stratigraphic

Nomenclature. This minority-view proposal points out that radiometric age dating has provided a time scale based on years that is essentially independent of the so-called "relative time scale" based on presence of fossils and physical stratigraphic evidence. It suggests that the term "chronologic" be restricted to the radiometric time scale and that the term ***geochronologic*** be applied to the traditional time scale of eras, periods, and epochs. This proposal also suggests that the rock bodies (systems and series) representing the geochronologic units be called ***chronostratigraphic units.*** Chronostratigraphic units might then be defined by using biostratigraphic units such as zones, as well as any other pertinent criteria. Their boundaries would not have to coincide with biostratigraphic boundaries, but commonly would. The time span of each geochronologic unit would be represented by widespread rocks (the chronostratigraphic unit), plus all the gaps in these rocks owing to nondeposition or erosion. The type section is maintained in this scheme, but its limitations are recognized. In a sense, the type section in this scheme is like the type specimen of a species, the chronostratigraphic unit is like a hypodigm (page 113) and the geochronologic unit is like a species. This analogy should be clear if the reader understands the taxonomic concepts presented on pages 113 and 114.

Accuracy of Correlation

As implied by the controversy over designation of systems, series, and stages as time-rock units, there is considerable debate regarding the accuracy of correlation by fossils.

As we have seen, faunas and floras of past times tended to be segregated into somewhat isolated geographic realms, just as they are today. The relatively high degree of biotic uniformity within provinces makes correlation more accurate within provinces than between them. Correlation between provinces is commonly used to establish large-scale, approximately synchronous "surfaces" for reconstruction of paleogeography for brief intervals of geologic time. Often, two or more continents are included. Because paleogeographic maps typically portray large areas, they seldom provide local detail. Correlations used for their construction need not be nearly as accurate as those used for local environmental studies.

Most intercontinental correlations are based on index fossils, rather than on comparison of large fossil faunas or floras. Commonly, the index fossils in the two areas being compared belong to two species of the same genus; sometimes they belong to the same species. If Simpson's estimate of 0.5–5 million years for average species duration is correct (Simpson, 1952), this range of values establishes the approximate accuracy of correlations based on single fossil species, assuming no other evidence is applied to a given problem. Actually, many marine invertebrate species have lived for longer than 5 million years (page 323). An idea of the accuracy of correlation based on genera, rather than species, is given by the estimate of 8 million years for duration of Cenozoic carnivore genera and 78 million years for Cenozoic bivalve genera

(Simpson, 1953). Certainly, there has been much variation in the duration of genera and species throughout geologic history. Some higher categories have been characterized by much longer mean duration of their component genera and species than others. It is obvious why one prime requisite of a good index fossil is a relatively short stratigraphic range. The statistical value of using several taxa for a single correlation should also be readily apparent.

Many intercontinental correlations are accomplished in a series of two or more steps, one biota being judged to have existed contemporaneously with a second, the second with a third, and so on. Thus, the third biota is approximately correlated with the first, though the two may have few (or no) species or genera in common. Each added step in a sequence of equidistant correlations doubles the inaccuracy of the total correlation, all other factors being equal. A great deal of subjectivity enters into the making of most correlations in choice and weighting of both the taxa and the events (such as first appearances and last appearances) that are used.

The statements of many workers suggest that system boundaries traced around the world are nearly synchronous surfaces. Other workers, such as Weller (1960, page 562), are of a different persuasion:

> Because of more or less effective isolation, the organic relations between different faunal or floral provinces are likely to be more remote than those connecting different parts of a single province. Consequently, comparisons and correlations must be made on the basis of more general evolutionary changes, and the results are less precise. Just what allowance should be made on account of these less direct relationships is uncertain. . . . On the whole, interprovincial correlations based on paleontology are probably not accurate to much less than one-quarter of a geologic period.

For most correlations within a single biogeographic province, stratigraphic sections are separated by intervals of a few kilometers (seldom exceeding 100 kilometers) and the total area of study has dimensions measured in tens or hundreds of kilometers (seldom exceeding 2,000 kilometers). Correlation on this scale can often be based on biotas possessing many common faunal or floral elements. The required accuracy of correlation is, however, also much greater than for correlation between provinces. Within a single province, extremely accurate correlation is often desired for local paleoecologic studies. Rates of species evolution is too slow for changes in fossils to be of value in most such local studies. Other stratigraphic features, such as nearly isochronous rock units, may prove to be more useful in correlation. Beds formed by volcanic ash-falls or by other apparently widespread deposition over short lengths of time may be especially valuable and are commonly referred to as ***key beds.***

In fact, under certain circumstances, the presence of similar biotas in rocks of separate areas may indicate *different* ages for the rocks. For example, suppose a worker identifies a certain fossil fauna as a marine intertidal community that lived in a narrow belt along the ancient shoreline. Discovery of this biota in two isolated outcrops 300 kilometers apart along an axis perpendicular to the known shoreline orientation would strongly suggest different ages for the two outcrops.

Some workers have expressed hope that improved radioactive dating techniques and increased numbers of absolute age determinations may provide much better correlation accuracy than has been attained with the use of fossils. Unfortunately, this sanguine outlook is largely unjustified except for very young rocks.

The estimated ***precision*** of a radiometrically determined date is commonly listed as a standard error (a "plus-or-minus" value following the date). Lack of precision is the result of both statistical variation among samples of the same origin and errors in measurement. In addition, most measured dates are subject to considerable errors in ***accuracy,*** primarily because of loss of the measured daughter product of radioactive decay through weathering, heating, or alteration of the radiometric solution, for example.

The absolute geologic time scale shown in Figure 9-16 is based on selected age determinations made from samples that are believed to have remained as virtually closed systems. The standard errors attached to particular values are therefore the approximate total errors for most determinations. These errors are typically from 1 to 2 million years for Cenozoic dates but increase for older dates, averaging about 10 million years for dates of lower Paleozoic rocks. We can assume that fossil correlation is also poorer for older rock units because of poorer preservation and less widespread rock exposures. Whether correlation based on fossil evidence or radiometric age dating is more reliable varies from study to study.

There are also difficulties in attempting to relate radiometric age determinations to established biostratigraphic subdivisions. One major problem is that fossils are seldom directly datable by radiometric methods. Many measured dates are for minerals formed at the time of deposition of sediments containing fossils, but others are for igneous rock units whose stratigraphic relationships only indicate their approximate time relationships to fossil-bearing sedimentary rocks. To arrive at the average dates shown in Figure 9-16, Harland et al. (1964) chose only measured dates that represent minerals whose origin can be closely fitted into a biostratigraphic framework.

Additional problems arise through poor biostratigraphic correlation of those fossils that are closely associated with radiometrically dated minerals. Ideally, it would be desirable to obtain the absolute ages of all system and stage boundaries in type areas. Because absolute age determinations are restricted to certain mineral occurrences and geologic conditions, however, this goal can seldom be attained. The dates of most system and stage boundaries of the time scale of Figure 9-16 have been estimated from age determinations made far from the type locations of the various systems and stages.

Radiometric age determinations are of little value in local correlations, especially for rocks well back in the stratigraphic record. Most local environmental and paleoecologic studies require much greater precision than is provided by radiometric measurement, even in sediments of Tertiary age, where nearly all measurements are subject to errors of 0.5 million years or more.

There is some hope that quantitative biostratigraphic techniques of the type introduced by Shaw (1964) may improve correlation accuracy in the future. The primary roadblock is the extremely detailed stratigraphic measurement and fossil collection required. An astronomical number of man-hours would be required for

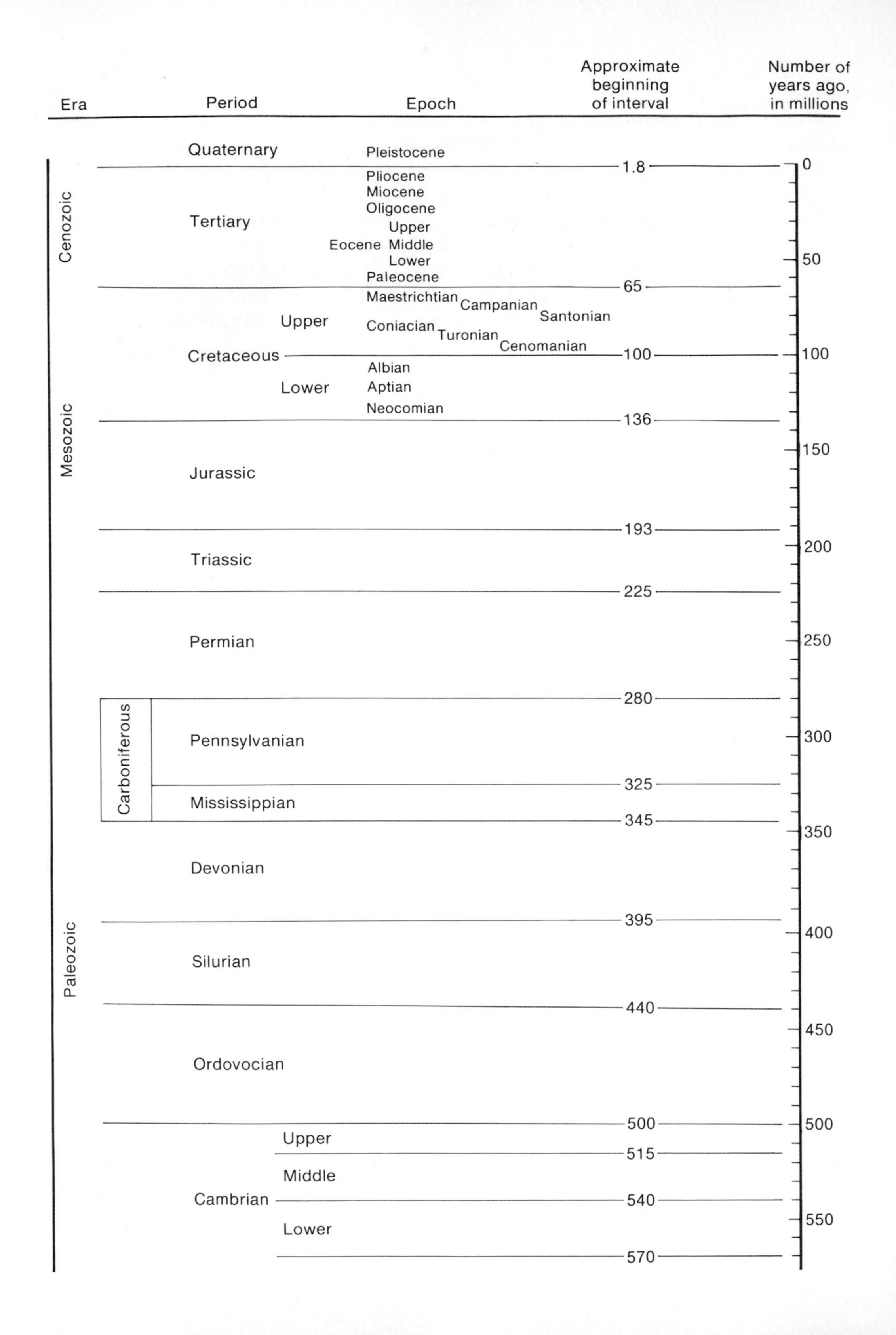
Era
Period
Epoch
Approximate beginning of interval
Number of years ago, in millions
Cenozoic
Mesozoic
Paleozoic
Quaternary
Pleistocene
Tertiary
Pliocene
Miocene
Oligocene
Eocene Upper
Middle
Lower
Paleocene
Cretaceous
Upper
Maestrichtian
Campanian
Santonian
Coniacian
Turonian
Cenomanian
Lower
Albian
Aptian
Neocomian
Jurassic
Triassic
Permian
Carboniferous
Pennsylvanian
Mississippian
Devonian
Silurian
Ordovocian
Cambrian
Upper
Middle
Lower
1.8
65
100
136
193
225
280
325
345
395
440
500
515
540
570
0
50
100
150
200
250
300
350
400
450
500
550

FIGURE 9-16
Absolute geologic time scale. (Data largely from Harland et al., 1964.)

extensive application of Shaw's method, especially for the establishment and correlation of many composite standard sections.

Furthermore, some workers refuse to accept the validity of quantitative approaches to correlation. Their arguments are similar to those of taxonomists who attack the validity of numerical taxonomy (page 142). The question raised is whether it is valid to give equal statistical weight to all fossil taxa in correlation (or to all measured morphologic characters in taxonomy). Just as some taxonomists argue for the special utility of certain "key characters" in the classification of taxonomic groups, some stratigraphers argue for the special utility of certain index fossils in correlation (see, for example, Jeletsky, 1965). The opposing argument claims that if enough measurements are used, the less meaningful data merely dilute, but do not invalidate, the statistical conclusions. Perhaps the real question is whether the necessary quantities of data can be gathered economically enough to justify widespread use of statistical approaches.

Biostratigraphers are constantly striving to improve correlations and correlation techniques. The assessment of correlation error in this chapter is not meant to belittle the validity or utility of our existing stratigraphic framework, but to alert the student to the potential pitfalls of biostratigraphic correlation and, perhaps, to stimulate him to contribute improvements to existing correlations and correlation techniques.

Supplementary Reading

American Commission on Stratigraphic Nomenclature (1961) Code of stratigraphic nomenclature. *Bull. Amer. Assoc. Petrol. Geol.*, **45**:645-665. (Outlines principles and practices for consistency in classifying and naming stratigraphic units.)

Berry, W. B. N. (1968) *Growth of a Prehistoric Time Scale.* San Francisco, W. H. Freeman and Company. 158 p. (History of the development of the relative geologic time scale.)

Donovan, D. T. (1966) *Stratigraphy.* Chicago, Rand McNally, 199 p. (A concise summary of stratigraphic principles with reference to their historical development.)

Dunbar, C. O., and Rodgers, J. (1957) *Principles of Stratigraphy.* New York, John Wiley & Sons, 356 p. (Contains chapters encompassing biostratigraphy, with many examples from geologic literature.)

Eicher, D. L. (1976) *Geologic Time.* Englewood Cliffs, N. J., Prentice-Hall, 150 p. (A concise, well written treatment of the elements of stratigraphy.)

Krumbein, W. C., and Sloss, L. L. (1963) *Stratigraphy and Sedimentation,* 2nd Ed. San Francisco, W. H. Freeman and Company, 660 p. (A popular text emphasizing principles and incorporating the Code of Stratigraphic Nomenclature as an appendix.)

Matthews, R. K. (1974) *Dynamic Stratigraphy.* Englewood Cliffs, New Jersey, Prentice-Hall, 370 p. (A case-study introduction to modern stratigraphy.)

Weller, J. M. (1960) *Stratigraphic Principles and Practice.* New York, Harper & Brothers, 725 p. (Includes useful discussions of practical stratigraphic techniques).

CHAPTER 10

Paleoecology

The word "ecology" has been flagrantly misused in recent years, during the surge of concern about the quality of our modern environment. To the biologist, ecology is the study of interrelationships between organisms and their environment. Man is only one organism among many, and ecologic studies consider many systems in which he plays at most a minor role. Paleoecology is simply study of the ecology of fossil species.

The reasons for studying paleoecology are many. A major goal of geology is to unravel the history of the earth and its inhabitants. Although most detailed paleoecologic analyses are limited to short geologic time intervals and geographic areas whose boundaries are measured in meters, kilometers, or tens of kilometers, the accumulation of information from many such analyses gives us a general history of life and environments on earth. Because there are gaps in the stratigraphic record and because fossils do not provide complete biologic information, we can never reconstruct this history completely. Nevertheless, paleoecology can provide us with considerable knowledge about the history of life on earth, and we are only in the beginning stages of exploring its possibilities.

A major limitation of paleoecology is our inability to observe, sample, and measure most features of ancient environments directly. To establish frameworks for analysis, we must therefore rely on partial reconstruction of environments by stratigraphic and sedimentologic techniques. Another limitation is imposed by the incompleteness of the fossil record, a fact that was discussed in Chapter 1. Both kinds of limitation will be treated later in this chapter, after the introduction of various biological concepts.

As discussed in Chapter 1, marine animals, most of which are invertebrates, inhabit a vast depositional basin in which an organism's remains are generally more likely to be preserved than on land. The majority of terrestrial organisms are insects, which are not easily preserved. In order to be preserved, the remains of most preservable terrestrial species (chiefly plants and vertebrates) must accumulate in low-lying depositional areas, such as swamps, streams, or lakes. Some species live in or near the environments in which they are fossilized, but others are transported great distances. Nonetheless, many terrestrial fossil biotas are of great paleoecologic importance. Although often transported before burial, plant remains offer valuable evidence of past climates. Because life habits of vertebrates are readily interpreted, well-preserved assemblages of bones and teeth offer excellent opportunities for ecologic analysis. Toward the end of this chapter and in Chapter 12 we will examine some of the fruitful kinds of analyses that have been undertaken in the fields of terrestrial paleoecology and biogeography. Because shelled marine invertebrates have been the subjects of most paleoecologic studies, however, they will be emphasized in our discussion.

Fundamental Ecologic Principles

The largest unit of study considered in ecology is the ***ecosystem,*** which consists of a chosen portion of physical environment plus all of the organisms contained in it. The ecosystem includes all the interactions—chemical, physical, and biologic—that occur within the chosen physical boundaries. thus we mght consider as an ecosystem the biosphere or, just as reasonably, a tiny puddle of rainwater containing two or three protozoan species.

The ***habitat*** is the environment in which an organism lives. It might be a rocky seashore, a grassland, or for a parasite such as a flea, a host species such as a dog. There may be more than one habitat in an ecosystem. The ***ecologic niche*** is often defined as the organism's position in the habitat, including its way of life and the role it plays in the ecosystem. Some workers prefer to define "ecologic niche" in terms of the features of the environment that permit a species to exist. What we are really concerned with is the interaction between a species and its environment. The two types of definition are not fundamentally different, but simply differ in their emphasis. The first stresses the physiologic and behavioral adaptations of the species to the environment, and the second stresses the environmental limits of adaptation of the species. The first commonly makes description of a particular niche easier, but the second commonly facilitates comparison of two or more similar niches. Most habitats are occupied by several species, each with its own ecologic niche. Usually each species is represented by two or more individuals, which constitute a ***population*** (Chapter 4). Populations of two or more species occupying a habitat are commonly referred to as a ***community.*** Commonly a community is named for one or two of its most conspicuous and abundant species. Familiar examples might be a spruce-fir forest community of northern latitudes and a barnacle-mussel community along a rocky ocean shore. Just as there may be more than one habitat in an ecosystem, there may be more than one community. The problem of rigorously defining the term "community" in ecology remains unsettled: we will discuss it in more detail later in the chapter.

Within any ecosystem many constituents, both living and nonliving, interact. Among the most fundamental interactions are those in which energy and materials are transferred. The idealized pattern for the flow of materials through a simple but typical ecosystem is shown in Figure 10-1. In most communities the organic compounds are synthesized from the environment by ***producers*** in the form of photosynthetic plants. These are consumed by ***herbivores,*** some of which are devoured by ***carnivores. Parasites,*** which feed on living organisms without necessarily killing them, and ***scavengers,*** which feed on dead organisms, may enter into the system in a variety of ways. The sequence of species from producers through unpreyed-upon carnivores in any ecosystem is termed a ***food chain*** or ***food web.*** Organic materials of producers, herbivores, and carnivores not assimilated by species higher in the food web are broken down by organisms called ***decomposers,*** which are chiefly bacteria. Finally other organisms called ***transformers,*** also mostly bacteria, chemically alter

FIGURE 10-1
Diagram of the flow of materials through a typical ecosystem. (Modified from Clarke, 1965.)

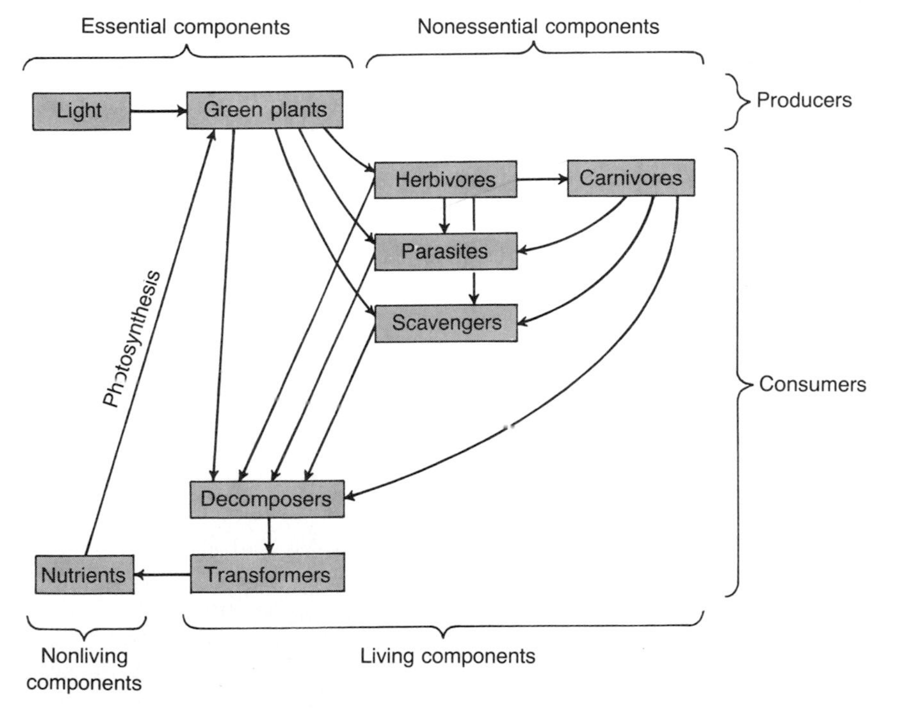

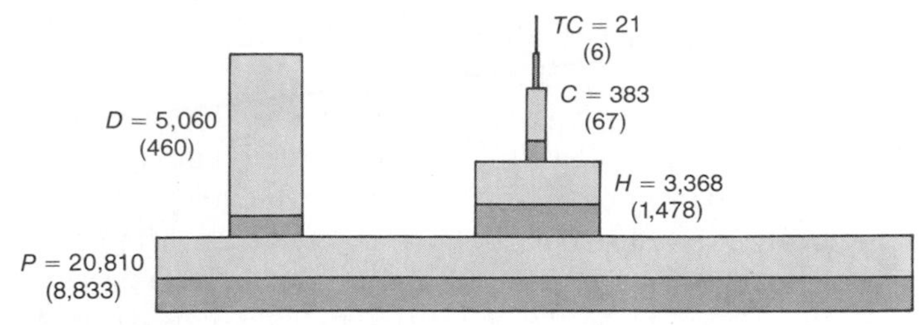

FIGURE 10-2
Energy pyramid for an ecosystem at Silver Springs, Florida. Trophic levels: P = producers, H = herbivores, C = carnivores, TC = top carnivores, D = decomposers. Each bar represents the total energy flow through a given trophic level. The darker portion of each bar represents energy locked up in biomass in the area studied; the lighter portion represents energy lost through respiration or movement downstream and out of the study area. (From Odum, 1959.)

certain decomposition compounds to render them utilizable once again by producers. Bacteria are thus extremely important in effecting completion of the cycle; without them, ecosystems as we know them could not exist.

It should be understood that the cycle shown in Figure 10-1 is a cycle of materials. Energy, though also moving clockwise through the cycle, is not recycled. The amount of living matter, including stored food, in the ecosystem or in any part of the ecosystem at any time is termed the ***biomass.***

Flow of materials or energy through a food chain or web can be represented by a simple diagram. Figure 10-2 shows what is called an energy pyramid and represents rate of energy flow. The ultimate source of energy for most ecosystems is the sun, because most producers use solar energy to synthesize compounds necessary for life. Producers are fed upon by herbivores, which are fed upon by carnivores, which in turn are fed upon by "top carnivores." The slope of the pyramid represents the energy loss at each successive step, resulting primarily from inefficiency of metabolic systems and decay of unpreyed-upon individuals. A similar pyramid can be constructed for biomass within an ecosystem. Because most carnivores are larger than their prey, the biomass at the top of a food pyramid is not only smaller, but tends to be divided among fewer animals. Most food pyramids are formed of only three or four steps. The fewer the steps in a food pyramid, the less the energy loss tends to be.

Determining the flow of energy and materials through ecosystems is virtually beyond the reach of paleoecology. The paleoecologist can, however, reconstruct at least a partial picture of food chains for certain fossil assemblages. He can, for example, commonly recognize predatory species and their prey.

Diversity refers to the taxonomic richness of a community. This can be measured in several ways. The simplest technique is simply to count the number of species present. Other techniques give more weight to abundant species than to rare species (see, for example, Pielou, 1975). Different methods of measuring diversity often yield different relative values of diversity for two faunas. In addition, subjectivity is required in the choice of the area in which diversity is measured. What one worker may regard to be a single, heterogeneous habitat with high diversity, another worker may consider to be several distinct homogeneous habitats, each with rather low diversity. Comparison of diversities for distinct environments and biotas is therefore problematical. Measurement of diversity in the fossil record is even more difficult because species may be transported, mixed, or differentially preserved after death.

The Marine Ecosystem

The present-day marine ecosystem is used by most paleoecologists as a general model for interpretation of ancient marine sediments and faunas. One drawback in this approach is that climates have been considerably warmer than at present for most of the time period since the beginning of the Cambrian (Dorf, 1959). Also, continents are relatively emergent today; during most of the Paleozoic, Mesozoic, and Cenozoic, larger areas of continental crust were submerged beneath shallow epicontinental seas. The Baltic Sea and Hudson Bay are examples of modern epicontinental seas, but both are situated in cold northern latitudes.

We can begin our discussion by defining some terms that will be helpful in visualizing the marine ecosystem. Figure 10-3 is an idealized block diagram of the

FIGURE 10-3
Idealized block diagram of the edge of a modern continent. The vertical scale is exaggerated and distorted.

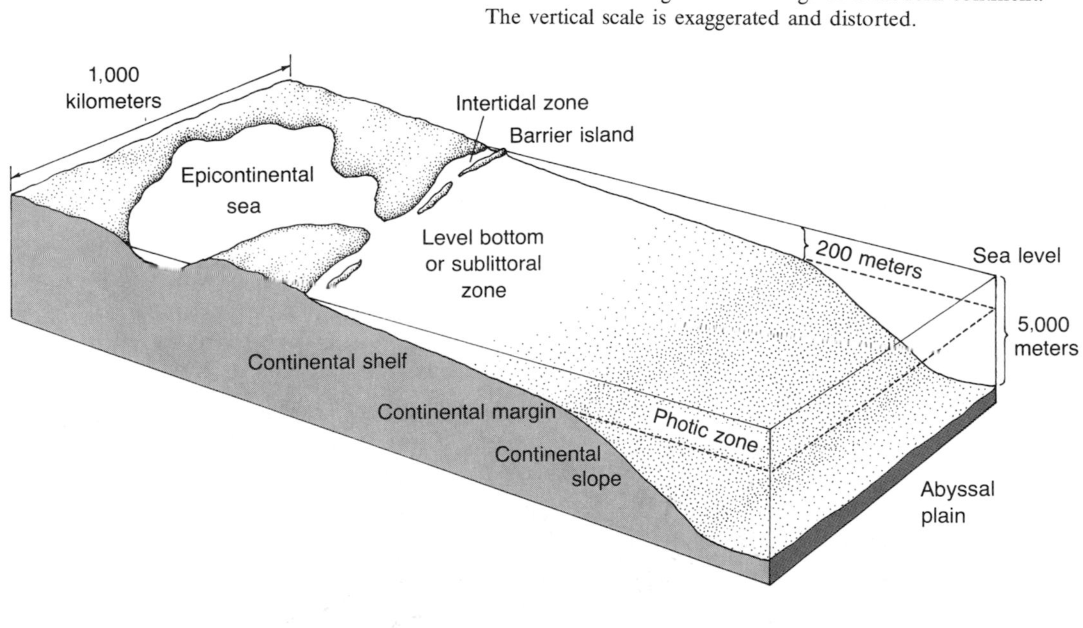

edge of a modern-day continent. The submerged border of the continent is called the ***continental shelf.*** The continental margin lies at a depth of about 200 meters. From it the ***continental slope*** extends to a depth of about 5,000 meters, where it reaches the floor of the ocean basin.

The marginal marine region alternately covered and uncovered by the tides is called the ***intertidal,*** or ***littoral, zone;*** the continental shelf surface is called the ***sublittoral zone*** or ***level bottom*** and the surface of the continental slope forms the ***bathyal zone.*** The ocean floor forms the ***abyssal plain,*** which is in places interrupted by deep trenches, towering submarine mountain chains, and other less conspicuous topographic features. The ***photic zone*** is the portion of water that is penetrated by light. The lower limit of the photic zone varies from place to place, depending primarily on water clarity, but usually coincides more or less with the margin of the continental shelf. Only the upper half, that is, the upper 100 meters, has sufficient illumination for substantial photosynthesis. In Figure 10-3 an arm of the ocean extends inland to form a shallow epicontinental sea. Like the continental shelf farther offshore, it is floored by what is called a sublittoral or level bottom substratum. Most marine deposits we now encounter on continents were deposited in epicontinental seas of this type; few were deposited at abyssal depths.

Marine organisms are commonly classified according to a simple scheme, depending on where they live and whether they are capable of self-propulsion (Figure 10-4). Bottom-dwellers are called ***benthos*** (from which is derived the adjective ***benthonic,*** or ***benthic***). They may be ***epifaunal,*** which means they live *on* the substratum, or ***infaunal,*** which means they live *in* the substratum. Benthonic forms capable of locomotion are called ***vagile*** and immobile forms are called ***sessile.*** Organisms living in the water above the bottom are described as ***pelagic,*** those whose major movements are accomplished by swimming being ***nektonic*** and those that are transported primarily by waves and currents, ***planktonic. Phytoplankton*** are plants and ***zooplankton,*** animals. Figure 10-5 shows a simplified version of the food cycle in the oceans. Food webs in the marine ecosystem tend to be relatively uncomplicated. The principal producers in the modern-day seas are single-celled phytoplankton, mainly diatoms and dinoflagellates. In offshore waters, phytoplankton are the sole photosynthetic producers, but on the continental shelf, where the photic zone reaches the sea bottom, benthonic plants, including marine grasses, augment food production.

The primary consumers in offshore waters are small zooplankton. Most of the adult organisms among the zooplankton of modern seas belong to two crustacean groups. In addition, a large percentage of the zooplankton in many areas consists of tiny larvae of invertebrates that are benthonic in their adult stages. On the ocean floor benthonic invertebrates are the primary consumers. Metabolic and decay products of plants apparently form much of the food of both zooplankton and primary benthonic consumers. The absence of plants on the abyssal plain suggests that the animals there feed for the most part on inanimate organic matter.

The biomass pyramid in many marine settings tends to be inverted (Figure 10-6). The reason frequently given to explain why zooplankton typically outweigh phytoplankton is that there is a rapid turnover of phytoplankton (a high *rate* of

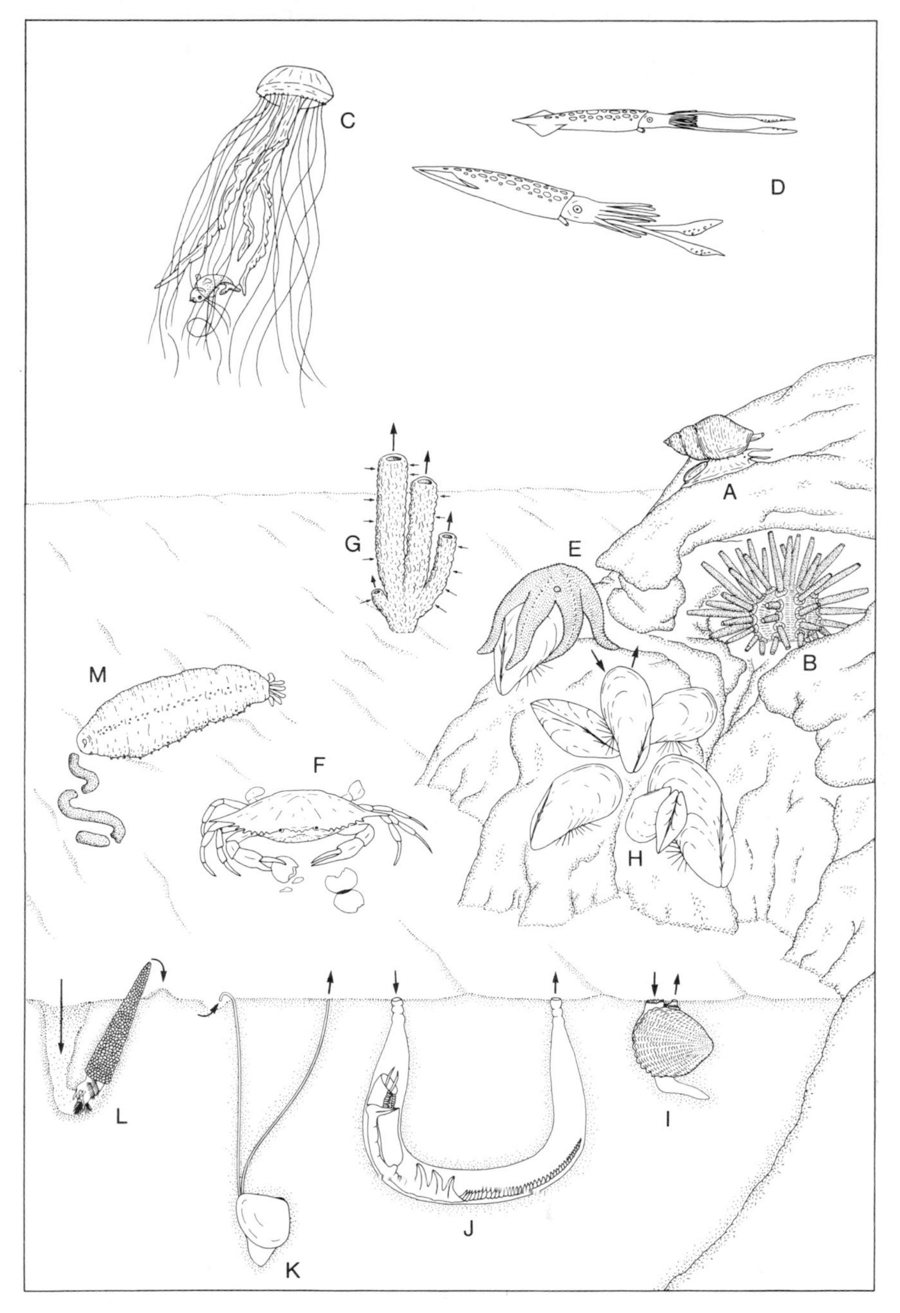

FIGURE 10-4
Life habits of marine invertebrates. A: A snail (gastropod mollusc). B: A sea urchin (echinoid echinoderm). C: A jellyfish (scyphozoan coelenterate). D: Squids (cephalopod molluscs). E: A starfish (asteroid echinoderm). F: A crab (crustacean anthropod). G: A sponge (poriferan). H: Mussels (bivalve molluscs). I: A cockle (bivalve mollusc). J: A chaetopterid worm (polychaete annelid). K: A macomid clam (bivalve mollusc). L: A trumpet worm (polychaete annelid). M: A sea cucumber (holothurian echinoderm). Animals C and D are pelagic (C is planktonic and D is nektonic).

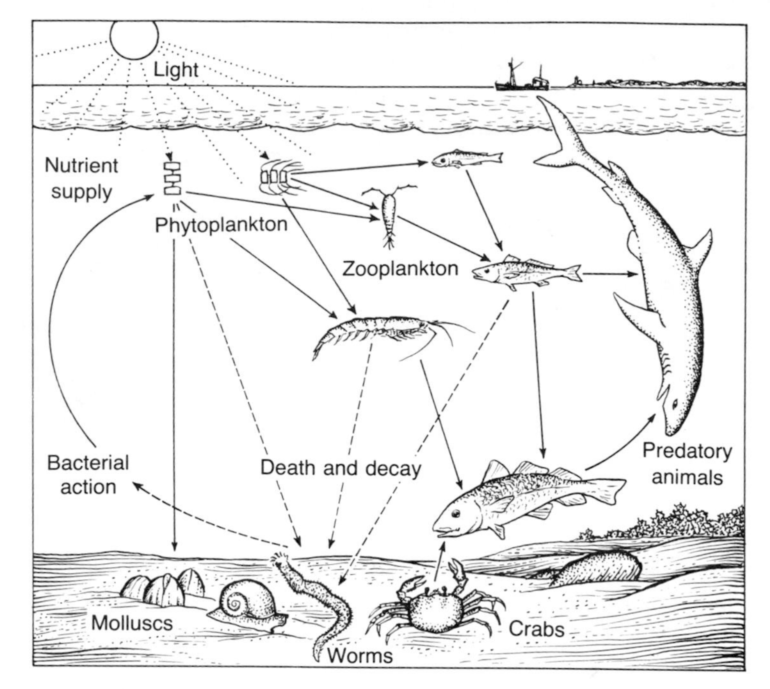

FIGURE 10-5
The food cycle of the oceans. (From Clarke, 1965.)

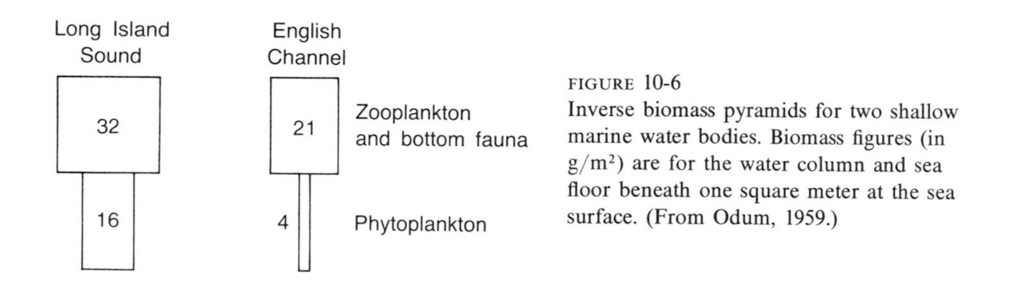

FIGURE 10-6
Inverse biomass pyramids for two shallow marine water bodies. Biomass figures (in g/m^2) are for the water column and sea floor beneath one square meter at the sea surface. (From Odum, 1959.)

production). Some workers, however, believe that the metabolic products of phytoplankton constantly being released into the water form a large proportion of the food used by zooplankton and benthonic consumers.

Carnivores feeding on zooplankton are largely nektonic fishes. In shallow water, zooplankton are joined by large numbers of benthonic invertebrates, which also feed on phytoplankton. Benthonic invertebrates also live at bathyal and abyssal depths, but in less abundance because of the reduced food supply.

Marine paleoecologists studying local faunas concern themselves primarily with benthonic organisms because species living on or in the substratum are much more likely to be preserved with little or no post-mortem transport than are pelagic species. Pelagic species must be transported to the bottom from their native habitat to be preserved at all, and the substratum on which they fall may bear no relation to the environment in which they lived.

Hunt (1925) proposed a classification of benthonic consumers that has been widely adopted (see Figure 10-4 for examples). He recognized three feeding groups of soft-substratum dwellers. ***Carnivores*** are "animals which feed mainly upon other animals, whether living or as carrion." (Because carrion-eating or "scavenging" is engaged in by most predators and because most scavengers eat some live animals, strict separation of scavengers and predators is unwarranted.) ***Suspension feeders*** are "animals which feed by selecting from the surrounding water the suspended micro-organisms and detritus." ***Deposit feeders*** are "animals which feed upon the detritus deposited on the bottom, together with its associated micro-organisms." To Hunt's three categories we might add a fourth, ***grazers,*** which feed by selectively removing organic surface films (chiefly algal coatings) from the substratum. Most grazers inhabit hard substrata; none were present in the soft-bottom fauna that gave rise to Hunt's classification.

Life Habits

We have been concerned with limiting factors and habitats of living and fossil benthonic species. We will now consider the life habits or modes of life of species within their respective habitats. For fossils only certain types of information may be within our grasp: What was the organism's orientation in life? Did it move, and if so, how? Was it attached to the substratum or free-living? How did it obtain food? How did it reproduce?

DIRECT EVIDENCE THROUGH PRESERVATION

The most direct way to learn the life habits of extinct fossil species is to observe them preserved in the midst of some life activity. An example is shown in Figure 10-7. Many such examples represent unique preservational circumstances, such as sudden, catastrophic burial.

Inference of life habits from mode of preservation frequently requires ingenuity. An example is the analysis by Grant (1975) of the feeding mechanism of richthofeniacean and lytoniacean brachiopods. Such brachiopods have an unusual shape. The cup-like ventral valve is cemented to the substratum. The dorsal valve is reduced and flattened, to form a "lid" that lies recessed within the ventral valve (Figure 10-8). Grant evaluated the possibility suggested by Rudwick (1961) and other authors that

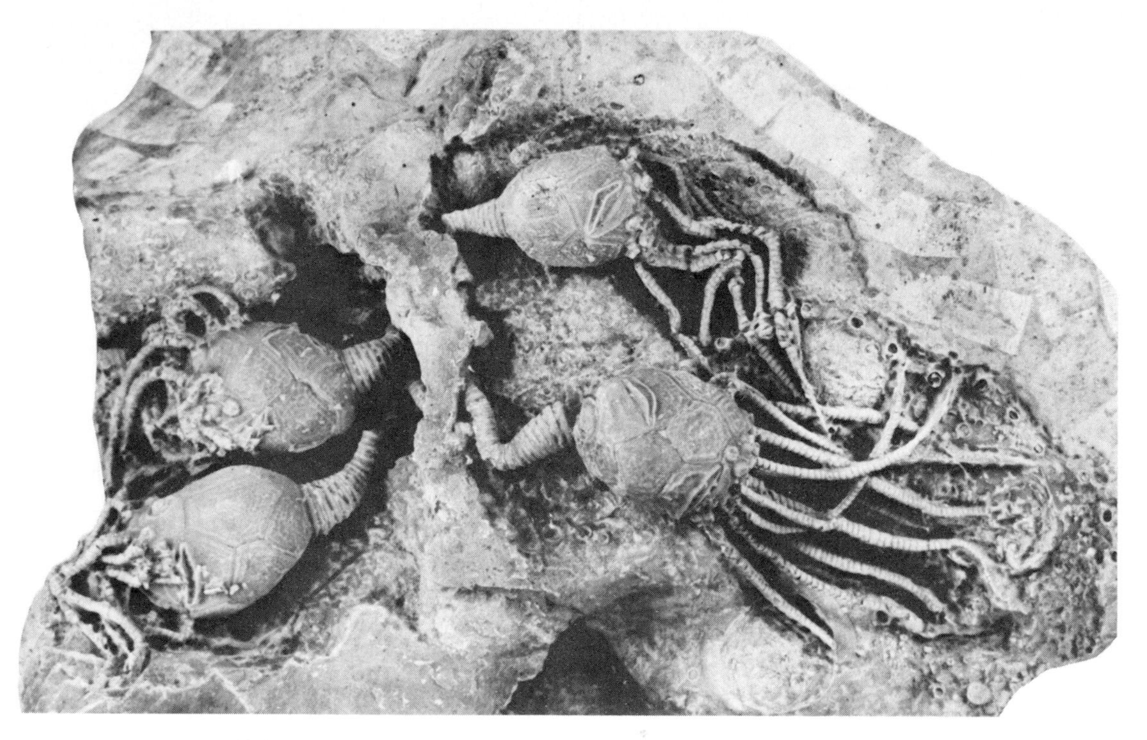

FIGURE 10-7
The Devonian cystoid *Adocetocystis williamsi* preserved in life position, attached to an ancient erosion surface within the Shell Rock Formation, Iowa (×1.3). (From the first published report of fossil cystoids unequivocally attached by stems—Koch and Strimple, 1968.)

these brachiopods fed by clapping the lid-like dorsal valve to bring water and food into the mantle cavity. This method of feeding would have been quite unusual because in most brachiopods food-laden water is pumped through the mantle cavity by ciliary activity on the lophophore. Furthermore, Grant (1972) found evidence suggesting that a normal lophophore may have been present in the richthofeniacean and lytoniacean brachiopods. This led him to question the proposed clapping mechanism because it seemed unnecessary. Moreover, the lophophore would have been attached to the dorsal valve, making clapping of this valve difficult if not impossible. Grant (1975) discovered a single specimen of a richthofeniacean in which another kind of brachiopod had been living on top of the lid-like valve; the other brachiopod was trapped beneath the meshwork of the ventral valve that formed a kind of sieve-like structure above the dorsal valve. The encaged animal must have passed through the meshwork as a larva and then grown to large size. Evidence that the intruder matured while its richthofeniacean host was still alive would indicate that the host did not feed by clapping its dorsal valve. Effective clapping would have been impossible with the

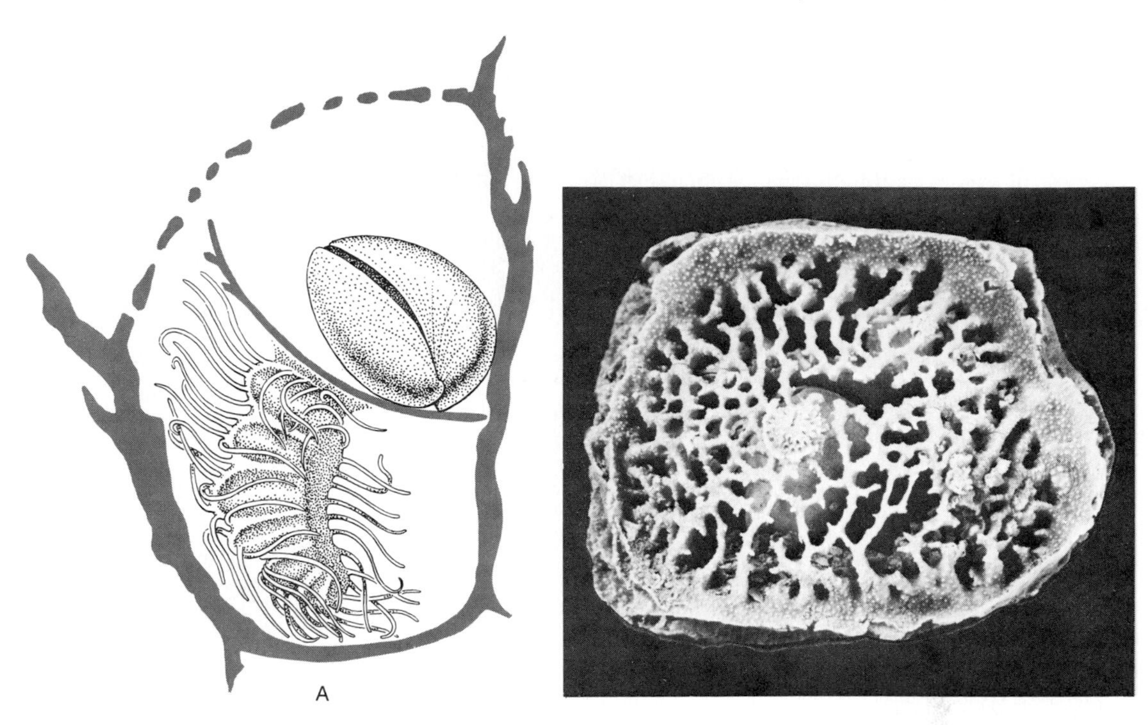

FIGURE 10-8
A: Cross-sectional diagram of a specimen of the richthofeniacean brachiopod genus *Hercosestria* in which a brachiopod of the genus *Composita* has been preserved in life position. The *Composita* lived between the lid-like dorsal valve and the meshwork of the ventral valve of the *Hercosestria*. B: Top-view photograph of the specimen represented in A. The *Composita* can be seen through the meshwork. (From Grant, 1975.)

intruder stationed as it was, and if the host had fed by clapping, the process would almost certainly have killed the intruder. Grant also noted that the dorsal valve of the richthofeniacean was curved downward in an unusual way (Figure 10-8). It apparently grew abnormally because of the presence of the encaged animal. His conclusion was that this richthofeniacean, and presumably all others, pumped water by ciliary activity in the normal brachiopod fashion, without clapping the dorsal valve.

Intimate species associations of the sort just described between two epibiont brachiopods offer many kinds of useful information. ***Epibionts*** are organisms attaching to other organisms, ***epiphytes*** being plants with this habit and ***epizoans,*** animals. The chief problem in using epibionts as indicators for the life habits of their host species is one of establishing whether attachment took place before or after death of the host. As in the example above, both the location and orientation of epibionts may be useful in this regard. Regions of current flow are the only parts of the shells of some aquatic animals that are exposed above the sediment surface in life; epibionts living on these parts of a host therefore indicate the orientation of the live host species.

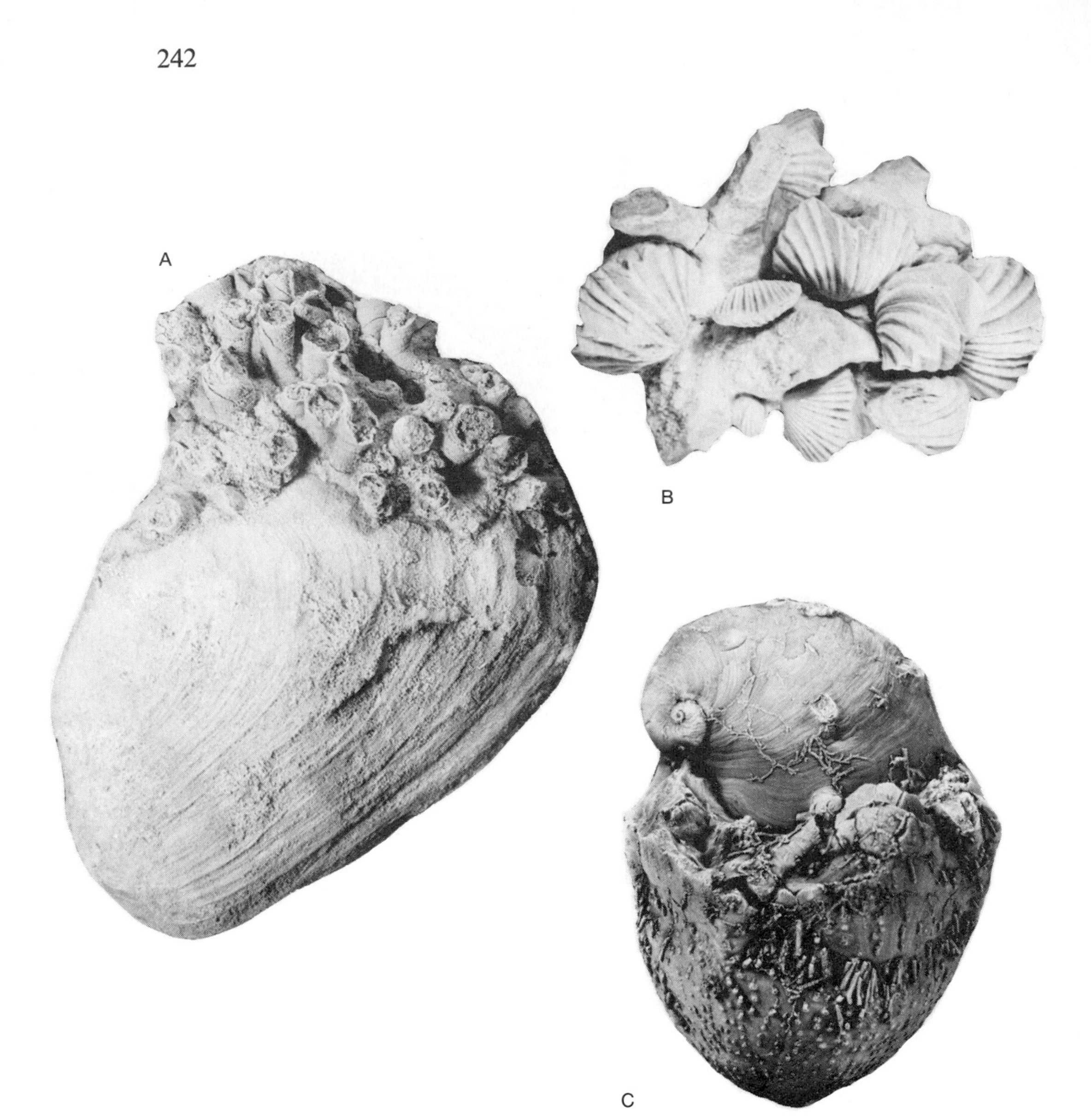

FIGURE 10-9
Preservation of intimate species associations. A: Specimen of *Modiomorpha concentrica* from the Devonian of Indiana, with epibionts encrusting the posterior region (×1). The lower limit of epibionts on the shell probably indicates the position of the sediment-water interface when the animal was alive. B: Three species of Ordovician brachiopods attached to a branching bryozoan colony (×1.4). C: Platycerid snail attached to the anal region of the calyx of the Mississippian crinoid *Arthracantha carpenteri* (×2). (A, Photo courtesy of John Pojeta, U.S. Geological Survey; B from Richards, 1972; C, Photo courtesy of Porter M. Kier, U.S. National Museum.)

Epibionts on the *Modiolus*-like mussel of Figure 10-9,A show that it lived in the endobyssate manner that we would predict from its shell form (pages 192–195). The Late Ordovician brachiopods that are shown in Figure 10-9,B lived attached to branching colonies of trepostome bryozoans. *In situ* preservation shows that members of the gastropod family Platyceratidae, which range from Ordovician to Permian, attached to the calyces of stalked crinoids (Figure 10-9,C). The apertures of platyceratid snail shells fitted over the anal openings of crinoids, and the snails fed on fecal material.

The manner in which an organism becomes fossilized sometimes indirectly reveals its life habits. Seilacher et al. (1968) found calyces of the enormous crinoid *Seirocrinus* preserved beneath its stems, with the arms spread radially, on huge slabs of the Jurassic Posidonia Shales. (A slab of the Shale is shown in Figure 10-10,A, on the following page.) Disentangling of the stalks showed that the parts farthest from the calyces were uppermost and were the last to settle. These observations and supporting evidence from functional morphology led to the conclusion that in life the crinoids must have dangled calyx-downward from a large floating object. It should be noted that these interpretations have been questioned by Rasmussen (1976), who believes that other kinds of evidence suggest a benthic mode of life for *Seirocrinus.*

HOMOLOGY

A less direct approach to understanding life habits is to infer them by homology (page 178). It has often been said, for example, that were it not for the existence of living turtles, we would have no concept of the life habits of fossil turtles. We might suspect that the turtle shell served a protective function for some bizarre creature, but we certainly would never be able to deduce from their fossil remains the differences in life habits of a largely terrestrial box turtle and a fresh-water snapping turtle. Certainly the deductions that have been made about the predatory, nektonic habits of ammonites could never have been made without information gained from study of the Recent *Nautilus.* Many years of endeavor would probably have been required simply to ascertain that ammonoids were cephalopods. Only recently have preserved cephalopod jaws or "beaks" been found associated with fossil ammonities.

FUNCTIONAL MORPHOLOGY

Functional morphology, a subject discussed more fully in Chapter 8, represents a major source of evidence for the reconstruction of life habits, when coupled with inferences about habitat derived from sedimentologic evidence. An example is the study by Grant (1966) of the brachiopod genus *Waagenoconcha,* found in the Permian of Pakistan. Juvenile animals bear convergent attachment spines near the beak that were apparently used for clinging to stalk-like benthonic plants, sponges, or bryozoans on which the larvae apparently settled (Figure 10-11). Adults lived free on fine-

A

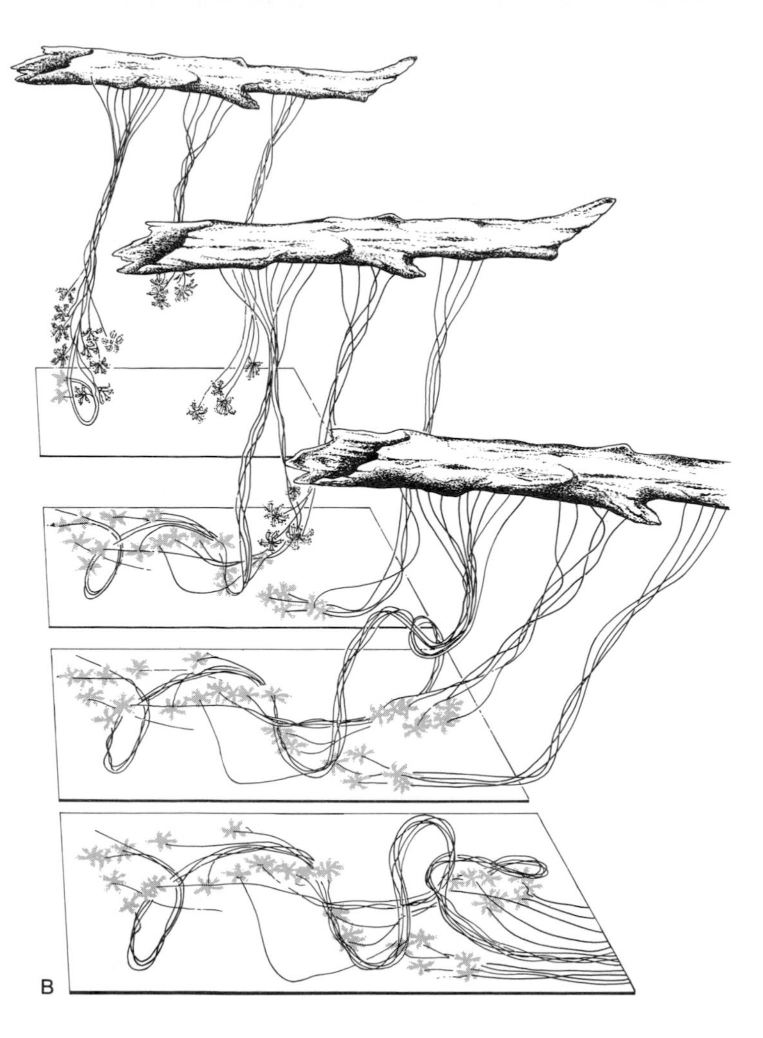

B

FIGURE 10-10
Reconstruction of the life habits of the giant Jurassic crinoid *Seirocrinus.* A: A slab of Jurassic Posidonia Shale on which are preserved a number of specimens of *Seirocrinus.* The slab is mounted on the wall of the Geological Institute of the Tübingen University in Germany. B: A series of reconstructions depicting, from top to bottom, the manner in which the group of crinoids of *A* are inferred to have settled to the bottom, apparently from pendant life positions in which they were attached to a log or some similar floating object. (From Seilacher, Drozdzewski, and Haude, 1968. Copied with the permission of The Palaeontological Association.)

grained sediment. From direct preservational evidence and from the great weight of the structure, Grant concluded that the spiny ventral valve lay undermost. The spines apparently served primarily to support the suspension-feeding animal on a soupy mud bottom, though Grant suggested that they also served an anchoring function. Like snowshoes, they prevented a dense object from sinking into a soft underlying medium. The plight of the juveniles, had they not been attached to plants, is predictable. Their small size would scarcely have elevated their commissure line above the mud. Any major disturbance of the bottom would probably have clogged their feeding mechanisms, interrupted normal activities, and led to the early demise of most individuals.

FIGURE 10-11
Reconstruction of the life habits of the productoid brachiopod *Waagenoconcha abichi* from the Permian of Pakistan. Spat are attached to idealized algae. (From Grant, 1966.)

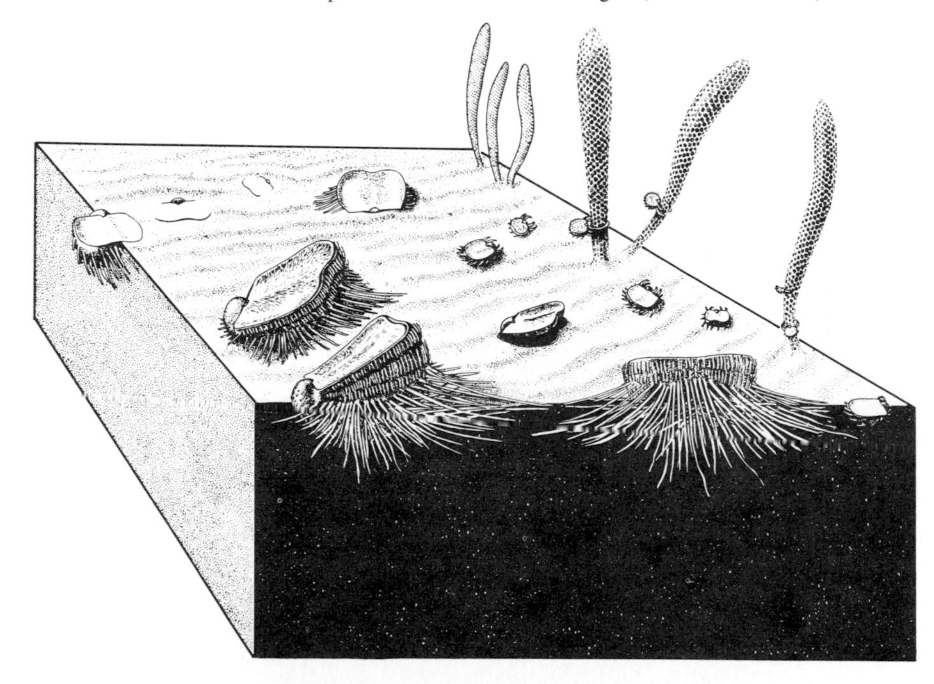

FIGURE 10-12
The trace fossil *Helminthoidea labyrinthica,* found in Cretaceous and Eocene sediments in the Alps and Alaska. (From Seilacher, 1967, after R. Richter.)

EVIDENCE OF BIOLOGIC ACTIVITY

Even when not represented by skeletal parts, many organisms have left fossil evidence of their presence and way of life in the form of tracks or trails. The Germans, who have contributed most to the study of these features, call them *Lebensspüren* (traces of life). A trace fossil was produced by the movements of an organism and can therefore reveal much about the organism's behavior. Formation of many types of trace fossils can be studied in the Recent, both in nature and in the laboratory.

Some Lebensspüren mark resting places of animals, others represent more or less permanent shelters (Figure 10-12), and some are merely crawling-tracks.

One pioneer student of trace fossils, Rudolf Richter, analyzed the form shown in Figure 10-12. It was produced by a worm-like deposit feeder that moved systematically in a special pattern through fine-grained deposits, to ingest sediment without passing the same material through its system twice. Four "rules" were necessary to produce the observed pattern. The animal had to: (1) tunnel horizontally through a single sediment layer, (2) make a U-turn after moving a certain fixed distance, (3) move close to one or more previous segments of its tunnel system, and (4) maintain a certain minimum distance from previous tunnel segments. The two final "rules" tended to produce a pattern in which segments of the tunnel system were separated by uniform distances. The "rules" must have taken the form of genetically coded instructions. The required instructions can be programmed for a computer to simulate the observed pattern (Raup and Seilacher, 1969). Presumably, animals in nature carry out their genetic instructions by responding to chemical and tactile stimuli in the surrounding medium.

Interesting stories are also told by tracks of terrestrial vertebrates. Much of the thick Triassic redbed sequence of the Connecticut Valley region was deposited in swampy lowland areas. Tracks on bedding surfaces reveal the presence of several dinosaur species (Figure 10-13).

FIGURE 10-13
Triassic dinosaur tracks from Rocky Hill, Connecticut. The animal walked from a firm mud bank, where rain prints are visible (arrow at extreme left points to shallow track), into a pond or stream whose soft bottom-mud was covered by ripple marks (note the two deeper tracks). (Courtesy of J. H. Ostrom.)

Trace fossils are especially important in providing evidence of past life where no skeletal remains are found. Connecticut Valley Triassic rocks yield virtually no dinosaur bones, yet abundant tracks attest to the existence of a diverse dinosaur fauna. The warm humid climate of the region in which the redbeds formed was apparently unfavorable for skeletal preservation, despite the fact that the region was a depositional basin receiving thousands of feet of sediment over a period of several million years. Otherwise barren marine sedimentary rocks also commonly show evidence of abundant benthonic life through the presence of trace fossils.

Limiting Factors

Our earlier discussion of ecologic concepts was largely restricted to present-day ecosystems. In this and the following sections we will provide examples to show how ecologic data and principles can be applied to the fossil record. We will begin with a discussion of ***limiting factors:*** those physical, chemical, and biologic properties of the environment that limit the distribution and abundance of particular species. Distribution is sometimes considered to be an aspect of abundance, in that areas where a species is absent are, in effect, areas where its abundance is zero.

TEMPERATURE

Temperature is one of the most important limiting factors in nature. Its primary effect, however, is on large-scale geographic distribution, which will be treated separately in Chapter 12. Microclimatic temperature may also have significant effects, especially where there are sharp temperature gradients or discontinuities. For example, in the intertidal zone, which is a highly unstable band of environment between land and sea, temperature is one of several conditions that fluctuate markedly with the tide. Most local trends in temperature coincide with trends in other limiting factors like water depth and light intensity. In fact, close correlations between trends in two or more limiting factors are common in nature, as will become apparent in the following discussion. Our separate consideration of these factors is merely a matter of convenience.

OXYGEN

When paleontologists collect fossils from sediments deposited in epicontinental seas (see Figure 10-3), they usually discover trends in species diversity that are the opposite of those that we have described for present-day transects from shallow water

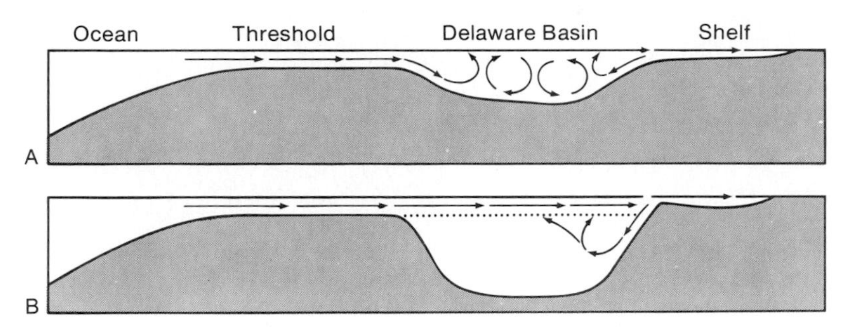

FIGURE 10-14
Reconstructed hydrography of the Delaware Basin. A: Hydrography during Leonardian times, when the Victorio Peak and Bone Spring Formations (which are shown in Figure 10-15,C) were being deposited. B: Hydrography during late Guadalupian times, when the Capitan Limestone was forming. At this stage, the lower portions of the basin were stagnant. (From Newell, et al., 1953.)

into the deep sea. The floor of the deep sea, located between continental land masses, lies at depths greater than about 5,000 meters. Few epicontinental seas have been deeper than a few hundred meters, and except in very cold climates, their deeper waters have not been as cold as the present deep sea. Their deeper waters may also have been less constant in temperature or salinity. Unfortunately, we live at a time when continents are relatively emergent so that we have few present-day epicontinental seas to use as models for reconstructing the nature of ancient ones. Study of large lakes and inland seas has nonetheless been of value, and one of its contributions has been to illustrate the importance of oxygen as a limiting factor. Oxygen is constantly renewed to the deep sea by worldwide circulation patterns. Cold surface waters sink downward at the poles and pass equatorward along the sea floor. In contrast, unless shallow, landlocked water masses are stirred by waves to considerable depths, they become stratified and their deep layers develop a deficiency of oxygen. This condition excludes many kinds of benthos and seems to account for decreased diversity of fossil species away from ancient shorelines, toward the interiors of shallow basins.

An example of the limiting nature of oxygen was discussed by Newell et al. (1953) in a study of Permian rocks of the Delaware Basin region of Texas. A reconstruction of the hydrography of the basin is shown in Figure 10-14. This basin persists today with much the same topography that it had in Late Permian times (see the diagrams that comprise Figure 10-15 on the following three pages). Around its margins enormous

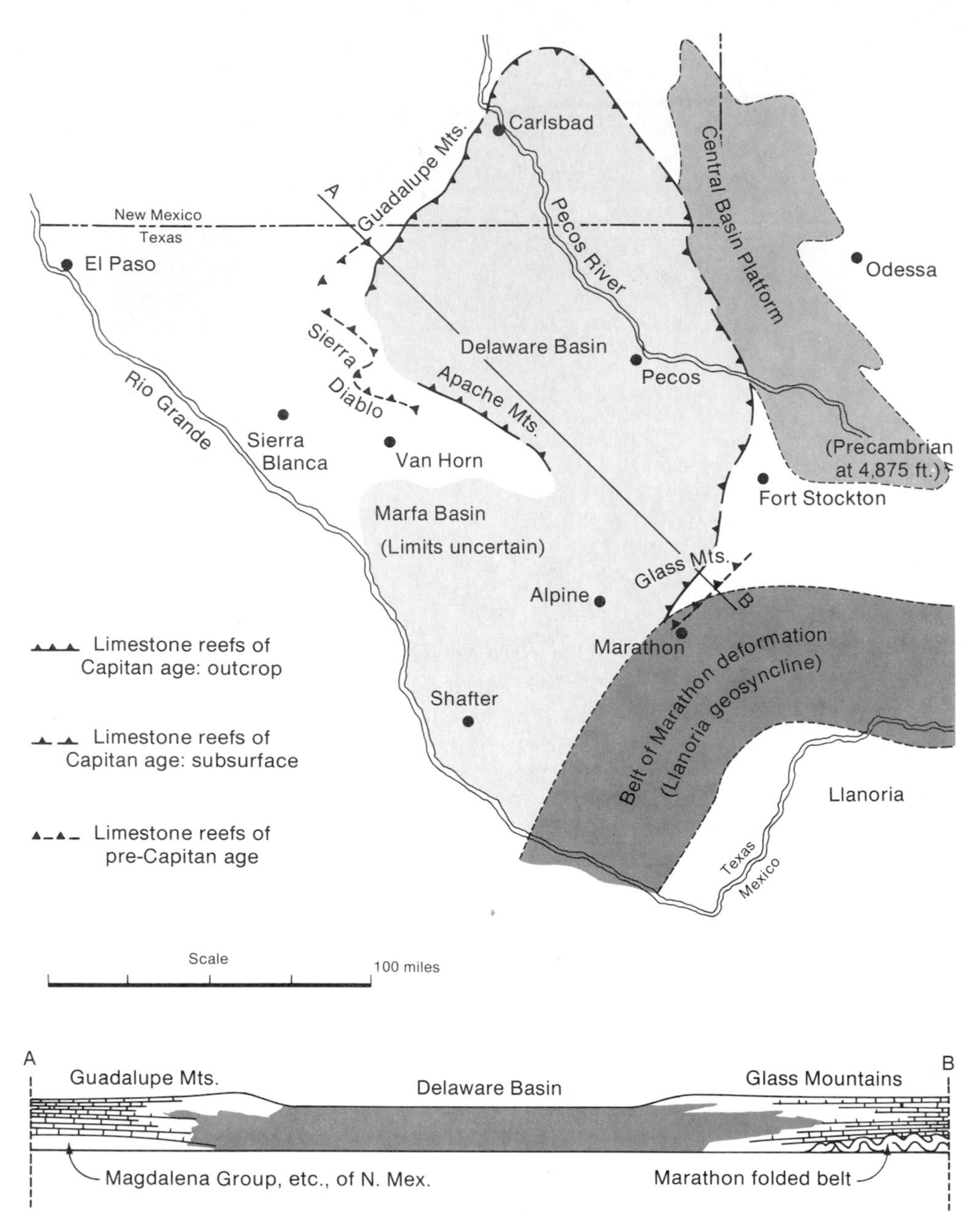

FIGURE 10-15,A
Geology of Permian deposits of the Delaware Basin, Texas: Map showing the distribution of reef-like deposits around the edge of the Delaware Basin.

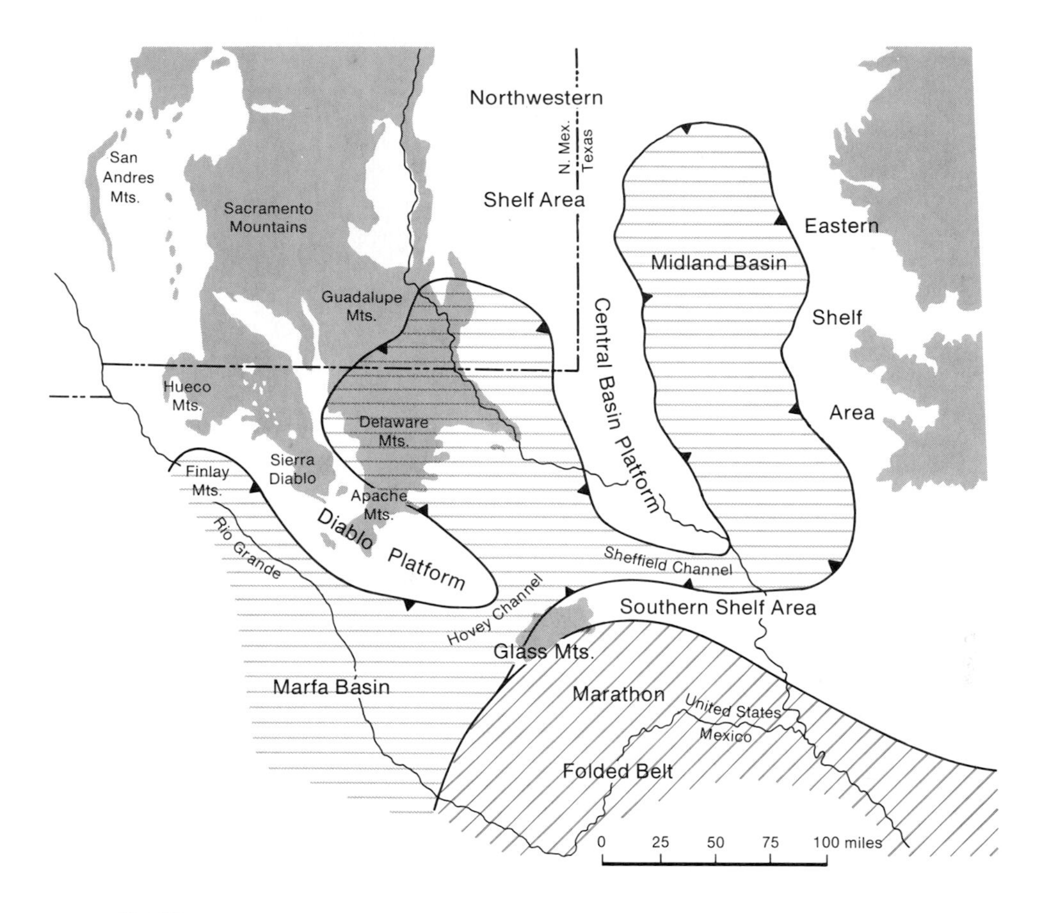

FIGURE 10-15,B
Main structural features of the basin and surrounding areas. Note the presence of Hovey Channel at the southern end of the basin (lower center of map). It is the passage above the threshold in Figure 10-14, through which the basin was connected to the ocean.

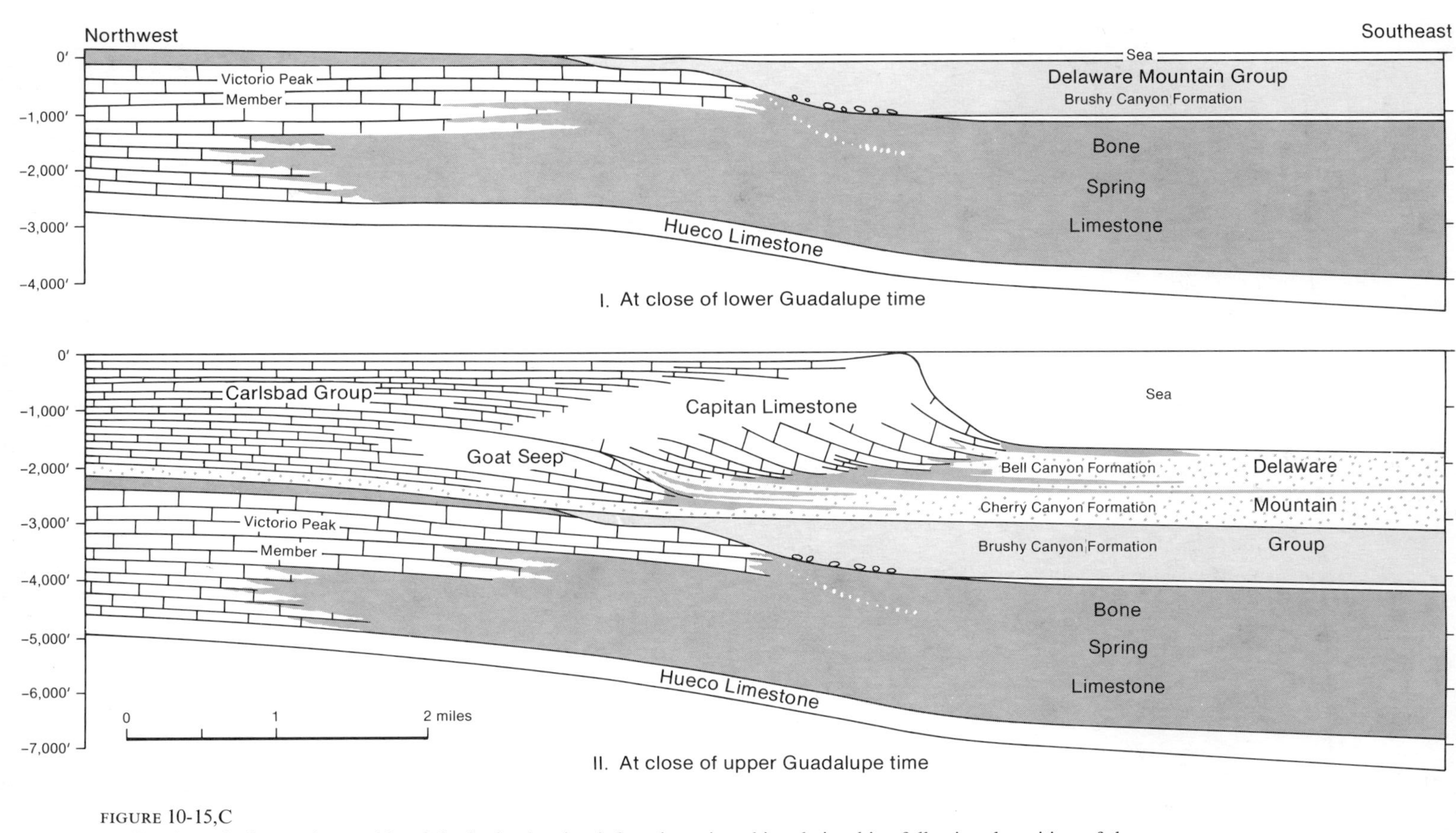

FIGURE 10-15,C
Profiles through the northwest side of the basin showing inferred stratigraphic relationships following deposition of the Brushy Canyon Formation (1) and following deposition of the reef-like Capitan Limestone (II). The body of water labelled "sea" occupies the Delaware Basin. (Parts A, B, and C all from Newell, et al., 1953, after King, 1948.)

thicknesses of carbonate sediments were laid down, largely through the activities of marine life. As a result, in the later stages of development of the basin (Figure 10-15,C-I), there was great topographic relief at the basin margins. Much debris, including large carbonate blocks, periodically slumped from the shallow marginal banks into the basin. The resulting talus deposits now interfinger with basin deposits, which consist mostly of dark sandstones and siltstones associated with limestones. In many beds of the basinal deposits, the grain size of sediment grades upward from coarse to fine. Part of the coarse material consists of skeletal debris from organisms that lived around the basin margin. Clearly the graded beds were deposited by dense, turbid flows that descended from the basin margin. It appears that the final configuration of the Delaware Basin was developed by the following sequence of events.

Early in its history the basin was relatively shallow and its waters were well mixed and well oxygenated (Figure 10-14). When the Victorio Peak and Bone Spring lithologic units were deposited, the basin was inhabited by nuculoid bivalves (such as are shown on page 269), gastropods, siliceous sponges, and the brachiopod *Leiorhynchus*. All of these are preserved in the Bone Spring Limestone. Subsequently, however, sea level rose rapidly and, though accumulation of organically produced limestone at the basin margin kept pace, infilling of the basin floor lagged behind. The basin deepened to perhaps 600 meters (Figure 10-15,C-II). Its margins also steepened as the Capitan Limestone, which was formed by organisms that could not grow above sea level, accreted seaward. The basin connected with the open sea through a channel to the south (Figure 10-15,B). This channel seems to have had a shallow threshold, however, and the deepening of the basin just described led to a condition in which normal circulation failed to reach the deeper basin waters or the basin floor. (Figure 10-14). Oxygen depleted by respiration and other oxidational processes was not renewed effectively, and nearly all types of benthos were excluded. The kinds of benthos listed earlier for the Bone Spring Limestone are found only in those deposits of the Delaware Mountain Group (Figure 10-15,C) that formed near the basin margin, at depths shallower than the central basin floor. Numerous sponge spicules and scattered occurrences of the brachiopod *Leiorhynchus* provide the only evidence of benthic life in the basin center, and the sponge spicules may have washed in from shallower depths. Otherwise, only plant spores and the skeletons of pelagic ammonites and radiolarians that lived in the shallow, oxygenated waters accumulated here.

Rhoads and Morse (1971) have provided a general view of the effects of dissolved oxygen on the depth distributions of benthos of marine basins. In water bodies like the Black Sea, diversity of species drops off markedly as levels of dissolved oxygen decline with depth. Furthermore, heavily calcified taxa, which paleontologists are most likely to discover, disappear before soft-bodied organisms (Figure 10-16,A), and small organisms tend to extend to greater depths than large organisms (Figure 10-16,B).

Reduction in dissolved oxygen with increased water depth has been invoked to explain the decline in diversity of species in an offshore direction within the Late Cretaceous Interior Seaway of the Western United States (Rhoads et al., 1972) and to explain a similar trend in an offshore direction from the Devonian Catskill "Delta" of New York State (Bowen et al., 1974).

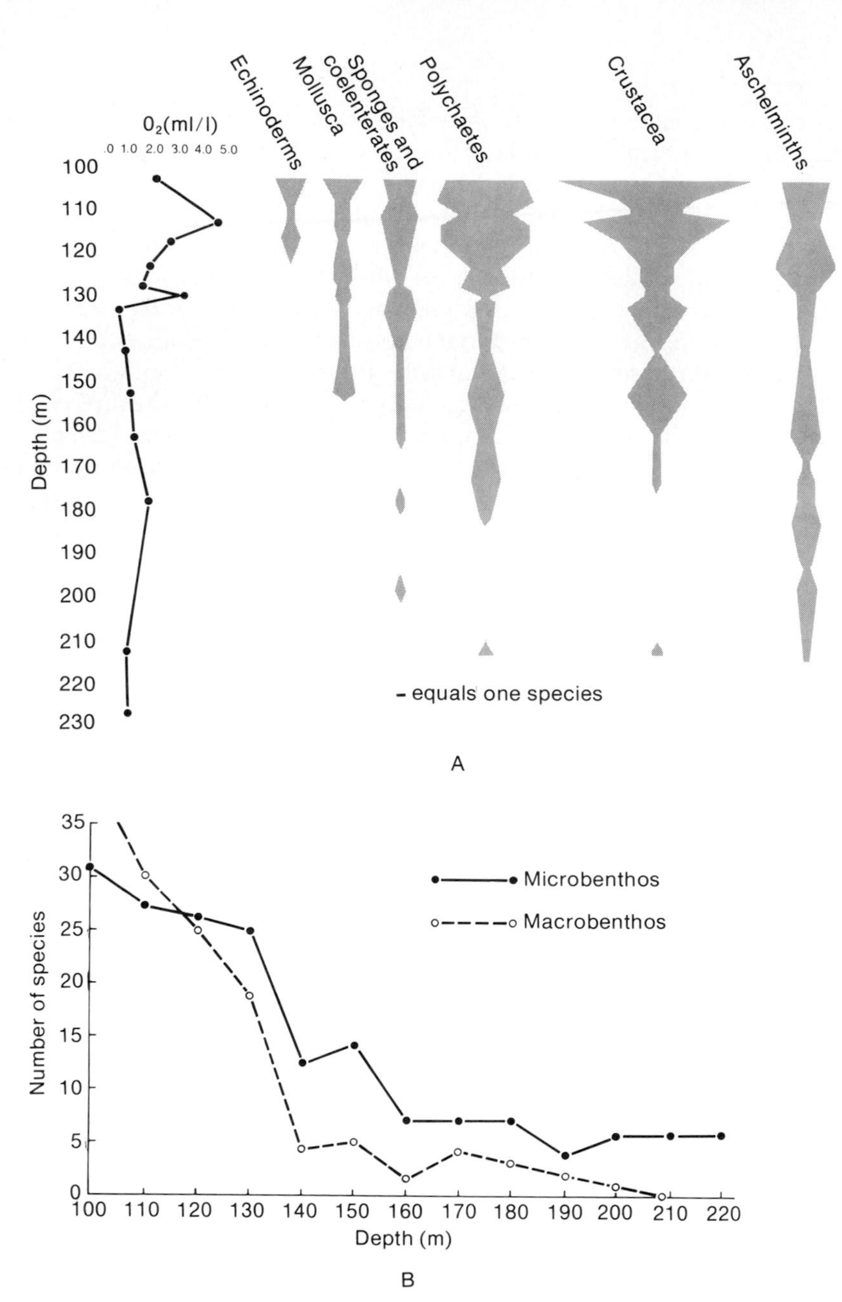

FIGURE 10-16
Relation between species abundance and depth in the Black Sea. A: Decline of major benthic invertebrates with increasing depth and decreasing concentration of dissolved oxygen. B: Decline in total abundance of large and small benthic animals with depth. (From Rhoads and Morse, 1971, after Bacescu, 1963.)

WATER DEPTH

Although it is one of the most useful variables for reconstruction of environments of the past, water depth is very difficult to determine, even in an approximate way, from marine fossil assemblages. Ancient water depth is important to know because it defines the configuration of marine basins and contributes to our picture of paleogeography for particular intervals of geologic time. Water depth affects organisms most directly through hydrostatic pressure, but exerts relatively little direct control over the distribution of marine life, except over larger depth intervals than are commonly represented in paleoecologic studies. Furthermore, water depth has little effect on solid skeletons, so that adaptation to pressure is not directly correlated with skeletal strength. Rapid pressure change does affect soft tissue. Most living animals brought rapidly upward from the abyssal plain in bottom-sampling devices suffer from the rapid decrease in pressure. Although many taxonomic groups have diversified to inhabit a variety of water depths, taxonomic and functional morphologic relationships to minor variations in water depth are rare.

Indirectly, water depth may affect species distribution through such depth-related parameters as light intensity, salinity, temperature, food supply and, as we have seen, dissolved oxygen. For example, many living benthonic species inhabit deeper water toward the equatorward part of their geographic range than toward the poleward part because the shallow waters toward the equator are too warm for them.

Because no food may be produced by photosynthesis below the photic zone, food supply decreases markedly with depth below about 200 meters. It has long been recognized that the abundance of organisms decreases downward across the continental slope to the abyssal plain. Sanders, Hessler, and co-workers (1965; 1967) at the Woods Hole Oceanographic Institution confirmed this trend in a large-scale study with modern sampling devices, but have demonstrated greater abundance of benthonic life in the deep sea than was previously recognized. In addition they found taxonomic diversities in the deep sea to be much higher than suggested by earlier workers. It is difficult to compare these diversities with those of shallower regions because faunal densities in the deep seas are quite low and environmental homogeneity cannot be determined easily; nevertheless, numbers of species have been compared with those of shallow tropical sea bottoms. A sidelight of the Woods Hole study is the demonstration of a marked dominance of deposit feeders in deep-sea deposits. The sparse concentration of suspended food makes suspension feeding relatively unprofitable.

Light intensity is obviously important in controlling plant distribution in the marine realm, as suggested by the very presence of the photic zone, in which light intensity decreases with depth to zero. Light toward the red end of the wavelength spectrum is more rapidly absorbed than light toward the blue end. While most plants can apparently utilize some light of all wavelengths, different species have different wavelength optima. Among the marine algae, the taxonomic group called the "green algae" tends to absorb radiation preferentially at the red end of the spectrum and is therefore largely restricted to intertidal and shallow subtidal depths. "Red algae" extend much deeper into the subtidal realm. Both groups contain species that secrete calcium carbonate skeletons and that are represented in the fossil record. The presence of untransported fossil green algae generally indicates ancient water depths of less than 60 or 70 meters.

For marine paleoecology one of the most important ecologic effects of light intensity is on the depth distribution of reef-building corals. Among present-day corals (Scleractinia, sometimes called hexacorals), there are two ecologic groups. Ecologic differences between the two groups are related to the presence of certain algae, called ***zooxanthellae,*** in the soft tissues of one group. There is evidence that the algae aid these corals by assimilating metabolic wastes, by helping in secretion of the calcareous skeleton, and by serving as a source of food. Large, massive colonial skeletons are restricted among hexacorals to those containing zooxanthellae. This group, called ***hermatypic*** ("mound-variety"), are the frame-builders of tropical coral reefs. Species lacking zooxanthellae are called ***ahermatypic*** corals. The light requirements of zooxanthellae restrict hermatypic corals to water shallower than about 100 meters. They do not build reefs effectively below about 70 meters, however, and certain species and growth forms are restricted to much shallower depths. Coral reefs are therefore useful indicators of shallow-water depositional settings. Present-day reefs are formed by large numbers of coral species, in association with calcareous algae and a rich invertebrate fauna (Figure 10-17). Deep-water coral "banks" are built by a few ahermatypic species in association with sparse invertebrate faunas and no calcareous algae. On this basis, Teichert (1958) concluded that what were formerly thought to have been coral reefs in the Late Triassic of southern Alaska, comprising skeletal material from only eight coral species, must actually have been comparable to the present-day deep and cold-water coral banks off the coast of Norway.

We must be careful in applying the limiting factors of present-day coral reefs to fossil reefs. Enough Late Triassic and Early Jurassic taxa survive today to warrant the assumption that the earliest (mid-Triassic) reef-building hexacorals contained zooxanthellae and had approximately the same depth limits as present-day forms. Major Paleozoic reef-building groups, such as tabulate corals and stromatoporoids, were taxonomically distinct from present-day hexacorals. We may never know whether they lived in association with zooxanthellae, but the bulk of independent evidence suggests that Paleozoic reef and reef-like structures, which exhibit high species diversity, grew in shallow water. Many such structures restricted water circulation in back-reef areas, suggesting that they grew to, or nearly to, sea level.

Fossil sponges have been shown to be useful indicators of water depth, though they are abundant in relatively few sedimentary rocks. Sponges are sessile, epifaunal suspension feeders (Figure 10-4,G). Living sponges of the class Calcarea, in which the supporting skeletal elements (called spicules) are calcareous, are most common on the sea bottom at depths of less than 100 meters, although a few species live in deeper

FIGURE 10-17
A coral reef knoll in a few meters of water near North Rock, Bermuda. The scale is 50 centimeters long. The large head of brain coral at the lower right and the one at the left center belong to different species of the genus *Diploria.* The sheet-like encrusting coral colony below the scale belongs to the genus *Montastrea.* The branching, tree-like forms are alcyonarian sea whips and sea fans. (Photo supplied by Peter Garrett.)

water. Living sponges of the class Hexactinellida, in which the spicules are siliceous, are most common in the depth range of 200–300 meters; few species therefore occur on continental shelves and almost none at depths of less than 10 meters (Reid, 1968). Finks (1960) has shown that the same ecologic relationships existed as long ago as the late Paleozoic. In the Permian rocks of Texas he has recognized shallow-water "shelf" sponge faunas, in which species of the Calcarea abound, and deep-water "basin" sponge faunas, in which the Calcarea are lacking but hexactinellids are numerous. The Delaware Basin faunas (page 253) form an example.

In general the sharpest ecologic break in the bathymetry of the oceans lies between the intertidal and subtidal zones. Species in the intertidal zone are subjected to rapid and extensive fluctuations in the physico-chemical environment, especially fluctuations in temperature and salinity. On soft bottoms many intertidal species (especially species of annelid worms, crustaceans, and bivalve molluscs) have become deep burrowers for protection. Evidence that such species lived in a certain place may be provided in the fossil record by preserved burrows. Figure 10-18 shows the deep burrow of the crustacean *Callianassa,* which lives today along intertidal sandy shores and has left a record of burrows extending back at least into the Cretaceous. Throughout its geologic range the genus is an excellent indicator of intertidal deposits. Seilacher (1964) and Rhoads (1967) have contrasted deep intertidal burrows with the shallow horizontal burrows of deep-water species, whch have less to gain from deep burial. Seilacher suggests that most deep-sea burrows belong to deposit feeders, as we discussed earlier (page 255), and follow patterns that permit systematic coverage of sediment layers containing food (Figure 10-12).

The foraminiferans were one of the first taxonomic groups to be widely employed as indicators of ancient water depth. Natland (1933) discovered that species of benthic foraminiferans living off the California coast are found in zones parallel to the shoreline. Most of the species are also found in Pleistocene and older Cenozoic sediments of California. It has therefore been possible to estimate water depths of these deposits and, from them, relative sea level changes during the Cenozoic. No one knows why foraminiferans show such distinct depth zonation. It may be that, as single-celled organisms, they are particularly sensitive to water pressure. Foraminiferal faunas also exhibit other trends along sampling transects perpendicular to shorelines. Some of these are depicted in Figure 10-19.

Determination of water depth of deposition for ancient sedimentary deposits and fossil assemblages must be approached through many lines of evidence. Frequently the final designation reflects general distance from shore (such as "nearshore" or "offshore") rather than actual water depth. Intertidal deposits can often be recognized from physical and chemical evidence, such as mudcracks and fine-grained dolomite, as well as from biologic information. Regional stratigraphy and associated evidence commonly permit recognition of shallow-water lagoonal deposits. Sandy-beach deposits may be recognized by a variety of characteristics, including sediment grain size, bedding features, and low species diversity. Deep-sea deposits may also be

A

B

FIGURE 10-18
Callianassid decapod crustaceans. A: A living *Callianassa* exposed in its burrow. B: External surface of a Recent burrow (×1). C and D: Similar burrows found in the Cretaceous Fox Hills Sandstone of the Denver Basin. (Photos provided by R. J. Weimer and J. H. Hoyt, A, B, and D from Weimer and Hoyt, 1964.)

C

D

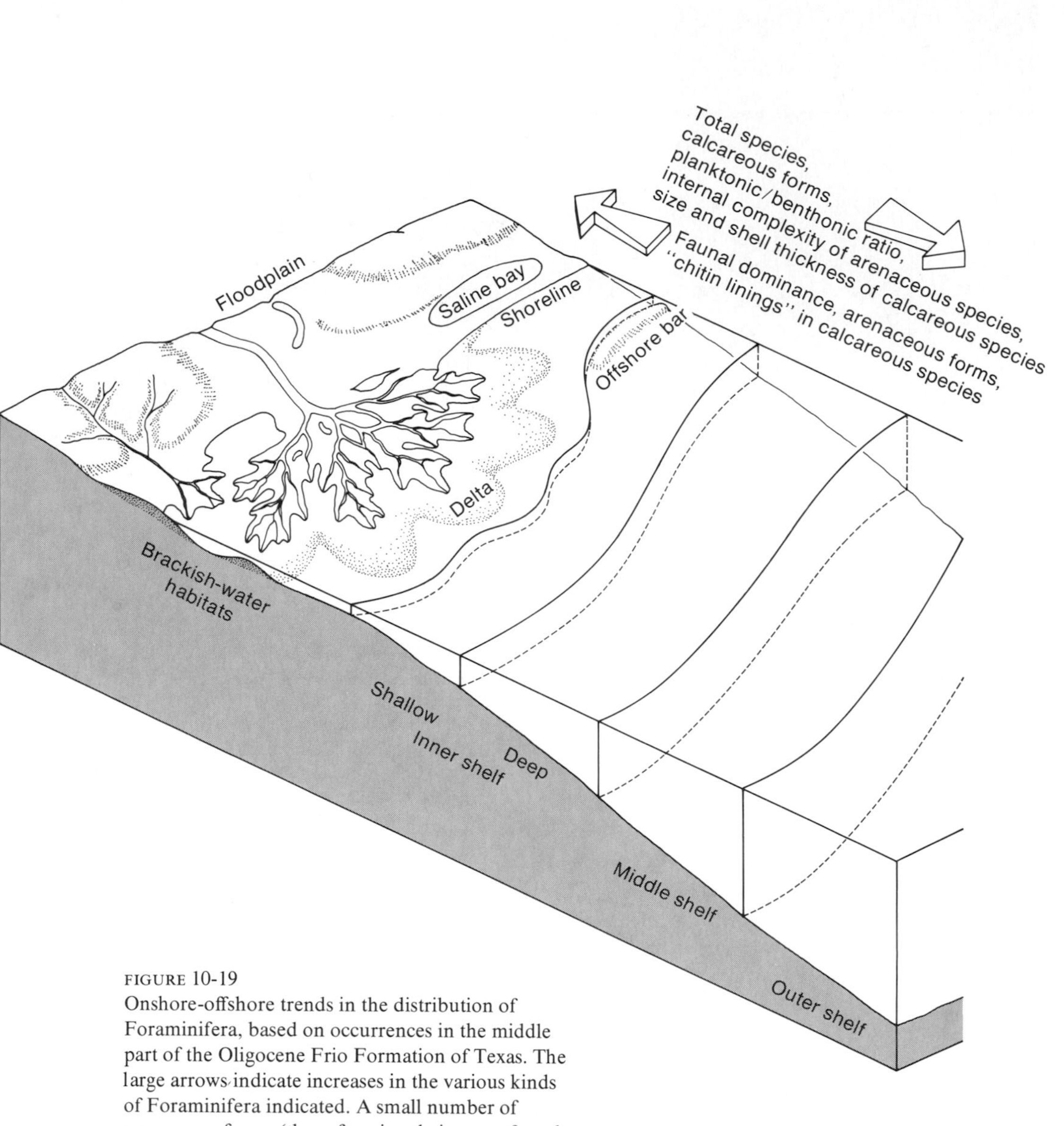

FIGURE 10-19
Onshore-offshore trends in the distribution of Foraminifera, based on occurrences in the middle part of the Oligocene Frio Formation of Texas. The large arrows indicate increases in the various kinds of Foraminifera indicated. A small number of arenaceous forms (those forming their tests of sand grains) dominate nearshore habitats. Calcareous species, and especially planktonic varieties, increase in an offshore direction. (From Gernant and Kessling, 1966.)

identified by certain sedimentologic and biologic features. Fine resolution, however, is seldom possible. Unfortunately it is especially difficult to estimate depth of deposition for level-bottom sediments and faunas of epicontinental-sea and continental-shelf deposits, which form much of our sedimentary rock record.

SALINITY

Salinity, which is a measure of the dissolved salt content of natural waters, is usually measured in parts per thousand. The value for undiluted seawater is about 35 parts per thousand. Table 10-1 shows the names generally applied to natural waters of various salinities.

TABLE 10-1
Classification of Salinities in Natural Waters

Descriptive term	*Salinity in parts per thousand*
Freshwater	0–0.5
Brackish water	0.5–30
Seawater	30–40
Hypersaline water	40–80
Brine	>80

SOURCE: Kinne, 1964

The effects of salinity on organisms vary from species to species, with respect to both the mean and the range of salinities tolerated. A species with a narrow tolerance range is said to be ***stenohaline,*** and one with a broad range, ***euryhaline.***

The most striking ecologic effect of salinity is on local species diversity. The largest number of aquatic species are marine forms. Many of these are stenohaline, often living in offshore areas where they are never subjected to appreciable salinity changes. A smaller number of aquatic species are stenohaline freshwater forms. Relatively few species can live in brackish *or* hypersaline water, but many species that tolerate either tolerate both. Brackish and hypersaline water is found in nearshore bodies of water (such as bays and lagoons), which do not undergo thorough or rapid-enough mixing with the open ocean to reach a stable salinity of about 35 parts per thousand (the Baltic Sea, Figure 10-20, is a good example); instead, salinity fluctuates with tidal movements, floods, storms, and seasonal climatic changes. Inhabitants of such unstable and stressful nearshore environments must be euryhaline. Apparently there have been relatively few euryhaline species at any time in geologic history. The few species that live in most brackish and hypersaline environments usually are present in great abundance because of reduced interspecific competition for ecologic niches.

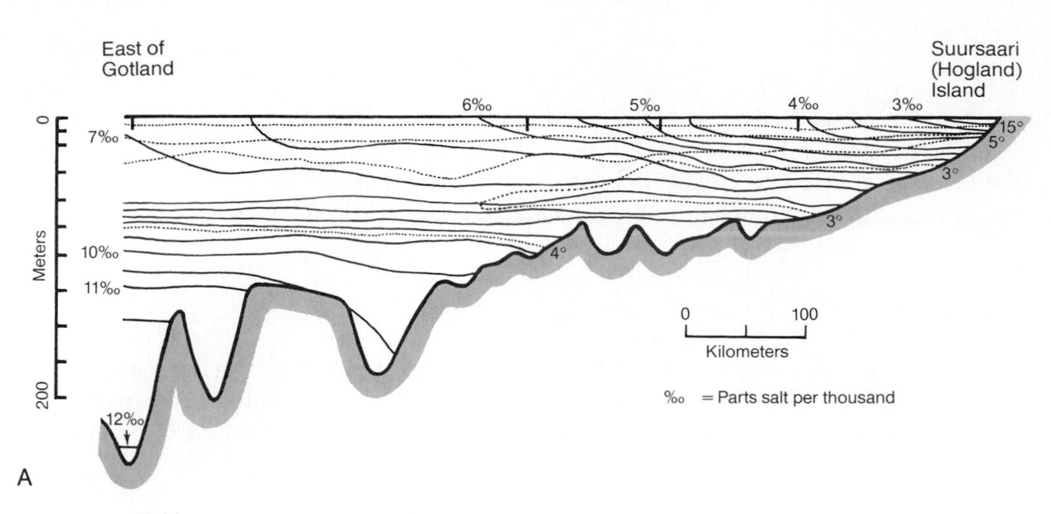

FIGURE 10-20
Relation of marine species diversity to salinity in the Baltic Sea. A: Schematic cross-section of the Baltic Sea from inner reaches of the Gulf of Finland to Gotland Deep, July, 1933. Solid contours show salinity in parts per thousand (‰) and dotted contours show temperature in degrees Centigrade. (From Jurva, 1952.) B: Molluscan percentages and salinity gradient along a slightly different transect. (From Sorgenfrei, 1958.)

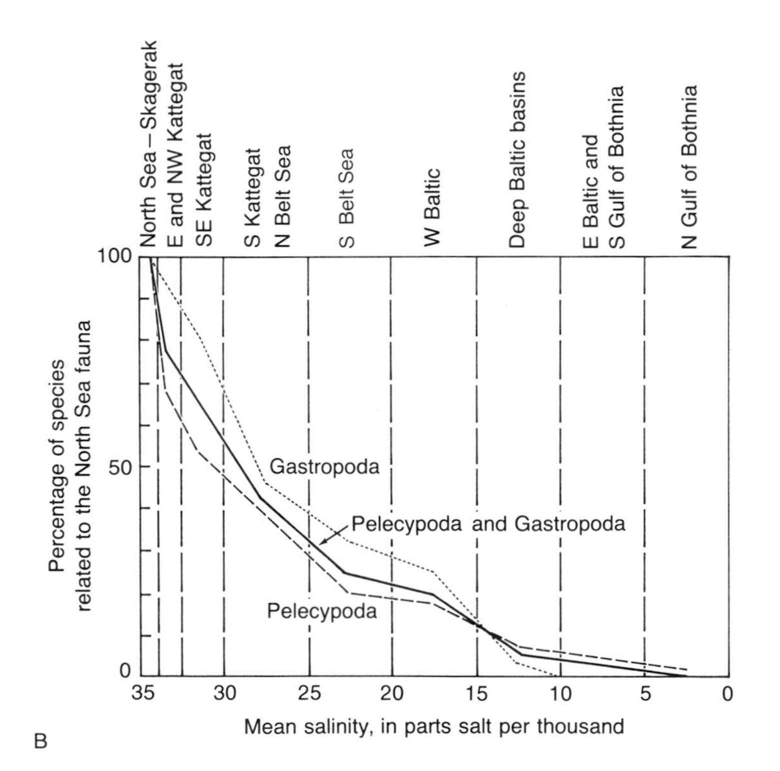

Most brackish-water species have evolved from marine, rather than from freshwater, forms, and few are capable of living in freshwater, though many can tolerate normal marine salinity.

Lowered species diversities within brackish and hypersaline environments can commonly be recognized in the fossil record on the basis of stratigraphic evidence documenting nearshore depositional environments. Especially important are contiguous terrestrial, lake, or river deposits, which can often be recognized by fossil content, sedimentology, and regional stratigraphy.

Most hypersaline water bodies exist in warm, arid climates. Their sediments commonly include minerals such as halite, gypsum, and anhydrite, that were formed by chemical precipitation due to evaporation. Such sediments are sometimes associated with dune deposits. Nearby river delta deposits, plant remains, or lake beds may be diagnostic evidence for brackish rather than hypersaline conditions.

Extrapolating backwards in time from a single living species to a related fossil species with respect to salinity tolerance is dangerous. One genus that appears to have been euryhaline for many millions of years is the inarticulate brachiopod *Lingula* (see Figure 10-37), which today lives in nearshore habitats, often in brackish water. The fossil record of this "living fossil" extends back at least to the Ordovician Period. A wealth of evidence suggests that *Lingula* occurs throughout its stratigraphic range commonly in nearshore faunas, which are characterized by low species diversity. Ostracodes of the family Leperditidae have been found frequently in Ordovician, Silurian, and Devonian sedimentary rocks that offer evidence of having been deposited in shallow, hypersaline aquatic environments.

Oxygen isotope determinations are sometimes used to estimate salinities for depositional environments of late Cenozoic sedimentary units. Isotopes O^{18} and O^{16} occur in calcium carbonate shells secreted by aquatic animals in ratios that differ from the ratios of those isotopes in the waters in which the animals live. The degree of difference between the ratios varies with temperature and has been measured in present-day seas. If isotope ratios are measured for well-preserved fossil shells and if temperature at the time of deposition is approximately known, then isotope ratios of the contemporary seawater can be calculated. Freshwater from land that mixes with shallow seawater and reduces its salinity normally has a distinct O^{18}/O^{16} ratio. Because mean local temperatures are reasonably constant, variations in the O^{18}/O^{16} ratios of fossil shells of shallow-water deposits may chiefly reflect degree of dilution by freshwater, which is to say, salinity. Studying Pliocene deposits of the Kettleman Hills, California, Stanton and Dodd (1970) found a good correlation between relative salinities determined by the O^{18}/O^{16} method and those estimated from the composition of the faunas, which contain many living species and genera (Figure 10-21,A). Absolute salinities could not be determined here because the O^{18}/O^{16} ratio of the diluting freshwater was unknown. The boundaries between fresh, brackish, and normal marine conditions were therefore determined subjectively, based largely on faunal evidence. Isotopic estimation of salinities is limited by conditions of preservation. Shells older than a few million years are generally so greatly altered by the

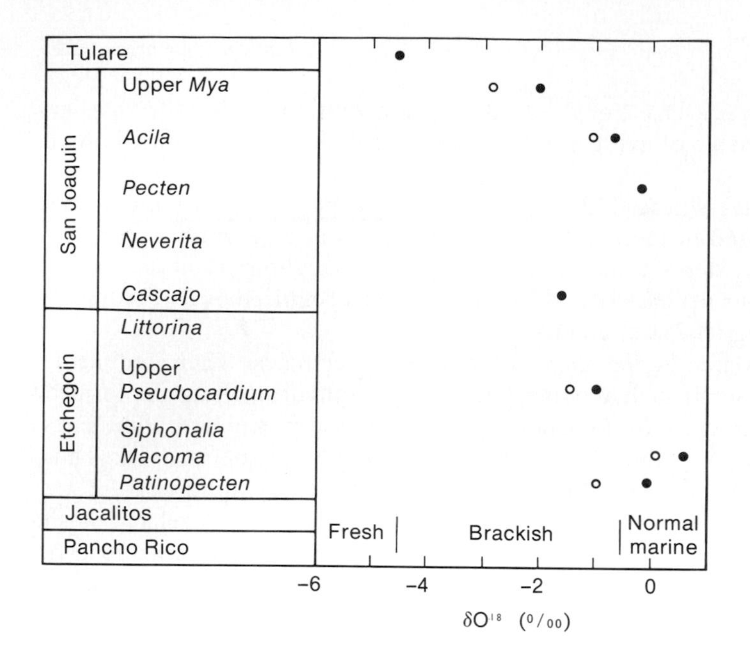

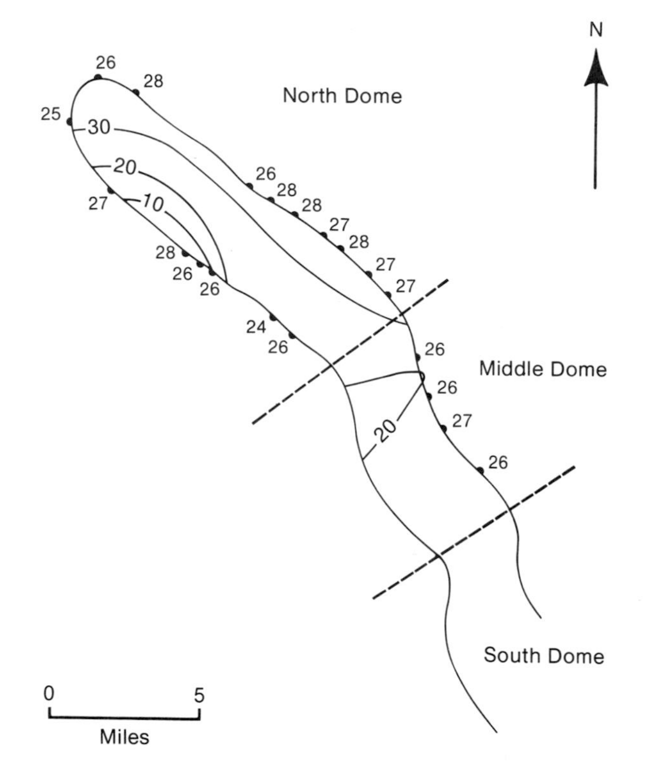

FIGURE 10-21
Variations in salinity of depositional environments of Pliocene deposits in California. A: Change in salinity through time. Formation names and biostratigraphic zones are shown at the left. The scale labelled δO^{18} records the O^{18}/O^{16} ratio determined for fossil bivalve shells. Salinity decreases with an increase in δO^{18}. B: Geographic distribution of mean number of ribs on single valves of *Anadara* in the *Pecten* zone of A. Contours are for estimated salinities, measured in parts per thousand (‰). (A from Stanton and Dodd, 1970; B from Alexander, 1974.)

removal and addition of chemical components that accurate results cannot be obtained. Even very young fossils must be exceptionally well preserved for measurements to be meaningful.

Sometimes the salinity of a shelled marine species' habitat is correlated with shell form. With decreasing salinity, the European cockle *Cardium edule* grows a thinner shell with fewer ribs. Working in the same Pliocene units as Stanton and Dodd, Alexander (1974) found that the bivalve *Anadara* exhibited the same kind of correlation between environmental salinity and rib number (Figure 10-21,B). To establish this correlation, he used salinity values estimated by Stanton and Dodd for the brief interval of time represented now by a biostratigraphic unit, the *Pecten* zone. For reasons outlined in the preceding paragraph, absolute values of salinity are only approximate, but a relative westward increase is apparent, and rib number for *Anadara* averages about 26 on the west side of the basin, compared to 27 on the east. We must always be cautious in applying a relationship such as that observed between rib number and salinity to the fossil record. For some extinct species, an inverse correlation may have held!

Hudson (1963) has related faunal distribution and diversity to salinity for the Jurassic Great Estuarine Series of England. He has delineated several faunal assemblages, each representing a group of species that tend to occur together in certain rock types. Hudson's conclusions are summarized in Figure 10-22. The assemblages do not form a simple stratigraphic sequence and they have not been traced laterally to determine how they may intergrade, but two independent lines of evidence have pointed to Hudson's interpretation. One is the salinity tolerances of certain of the fossil genera that survive today. *Unio,* a freshwater clam, and *Viviparus,* a freshwater snail, are both abundant in many regions today. *Mytilus* and *Modiolus* are common mussels in present-day seas, inhabiting brackish and normal marine waters, but most *Modiolus* species cannot tolerate salinities as low as those tolerated by *Mytilus* species. *Liostrea* is similar to Recent oyster species that occupy restricted, low-salinity bays. Hudson assumes there has been no major change in the absolute salinity of the ocean since the Jurassic. Arrangement of still-living genera in a sequence according to their present-day salinity tolerances (and inclusion of associated fossil species) produces the pattern shown in Figure 10-22. The scale of approximate salinities, which is determined by comparison with living relatives, yields important supporting evidence. The minimum species diversity falls in the lower brackish range (5-9 parts per thousand on the chart), and the maximum diversity falls in the higher brackish and marine range. Taxonomic diversity data suggest that the genera with living representatives have not tended to alter significantly their salinity preferences in the course of about 150 million years.

Certainly some aquatic groups, throughout their geologic history, have been largely stenohaline and have lived in normal marine water, while others have tended to be euryhaline. It is generally thought, for example, that most corals, cephalopods, articulate brachiopods, and echinoderms have been stenohaline-marine. Ostracodes, bivalves, and gastropods have tended to inhabit both freshwater and marine

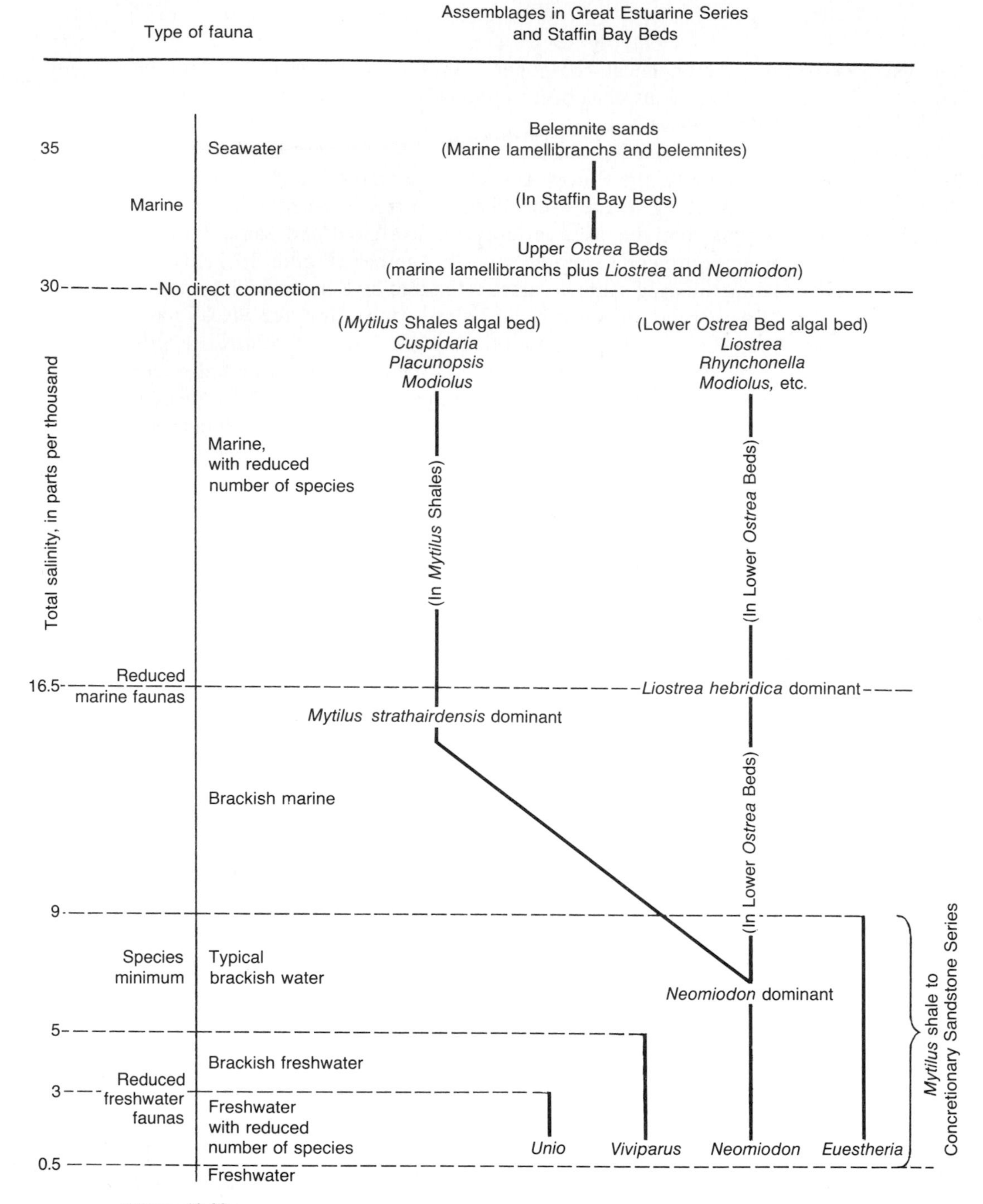

FIGURE 10-22
Salinity-determined faunal assemblages in the Great Estuarine Series of the British Jurassic. (From Hudson, 1963.)

environments and many species of these groups have been euryhaline. Arenaceous foraminiferans today are especially common in nearshore, low-salinity environments and are thus widely used as paleoecologic indicators.

From our knowledge of *Lingula,* leperditid ostracodes, and the genera discussed by Hudson, we might argue that all species of a genus have approximately the same salinity tolerance. Still we must be wary of oversimplification. Numerous living genera contain species with a variety of salinity tolerances. It is certainly not safe to make generalizations about salinity tolerances within a family. Many mollusc families, for example, show considerable variation. Some, such as the Ostreidae (oysters), Mytilidae (mussels), and Neritidae (intertidal snails), have demonstrated a general tendency to produce brackish-water taxa, but all have also produced stenohaline marine forms. Strict taxonomic arguments should be combined with species diversity data and sedimentologic and stratigraphic arguments in any attempt to determine approximate salinities of ancient depositional environments. In most cases only crude conclusions can be reached even when all approaches are combined. We are doing well simply to distinguish between fossil assemblages from freshwater, brackish water, marine, and hypersaline environments. The further back in the fossil record we venture, the poorer is our resolution.

SUBSTRATUM

Soft substrata are the preservation medium for most fossils. They are especially important in paleoecology because, excepting the fossils themselves, the character and distribution of sediments are the sole source of information for reconstructing ancient habitats.

The fabrics of sedimentary rocks commonly reflect depositional processes, and the geometric configurations of sedimentary rock units commonly reflect depositional settings. Sedimentary particles range in grain diameter from clay size (<0.004 mm) through silt size (0.004–0.0625 mm) and sand size (0.0625–2 mm) to pebbles, cobbles, and boulders. Mud includes clay-size and silt-size particles. Grain size commonly reflects the degree of wave or current agitation in the depositional environment. Fine-grained sediments are usually deposited in quiet-water areas; well-sorted, coarse-grained deposits are most common in areas of strong wave and current activity. Most marine sediments are one of two types: (1) terrigenous sediments are land-derived and consist primarily of silicate fragments, especially of quartz, feldspar and clay particles, and (2) calcium carbonate particles are primarily skeletal debris and aragonite needles precipitated by algae or inorganic processes. As yet, terrigenous and carbonate sediments have not been shown to have differing effects on the distribution of benthonic organisms: the main effects of substratum are associated with ***grain size.***

One of the most important relationships between substratum and the distribution of benthonic organisms concerns feeding mechanisms. Sanders (1956) has shown that deposit feeders are more abundant than suspension feeders in muddy sediments, whereas the reverse is true in sandy sediments. Sanders has explained the occurrence of deposit feeders in muddy sediments on the basis of food supply. Fine particulate organic matter for deposit feeding settles to the bottom in quiet-water areas where mud also accumulates, but is winnowed out of coarse-grained sands that are subjected to greater agitation. Rhoads and Young (1972) have shown that the loose surface layer of soft muds tends to clog feeding apparatuses of many suspension feeders. For this reason most suspension-feeding species are rare in muddy substrata.

Relationship between sediment type and feeding mechanism can be seen well back into the Paleozoic record, in which, for example, nuculoid clams (Figure 10-23) are found in many mudstones and shales. Modern nuculoid clams depend largely on deposit feeding for food gathering and live predominantly in muddy sediments. Because basic morphologic features of the superfamily Nuculacea are associated with the group's deposit-feeding mechanism, homology as well as position in the stratigraphic record suggests that most nuculoid clams of the Paleozoic were deposit feeders.

Substratum is also important relative to method of attachment and locomotion of benthonic animals. We will consider four common types of substratum in this light. Rocky bottoms are primarily sites for the attachment of epifaunal species. Only a few infaunal species bore into rock or nestle in pre-existing crevices. Because rocky surfaces are seldom horizontal and are commonly scoured by currents and waves, attachment is necessary for most species that colonize them. However, because they usually represent areas of erosion rather than deposition, large rocky surfaces are seldom preserved in the stratigraphic record. More commonly preserved are solid-surface microhabitats, in the form of cobbles, boulders, or shell debris.

In an exceptionally detailed study of the Cretaceous Chalk of Denmark, Surlyk (1972) analyzed the ecology of brachiopods attaching to shell debris. The Chalk is weakly lithified. Surlyk disaggregated about 150 five-kilogram samples of the Chalk and obtained about 50,000 brachiopods belonging to 43 species. It appeared that neither on the quiet sea floor of deposition nor after entombment in the Chalk sediment were many shells moved or destroyed. In other words, preservation of the fauna was truly exceptional. The Chalk is largely of biogenic origin, consisting primarily of coccoliths, which are tiny calcium carbonate plates of planktonic algae. The plates are composed of the mineral calcite, which is less soluble in acidic waters near the earth's surface than is aragonite, the other form of calcium carbonate commonly secreted by organisms. This largely accounts for the weak lithification of the Chalk and the excellent state of preservation of fossils.

Surlyk found that the center of the vast basin in which the coccoliths and other small particles settled was generally devoid of large sedimentary particles, except for the skeletal remains of certain benthic creatures. Brachiopods, which required hard substrata for temporary or permanent attachment by a pedicle, seem to have depended heavily on the death of bryozoan colonies as a source of skeletal debris.

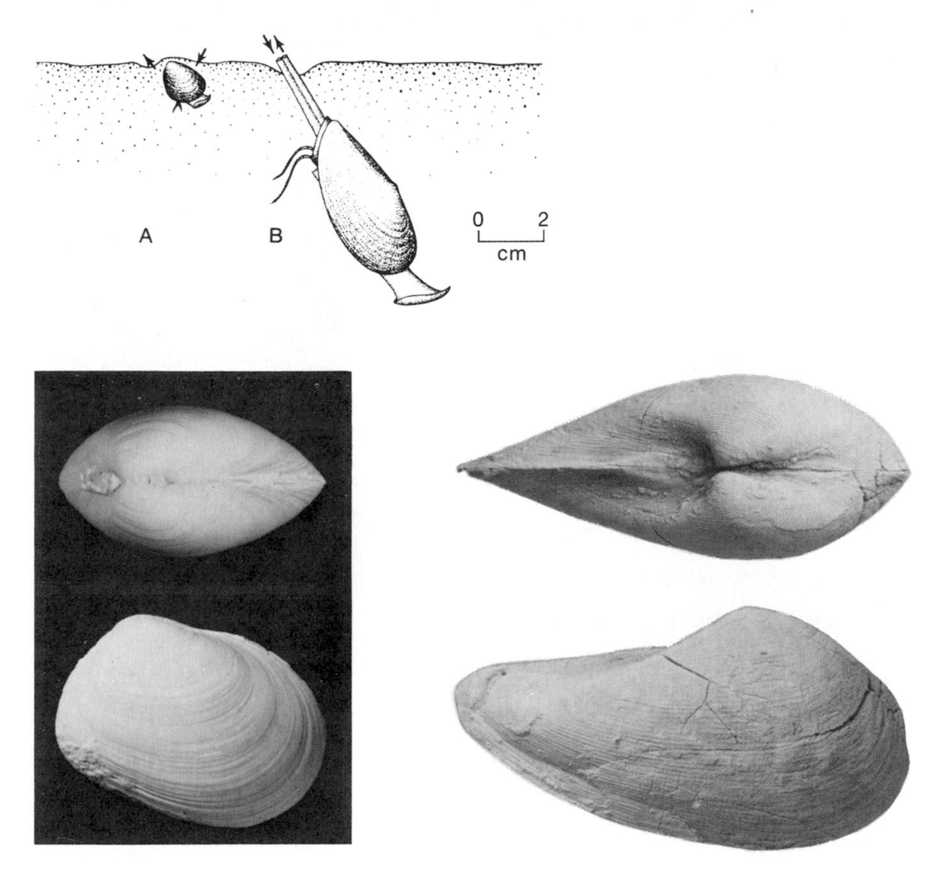

FIGURE 10-23
Nuculoid clams: A, *Nucula,* and B, *Yoldia,* both feeding by appendages of the lip region called palp proboscides; C, *Clinopisthia* (dorsal and right lateral views), a Carboniferous genus whose life habits were probably similar to those of *Nucula;* D, *Polidevcia* (dorsal and right lateral views), a Carboniferous genus whose life habits were probably similar to those of *Yoldia.* (A and B from Stanley, 1968; photographs of C and D courtesy of A. L. McAlester and E. G. Driscoll.)

Surlyk found a strong correlation between abundance of brachiopods and abundance of bryozoans. He was also able to group the brachiopod species into four life-habit groups by means of homology and inferences based on functional morphology (Figures 10-24 and 10-25). Species having different life habits apparently had different population dynamics that reflected their differing substratum requirements.

I. Attached to the substrate by means of a pedicle: 24 species.

(a) Minute forms able to use very small substates: 20 species

Terebratulina faujasii (Roemer) — 4.6 mm

Terebratulina longicollis Steinich — 5.2 mm

Terebratulina subtilis Steinich — 2.5 mm

Rugia tenuicostata Steinich — 3 mm

Rugia acutirostris Steinich — 2.5 mm

Rugia tegulata Surlyk — 2.5 mm

Rugia spinosa Surlyk — 1.3 mm

Gisilina gisii (Roemer) — 5.5 mm

Gisilina jasmundi Steinich — 4.5 mm

Dracius carnifex Steinich — approx. 4 mm

Aemula inusitata Steinich — 3.6 mm

Scumulus inopinatus Steinich — 2.5 mm

Argyrotheca bronnii (Roemer) — 5 mm

Argyrotheca coniuncta Steinich — 3.7 mm

Argyrotheca n. sp. aff. *coniuncta* — approx. 4 mm

Argyrotheca hirundo (Hagenow) — 4 mm

Argyrotheca obstinata Steinich — 4 mm

Argyrotheca stevensis (Nielsen) — 4 mm

Argyrotheca bronnii s.l. (late form) — 5 mm

Dalligas nobilis Steinich — approx. 5 mm

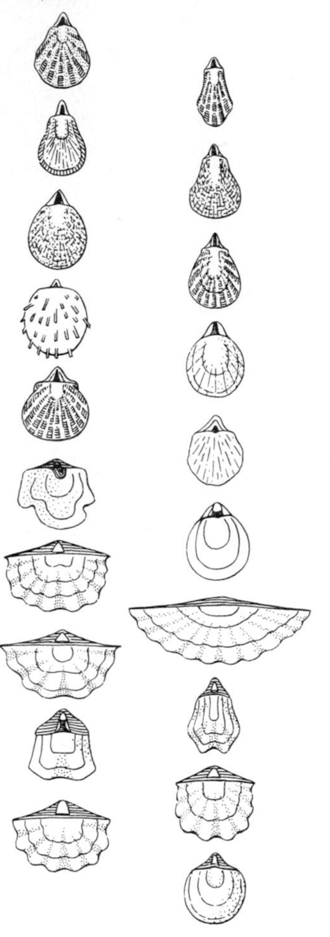

(b) Medium to very large sized forms confined to large, hard substrates: 3 species

Neoliothyrina obesa Sahni — 70 mm

Neoliothyrina fittoni (Hagenow) — 15 mm

Kingena pentangulata (Woodward) — 17 mm

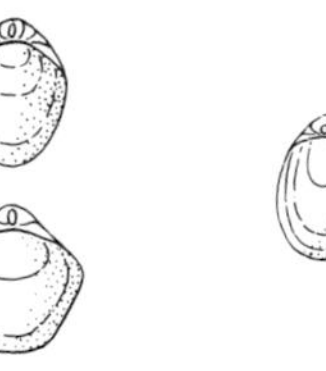

(c) Attached directly to the sediment: 1 species

Terebratulina chrysalis (Schlottheim) — 22 mm

II. Secondarily free-living forms, medium to large sized: 10 species

Cretirhynchia limbata (Schlottheim) — 12 mm

Cretirhynchia retracta (Roemer) — 25 mm

Cretirhynchia sp. — approx. 30 mm

Carneithyris subcardinalis (Sahni) — 45 mm

Terebratulina gracilis (Schlottheim) — 13 mm

Trigonosemus pulchellus (Nilsson) — 20 mm

Gemmarcula humboldtii (Hagenow) — 20 mm

Magas chitoniformis (Schlotteim) — 11 mm

Meonia semiglobularis (Posselt) — 4 mm

Thecidea pappilata (Schlottheim) — 10 mm

III. Burrowing forms: 1 species

Lingula cretacea Nilsson — 8 mm

IV. Attached to the substrate by cementation: 8 species

(a) Attached to the very small substrates: 2 species

Isocrania costata (Sowerby) — approx. 7 mm

Isocrania barbata (Hagenow) — approx. 7 mm

(b) Confined to large, hard substrates: 6 species

Ancistrocrania tubulosa (Nielsen) — 8 mm

Crania antiqua Defrance — 14 mm

Crania aff. *craniolaris* (Linnaeus) — 7 mm

Vermiculothecidea vermicularis (Schlottheim) — 12 mm

Thecidea recurvirostra Defrance — 7 mm

Bifolium wetherelli (Morris) — 4 mm

FIGURE 10-24
Brachiopods of the Danish Chalk grouped according to life habits. (From Surlyk, 1972.)

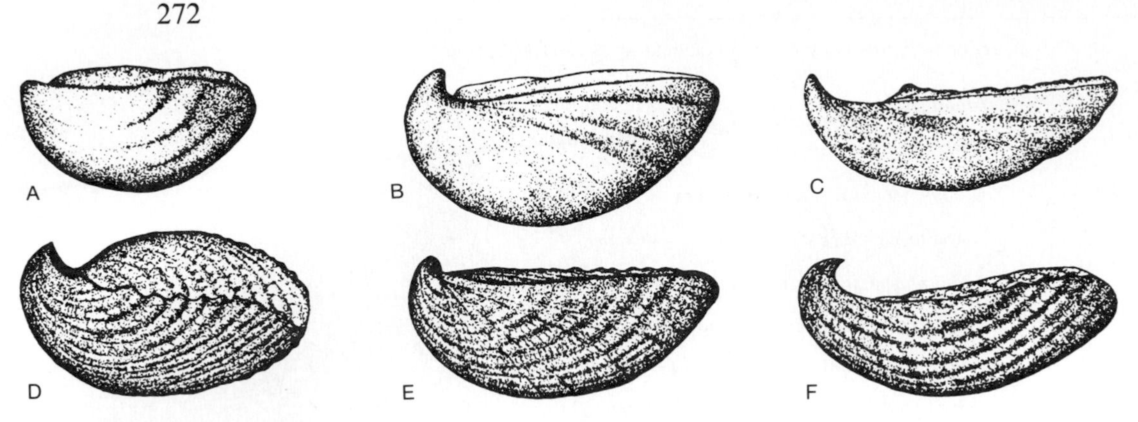

FIGURE 10-25
Lateral views of secondarily free-living, hemispherical brachiopods of the Danish Chalk. A: *Meonia semiglobularis.* B: *Magas chitoniformis.* C: *Thecidea pappilata.* D: *Gemmarcula humboldtii.* E: *Terebratulina gracillis.* F: *Trigonosemus pulchellus.* (From Surlyk, 1972.)

Because of the remarkably complete preservation of the Chalk populations, Surlyk was able to uncover these differences using techniques like those discussed in Chapter 4. He found that species of Group I, which were very tiny, attached to the substratum by means of a pedicle. Juvenile mortality was very great, apparently because the small size of the young brachiopods made them vulnerable to burial in the fine sediment. Species of Group II lived free on the surface of the sediment as adults. Their larger adult size made this mode of life feasible by reducing the likelihood of burial or clogging by sediment (see page 245 and Figure 10-11). These species did, however, face imminent burial or clogging early in their ontogeny, and the growth curves of these species offer evidence that the animals grew rapidly throughout their juvenile stage, perhaps as a special adaptation to enable a large fraction of individuals to gain the protection afforded by large size.

Not all occupants of hard substrata settle upon all types of surfaces. Thayer (1974) showed this for several taxa in the Devonian Genesee Group of New York State. Perhaps the most interesting of his findings was that a species of worm believed to have been a spionid polychaete bored preferentially into spiriferid brachiopods, which have impunctate shells. This observation was taken to support the suggestion of Owen and Williams (1969) that brachiopod punctae (perforations of the shell that pass from the interior almost to the surface) housed fleshy protuberances that secreted substances repellant to boring organisms.

Of the types of soft substrata, shifting sands are the least easily colonized by benthonic organisms. They tend to be inhabited by a few, highly mobile infaunal species that can reburrow rapidly when exhumed by currents or waves. That fossil remains are absent from or very scarce in many beach deposits of the geologic record reflects not only unfavorable conditions for preservation, but also unfavorable conditions for life. A few kinds of extinct organisms have characteristically occupied shifting sands. The bivalve genus *Tancredia* (Figure 10-26) has been found in sandy shoreline deposits of the Jurassic and Cretaceous (Speden, 1970). Its form is that of a rapid burrower (Stanley, 1970) and, as would be expected, few other fossils are found with it. One type that is found is the corncob-like burrow of callianassid crustaceans (Figure 10-18). As discussed earlier in this chapter, callianassids also often occupy shallow, sandy beaches. In the Paleozoic the pentamerid brachiopods (see Figure 10-29) commonly inhabited somewhat unstable, sandy substrata. Their habitat preferences will be discussed more fully in the section of this chapter entitled "Soft-Bottom Communities."

Muddy bottoms in quiet-water areas tend to be soft and soupy. Their shelled infauna consists largely of small animals, many of which have thin shells. Large, thick-shelled species would sink into soft mud; the feeding and respiratory mechanisms of many would be unable to function properly. Because of problems of attachment, flotation, and clogging of feeding and respiratory mechanisms, few epifaunal species inhabit soupy muds.

Substrata composed of mud-sand mixtures usually differ from shifting sand and quiet mud bottoms in being both stable and firm. Commonly such substrata are inhabited by a wide variety of species. Animals that form permanent dwelling burrows (tubes or channels) and do not cement the walls of their burrows are largely restricted to muddy sand because of its cohesiveness.

FOOD

Earlier in the chapter we presented a classification of modes of feeding in the marine ecosystem. Quite a different classification is necessary for terrestrial animals. Terrestrial herbivores naturally display different tooth and jaw structures from those of terrestrial carnivores. The main kinds of large herbivores are ***grazers*** (grass and ground plant eaters), ***browsers*** (leaf eaters), ***granivores*** (seed eaters), and ***frugivores***

FIGURE 10-26
Tancredia, a Mesozoic genus of rapid-burrowing bivalves. The anterior (left-hand portion) is elongated, reflecting the presence of a large foot and a pointed shell, for rapid penetration of substrata.

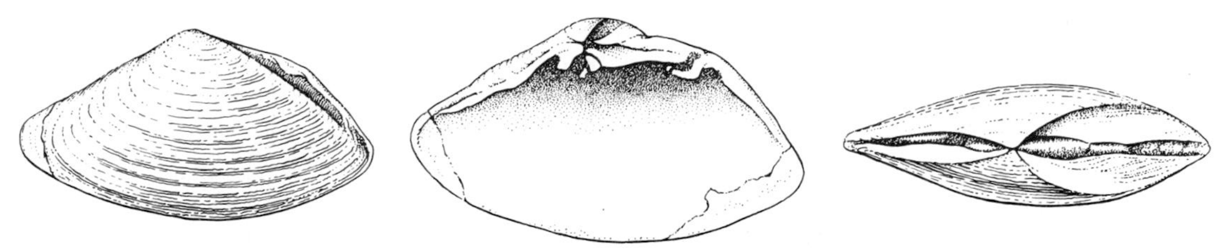

(fruit eaters). On a smaller scale, some insects suck nutritional substances from plants or mine their leaves. The striking difference between plant consumption in the terrestrial and marine realms is that in the latter most plant food is in the form of small particles when consumed. Single-celled algae and bacteria represent small particles even before they are broken down, and larger plants are often broken apart and decomposed by bacteria to form organic detritus that is eaten as it floats in the water or after it becomes buried in sediment. Some organic detritus forms in the ocean and some is washed there by rivers and streams.

The distribution of deposit-feeders and suspension-feeders in relation to sediment grain size, discussed in the previous section, exemplifies the way in which limiting factors tend to be interrelated. In a later section of the chapter we will discuss the degree to which food may limit population sizes for various types of animals.

Spatial Distribution of Populations

We have already noted that populations tend to be distributed in characteristic spatial patterns.

Even within areas a few meters square, many benthonic marine species tend to be aggregated in clusters of individuals. There are two main causes for this type of distribution pattern: inhomogeneities in the physical environment and the reproductive and social behavior of the species. The second of these causes appears to be the more important. A striking illustration is provided by species of marine grass that carpet shallow-water areas in many parts of the world (Figure 10-27), propagating partly by means of runners that spread laterally beneath the sediment surface. Once an individual plant grows from a seed in a previously barren area its runners may produce a discrete grass patch. Many submarine grass patches are separated by areas of barren sediment. Grass alters the bottom environment by stabilizing the sediment and commonly acts as a baffle-like trap for fine-grained sediments and also offers new substrata (plant stems and roots) for colonization by a variety of organisms. Certain species of clams, snails, sea urchins, and worms are largely restricted to the "grass-flat" habitat.

Mode of formation of local populations is another factor causing patchy distribution of benthonic invertebrates. Most species produce large numbers of larvae that spend variable periods of time (averaging about three weeks) as members of the plankton before settling to the bottom to metamorphose. There is much debate about the ability of larvae to choose suitable substrata for settling, and also about the effects of unfavorable benthonic conditions in delaying settling. At any rate, larvae of a given species often tend to be transported together, as a population, by currents and to settle in one area. Often, too, they are gregarious when settling and are attached to adult animals of their species, which have obviously enjoyed some degree of success where they settled. Thus, successful settlement is often in clumps whose distribution appears to be unrelated to basic inhomogeneities in the botton environment.

FIGURE 10-27
Marine turtle grass in the Mediterranean Sea off the coast of Spain. Note the roots exposed at the margin of the area where currents have scoured out the bottom. Such scoured areas usually support fewer benthonic animal species than are found in adjacent grassy areas. (Photograph supplied by E. A. Shinn.)

Benthonic species that lack a planktonic larval stage are even more likely to be found in patchy distributions. The ontogeny of some such species includes "brooding" of eggs and larvae and release of juveniles near the parent.

Finally, some vagile species tend to aggregate as adults, apparently in areas in which food supply or other factors make the environment especially suitable for habitation.

Some aggregation of benthonic species is caused by inhomogeneities of the physical benthonic environment. Substratum, rather than salinity and temperature, is especially important here. The common mussel *Mytilus* attaches to hard objects by a byssus. Even in sandy areas a few "pioneer" mussels can settle on scattered shell fragments or pebbles. Their shells commonly form attachment sites for additional mussels. The result may be an aggregated distribution of mussels over large areas (Figure 10-28). Clumps of Paleozoic brachiopods seem to have formed in the same way, as documented by Ziegler and co-workers for certain Silurian groups. Accumulation of shells of dead brachiopods nucleated colonies of several species, some of

FIGURE 10-28
A: Clumps of the mussel *Mytilus edulis* on a broad intertidal sand flat at Barnstable Harbor, Massachusetts. B: Closeup view.

which attached to the shells of other living individuals. The Paleozoic brachiopod genus *Pentamerus* left numerous colonies of closely packed individuals preserved in life position (Figure 10-29). It is uncertain whether adults were attached by ***pedicles*** (fleshy stalks possessed by many brachiopod genera) or rested free on the sediment. At any rate their orientation with the pedicle region directed downward suggests that they did not attach to each other, but simply aggregated during larval settlement. (The adult brachiopod had no means of locomotion.)

In environments in which patchy marine populations have been preserved in place, their fossil assemblages may retain the original distribution pattern. A common modifying factor is time. In areas in which the location of plankton, and hence that of the larval stage of the organisms, shifts from year to year and sedimentation rates are

FIGURE 10-29
Undersurface of a cluster of the brachiopod species *Pentamerus oblongus* preserved in life position, Silurian Red Mountain Formation, Alabama (×0.45).
(From Ziegler, Boucot, and Sheldon, 1966.)

low, the record of preservation over many years may tend to even out inhomogeneities in the distribution of single year classes.

It is important to understand that while there must always be environmental factors explaining the success of a certain species in a particular setting, the absence of the species in other similar settings may be largely the result of chance, principally because of the vagaries of larval transport and the transient, short-lived nature of many local populations. A uniform depositional environment does not necessarily support a uniformly distributed population or produce a uniformly distributed fossil assemblage.

Fossil Communities

We have considered life habits and local distributions of populations belonging to single species. Commonly these topics are united under the heading ***autecology,*** the study of the interactions between single species and their environments. The study of interrelations between two or more species and their environment is called ***synecology.*** Actually, part of the environment of any species is biologic, consisting of other species. Autecology and synecology are, therefore, not clearly separable, but represent different emphases in the study of ecology.

Consideration of species living in association with each other has led to the concept of the biologic ***community.*** Most present-day workers speak of a community as any natural assemblage of species living in a given area, and this definition will be adopted here.

Since it has long been recognized that certain groups of plants and animals tend to occur together in nature, some workers have specified that to be a community, an assemblage must recur in many areas with little variation in species composition. We will instead consider such an association to be a special type of community called a ***recurrent community.***

Finally, some ecologists have suggested that the label "community" should imply interaction and interdependence among component species. In some instances, strong interaction among neighboring species is obvious. In others, evidence for interaction of certain species pairs is lacking. It is generally agreed that interaction should be excluded from the definition of "community."

LIMITING FACTORS AND SPECIES INTERACTIONS

The ways in which species interact are in part determined by the kinds of limiting factors that affect them. In this light it is useful to divide the marine limiting factors we have considered into two groups—those that organisms consume or use up (dissolved oxygen, living space, and food) and those that they do not (depth, salinity, and temperature). The depletable factors are often referred to as ***resources.*** Species that vie for a resource that is in limited supply are said to compete, and ***competition*** is a very important kind of interaction in many organic communities. The phrase "in limited supply" is very important. A resource that is supplied as rapidly as it is used up, or more rapidly, cannot be the object of competition. Populations that use such a resource must be kept in check by some other resource or environmental condition. Dissolved oxygen may be the object of competition in certain oxygen-deficient basins, but most marine waters are well circulated and supplied with a superabundance of atmospheric oxygen. Area or volume of substratum and food are more commonly limiting factors for marine benthos. Area of substratum is a potentially limiting resource for epifauna and volume of substratum, for infauna.

Under certain circumstances it may happen that no resource limits the local abundance of a population. How can this be? For one thing, severe predation may keep abundance of individuals at such a low level that all resources are superabundant. Disease or parasitism, which are often grouped with predation in studies of theoretical ecology, sometimes perform the same kind of function, as may lethal agents of the physical environment. The situation is not always simple. Factors limiting population size vary among different kinds of organisms and among different kinds of habitats. They may even vary from place to place for a particular kind of organism within a single habitat. Sometimes there is a seasonal alternation of two limiting factors. Territorial behavior is important in this regard. All space available to a local population of mobile animals may be divided up into territories, which means that space is limiting. Food may well be superabundant most of the time, which raises the question of why large territories are maintained. The answer often seems to be that at times when food is especially scarce, individuals that maintain large enough territories to survive are favored by natural selection.

What generalizations can be made about the distribution of limiting factors in the marine ecosystem? At present, not enough studies have been undertaken for us to give a complete answer. We will offer some examples of the kinds of conclusions that have been reached. Soft-substratum habitats and organic reefs are of special paleontologic interest and will be dealt with in separate sections that follow. To provide a conceptual framework, we can focus first upon planktonic systems and hard-substratum habitats other than organic reefs.

Planktonic systems are especially difficult to study because they are constantly in motion. Recent phytoplankton are preyed upon quite extensively (see Figure 10-6), but when certain species of planktonic algae are in spring or fall bloom (times of rapid increases in biomass), they are apparently limited by the supply of nutrients (especially by supplies of nitrogen or phosphorus). The blooms are initiated in part by the stirring of water masses, which brings nutrients up to the photic zone. In areas where blooms are waning or have not recently developed because nutrients are lacking, zooplankters may multiply sufficiently to limit algal densities. Zooplanktonic organisms themselves are heavily preyed upon by fishes and other creatures, including other zooplankters, and appear sometimes to be limited by predators. This condition can produce a delicate balance, because the degree to which zooplankton are predator-limited will influence the degree to which they limit the abundance of phytoplankton. In the ocean, as in many ecosystems, top carnivores (see Figure 10-2) tend to be limited by food because they are not preyed upon.

Ecosystems of rocky intertidal surfaces are convenient to study and have been widely investigated. Most species attaching to hard surfaces are suspension feeders. These may compete for food, especially if they occupy crevices or cavities where water flow is restricted. Competition for space appears to be more common, however, which is to say that as populations grow, space is often depleted before the rate of food consumption exceeds the rate of food supply. Often, however, neither food nor space is limiting. Paine (1966) demonstrated this for one rocky intertidal community by removing the dominant predators. Before this modification, numerous species of multicellular algae, barnacles, and other kinds of organisms had naturally coexisted with the mussel *Mytilus.* When predators were excluded, *Mytilus* overgrew and killed off the other species: space became a limiting resource monopolized by one kind of organism. In many benthic settings space would be limiting if predation or physical disturbance were not so severe. In other words, predation commonly increases local diversity by preventing competitive exclusion. Dayton (1971) has shown that for a typical rocky intertidal surface in California, competition for space is a minor limiting factor. There, high mortality is more significant as a limiting factor because organisms are damaged by violent wave action and wave-tossed debris or perish from desiccation.

A basic question about the composition of any association of species in nature is whether it may be expected to change toward an equilibrium composition. In other words, as a vacant habitat is colonized by whatever species have access to it, will the association of species tend to alter toward a particular final, or ***climax,*** community regardless of the order in which the species become established? Such a transition is known as an ***ecologic succession.***

Where competition is a dominant factor and a habitat is stable over a long period of time, succession is sometimes observed. As an example, in forests of central New Hampshire, beech trees dominate in the climax forest (Forcier, 1975). The dense shade that they produce tends to prevent the establishment of yellow birch and sugar maple seedlings. In areas made barren by wind or other physical disturbance, however, yellow birch tends to invade rapidly. It is a ***pioneer species*** that survives by virtue of its ability to colonize rapidly. Such a species is sometimes called an ***opportunistic species.*** Although they are effective invaders, opportunistic species typically are poor competitors. Seedlings of yellow birch do not colonize effectively in the shade of adults of the same species. Seedlings of sugar maple are more successful, and as they grow into tall trees, yellow birch is shaded out and dies. Beech seedlings are superior competitors under the sugar maple canopy, and they, in turn, overgrow and shade out sugar maples to form the ***climax community.*** Frequent disturbance in the form of destruction by wind opens areas for invasion by yellow birch and the cycle is repeated. More severe disturbance, such as burning by forest fires, can carry the succession back to an even earlier stage, in which other kinds of vegetation invade before yellow birch.

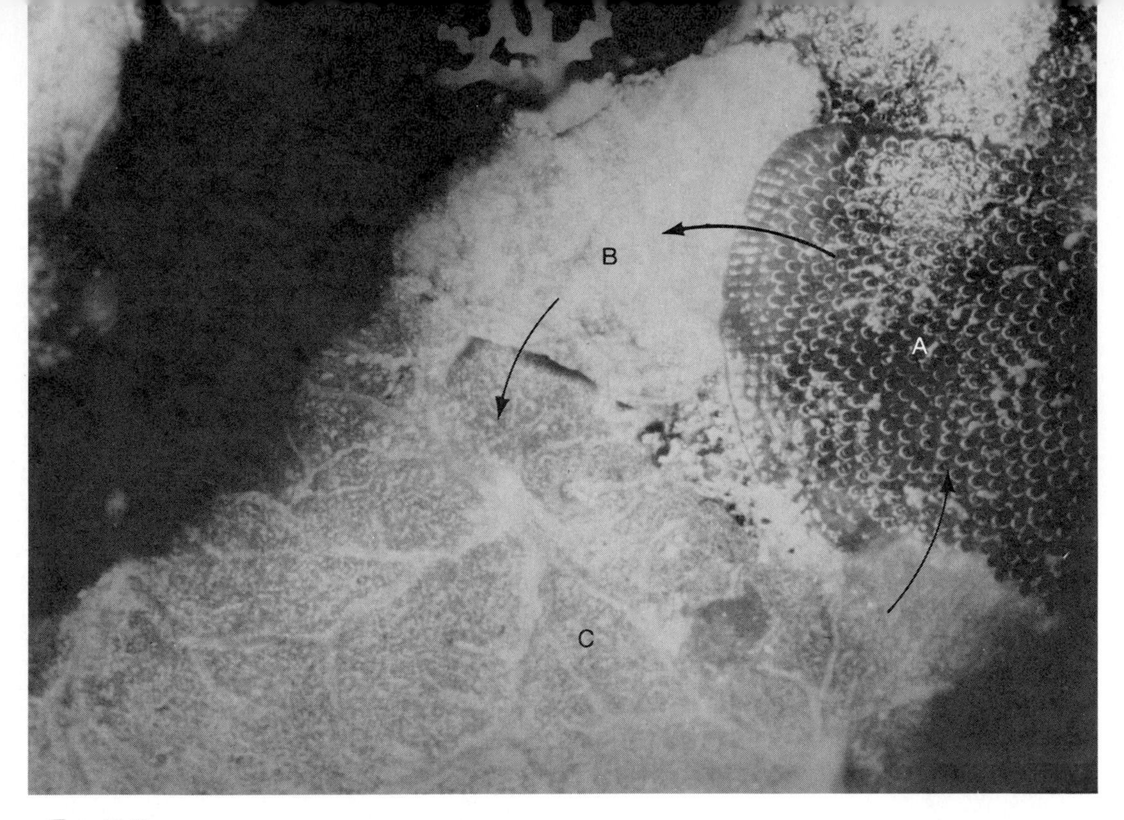

FIGURE 10-30
A competitive network of colonial organisms in the form of cyclical overgrowth on the dead undersurface of a tabular coral colony, at a water depth of about 15 meters at Rio Bueno, Jamaica. Species A (the bryozoan *Steganoporella magnilabris*) is overgrowing species B (the bryozoan *Stylopoma spongites*). Species B is overgrowing species C, an unidentified sponge, and species C is overgrowing species A. (Photo courtesy of Jeremy B. C. Jackson.)

Where competition is less important or where predation or environmental change frequently disturbs communities, a climax association is seldom attained. Sutherland (1974) has summarized evidence that historical factors play an important role in many habitats. The timing and success of recruitment (establishment of juvenile organisms) and the somewhat haphazard effects of predation and other agents of mortality often prevent any stable climax community from forming. Jackson and Buss (1975) have recognized another complicating factor. Studying epifaunal associations of sponges, bryozoans, corals, and other organisms that occupy hard substrata in the tropics, they found that there is no simple hierarchy of competitive dominance. For example, species A may dominate in competition with species B, which may dominate in competition with species C, and yet C may dominate in competition with A. This happens because competition advances by a variety of mechanisms, including overgrowth and toxic destruction of neighbors (Figure 10-30). As a result, it takes a very long time for any single species to take over a given surface, and usually some sort of disturbance will prevent this from happening. As bare spaces appear, new species with varying competitive relationships are added, and the system tends to remain in a constant state of flux.

In the past few years a number of paleoecologists have undertaken what are commonly called trophic analyses of assemblages of fossil benthos. Their procedure has been to assign species of an assemblage to feeding categories (for example, suspension feeders that feed just above the sediment surface) and then to rank the species according to relative abundance. It is alleged that dominant species of the assemblage should all fall within different feeding groups, and that each dominant species with an assemblage was the superior competitor within its feeding group. The ecological information of the preceding paragraphs shows that analyses of this type are largely unjustified. Many benthic populations are limited by space, predation, or disturbance rather than by food supply. Also, we do not know the precise feeding habits of many fossil species, and we seldom know their relative abundances. In fact, many species are not preserved at all.

Having introduced some basic concepts of the study of benthic communities, we will now turn to some actual examples in the fossil record.

ORGANIC REEF COMMUNITIES

Marine paleoecology emerged as a distinct branch of paleontology during the 1950's. Many of the early marine paleoecologic studies dealt with organic reefs, partly because of commercial interest in ancient reefs as traps for petroleum and partly because of scientific curiosity about reefs themselves.

The term "reef" has been defined and redefined by geologists. The most important characteristic included in most definitions is the presence of a rigid organic framework that stands, or once stood, above the adjacent ocean bottom. Organic reefs today are produced largely by frame-building hermatypic corals in association with encrusting calcareous algae and many minor constituents, and are among the most spectacular organic communities in the beauty and diversity of their component species (Figure 10-17).

Ancient reefs offer three major advantages for paleoecologic analysis: (1) most of the reef-forming species have rigid skeletons that are readily preserved; and (2) the skeletons, in forming a rigid framework, binding the framework together, or being trapped in it, are commonly preserved without post-mortem transport, many being preserved in life position; and (3) reefs form a continuous record of community composition through time. Reefs are also important geologically in that they alter the marine environment, often in neighboring areas as well as where they grow. Most reefs grow upward to, or very near to, sea level (page 256) and shelter back-reef areas from waves and currents, permitting fine-grained sediments to accumulate. Commonly zonation in species composition is evident across a reef from the windward to the leeward side, because the reef's obstruction of waves and currents produces an "energy gradient" (Figure 10-31).

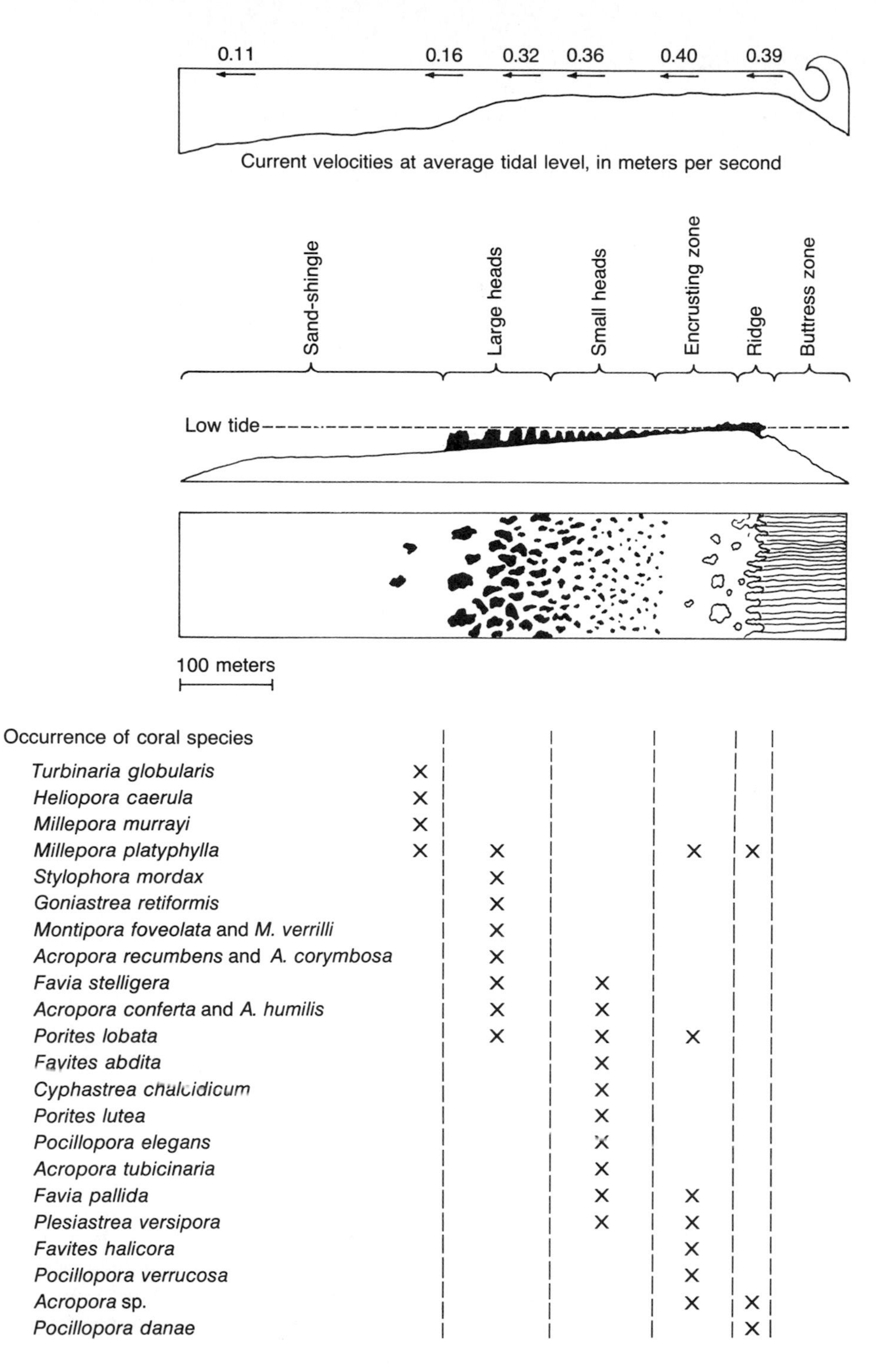

Occurrence of coral species	Sand-shingle	Large heads	Small heads	Encrusting zone	Ridge	Buttress zone
Turbinaria globularis	×					
Heliopora caerula	×					
Millepora murrayi	×					
Millepora platyphylla	×	×		×	×	
Stylophora mordax		×				
Goniastrea retiformis		×				
Montipora foveolata and *M. verrilli*		×				
Acropora recumbens and *A. corymbosa*		×				
Favia stelligera		×	×			
Acropora conferta and *A. humilis*		×	×			
Porites lobata		×	×	×		
Favites abdita			×			
Cyphastrea chalcidicum			×			
Porites lutea			×			
Pocillopora elegans			×			
Acropora tubicinaria			×			
Favia pallida			×	×		
Plesiastrea versipora			×	×		
Favites halicora				×		
Pocillopora verrucosa				×		
Acropora sp.				×	×	
Pocillopora danae					×	

FIGURE 10-31
Cross-section of the windward reef of Eniwetok Atoll, showing current velocity gradient and species zonation. (From Odum, 1959, and Odum and Odum, 1955.)

Many reefs grow for thousands of years, forming their own fossil record. Often the reef community changes as upward growth approaches sea level. Lowenstam (1950) studied Silurian reefs in the Great Lakes region of the United States that were formed by groups of organisms which are now extinct. The reefs were mound-like structures in which ecologic succession occurred during growth. As shown in Figure 10-32, the quiet-water stage, built by the pioneer community, consisted of relatively few groups of organisms. As the reef grew upward into rough water, the faunal composition changed and diversified; more robust species took over. A climax community formed the final wave-resistant stage near sea level. Other examples of succession are illustrated in Figure 10-33. These were observed by Kauffman and Sohl (1974) in reef-like frameworks formed by Caribbean rudists. Rudists are unusual, coral-like bivalve molluscs that lived during the Jurassic and Cretaceous Periods. It is easy to understand why reef communities often exhibit ecologic succession. Their builders tend to have durable skeletons that form relatively stable structures. As these structures grow, they alter the environment. Disturbance by predation and water movement is often insufficient to prevent intense competition for space.

On Recent organic reefs, species of corals, sponges, and calcareous algae compete with each other. Recent corals grow rapidly, partly because they secrete porous skeletons (as did rudists) and partly because they are infested with algae called zooxanthellae. The coral-algae association is termed ***symbiotic*** because it is mutually beneficial. The algae make use of coral waste products and benefit the coral by serving as partial sources of food and as removers of CO_2 from coral tissues; removal of CO_2 aids in secretion of the coral skeleton. Corals derive food by predation upon small organisms, especially zooplankton. Competitive interactions of coral species with each other and with other taxa are coming to be well known. Rapid growth into dead spaces on the reef is one effective mode of competition, and so is overgrowth of neighboring corals. Many coral species extrude gastric filaments and actually digest their neighbors (Lang, 1971). Porter (1974) has reported interesting competitive patterns that illustrate some concepts presented in our earlier discussion of community structure. In some areas corals display a hierarchy of competitive dominance. Where predation upon corals happens to be minimal, diversity of coral species is also minimal. Superior competitors monopolize space, and most of the reef is occupied. Where predation is extensive, competition is reduced, only a small percentage of space is then occupied, and many species coexist. In other areas, it happens that no species excel at all competitive strategies. Here, as a result of competitive balance, diversity tends simply to increase as the coral cover increases.

We know rather little about competitive interactions on ancient reefs, but Kauffman and Sohl (1974) note that in the Cretaceous, rudists seem to have competed successfully against corals for niches on reef-like structures. Reef corals had flourished in the Jurassic, but they are largely absent from large reef-like structures of the Cretaceous, occurring only in small peripheral patches. After the total extinction of rudists at the end of the Cretaceous, corals again emerged as dominant reef builders.

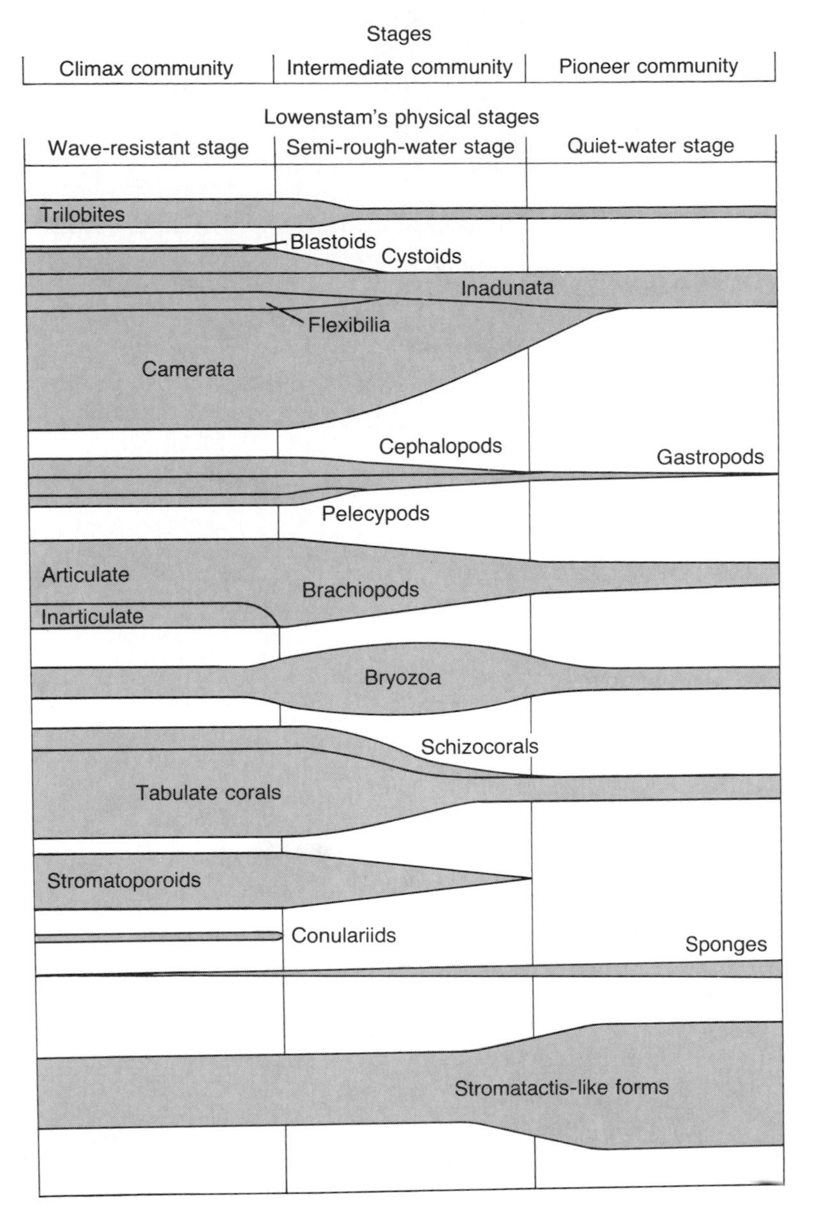

FIGURE 10-32
Three stages in the ecologic succession recognized by Lowenstam for Silurian reefs of the Great Lakes region of the United States. (From Nicol, 1962.)

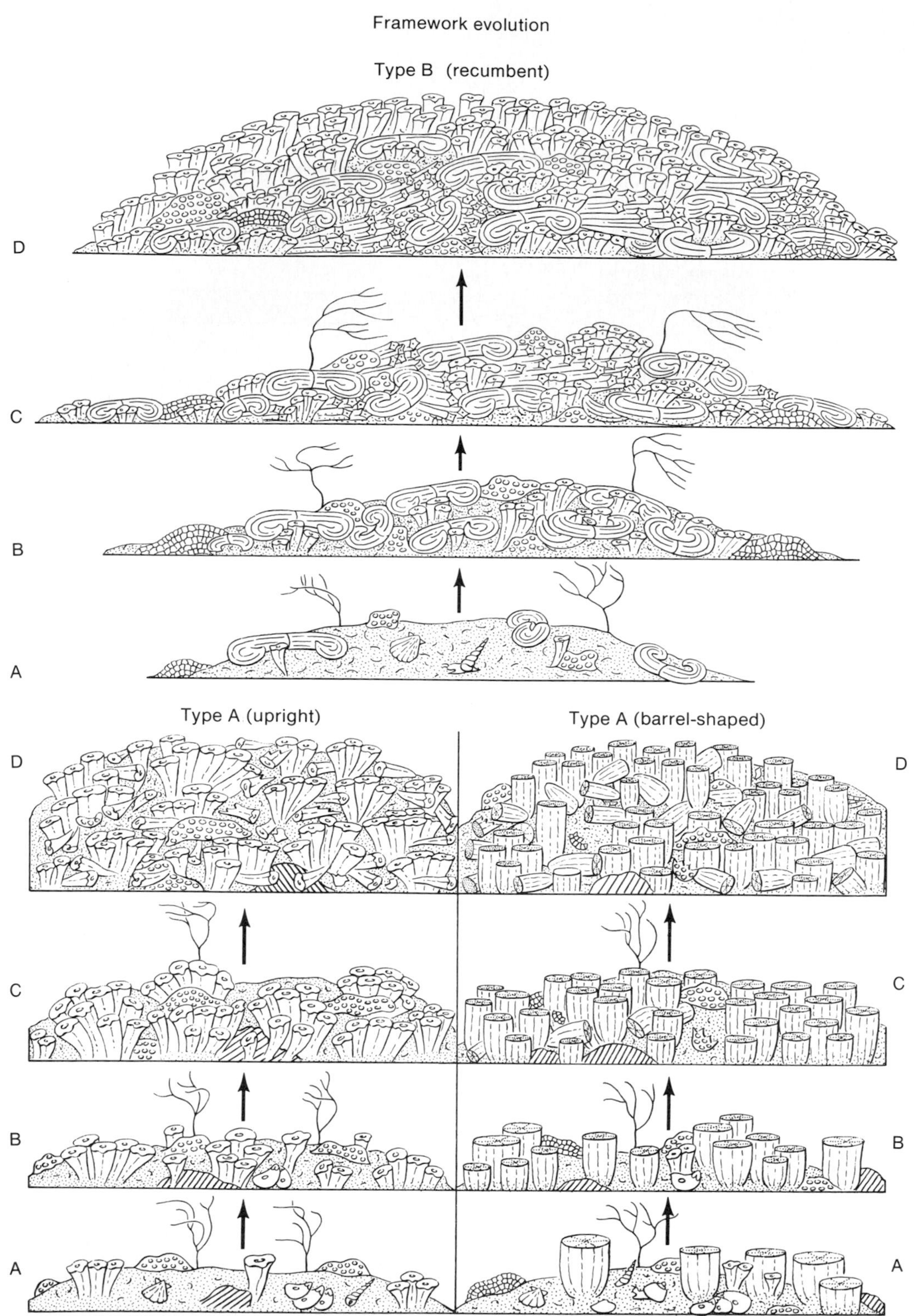
Framework evolution
Type B (recumbent)
D
C
B
A
Type A (upright)
Type A (barrel-shaped)
D
C
B
A
D
C
B
A

FIGURE 10-33
Diagrams illustrating sequences by which rudist bivalves built bioherms in the Cretaceous of the Caribbean region. Type A frameworks began with upright rudists, some of which would have been large, barrel-shaped forms (lower-right sequence). Type B frameworks began with large, recumbent rudists having two coiled valves of nearly equal size. (From Kauffman and Sohl, 1974.)

Despite their prominance, rudist frameworks did not form barriers as effective as those of coral reefs, apparently because cementing organisms like coralline algae were unable to grow on the shells of live rudists and bind them into a durable structure.

SOFT-BOTTOM COMMUNITIES

The structure of Recent communities occupying soft substrata varies greatly from place to place.

Deposit-feeding species typically exploit a reservoir of organic-rich sediment and might be expected to compete with each other for food. Levinton and Bambach (1975) suggested that in both Recent and Silurian deposit-feeding communities, species are vertically stratified within sediments because of competition for food or space; this condition may exist in the absence of extensive predation. It has also been observed, however, that deposit-feeding clams in shallow marine habitats increase in numbers enormously when protected from predation. Predation on young forms is normally so extensive here that populations never reach the size where they approach the limits of available food or space. Although the kinds of animals studied are typical of modern shallow-water deposit-feeding communities, the results may not apply to all similar communities of the present or past. We have already discussed the great diversity of the deep sea, where most herbivores are deposit feeders (pages 255). It has been suggested that this diversity of deposit-feeding species represents a condition in which niches have become finely partitioned through competitive interactions over long periods of time in a very stable environment (Sanders, 1968; Grassle and Sanders, 1973). There is experimental and photographic evidence, however, that predation by fishes is very extensive in the deep sea. This has given rise to the alternative argument that extensive predation reduces competition to the degree that many species can coexist (Dayton and Hessler, 1972).

Like deposit feeders, suspension feeders living in soft sediment are often extensively preyed upon. Part of the reason that herbivorous benthos are easily preyed upon is that most are slow-moving or immobile. Many predators concentrate upon young animals, which because of their small size are especially easy to consume (Thorson, 1966; Muus, 1973). Very high rates of reproduction and settlement of larvae compensate for high juvenile losses. Studying benthos in a tropical sea bottom covered by marine grass, Jackson (1972) concluded that intensity of predation increased in an offshore direction. For intertidal soft substrata, where fewer predators can live, abundance of benthic herbivores is sometimes limited by the rigors of the physical environment. Periodic desiccation and fluctuations in temperature and salinity may cause high rates of mortality.

When neither physical disturbance nor predation limits population densities, benthic herbivores may compete for space. One way in which space is monopolized is through feeding activity or movement of certain populations; such activities may disturb the sediment and prevent settlement or survival of larvae. Woodin (1976) has shown that this frequently happens in dense assemblages of worms. Adults may also compete with each other for space. Johnson (1959) found that suspension-feeding phoronid worms live in clusters in the intertidal zone but space themselves evenly, which apparently prevents mutual interference of feeding apparatuses. Vertical burrows assigned to the genus *Skolithos* (Figure 10-34) have been compared to phoronids of the type that Johnson studied. These fossil burrows are abundant in sandstones near the base of the Cambrian. They tend to be quite straight and are often several feet long! It is difficult to know what the spacial distribution of the burrows may have been in life because we cannot easily show that several found together in a rock were inhabited at the same time. Woodin (1974) found that a mobile burrowing species of polychaete worm living in an intertidal area along the coast of Washington

FIGURE 10-34
A block of Lower Cambrian Antietam Quartzite in which numerous vertical burrows of *Skolithos* are displayed. The bar at the lower right is 5 centimeters long.

State was limited by space. It was also excluded from areas occupied by immobile, tube-building worms that altered the substratum in such a way as to prevent normal movements.

Much effort in community paleoecology has been directed toward the recognition of ***recurrent benthic communities.*** The idea that such communities exist was developed by Peterson (1913), based on extensive studies in the North Sea. Peterson concluded that species tend to cluster into groups, each of which occurs in many separate areas on the sea bottom. Peterson emphasized that the recurrent communities he recognized were merely statistical entities. He avoided any implication that each of the recurrent communities existed because of ecologic interdependence among the species forming it. His view was that the species forming a recurrent community are simply species that have similar environmental requirements. Some workers have since taken issue with that concept, suggesting that just as no two benthonic environments are identical, no two communities should have identical species compositions. Indeed, many of the communities enumerated and named by Thorson (1957) are based only on assemblages obtained by limited subtidal sampling, with no regard for spacial distribution or recurrence. To test the concept of recurrence, Stephenson, Williams, and Cook (1972) analyzed Peterson's original data statistically. The communities delineated by their analysis differed with the kind of statistical procedures used. One kind of procedure produced a classification of communities resembling Peterson's; another did not. Many workers now believe that the concept of recurrent communities is valid for at least some soft-bottom environments, but more data than have been gathered will be needed for the concept to be evaluated definitively. For example, though some paleoecologic analyses have been premature in classifying fossil species rather rigidly into distinct communities, in contrast, relatively few studies of living benthic communities of the past decade have focused on the recognition of recurrent communities.

A pioneering study in the field of community paleoecology is that of Ziegler et al. (1968) for Silurian deposits of Great Britain. This study recognized five communities that were judged to be recurrent in space and time. They were recognized both on the basis of species composition and geographic distribution (Figure 10-35). The five communities intergrade and were judged to represent an approximate nearshore-offshore sequence. Although communities 1 and 2 or 4 and 5 may have several species in common, 1 and 4 or 2 and 5 may be mutually exclusive in species composition. In the study of Ziegler et al., the community nearest shore (community 1) was named the *Lingula* community (Figure 10-36). *Lingula* was described on page 263 as having been a predominantly nearshore euryhaline genus since the early Paleozoic. As seems typical for Paleozoic faunas, benthic molluscs also occur mainly nearshore.

By studying the sedimentology of the rocks containing the communities recognized by Ziegler, Bridges (1975) has superimposed an environmental framework on Ziegler's faunal distributions. The *Lingula* community apparently occupied a restricted environment in the protection of a peninsula. Studies in North America

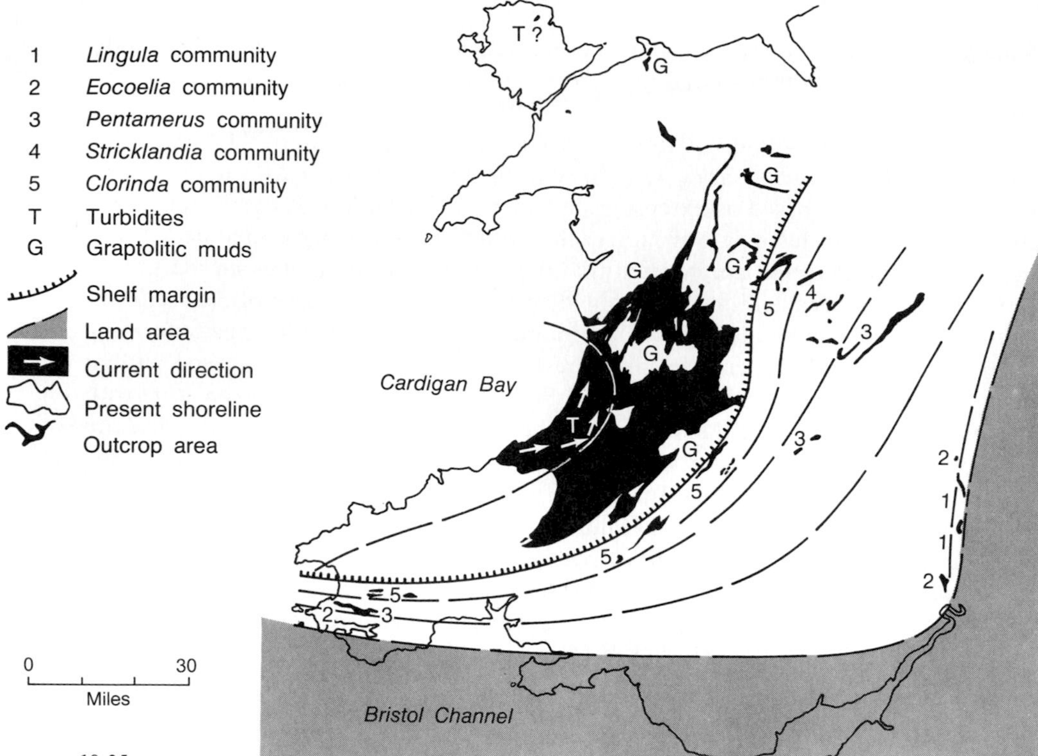

FIGURE 10-35
Early Upper Llandovery (Silurian) paleogeography of Wales, showing distribution of five recognized benthonic communities in zones parallel to the ancient shoreline. (From Ziegler, 1965.)

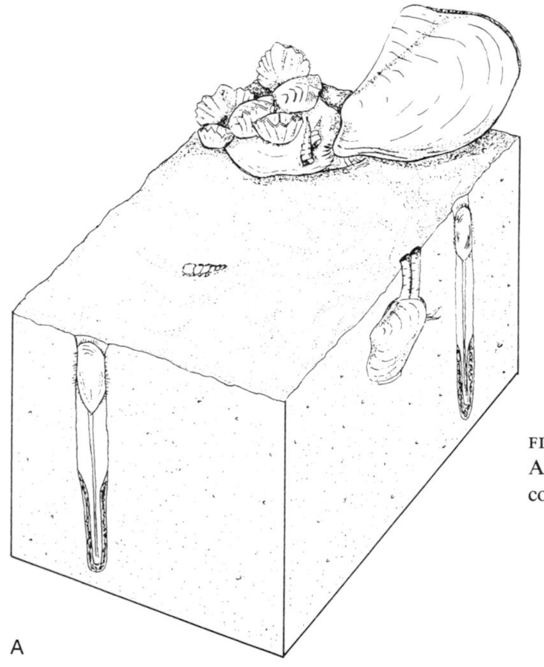

FIGURE 10-36,A
Artist's reconstruction of the *Lingula* community of Figure 10-35.

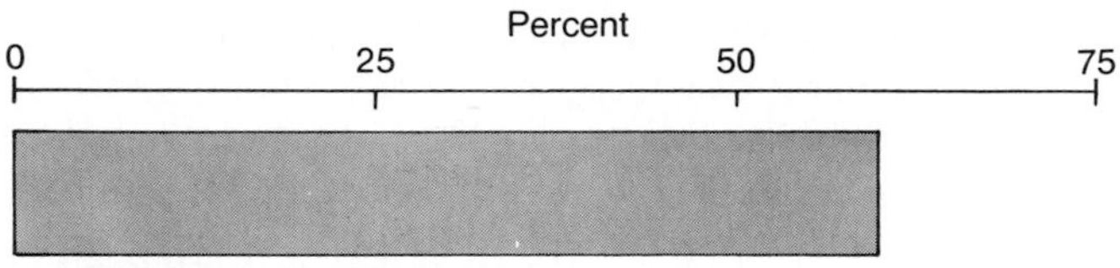

A: *"Camarotoechia" decemplicata*

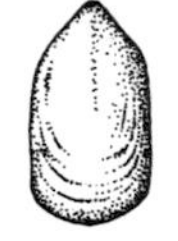

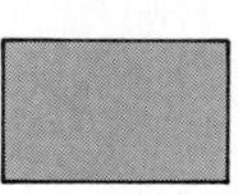

B: *Lingula* sp.

C: *Palaeoneilo rhomboidea*

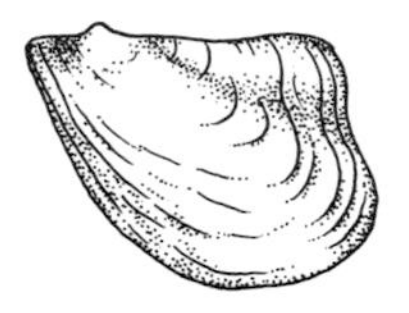

D: *Pteronitella* cf. *retroflexa*

E: *Cornulites* sp.

F: *Lingula pseudoparallela*

G: Others

B

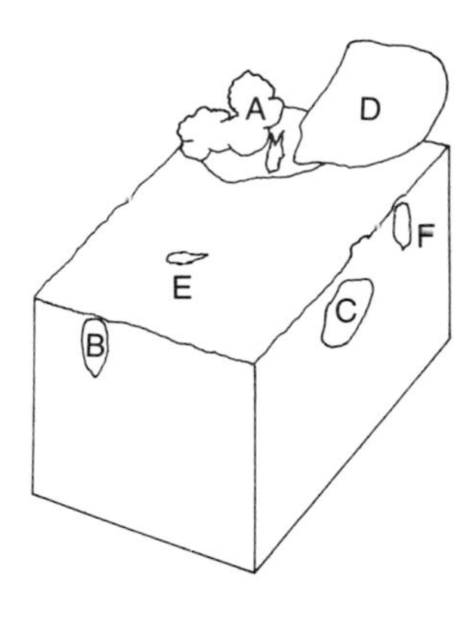

FIGURE 10-36,B
Relative species abundance in the *Lingula* community of Figure 10-35. (Both A and B from Ziegler, Cocks, and Bambach, 1968.)

corroborate this conclusion. Faunas of the Siluro-Devonian Kaiser Limestone of the Central Appalachians resemble the Welsh faunas studied by Ziegler and his co-workers, but include a greater variety of taxa. Faunal distributions in the Kaiser show clear evidence of control by environmental discontinuities (Makurath, 1977); barriers of sand and mounds held in place by bryozoans and crinoids separated lagoons from offshore areas, and broad intertidal areas fringed the shoreward margins of the lagoons. Echinoderms were virtually absent from these lagoons; they were probably excluded by reduced and fluctuating salinities (conditions which are not tolerated by most Recent echinoderms). Silurian pentamerid brachiopods have been found largely in normal marine sandstones, and Makurath has concluded that pentamerids, which range in age from Cambrian to Devonian, in general tended to inhabit somewhat unstable sandy substrata. Pentamerids often formed clusters (Figure 10-29), and thickening of the beak region of their shells would have helped to anchor the posterior and prevent dislodgement. Makurath suggests that the *Pentamerus* community of Ziegler's study (number 3 of Figure 10-35) occupied shifting substrata seaward of barriers that separated restricted lagoonal and perhaps intertidal associations of species (1 and 2 in Ziegler's classification) from offshore associations (4 and 5). A complex of barrier islands and lagoons occurs along many modern marine shorelines (see Figure 10-3), and the key to the recognition of such complexes in ancient rocks is often the vertical arrangement of sedimentary environments. The shifting of one depositional environment over another through geologic time results in characteristic stacking sequences of facies (as is diagrammed in Figure 9-9). The origin of the typical sequence, or ***vertical profile,*** observed in the Kaiser Limestone is shown in Figure 10-37. Recognition of this sequence, in conjunction with detailed study of

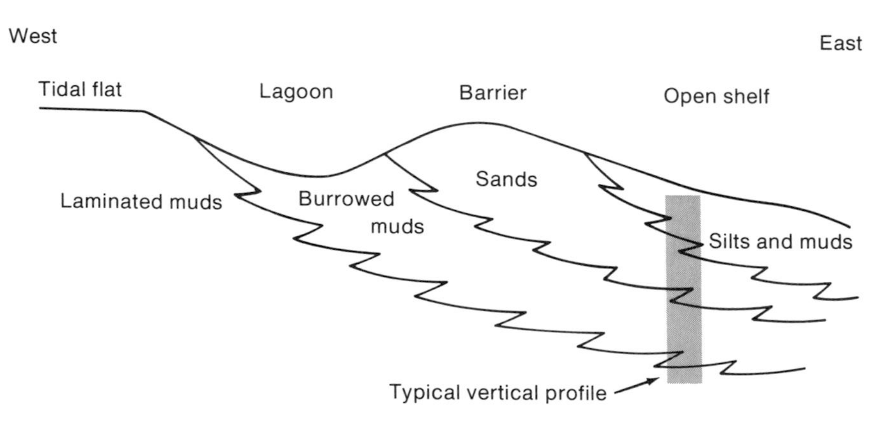

FIGURE 10-37
The origin of a typical vertical profile of sedimentary units in the Kaiser Formation of the Central Appalachians. This profile was produced by transgression, which shifted environments westward through time. (Modified from Makurath, 1977.)

sedimentary structures and lithologies, formed the basis for the environmental interpretation. Surprisingly, Makurath found little evidence of habitat specificity for *non*pentamerid brachiopods.

Using a similar approach, Walker and Laporte (1970) found that Middle Ordovician and Lower Devonian communities of New York State are remarkably similar with particular habitats. They recognized four habitats: supratidal (occasionally flooded), high intertidal, low intertidal, and subtidal.

Through study of sedimentary units in vertical profile, other kinds of complex shorelines can also be recognized. One of the most important is the deltaic shoreline, in which sandy, finger-like lobes of sediment build out over silty and muddy offshore deposits (Figure 10-38). Branching channels called distributaries, which divide up the river and carry water across the delta, form these lobes. Periodically old lobes are cut off by the shifting of distributaries and new lobes are formed. Being areas of heavy deposition, deltaic complexes tend to sink continuously. This condition, sometimes in combination with rising sea level, leads to burial of old lobes beneath new ones as distributaries—and sometimes the entire actively forming delta—shift back and forth. In this manner, "packages" of sediment that coarsen upward are stacked one on top of another in cyclical fashion (Figure 10-39). Different parts of a given cycle contain different faunas, reflecting the fact that shallow areas tending to have sandy substrata and fluctuating salinities support different assemblages of organisms than offshore muddy areas.

It is only by using sedimentologic and stratigraphic techniques to reconstruct ancient environments in detail that fossil communities can be meaningfully studied. Failure to do this has impeded progress in paleoecology to date. We remain largely in the inductive stage of the science, in which we must fit faunas into pre-established environmental frameworks in order to establish characteristic patterns of distribution. Especially in the Paleozoic, only a few fossil taxa and a few faunal characteristics can now be regarded as clear-cut indicators of certain habitats. Several of these have been discussed in this chapter, but all must be applied with caution.

TERRESTRIAL COMMUNITIES

Terrestrial floras have special value in the interpretation of ancient climatic conditions, and in Chapter 12 we will discuss them in this light. Terrestrial vertebrate communities, on the other hand, have been the subject of only a few paleoecologic studies. Nonetheless, some of these are especially interesting. Whereas it is commonly assumed that preservation of vertebrates is generally too poor to allow estimation of the relative abundances or biomasses of various kinds of animals, in fact there are numerous exceptions to this generalization, and certain well-preserved faunas have yielded remains of thousands of individuals. Furthermore, the potential for undertaking paleoecologic analysis of many of these is enhanced by the fact that

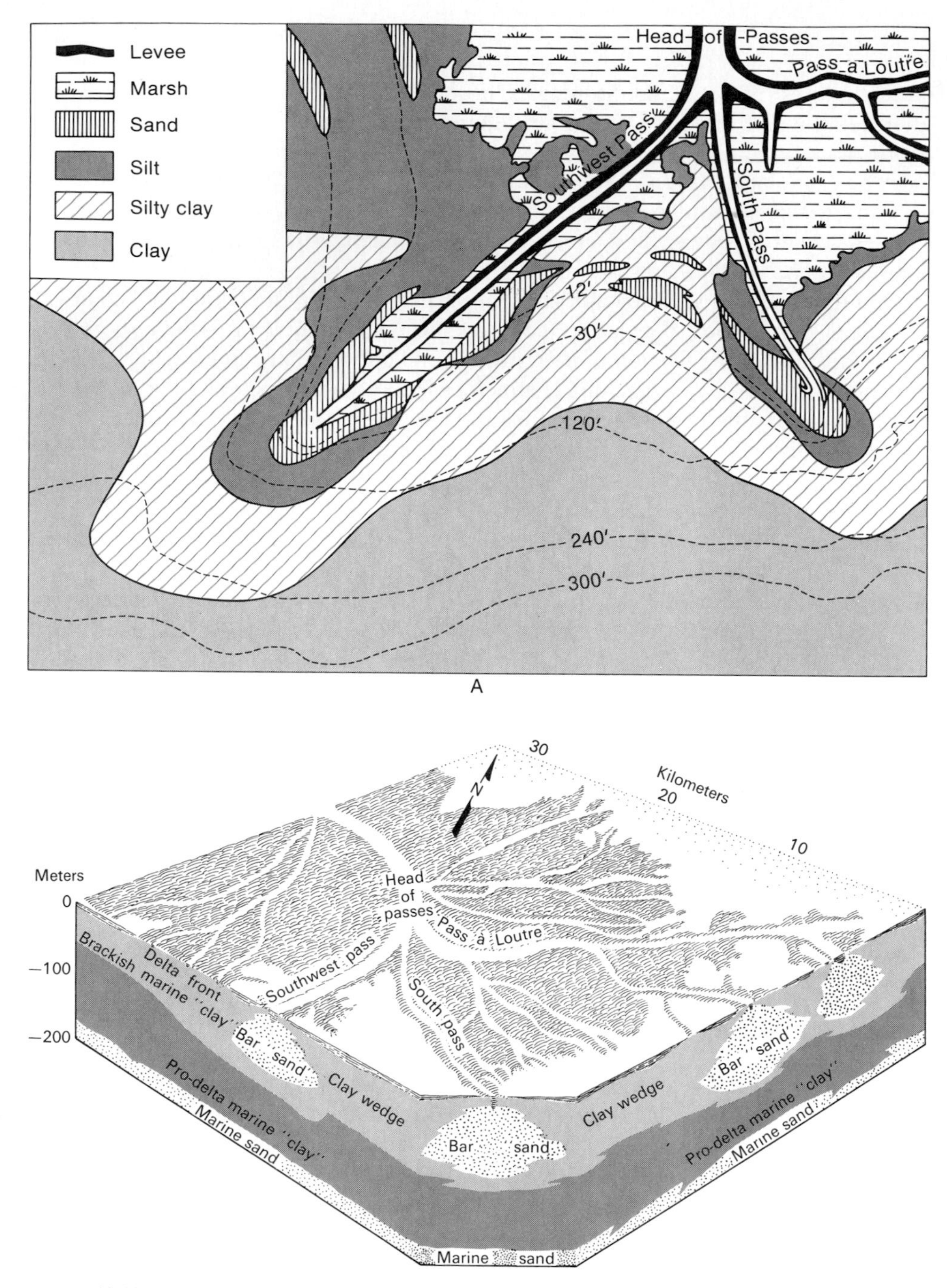

FIGURE 10-38

Deltaic deposits. A: Map of the distribution of sedimentary deposits of the Birdfoot subdelta of the Mississippi River. Submarine contours show depths in feet. B: Block diagram of the Birdfoot subdelta showing sections of the elongate sand bodies produced by the main distributaries. (After Fisk et al., 1954.)

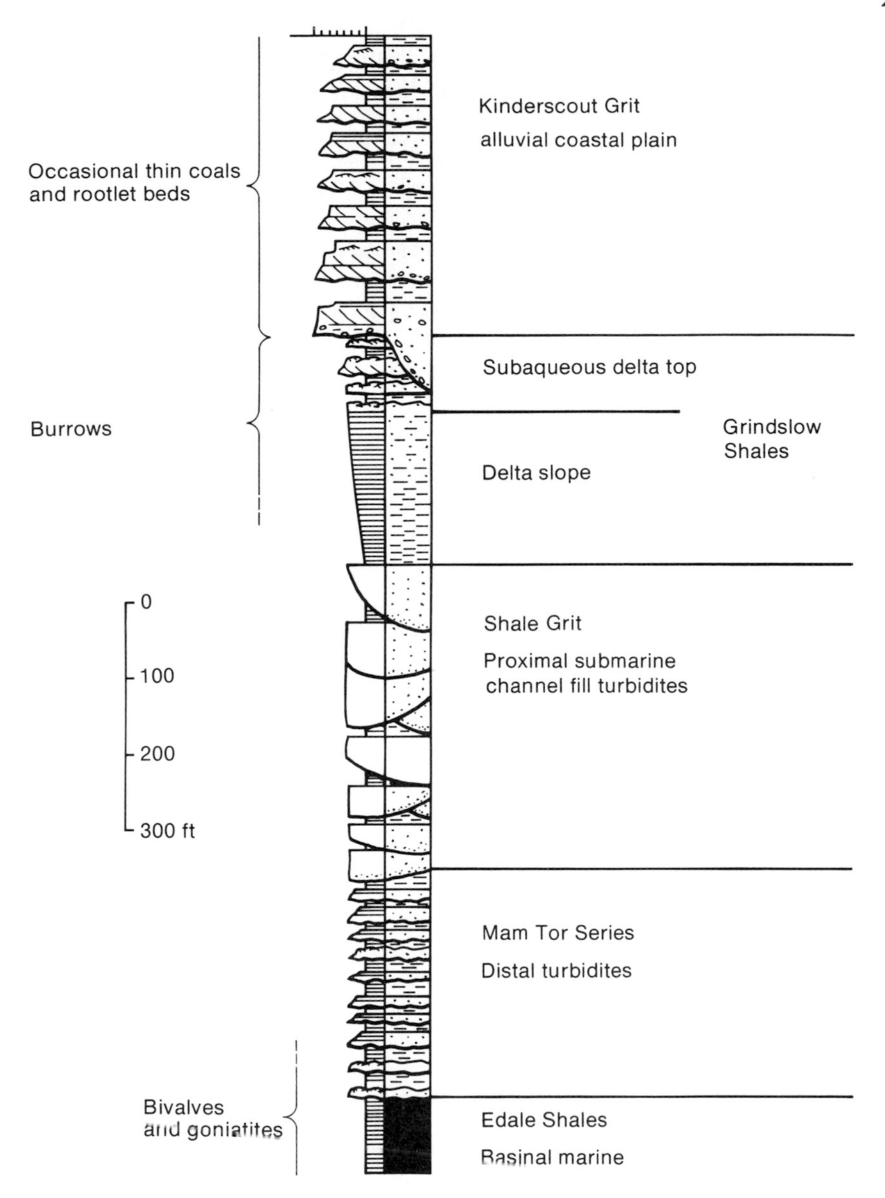

FIGURE 10-39
Generalized composite stratigraphic section of deltaic deposits in the Carboniferous of northern England. The sequence presents deposition, first in shallower marine water and finally on nonmarine terrain (toward the top). The sequence was produced when an ancient delta, or part of an ancient deltaic complex, grew out, or shifted, over a marine basin. (From Selley, 1970. Copyright © 1970 by Richard C. Selley. Used by permission of Cornell University Press and Chapman & Hall, Ltd.)

predator-prey relationships within vertebrate faunas are rather easily recognized, for both feeding habits and modes of locomotion are often readily discernable from skeletal morphology.

Making use of exceptionally well-preserved faunas, Bakker (1975) found that ratios of predator biomass to prey biomass suggest that dinosaurs were endothermic (warm-blooded) rather than ectothermic (cold-blooded). Endotherms (living birds and mammals) require large amounts of energy to maintain high body temperatures; annual caloric intake per unit of body weight is much greater for predatory endotherms than for predatory ectotherms. This explains why, in communities of living mammals, the ratio between predator biomass and prey biomass is lower than it is in communities of living lizards. For either group—lizards or mammals—the relationship holds regardless of animal size. Bakker estimated biomasses of fossil specimens from skeletal dimensions and found that the relationship is also apparent in well-preserved fossil faunas containing enormous numbers of individuals. For mammalian faunas of the Cenozoic, biomass estimated for predators turned out to be relatively low, as expected: it was only 3–5 percent as large as biomass estimated for associated prey species. For Early Permian communities of established amphibians and reptiles, percentages were much larger, and predator biomass ranged from about 40 percent to over 60 percent of estimated prey biomass. Late Permian and Triassic mammal-like reptiles display intermediate percentages that range from about 10 percent to 20 percent, apparently reflecting endothermy with lower heat production than that of modern mammals. The dinosaurs, which seem to have outcompeted mammal-like reptiles early in the Mesozoic, display very low ratios, like those of Recent mammals. Bakker cited this condition as evidence that dinosaurs were endothermic. This conclusion is corroborated by evidence from bone structure. Dinosaurs, advanced mammal-like reptiles, and living birds and mammals have bones that lack growth rings but are rich in Haversian canals, where rapid calcium phosphate exchange takes place, and blood vessels. Such bones seem to be characteristic of animals with endothermic metabolism.

Post-mortem Information Loss

It is convenient for us to divide the history of an organism from its birth to our discovery of its fossilized remains into time intervals, and to give different names to the study of the various intervals (Figure 10-40). In its strict sense, paleoecology concerns only the interval between the organism's birth and death. Although the names given here for the study of the later intervals have not been widely adopted, they do present a useful classification. Instead of "Post-mortem Information Loss," this section of the chapter might have been named "Taphonomy."

We have used "fossil assemblage" to mean a group of fossils (of one or more species) found together in the stratigraphic record. A further qualification should be added. In

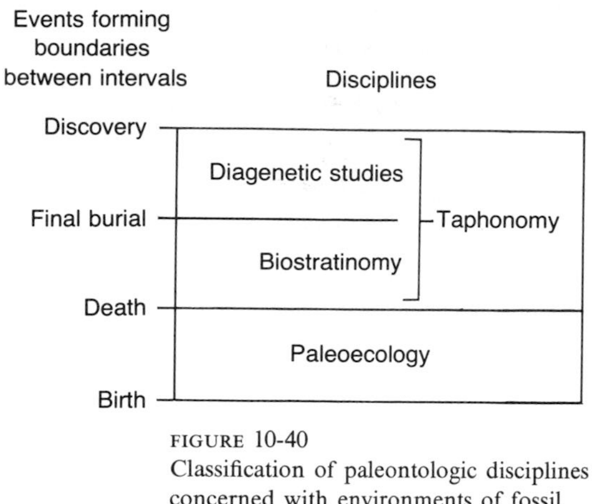

FIGURE 10-40
Classification of paleontologic disciplines concerned with environments of fossil organisms between their time of birth or hatching and discovery as fossils. (From Lawrence, 1968.)

order to be evaluated as a unit, the assemblage should be nearly homogeneous or uniformly heterogeneous in species composition. In the vertical dimension assemblages are usually restricted to a single stratigraphic bed or bedding plane but may extend through several beds. Horizontally, they may extend for centimeters, meters, or (rarely) kilometers. Vertically and horizontally, an assemblage may give way, abruptly or gradually, either to barren strata or to other assemblages. The homogeneous, or uniformly heterogeneous, composition of a fossil assemblage results from its unique ecologic and preservational history.

Many systems of classification have been used to distinguish between various types of fossil assemblages. One such system is shown in Figure 10-41. We will refer to a fossil assemblage composed entirely of species belonging to a single community and preserved in the environment where they lived as a ***life assemblage.*** If a fossil assemblage is composed entirely of species transported from the environment where they lived, even if they lived together, we will refer to the fossil assemblage as a ***death assemblage.*** Assemblages that contain species that lived in two or more habitats will be termed ***mixed death assemblages.***

Two items of information are generally sought in studying the taxonomic composition of a fossil assemblage. The first is what species are present. The second is the relative abundance of these species at the time when they lived. Studies on invertebrates, vertebrates, and plants have indicated that the second item can seldom be determined even approximately.

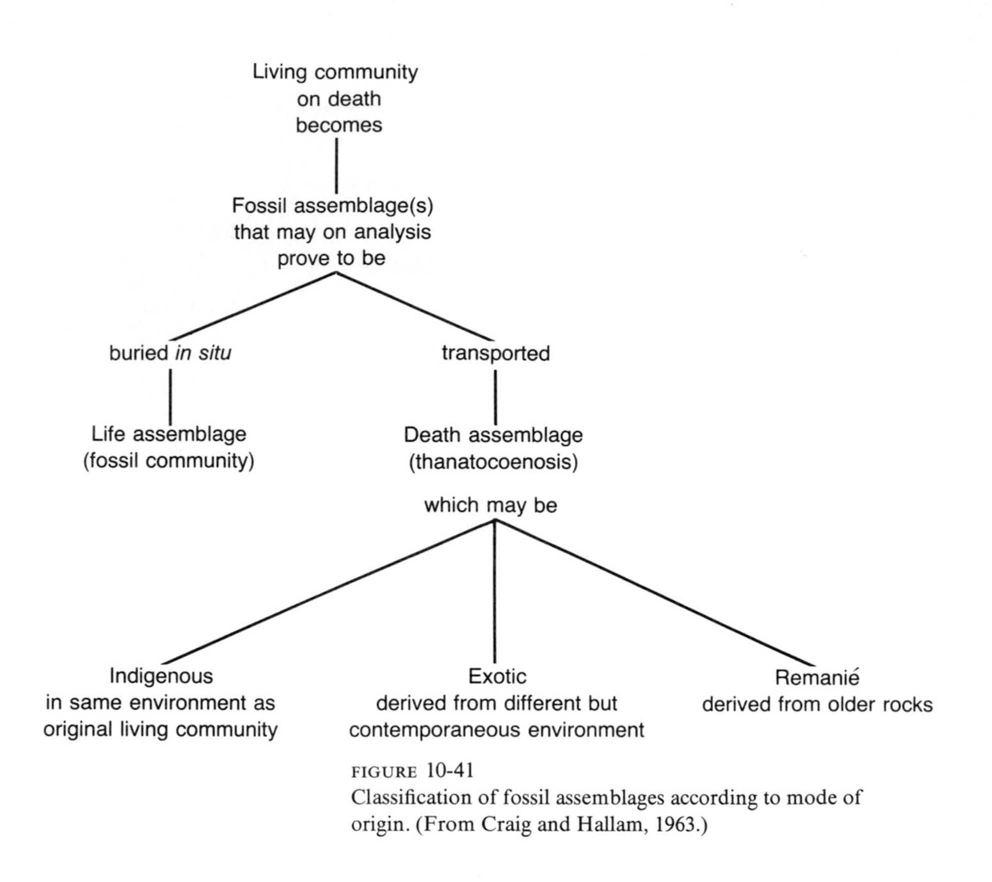

FIGURE 10-41
Classification of fossil assemblages according to mode of origin. (From Craig and Hallam, 1963.)

As discussed in Chapter 4, many criteria have been used to determine whether fossils are preserved where they lived. Preservation in life position is a sure clue for ruling out post-mortem transport. An attribute that makes both coral reefs and trace fossils especially valuable in paleoecologic interpretation is that they represent in-place preservation. Occasionally, other kinds of communities or populations may suddenly be buried in a blanket of sediment with the result that many individuals are preserved in life position.

That fossils in marine benthonic settings were subjected to post-mortem transport can often be recognized from the fossils themselves. Among the most important effects of transport are disarticulation, breakage, wear, and size sorting. Johnson (1960) analyzed these effects for three preservation models, which are presented in Box 10-A. The table lists the preservation features postulated for each model. The models, of course, represent positions in a spectrum. Most real fossil assemblages differ from all three, but many resemble one more closely than the other two.

BOX 10-A Preservation Models

MODEL I: A community lives in a restricted area of the sea bottom below the low-tide mark but above the maximum wave base. The water mass moves over the bottom at a low velocity which occasionally increases so that it becomes erosional. The substrate is composed of clastic sediments. Intermittently, small amounts of similar sediments are introduced into the area together with minor amounts of the durable remains of organisms. The abundant and diverse life consists of soft-bodied organisms and organisms bearing hard parts. Continually and under the conditions of normal mortality, elements of this community die and become potential fossils. These decompose; some are buried after varying periods of exposure. Then, suddenly, the entire community is buried and killed by the rapid introduction of a large amount of sediment not unlike the material of the former substrate. The sedimentary body containing the fossil assemblage is gradually compacted as a result of further accumulation of sediment at the site.

MODEL II: A community lives in an environment similar to that described for Model I. Continually and under the conditions of normal mortality, elements of the community die and become potential fossils. These decompose; some of the remains are carried away, whereas others of durable composition are buried after varying periods of exposure. In time, the local environment changes and the community is eventually replaced by another of different composition. The fossil assemblage of interest becomes more deeply buried, and the sedimentary body becomes gradually compacted as a result of further accumulation of sediment at the site.

MODEL III: A community lives in a restricted area of the sea bottom below the low-tide mark but above the maximum wave base. The water mass moves at a moderate velocity over a bottom of clastic sediments consisting in large part of the durable remains of organisms. Frequently the velocity of the water mass is high enough to move sediment and organic detritus through the area. The hydrodynamic circumstances favor the accumulation of organic remains at the site. The sparse fauna consists of a few epifaunal species of scavengers, boring and encrusting organisms. Continually and under the conditions of normal mortality, elements of this community die and become potential fossils. These decompose; some of the remains are carried away, whereas others of durable composition are buried after varying periods of exposure. Eventually, the rate of accumulation of organic remains at the site decreases as changes occur in the source areas of the sediment and debris. In time, the zone containing the high concentration of durable remains is buried and gradually compacted as a result of further accumulation of sediment at the site.

(*continued*)

BOX 10-A *(continued)*
Relative Expressions of Features Developed under the Conditions of the Models

Only the most distinctive or most probable development is shown.

Feature	*Expression*		
	Model I	*Model II*	*Model III*
Faunal composition	Ecologically coherent assemblage of species	As in Model I	Not necessarily as in Models I and II
Morphologic composition	Delicate structures and heterogeneous shapes and sizes may be preserved. Suites of shapes represented are ecologically consistent	As in Model I	May consist only of the durable parts of the species present; may be homogeneous in shape and sizes
Density of fossils	Wide range of densities possible	As in Model I	High
Size-frequency distribution	Many species exhibit a size-frequency distribution conforming to an ideal distribution for an indigenous population	Some species as in Model I	Not as in Models I and II
Disassociation	High proportion of articulated remains; disassociated parts represented in appropriate relative numbers for a species	Moderate proportion of articulated remains; disassociated parts as in Model I	Not as in Models I and II
Fragmentation	Low proportion of remains are fragments	Moderate proportion of remains are fragments	High proportion of remains are fragments
Surface condition of fossils	Surfaces of preserved structures as in life	Various states of wear represented	As in Model II
Chemical and mineralogical composition	No general expectations are warranted by present knowledge in this area	As in Model I	As in Model I
Orientation	Some species may retain orientation as at time of death	Majority of fossils oriented with long axis parallel to bedding plane (some exceptions)	As in Model II (some exceptions)

Feature	Expression		
	Model I	*Model II*	*Model III*
Dispersion	Articulated remains of some species may retain pattern of dispersion as in life	Not as in life	As in Model II
Sediment structure and texture	Consistent with inferred tolerances of the fauna and with relatively quiet waters	As in Model I	Not necessarily consistent with inferred tolerances of the fauna: consistent with relatively turbulent waters

SOURCE: Modified from Johnson, 1960.

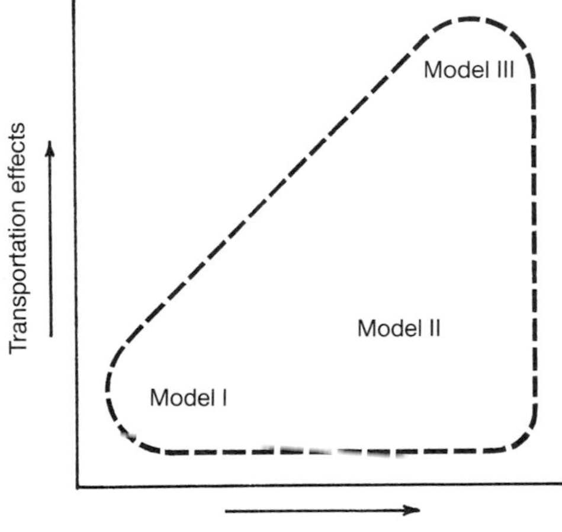

Hypothetical diagram representing the relative alteration of the fossil assemblages resulting from three modes of formation. Model I represents the sudden burial of a community; Model II, the gradual accumulation and burial of the remains of organisms living at the site of deposition; and Model III, an assemblage comprised almost entirely of transported remains. Exposure effects include degree of abrasion, solution, fragmentation, decomposition, encrustation, and boring. Transportation effects include spurious association of species and size sorting. Most fossil assemblages probably occur within the area of the dotted line. Assemblages formed in quiet waters should cluster near Models I and II while those of turbulent water should resemble Model III.

Supplementary Reading

Ager, D. V. (1963) *Principles of Paleoecology.* New York, McGraw-Hill, 371 p. (A documentary treatment with a valuable bibliography.)

Eltringham, S. K. (1971) *Life in Mud and Sand.* New York, Crane, Russak, 218 p. (A summary of organism-sediment interactions.)

Hecker, R. F. (1965) *Introduction to Paleoecology.* New York, Elsevier Scientific Publications, 166 p. (A translation from Russian dealing primarily with practical methodology.)

Hedgpeth, J. W., ed. (1957) *Treatise on Marine Ecology and Paleoecology, 1, Ecology.* Boulder, Colo., Geological Society of America. Memoir 67, 1296 p. (A comprehensive reference volume with chapters and bibliographies on many aspects of biologic oceanography.)

Imbrie, J., and Newell, N. D. (1964) *Approaches to Paleoecology.* New York, John Wiley & Sons, 432 p. (A collection of papers by authorities on various aspects of paleoecology.)

Kinne, O. (1970–1975) *Marine Ecology.* New York, John Wiley & Sons, Vol. 1 (3 parts), 1970, 1971, 1972; Vol. 2 (2 parts), 1975. (A comprehensive treatise on physiologic ecology in the marine realm.)

Ladd, H. S., ed. (1957) *Treatise on Marine Ecology and Paleoecology, 2, Paleoecology.* Boulder, Colo., Geological Society of America. Memoir 67, 1077 p. (Chapters emphasizing the regional biostratigraphic approach and bibliographies on paleoecology of fossil groups.)

Laporte, L. F. (1968) *Ancient Environments.* Englewood Cliffs, N.J., Prentice-Hall, 116 p. (A compact summary of biologic and physical approaches for reconstructing environments of the past.)

Moore, H. B. (1968) *Marine Ecology.* New York, John Wiley & Sons, 493 p. (A summary of habitats, limiting factors, and biologic adaptations in the sea.)

Schäfer, W. (1972) *Ecology and Paleoecology of Marine Environments.* Chicago, University of Chicago Press, 586 p. (A summary of many years' research on animal-sediment relationships, Lebensspüren formation, and fossilization; translated from German.)

Selley, R. C. (1970) *Ancient Sedimentary Environments.* Ithaca, N.Y., Cornell University Press, 237 p. (A useful summary of sedimentologic and stratigraphic approaches to the interpretation of ancient depositional environments.)

Valentine, J. W. (1973) *Evolutionary Paleoecology of the Marine Biosphere.* Englewood Cliffs, N.J., Prentice-Hall, 511 p. (A general review of ecological and evolutionary theory as applied to marine life.)

CHAPTER 11

Evolution and the Fossil Record

The traditional role of paleontology in the study of organic evolution has been to offer evidence about large-scale features of the history of life. These include phylogenetic relationships and rates, trends, and patterns of overall change. Sometimes such features are placed under the heading ***macroevolution,*** which refers to their magnitude, or ***transspecific evolution,*** which refers to the fact that they transcend species boundaries. Major deficiencies of the fossil record make it difficult for paleontologists to evaluate ***microevolution,*** which is minor evolutionary change within populations or species. Biologists can analyze microevolution more successfully, but even they suffer the limitation that their own life spans are not very long. Frequently they can observe populations through only a few generations. As it turns out, much biological study of microevolution is partly historical, taking into account changes in biological and physical geography in Pleistocene and Recent times.

As discussed in Chapter 1, the fossil record of species is very incomplete. For this reason much work in macroevolution has been undertaken by evaluation of data at the genus and family level, where the record is significantly better: data for genera and families are numerous enough to allow for meaningful analysis. Still, there are major problems in working with higher taxa. For one thing, a genus or family in one class or phylum may not be equivalent to a genus or family in another. Van Valen (1973b) has provided a useful introduction to this problem. There is no reason to believe that taxa of the same rank in different classes or phyla display comparable domains of morphologic, genotypic, or ecologic variability. Some kinds of organisms have a larger number of conspicuous morphologic traits than others, which may cause us to perceive them to evolve more rapidly (Schopf et al., 1975). Mean number of species per genus or family also differs from group to group. This factor is important in the comparison of longevities of genera and families because these higher taxa do not go extinct as single units. An average genus of bivalve molluscs survives about ten times as long as an average genus of mammals (Simpson, 1953; Van Valen, 1973a), but this is true partly because there are more than ten living species in an average genus of bivalves and only about three species in an average species of mammals. We must always bear in mind that the species is the basic unit of extinction.

Extinction

Natural selection, the primary process of microevolutionary change, was introduced in Chapter 5. Sometimes a species evolves sufficiently to survive in the face of environmental change. For this to happen, there must exist or quickly arise within the species enough genetic variability to produce individuals that are adapted to the new conditions. These individuals will be favored by the new kind of selection pressure. In other instances, a successful adaptive shift is impossible and extinction ensues. Environmental change of the type that we are discussing here can be viewed as a change in the shape of the adaptive landscape of the species (page 166). Later in the chapter we will consider the degree to which established species tend to undergo adaptive change. For the moment, let us simply review the external factors that may bring about such change. MacArthur (1972) pointed out that these are competition, predation, changes in the physical environment, and chance changes in population size. Because successful change is not always possible, these factors are also the basic agents of extinction. (It should be noted that the first three represent the kinds of ecologic limiting factors discussed in Chapter 10.)

Competition is most likely to lead to extinction when one or more species invade a new geographic region that includes the range of an inferior competitor. Barring complications, the inferior competitor may face extinction.

Predation is likely to cause extinction only if the predatory species feeds on a variety of prey species. If a predator depends only on one kind of prey, heavy exploitation of this prey will lead to decline of the predatory population by starvation, and this decline will give the prey species a chance to re-expand its population size, which, in turn, will lead to renewed population growth of the predator. Theoretically, the result will be a cyclical pattern, in which sine-wave-like curves describing the increase and decrease of the two populations through time will be out of phase (Figure 11-1). A predator that has several abundant sources of food can more easily prey a particularly accessible species to extinction. From the standpoint of theoretical ecology, a disease or a parasite acts in the same way as a predator. Unfortunately, these agents of

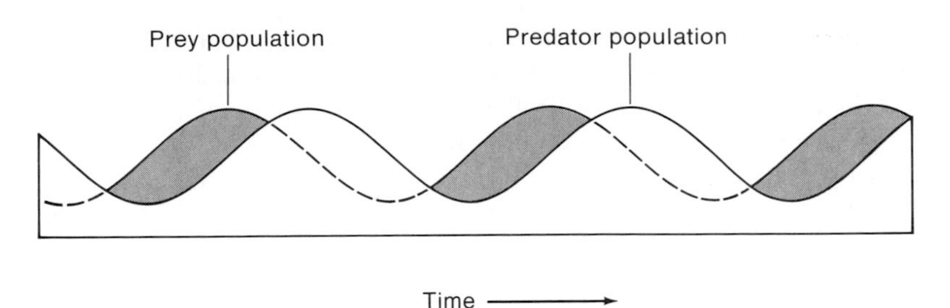

FIGURE 11-1
Oscillations in predator and prey abundance. (After Emlen, 1973.)

extinction seldom leave a direct imprint of their activities in the fossil record. Even recognition of macropredators is often impossible.

Conditions of the physical environment that can cause extinction are too varied to enumerate. On land various limiting factors often described under the heading of climate are of great importance. Comparable marine factors are temperature, salinity, and dissolved oxygen.

No single destructive agent will cause extinction unless its effects are widespread or the victim species has a narrow geographic distribution. This consideration leads to our discussion of the final source of extinction cited by MacArthur (1972): chance fluctuation in population size. Probability theory tells us that the role of chance in causing extinction must increase greatly at small population sizes. Let us pretend that a coin is tossed to determine the fate of every deme (see page 69) of a species through some interval of time. Heads will represent survival of a deme and tails, extinction. Then for a species having two demes the probability of obtaining 100 percent tails, which amounts to extinction of the species, is 25 percent. For a species with 100 demes, the chance of extinction is reduced to $(\frac{1}{2})^{100}$, which is an infinitesimal fraction. Chance factors include high rate of accidental death or low birth rate unrelated to environmental conditions. In most instances, chance factors probably operate in conjunction with other agents of destruction, finishing off a species after its populations are decimated by one or more environmental changes. The relationship between population size and likelihood of extinction by environmental change or chance is very important in any comparison of species longevities (Jackson, 1973). Species having large, widely dispersed populations should tend to survive longer, on the average, than smaller, more localized species. An example of such a species is the brachiopod *Terebratulina chrysalis.* It is one of the species of the Cretaceous Chalk of Denmark that Surlyk (1972) studied (see page 268). It was present in every one of the 150 samples of deep-water Chalk that Surlyk analyzed, occurring in all types of substrata. It is the most common brachiopod species of the Chalk and is also the only species of Surlyk's study that also occurs in adjacent nearshore facies of Sweden. Thus, in life it had both an unusually broad ecologic niche and an unusually great abundance. Surlyk noted that it belonged to a lineage that has survived to the present day with little change. Its living descendent, *T. retusa,* differs from it only in minor details of rib pattern. Part of the ecologic versatility of members of the lineage is shown by the variety of shapes that the pedicle of the living *T. retusa* can develop for attaching to various types of substrata (Figure 11-2).

FIGURE 11-2
Variation in pedicle morphology and mode of attachment in the living brachiopod *Terebratulina retusa.* (From Surlyk, 1972.)

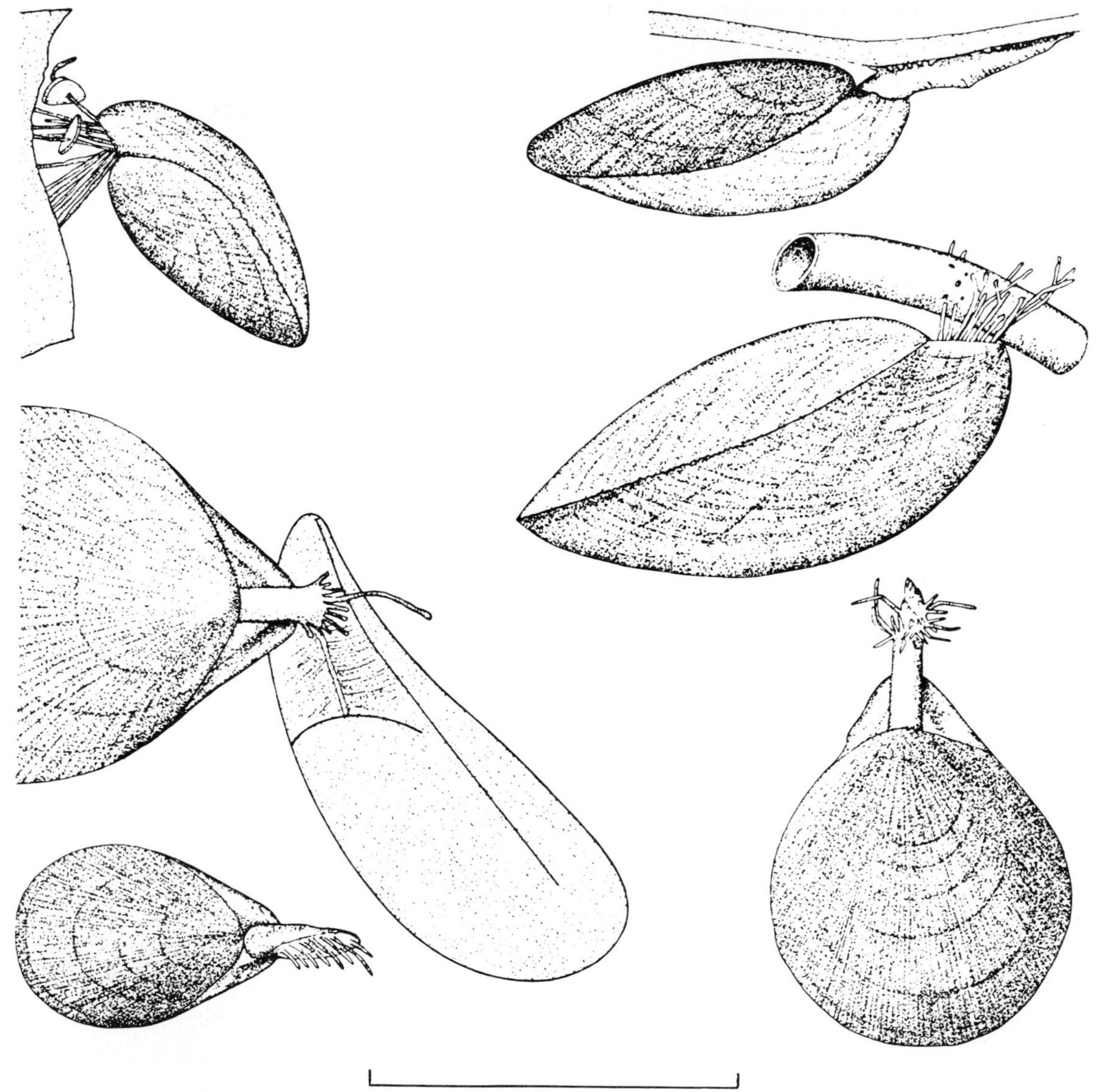

The generalization that species having large populations tend to survive for a long time has exceptions. For example, confinement of a species to a small geographic ***refugium,*** such as an island or lake, may prevent the population of the species from expanding to large size but still improve chances for survival. The reason is that this kind of confinement may provide isolation from competitors or predators.

Rates of Evolution and Extinction

The intensities at which both natural selection and extinction operate obviously vary widely in space and time. We must consider at least two conditions for each. One is the "background" condition, characterized by relatively low rates of evolution and extinction of the sort that exist within most groups of species at most times. The other is the less common condition in which rates are dramatically accelerated. ***Adaptive radiation*** and ***explosive evolution*** are terms commonly applied to episodes of rapid proliferation of new forms, and ***mass extinction*** is the term most commonly applied to episodes of rapid, large-scale demise of taxa.

The fossil record can be used to study rates of evolution in many ways. George Gaylord Simpson (1944, 1953) was a pioneer in this area, and several techniques of measurement have been added to those that he introduced. Simpson divided rates of evolution into two basic varieties. One of these is rate of increase in variety, of the sort that occurs during adaptive radiation. Rates of this type are normally measured by taxonomic criteria. Simpson termed them ***taxonomic frequency rates.*** The most common units of measure are numbers of new taxa per million years. We will return to a discussion of taxonomic frequency rates after discussing the other basic kind of rate, which we will call ***rate of progressive evolution.*** Here we measure amount of evolutionary change per unit time within a particular line of descent, as described in the following section.

RATE OF PROGRESSIVE EVOLUTION

In discussing the nature of species, we defined ***phyletic evolution*** as evolution within an established lineage (page 103). We also noted that phyletic evolution is virtually omnipresent because there is very little chance that gene frequencies within a species will remain constant from generation to generation. Even so, much phyletic evolution has been so slow that it is unrecognizable in the fossil record. When phyletic evolution is recognizable, its rate can be measured by using either morphologic or taxonomic criteria. We will consider these separately, beginning with morphologic criteria.

One way to evaluate amount of evolution per unit time is to plot a graph of morphologic measurements against stratigraphic position. This can be done for populations within a single lineage. If change is largely unidirectional, it produces what we term an ***evolutionary trend.*** Unfortunately, gaps inevitably exist in the stratigraphic record of lineages. Because of preservational gaps, we can never prove that an apparent trend, documented by several discrete faunas, is a real trend. Fossils appearing to document a trend will never represent every generation of fossil life. Furthermore, a choice must often be made as to what temporal and geographic components of a fauna will be considered a single population and represented by a single point on a graph. Fossils directly above any barren interval may represent an exotic fauna that migrated into the area of study. A sequence in which several successive populations show stepwise change in one direction is often judged as probable proof of an evolutionary trend. The case is strengthened if the range of values measured for successive populations overlap, even though the mean value shifts. The case is also strengthened if two or more types of morphologic change occur simultaneously. It is less likely that several successive speciation events or chance migrations would produce a chain of steps in a particular morphologic direction. Let us consider the simple case of change in some measurable feature like relative size of a given organ. If we postulate that increase or decrease is equally likely in each step as we pass stratigraphically upward, and if we have reason to believe that the first change should be in the observed direction, then the chance that a given number (N) of successive changes would be in one direction will be $(\frac{1}{2})^N$. If we have no way of predicting the direction of the initial step, we are said to have one degree of freedom and the probability will be $(\frac{1}{2})^{N-1}$. We can use slightly more sophisticated statistical tests to analyze the probability of net shifts in which every step is not in the prevailing direction. Figure 11-3 is a plot showing a net stratigraphic shift in the size of the proloculus (first chamber) of a Permian foraminiferan. One of the populations of Figure 11-3 has the same mean diameter as the preceding population at the level of precision at which the measurements were taken. Of the others, 24 show increases and only 8 show decreases. The Binomial Theorem predicts that if change were equally likely in either direction, there is a probability of 0.007 that such frequencies could arise by chance. This is the chance of flipping a coin 32 times and getting 24 heads. (For an elementary discussion of the Binomial Theorem, see Sokal and Rohlf, 1973.) Thus, it is highly unlikely that the net observed shift in morphology represents a succession of steps, in which each of the local populations disappears and is replaced by a new species invading from another area.

Given the high probability that Figure 11-3 documents a real evolutionary trend, we can measure the rate of the apparent phyletic evolution. It turns out that the average rate of increase in mean diameter is about 6.4 percent per million years. Another example is shown in Figure 11-4. Here several related lineages of Eocene condylarths exhibit apparent trends in tooth size, which is a rough measure of body size. Splitting of lineages is also documented.

Phyletic rates can be measured using taxonomic criteria, just as they can be measured using morphologic criteria, but only in a highly subjective fashion, because we can only subdivide most lineages into segments that we call species or subspecies by the arbitrary placement of taxonomic boundaries (page 110). The foraminiferan lineage depicted in Figure 11-3, for example, is subjectively divided into three subspecies (Hayami and Ozawa, 1975). The close tie between morphologic and taxonomic rates is indicated by the procedure implicitly followed here. In addition to

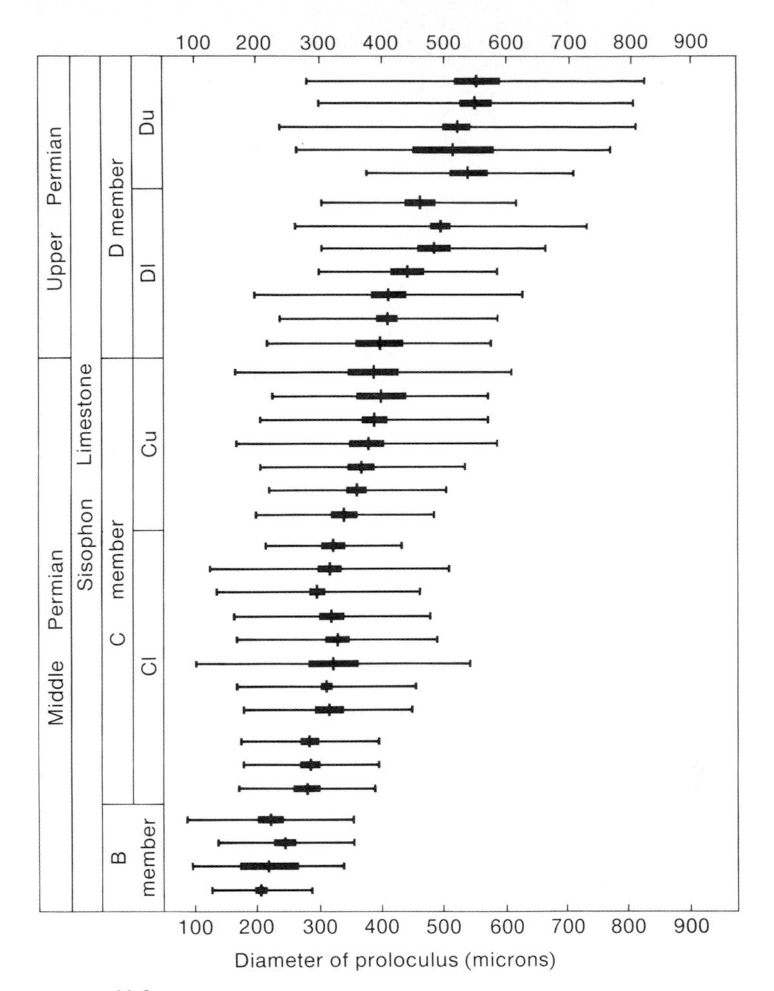

FIGURE 11-3
Shift through time in the size of the proloculus of the Permian fusulinid foraminiferan *Lepidolina multiseptata* from eastern Asia. For each stratigraphic level sampled, the mean value, its range with 95 percent confidence (thick bar), and two standard deviations (narrow bar) are plotted. The interval for which data are plotted spans about 15 million years. (From Hayami and Ozawa, 1975.)

the evolution shown in Figure 11-3, the lineage exhibits other minor changes in morphology. Knowing the typical degree of morphologic difference between closely related living species of foraminiferans, however, Hayami and Ozawa have judged that the lineage did not undergo enough morphologic change for young populations to be assigned to a species different from older populations. If a descendant species were recognized, exact placement of the interspecies boundary would, of course, be a matter of judgment. Other examples of apparent phyletic evolution are presented in Figures 11-5 and 11-6.

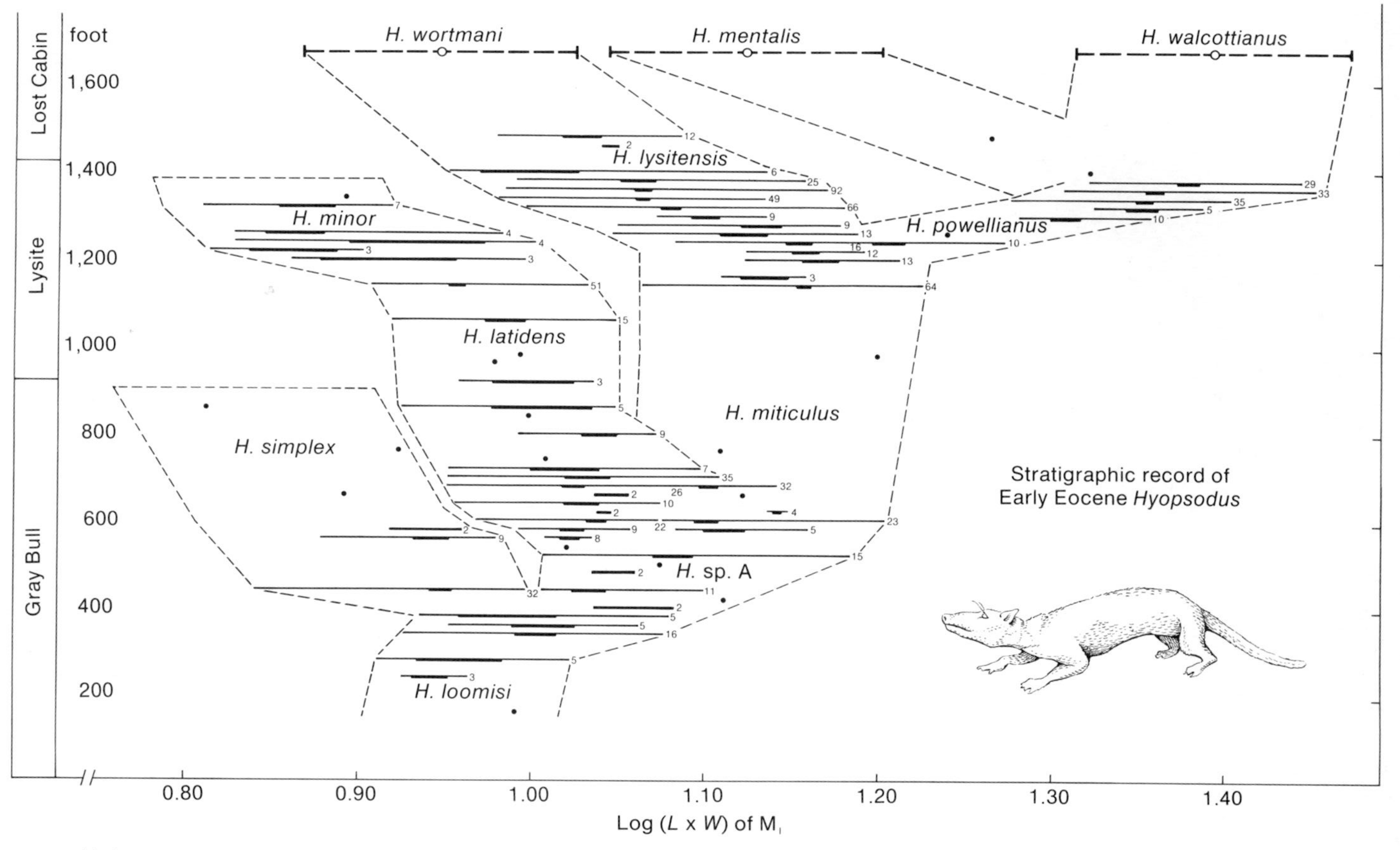

FIGURE 11-4
Stratigraphic occurrence of the condylarth genus *Hyopsodus* in the Early Eocene of northwestern Wyoming. On the horizontal axis is plotted the logarithm of the product of the length and width (in millimeters) of the first lower molar. The heavy bar represents the standard error for a sample, and the light bar, its range of values. Numbers to the right of bars indicate sample sizes (number of specimens). Solid dots represent single samples. Open circles and horizontal dashed lines at the top are means and expected ranges of species poorly represented in the stratigraphic section studied but well represented elsewhere. The stratigraphic section represents about 5 million years. (From Gingerich, 1974.)

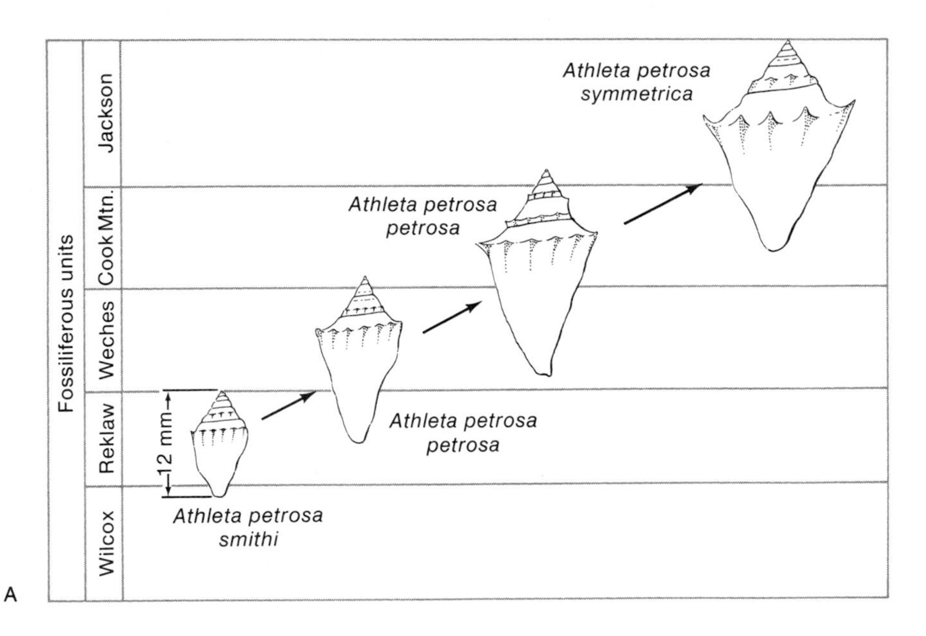

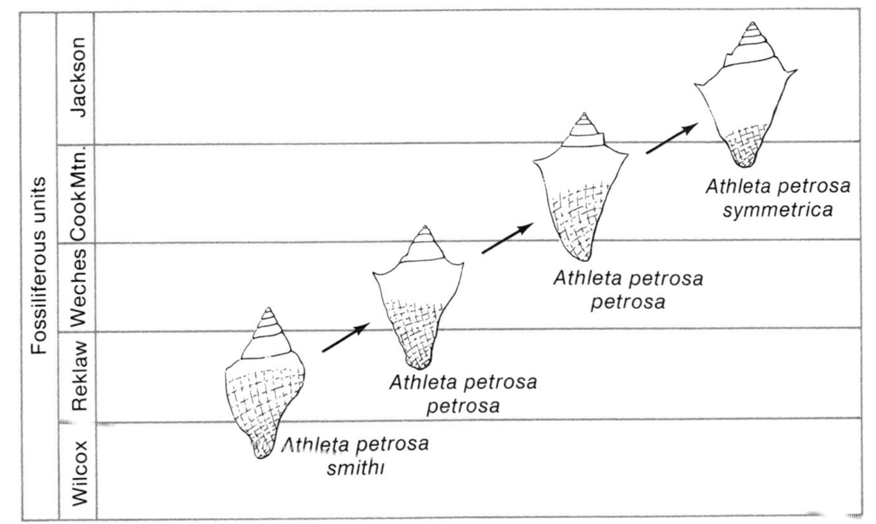

FIGURE 11-5
Evolution of *Athleta petrosa* in the Eocene of Texas. A: Changes in size, shape, and spinosity. B: Changes in the distribution of longitudinal and spiral ridges. The sequence of named subspecies spans approximately 5 million years. (Based on Fisher, Rodda, and Dietrich, 1964.)

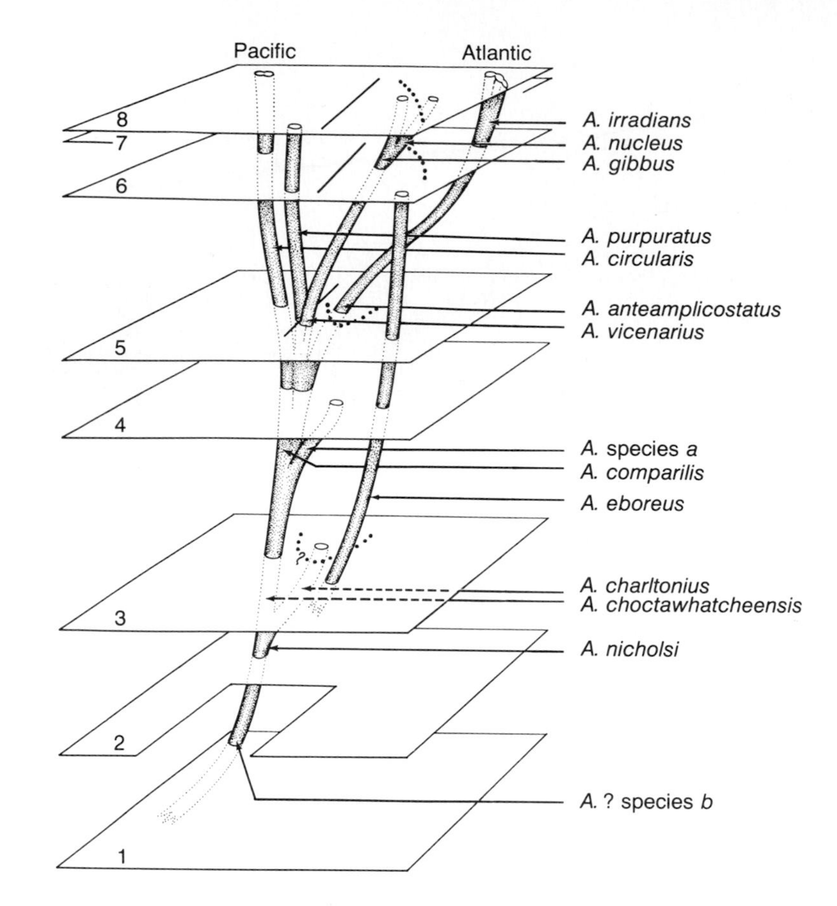

	Age	*Years ago*
1	Early Middle Miocene	19,000,000
2	Middle Middle Miocene	16,000,000
3	Late Middle Miocene	13,000,000
4	Late Miocene	8,000,000
5	Early Pliocene	6,000,000
6	Early Pleistocene	1,800,000
7	Late Pleistocene	250,000
8	Present	0

FIGURE 11-6
Cenozoic phylogeny of scallops of the genus *Argopecten* in North America. Numbers 1 through 8 identify chronologic planes spaced according to absolute time; each plane represents a stratigraphic unit that has contributed fossils and other data used to construct the phylogeny. The straight solid line on planes 5, 6, and 8 symbolizes the geographic barrier separating the Atlantic Ocean, Gulf of Mexico, and Caribbean from the Pacific Ocean after the Miocene. The curved dotted line on planes 3, 5, 6, and 8 symbolizes the ecologic barrier separating enclosed bay environments (upper right) from open marine environments (lower left) on the Atlantic side. (From Waller, 1969.)

The rate of phyletic evolution can be measured taxonomically in terms of the mean duration of all species that arise and disappear by phyletic evolution. Such species are sometimes termed ***chronospecies, successional species, paleospecies,*** or ***evolutionary species*** (page 110). You will note that such species in Figures 11-3, 11-4, 11-5, and 11-6 last for relatively long intervals of geologic time. Though one might wish to quibble with the placement of species boundaries in the lineages depicted in these figures, there is no question that the evolutionary rates that are documented are quite slow.

Whether rates of phyletic evolution should be measured by using taxonomic or morphologic criteria is a matter of judgment. Though not always reliable, taxonomic assignments and information on occurrences are often easy to obtain from the literature. On the other hand, morphologic measurements must often be made especially to suit a particular purpose. Both kinds of measurement offer problems for comparison of evolutionary rates within distantly related taxa. Morphologic criteria are easier to standardize, in that it is possible to use relative rates, such as percentage of change per unit time. On the other hand, different morphologic features of a single kind of organism may evolve at different rates. The choice of features then becomes important. The extreme case, in which some features change markedly while others remain relatively unchanged, is sometimes called ***mosaic evolution.***

In contrast to the ***phyletic rates*** of evolution that we have been discussing are what can be called ***phylogenetic rates,*** which consider the net rate of progressive change within the phylogeny of a higher taxon, such as a genus or family, without regard to the details of phylogeny. In effect, average morphologic condition within a group of closely related species is plotted against time. Sometimes a generic "lineage" is evaluated. The word lineage is placed in quotations here because it refers to a complex phylogeny that, for simplification, is treated as a single line of descent. For example, a sequence of three genera is plotted in Figure 11-7, to show the belief that some species of *Protoceratops* gave rise to the first species of *Monoclonius,* and that some species of the latter gave rise to the first species of *Triceratops.*The number of species belonging to each genus and the exact phylogenetic relationships of the species are ignored. Therefore the plot is cruder than those of Figures 11-3, 11-4, 11-5, and 11-6, but it also allows analysis of a poorer fossil record and spans a greater period of geologic time. An even longer time scale is considered in the plot of lungfish genera in Figure 11-8. Here genera of lungfishes existing during various intervals of geologic time were assigned scores reflecting degree of morphologic advancement. The scores were assigned subjectively, using modern lungfishes as standards for comparison. The score for each genus is an average of scores for more than 20 morphologic variables. The rate of overall change was relatively high at first but declined to a very low level toward the end of the Paleozoic.

While phyletic rates of evolution are often measured by using taxonomic criteria, it is generally inappropriate to measure phylogenetic rates in this way. The reason is that a new higher taxon generally does not arise from an entire ancestral taxon of the same rank. Many higher taxa probably arise from single species, but even those that are polyphyletic seldom arise from more than a small fraction of species of the ancestral

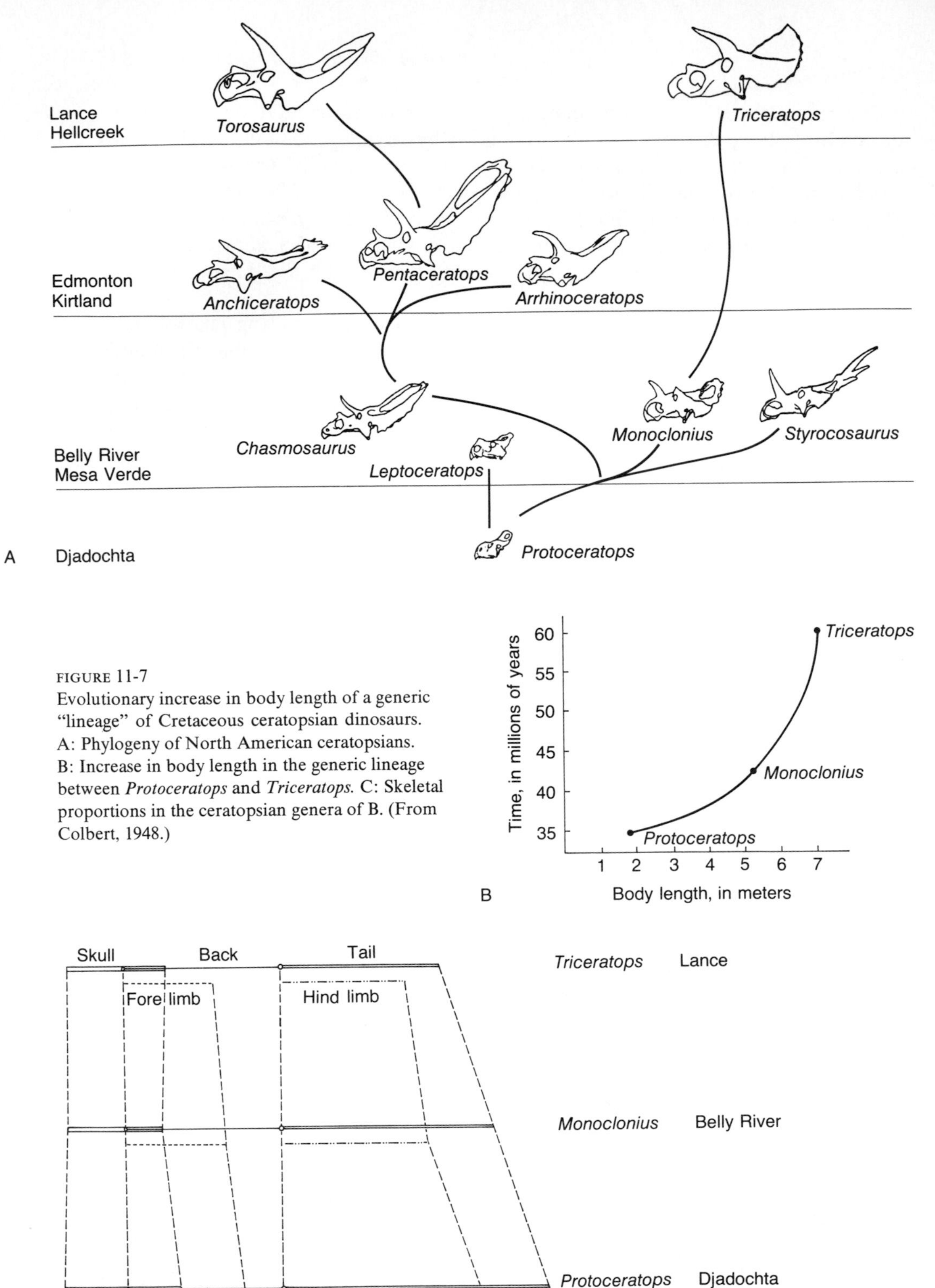

FIGURE 11-7
Evolutionary increase in body length of a generic "lineage" of Cretaceous ceratopsian dinosaurs. A: Phylogeny of North American ceratopsians. B: Increase in body length in the generic lineage between *Protoceratops* and *Triceratops.* C: Skeletal proportions in the ceratopsian genera of B. (From Colbert, 1948.)

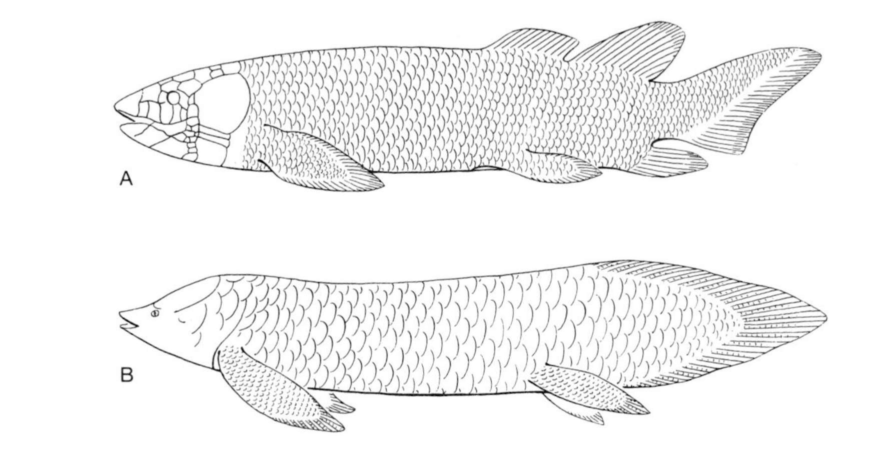

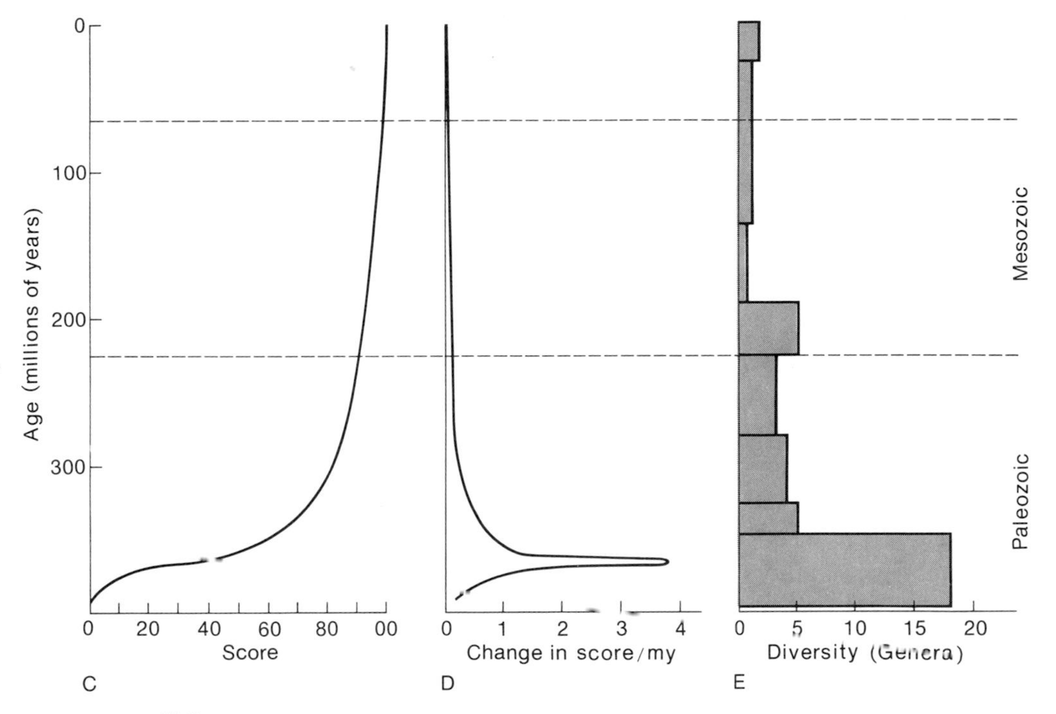

FIGURE 11-8
Evolutionary stagnation of the lungfishes. A: The Devonian genus *Dipterus* (×0.25). B: The living genus *Epiceratodus,* which may reach a length of 1.5 meters. C: Average scores for evolutionary advancement of genera plotted against time. D: Rate of change in average score plotted against time. E: Generic diversity. (A and B from Colbert, 1955, copyright © 1961 [reprinted by permission of John Wiley & Sons, Inc.]; C and D after Simpson, 1953, data from Westoll, 1949; E from Stanley, 1975.)

taxon. The result is that if we calculate the number of million-year periods per genus or family for any group of organisms, we are really calculating a rate of extinction by termination, not a rate of phyletic transition between taxa. For this reason, we will discuss this kind of measurement in a subsequent section in which rates of extinction are considered.

TAXONOMIC FREQUENCY RATES

Thus far we have considered rates of phyletic and phylogenetic evolution, that is, rates of directional evolution within particular lineages or closely related groups of lineages. It is also useful to measure the rate of diversification, which is to say, the rate at which totally new directions of evolutionary change are taken. These directions are embodied in new taxa. As mentioned earlier, the rates of appearance of taxa are called ***taxonomic frequency rates.*** For example, we can calculate the number of new genera or families per million years or the percentage of new taxa per million years. The percentage calculation is often useful because we measure rates for episodes of adaptive radiation during which the total number of taxa is increasing. This means that there is bound to be an increase in the number of new genera per million years, because the increase is basically geometric: the more genera that exist, the more potential ancestors there will be for the further addition of genera. Box 11-A illustrates the way in which this happens.

In measuring taxonomic rates, we must recognize the fact, pointed out earlier, that a genus or family in one class or phylum may not be equivalent to a genus or family in another. In fact, strict comparison is virtually impossible. Phylogeny tends to have a hierarchical pattern, in which clades occur on several levels (Figure 11-9). As we have seen in Chapter 6, both the recognition of the clades and their arrangement into higher taxa are highly subjective. Exactly what constitutes a genus or family within a large group may tend to reflect the structure of this hierarchy as well as the classifier's idea of how much morphologic variability should in general characterize a genus or family. There are two ways to introduce greater consistency into the measurement of taxonomic rates. One is to confine one's efforts to a coherent taxon, such as a class or order in which the families or genera to be compared have similar gross morphologies and are classified in a consistent way. Another is to use species as units of measure. Because of the poor fossil record of species (Chapter 1), this can only be done in special cases and in special ways, some of which we will now consider.

In particular, techniques exist for species-level analysis of Cenozoic taxa that have good fossil records. For a taxon of this sort that is in the midst of adaptive radiation we can obtain an estimate of doubling time for number of species if we know two things: time of origin of the taxon and number of living species within it. In Figure 11-10 we compare doubling time and increase in species diversity for an average radiating Cenozoic family of marine bivalve molluscs and an average radiating Cenozoic family of mammals. This comparison is based on the assumption that the families have arisen from single species. Polyphyly would alter the shapes of the plots slightly, but it would take a very high degree of polyphyly to alter them significantly. The fact that adaptive radiation is basically geometric accounts for the slender "tail" at the base of many empirical plots of taxonomic diversity through time. A comparison of Figures 11-10,A and 11-10,B should make apparent the great effect that doubling time has on overall amount of increase. Terrestrial mammals typically radiate much more rapidly than marine bivalves.

Mathematical treatment of the data used to construct Figures 11-10,A and 11-10,B can yield a percentage (or fractional) increase in number of species per million years.

FIGURE 11-9
The hierarchial pattern of clades in phylogeny. Three families, separated by the dashed lines, might be recognized in this hypothetical example. Smaller clusters of lineages might be designated as genera.

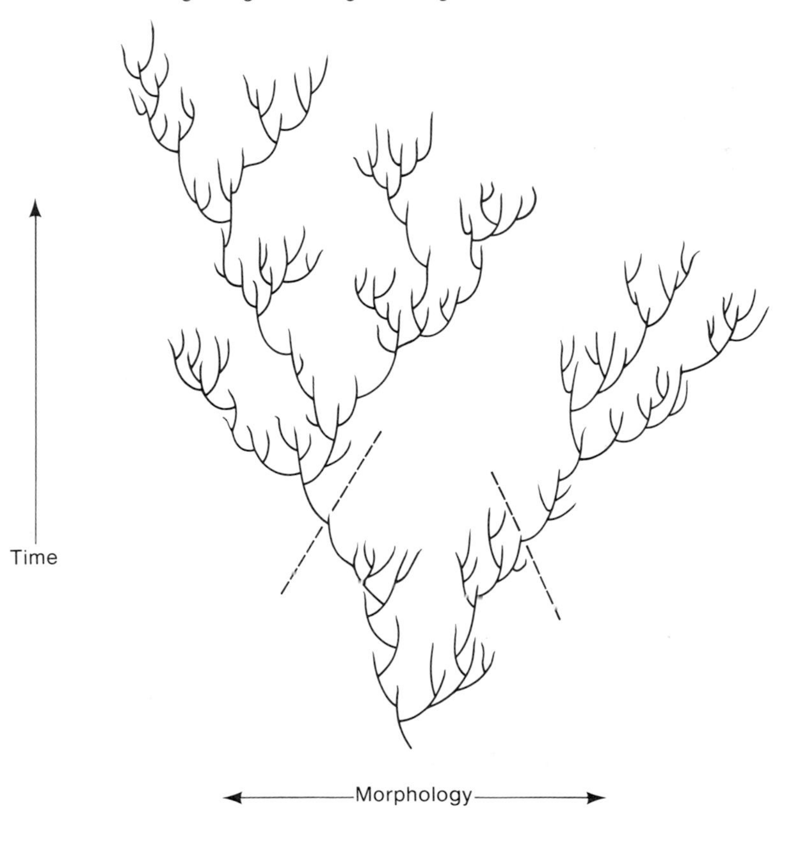

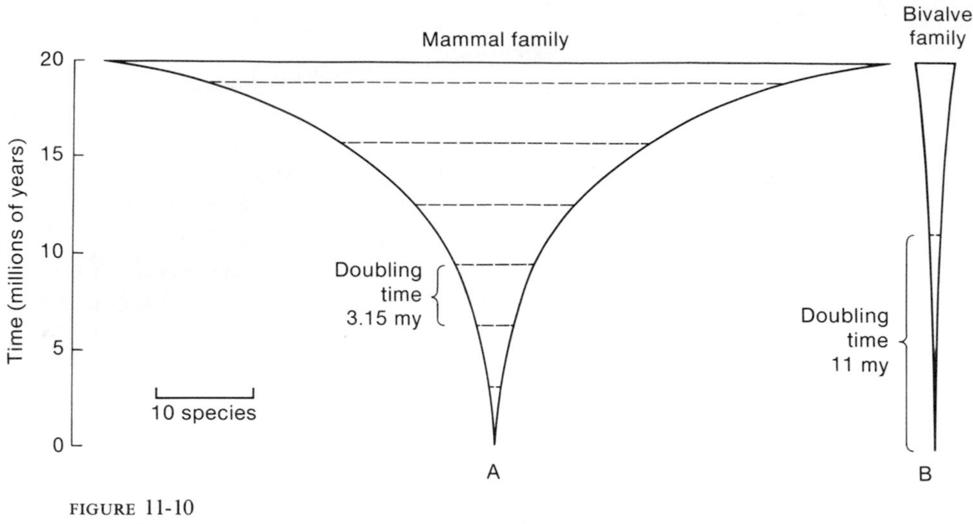

FIGURE 11-10
Diversification of an average Cenozoic family of mammals (A) and of marine bivalve molluscs (B). (Data from Stanley, 1977.)

This is the kind of "compound interest rate" discussed in Box 11-A. The method of calculation of this rate is available elsewhere for the reader who wishes to pursue this subject (Stanley, 1975; 1977). For simplicity here we can recall that net rate of increase is inversely proportional to doubling time (Box 11-A). Figure 11-10 shows that doubling time is about one third as long for mammals as for marine bivalves, which means that net rate of increase is about three times as high. As we showed in Box 11-A, the following relationship should hold:

Net rate of increase = Speciation rate − Rate of extinction of lineages
(or rate of formation of lineages)

We have estimated relative net rates of increase for mammals. If we can estimate relative extinction rates, then by adding the two, we can estimate relative speciation rates. This will be our next procedure. As would be expected from the contrast between Figure 11-10,A and Figure 11-10,B, we will find that rate of speciation is much higher for a typical adaptive radiation of mammals.

How can we calculate extinction rates? We can employ a short-cut method to make a crude estimate. All we need in order to make this estimate for a higher taxon is an estimate of the duration in geologic time of an average species. The rate of extinction will be approximately the reciprocal of the duration of an average species. In other words, if an average species of a higher taxon lasts 5 million years and species are

BOX 11-A Calculation of Taxonomic Frequency Rates

We will plot data for the initial adaptive radiation of ahermatypic (non-reef-building) scleractinian corals (Wells, 1956) in two ways (see the diagrams on the following two pages). Graph A shows how the total number of genera changed through time. Graph B shows that the number of new genera per million years also increased after the Jurassic. These graphs might seem to suggest that the adaptive radiation persisted unabated during the interval of time considered. Here we come to a crucial point, however. Graph B is misleading because, though the *number* of genera added per million years did not decline, the *percentage* of new genera added per million years did. This suggests that one or more factors, such as increased competition, heavier predation pressure, or deterioration of the physical environment, made the appearance of new genera less likely. If the percentage of increase per unit time had remained at the value of the first 60 million years, increase in number of genera would approximately have followed the dashed line of Graph C. This kind of increase is called ***geometric*** or ***exponential increase.***

Compound interest produces geometric increase of savings that are allowed to accumulate in a commercial bank. Geometric increase of any sort is characterized by a ***doubling time.*** One kind of geometric increase is illustrated in Graph D. One way of understanding why a doubling time exists is to recognize that the use of a doubling time is just one way of describing increase by a fixed percentage per unit time. An interest rate of 5 percent per year will result in an increase of 100 percent (a doubling) about every 14 years. An increase of 10 percent per year will result in doubling about every 7 years. In other words, multiplying the rate of increase by two cuts the doubling time in half: rate of increase and doubling time are inversely related. Readers who are familiar with the nature of radioactive decay will recognize that doubling time is similar to the half-life of a radioactive isotope. A half-life represents geometric decrease rather than geometric increase.

It is important to realize that evolutionary diversification is more complex than the simple increase shown in Graph D. This is true because taxa not only arise but also become extinct. Nonetheless, increase in diversity will occur if the percentage of new taxa added per unit time exceeds the percentage of taxa becoming extinct per unit time. Furthermore, if the two percentages are constant, the increase will still be geometric. Also the *net* percentage of increase will equal the first percentage minus the second. An analogy with banking is again appropriate. If money is deposited in an account yielding 5 percent per year and 3 percent of the balance is removed per year, the savings will increase at an annual rate of 2 percent. To summarize:

$$\begin{bmatrix}\text{Net rate}\\\text{of increase}\end{bmatrix} = \begin{bmatrix}\text{Interest rate}\end{bmatrix} - \begin{bmatrix}\text{Withdrawal}\\\text{rate}\end{bmatrix}.$$

The analogous relationship in adaptive radiation will be:

$$\begin{bmatrix}\text{Net rate}\\\text{of diver-}\\\text{sification}\end{bmatrix} = \begin{bmatrix}\text{Rate of origin}\\\text{of taxa}\end{bmatrix} - \begin{bmatrix}\text{Rate of}\\\text{extinction}\\\text{of taxa}\end{bmatrix}.$$

Geometric increase is an ideal phenomenon that will never be perfectly adhered to in any real adaptive radiation. Nonetheless, approximate adherence to it can permit us to use either *percentage* increase per unit time or doubling time as a rough measure of rate of diversifica-

(continued)

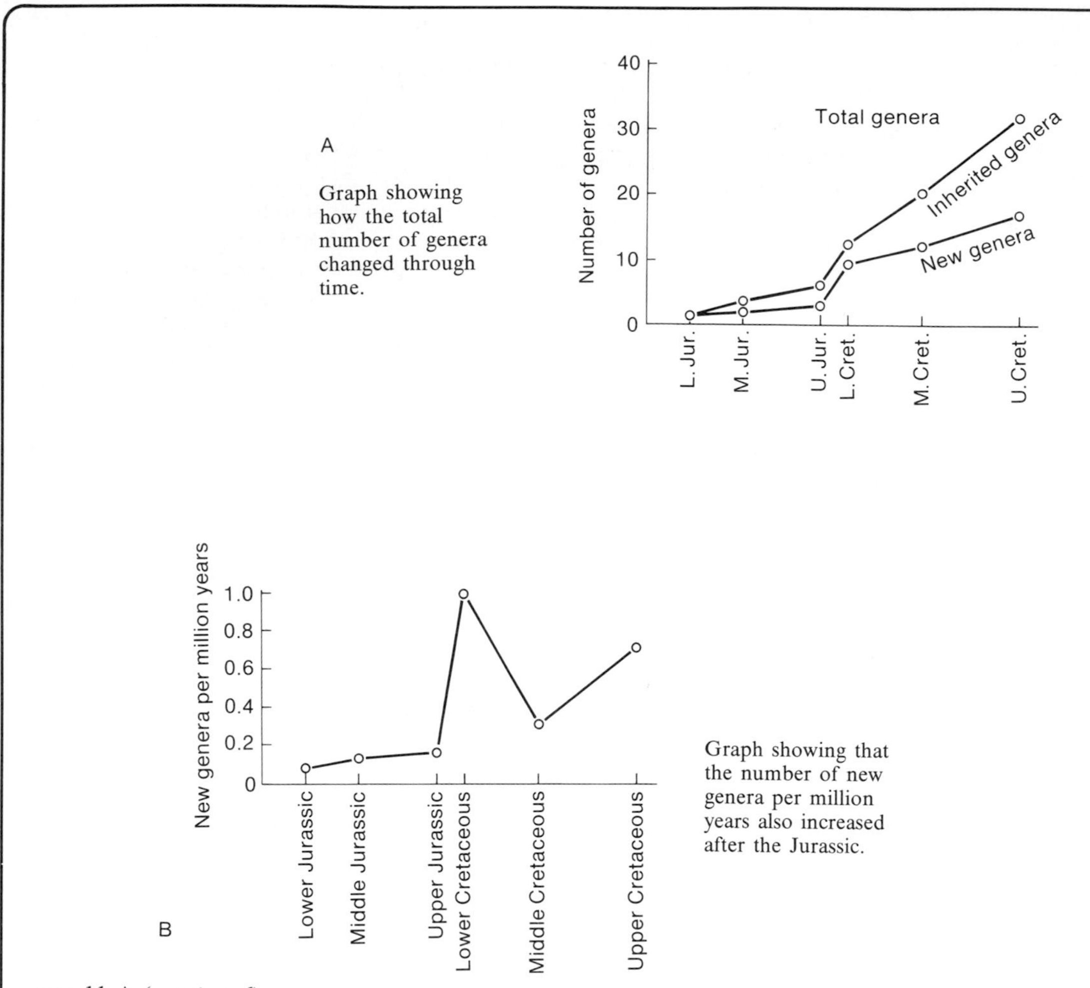

Graph showing how the total number of genera changed through time.

Graph showing that the number of new genera per million years also increased after the Jurassic.

BOX 11-A (*continued*)

tion. These measures are more meaningful than is the simple number of new taxa per unit time, which would be expected to increase as the number of taxa increases.

Clearly percentage increase per unit time will tend to decline after the initial stage of an adaptive radiation. One reason for this is crowding. If species compete with each other to some degree, fewer and fewer readily accessible ways of life will be available as diversification proceeds. Another reason, at least when we consider higher categories, is that increase in specialization of taxa through time may tend to reduce the likelihood that new higher taxa will arise. In general, doubling times should only be calculated for taxa that are in the early stages of adaptive radiation and whose rates of diversification have not slowed appreciably.

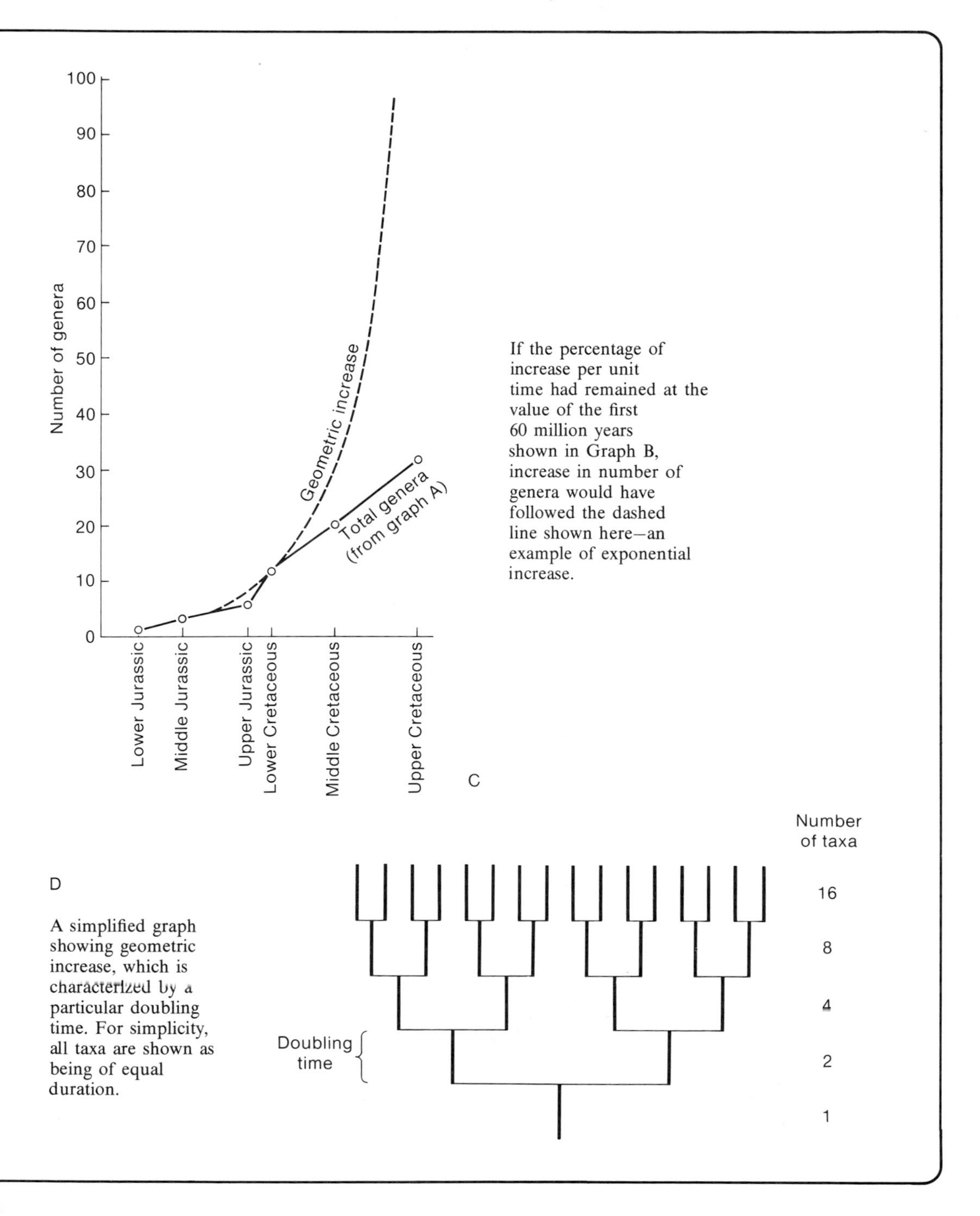

If the percentage of increase per unit time had remained at the value of the first 60 million years shown in Graph B, increase in number of genera would have followed the dashed line shown here—an example of exponential increase.

A simplified graph showing geometric increase, which is characterized by a particular doubling time. For simplicity, all taxa are shown as being of equal duration.

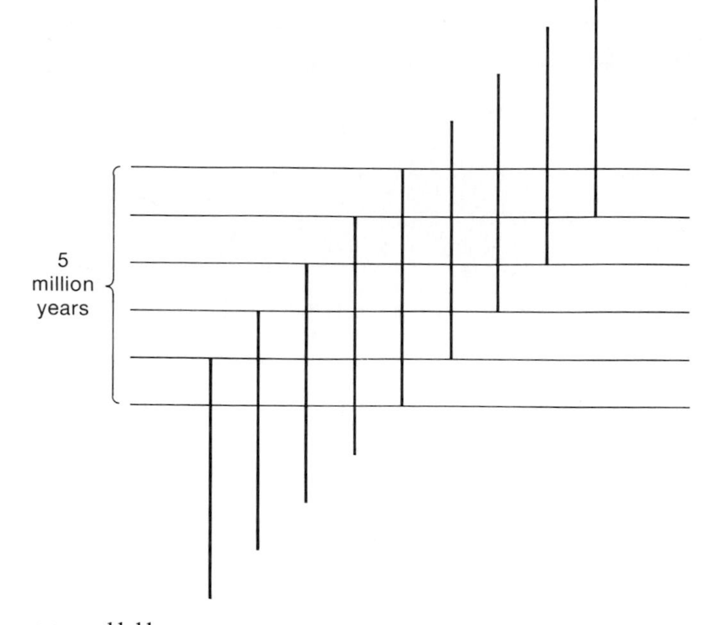

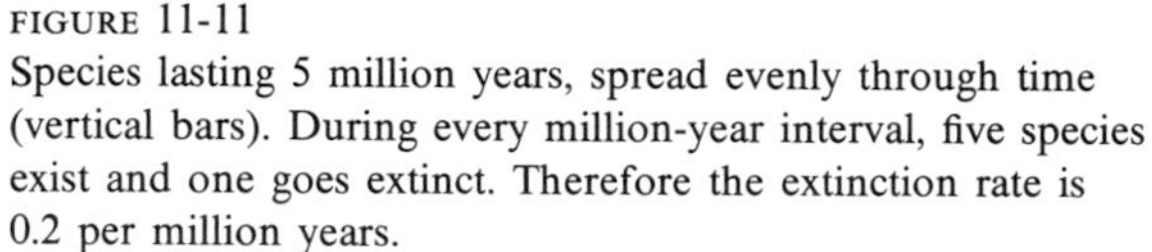
FIGURE 11-11
Species lasting 5 million years, spread evenly through time (vertical bars). During every million-year interval, five species exist and one goes extinct. Therefore the extinction rate is 0.2 per million years.

distributed evenly through time, about one in five will become extinct every million years (Figure 11-11). Another way of viewing this situation is to recognize that if an average species lasts 5 million years, it can be viewed as becoming one-fifth extinct every million years. Its individual extinction rate is, therefore, one-fifth per million years. If it is an average species, then its individual "extinction rate" will approximate the extinction rate of the entire taxon. Just how good an estimate it is will depend on the shape of the histogram of species durations for the entire taxon. At least it will be a crude estimate.

How do we determine how long an average species of a taxon lasts? There is a simple way of making a rough estimate, and this will suffice for our present discussion. If we look backward through time at fossil faunas of a taxon, we will find that successively older faunas contain fewer and fewer species that are alive today. Let us assume that we are dealing with a typical evolutionary interval for the taxon—one in which there were no major pulses of extinction, for example. If, in looking back through time, we stop at the point where faunas contain 50 percent living species, we will have an estimate of how long it should take for half of the species living at any time to become extinct. Complete faunal turnover (complete replacement of old species by new ones) would take about twice this long. As shown schematically in Figure 11-12, this turnover time will approximate the duration of an average species. We are, of course, oversimplifying the situation in Figure 11-12 by assuming that all species are of equal duration.

About 50 percent of bivalve mollusc species of the earliest Pliocene are still alive. The Pliocene Epoch began about 5 million years ago. This gives us an estimate of about 10 million years for the duration of an average species of bivalves. Applying the same technique to mammals, we obtain a much shorter estimated duration for an average species: between 1 and 2 million years. The reciprocals of these durations give

FIGURE 11-12
If all species are of equal duration and evenly distributed through time, faunal turnover time will equal an average species duration, or twice the age of faunas in which 50 percent of all species survive to the Recent.

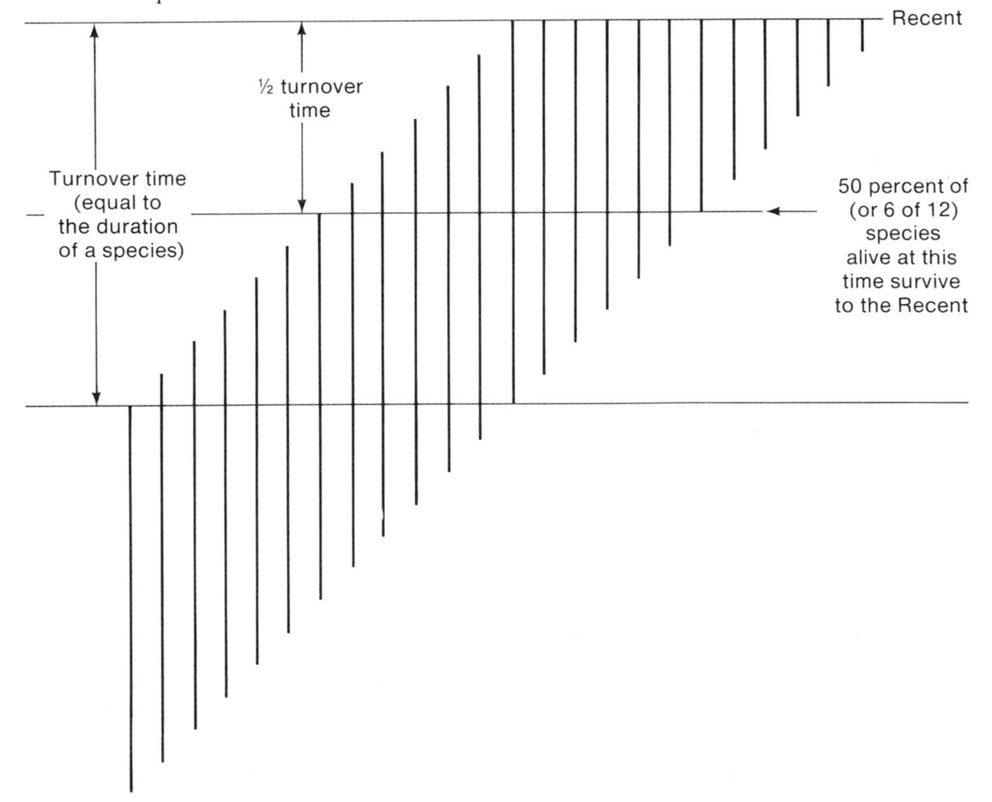

us crude estimates of extinction rates. These are 0.66 (66 percent) per million years for mammals and only 0.1 (10 percent) per million years for bivalves. In other words, our estimated extinction rate is more than six times as high for mammals. It is important to understand that the estimated rates include two kinds of extinction—extinction by termination of lineages and extinction by gradual phyletic transition of old species into new ones. We are interested here only in rate of termination of lineages. Probably the percentage of total extinction that is of this type is roughly the same for mammals and bivalves. In other words, rate of termination of lineages is probably at least five or six times as high for mammals.

To obtain an estimate of a speciation rate, we must add together our estimates of net rate of increase and extinction rate. Because both of the rates to be added are estimated to be several times as high for terrestrial mammals as they are for marine bivalves, we can estimate that a typical speciation rate in adaptive radiation will be several times as high for mammals. This difference is enormous because addition of species, like interest in a savings bank, is compounded. This, as we have seen, is the nature of geometric increase.

THE DISTRIBUTION OF EVOLUTIONARY RATES IN PHYLOGENY

We can now consider, from a paleontologic perspective, a question that was raised in Chapter 5. This is the question of whether most evolutionary change is related to the formation of new species. The alternative is that most change takes place as phyletic evolution within established species. We discussed earlier (page 107) the view of Mayr (1954; 1963) and others that homeostatic mechanisms tend to prevent established species from changing very rapidly. Mayr believes that rapid evolution tends to occur only in small populations, primarily in isolated ones that are becoming, or have just become, new species. The idea is that even though many speciation events produce only minor evolutionary divergence, others produce marked divergence, and the latter account for most large-scale change in the history of life.

Mayr (1954) also suggested that rapid change in speciation might explain the sudden appearance of evolutionary novelties in the fossil record. More recently various workers have argued that Mayr's scheme be applied to the fossil record in general. Some early Soviet contributions of this type are available in English translation. For example, Ruzhentsev (1964) suggested that most fossil species do not grade imperceptibly into one another but are separated by discontinuities representing sudden speciation events. Ovcharenko (1969) adopted this idea and reported an apparent example of rapid, divergent speciation for the Jurassic brachiopod *Kutchithyris* (Figure 11-13). Two common and geographically wide-ranging species of this genus show relatively consistent morphologies. *K. euryptycha* occurs stratigraphically above *K. acutiplicata.* They are different enough morphologically to have been placed in distinct genera by at least one author. In just one small locality, in a

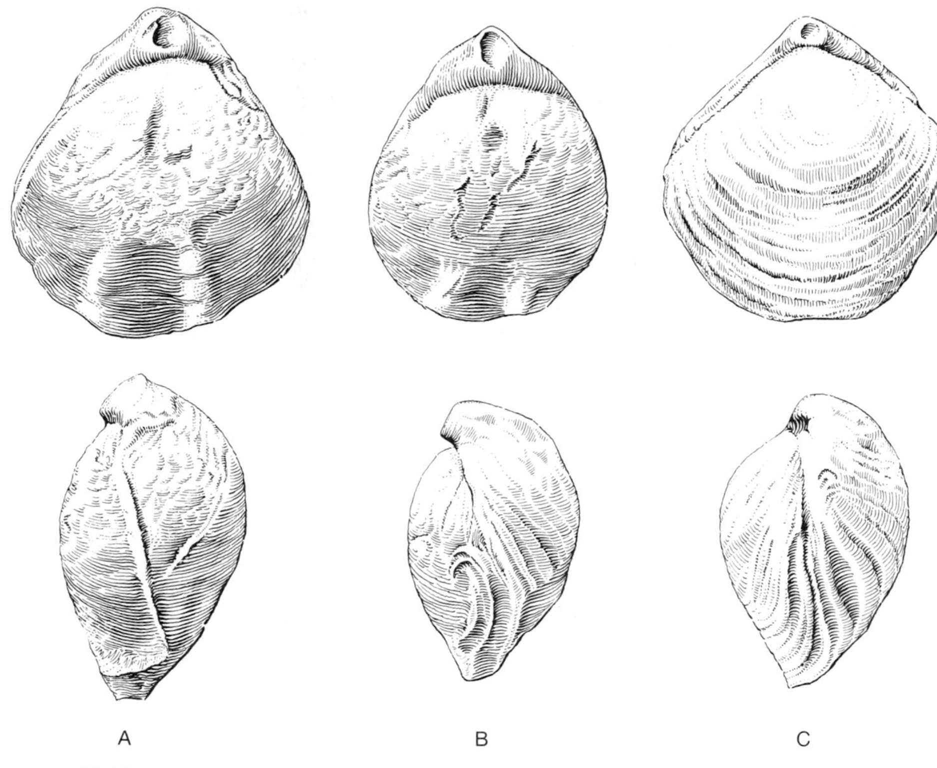

FIGURE 11-13
The Jurassic brachiopods (A) *Kutchithyris acutiplicata,* (C) the younger species *K. euryptycha,* and (B) a specimen of intermediate morphology and stratigraphic occurrence. (After Ovcharenko, 1969.)

sedimentary layer 1.0–1.5 meters thick, Ovcharenko found a transition from one species to the other. At the base of the layer only *K. acutiplicata* is found. In the upper part of the layer only *K. euryptycha* is found. Between these parts of the layer is an interval only 10 cm thick in which both forms appear, along with specimens of various intermediate morphologies. Here we seem to have a rare example of paleontologic documentation of rapid, divergent speciation by way of a small population. Several Soviet workers have adopted the view that the fossil record does not frequently document rapid phyletic trends. Their view represents acceptance of the arguments of biologists like Mayr (1954; 1963) and Grant (1963) that established species tend to evolve slowly. In other words, many lineages are wiped out before undergoing much phyletic change. The other side of the coin is the idea that much of the change that occurs in speciation is not recorded or discovered in the fossil record because it takes

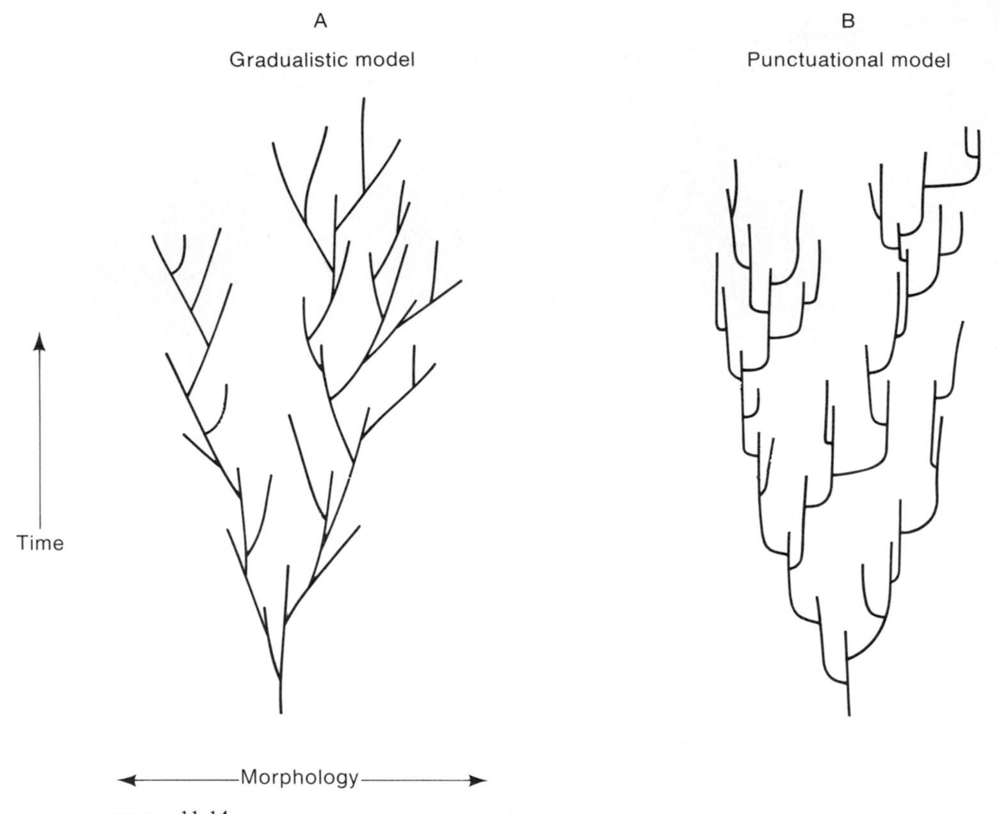

FIGURE 11-14
Phylogenetic trees based on (A) a gradualistic model and (B) a punctuational model.

place rapidly and in small, isolated populations. In North America, Eldridge (1971) independently suggested the same idea and provided a possible example within the Devonian trilobite genus *Phacops,* and Eldredge and Gould (1972) provided a general account of this view of evolution.

The distribution of rates of evolution in phylogeny is still being debated. If phyletic evolution and divergent speciation have accounted for nearly equal percentages of overall change, it matters little which mode of change has produced the larger percentage. On the other hand, if one mode of change has been clearly dominant, it is important to know which it is.

The idea that considerably more than 50 percent of evolution occurs through the proliferation of species has been termed the ***punctuated equilibrium model*** of evolution or the ***punctuational model.*** The name refers to the step-like or "punctuated" nature of the phylogenetic model (Figure 11-14,B). The opposing idea, that considerably more than 50 percent of evolution occurs as phyletic change within established species, has been termed ***phyletic gradualism*** or the ***gradualistic model*** (Figure 11-14,A). It represents the traditional view.

Several tests of the two models have been proposed (Stanley, 1975). One of these notes that if the punctuational model is valid, then taxa of a particular type would be expected to undergo little long-term change. These taxa are ones that last for long intervals of time without undergoing an appreciable amount of speciation. In fact, such groups are quite familiar and are commonly referred to as "living fossil" groups. These groups derive their name from the fact that they have persisted for long spans of geologic time with very little morphologic change. The important point is that taxa that have persisted for long intervals at low species diversities have almost invariably evolved very little. The lungfishes are a representative "living fossil" group. Figure 11-8 shows that the lungfishes ceased to evolve rapidly when they ceased to speciate rapidly. Figure 11-8,E is actually a plot for genera, but its shape obviously reflects rate of speciation. Other examples are the *Ginkgo* and *Metasequoia* trees, the mollusc *Neopilina* (see page 375), and the reptile *Sphenodon.* The punctuational model predicts the presence of such forms. If most evolution occurs by way of speciation, then groups that survive at low diversities will not change much, even over long periods of time. If, on the other hand, phylogenetic rates of evolution are largely due to phyletic change, then some "living fossil" groups, like other taxa, should produce distinctive new genera and families through phyletic evolution over periods of tens or hundreds of millions of years. There are many "living fossil" orders and classes, and yet none seem to show such amounts of change. In other words, the evolutionary histories of "living fossil" taxa seem to support the punctuational model.

Many new orders of mammals arose in the Paleocene and early in the Eocene, when mammals underwent dramatic modern expansion. Creatures as different as bats and whales arose from primitive ancestral groups during about 12 million years. Hardly any totally new morphologic features or modes of life have arisen in the Mammalia since the mid-Eocene, suggesting that in only about 20 million years the Mammalia evolved nearly all of the fundamental kinds of adaptations possible. An exception to this generalization, of course, is the evolution of humans. Major adaptive transformations in bivalve and gastropod molluscs have invariably required much longer. For example, the transition from burrowing clams through attached epifaunal forms (page 192) to free-living scallops took perhaps 150 million years (Stanley, 1972). The transition from burrowing clams to cemented epifaunal forms required about 300 million years (Newell and Boyd, 1970). The Gastropoda apparently did not develop advanced carnivorous habits until more than 400 million years after the group originated. Many other evolutionary innovations in both classes, including mantle fusion, which formed siphons for water flow, and the origin of many types of shell ornamentation also did not arise until the Mesozoic and Cenozoic Eras. This is, of course, but one comparison and does not by itself validate the punctuational model.

Perhaps the most effective way of evaluating the importance of phyletic evolution is to determine how long species survive in geologic time. The frame of reference here is the length of time it has taken for major evolutionary transitions to occur. If species are found to last a very long time in comparison to the time required for new families or orders to arise, then the gradualistic model may be difficult to justify. In other words, it will be difficult to claim that phyletic evolution is the source of most large-scale change. Rapid phyletic change within a lineage has to produce phyletic extinction at brief intervals, as each successive species of the lineage passes into the next. How long, then, do species tend to survive?

Although phyletic evolution has not frequently been documented with great care, all examples reported so far show relative stability for established species in comparison to rates of large-scale evolutionary change of the sort that occur in adaptive radiation. Figures 11-3, 11-4, 11-5, and 11-6 illustrate this point. For example, the condylarth species whose stratigraphic ranges are partially displayed in Figure 11-4 seem, on the average, to have survived for a million years or so. During the very five-million-year interval spanned by the sluggish episode of condylarth evolution shown in Figure 11-4, the numerous new families and orders of Cenozoic mammals, referred to earlier, were arising. It is not impossible that phyletic rates were dramatically accelerated to produce such change. It has been argued, however, that this prospect is unlikely because it would require that selection pressures acting upon a wide range of taxa living in many different kinds of habitats be greatly intensified for several million years.

On the other hand, speciation was obviously occurring at very high rates in the Mammalia of the early Cenozoic, as they spread into niches from which they had apparently been excluded, during the Mesozoic, by dinosaurs. This is, in fact, the sort of condition required for application of the punctuational model. If speciation events are almost instantaneous on a geologic time scale, then numerous successive speciation events can produce large-scale change in very little geologic time. This line of reasoning, labelled the Test of Adaptive Radiation, has been presented as an argument in favor of the punctuational model. Most invertebrate groups seem to exhibit slower rates of large-scale change than mammals, but their species also tend to last much longer. An average species of Cenozoic molluscs has survived 5 or 10 million years. This also is a very long interval of time in comparison to the rate at which large-scale evolution occurs in the Mollusca.

It could be argued that despite the slow rates of phyletic change observed in the fossil record, there may actually have existed in most groups at any time a significant fraction of short-lived species that have escaped our notice. In other words, perhaps a typical histogram for species durations in a higher taxon is strongly skewed to the right. Then mean duration may be quite long, though the mode may be close to zero. This would indicate the presence of many short-lived species that have perhaps undergone rapid phyletic change, one into another. This question can be evaluated for Plio-Pleistocene mammals using data that Kurtén (1968) has compiled for European species. It is actually possible to employ these data to produce a histogram of chronospecies durations that is not biased by the various problems that one normally encounters in plotting a histogram directly from reported stratigraphic ranges. Raup (1975b) and Sepkoski (1975) have analyzed these problems, the most obvious of which is that only segments of actual ranges are found in the fossil record. The technique for plotting a complete histogram (Stanley, 1977) will not be presented here. It will suffice to outline some of the major points.

Only about 12 percent of the living species of European mammals are not recognized in the fossil record of the Pleistocene. In fact, we can trace the evolutionary histories of nearly all existing lineages backward in geologic time for at least 200,000 years. Paleontologists studying late Cenozoic lineages of mammals tend to accept species designations that have been established for living faunas. Much is known about the biology of the living faunas and little disagreement exists about taxonomy at the species level. The paleontologist therefore uses the Recent as a taxonomic starting point for the study of an extant lineage. He traces the lineage backward through time until its members are different enough from living members to be recognized as a new chronospecies. Exactly where he draws a line to demarcate a new chronospecies will depend upon his particular species concept (see page 113). Of course, living chronospecies will "survive" backward into the past for different lengths of geologic time, depending on rate of evolution within the lineages to which they belong.

In considering the antiquity of living chronospecies, it is important to employ data that are treated consistently, preferably data that represent the species concept of a particular worker. The data that Kurtén (1968) has subjectively compiled for European mammals satisfy this criterion. What Kurtén's remarkably complete set of data reveals is that rather few living chronospecies have "survived" backward for less than about one-half million years, and virtually no living chronospecies has an antiquity of less than about 200,000 years. (We ignore here species that have formed recently by branching.) The sample of chronospecies is so complete that the results represent compelling evidence that phyletic evolution within late Cenozoic Mammalia has been extremely slow.

To assess in more detail the meaning of these chronospecies longevities it is instructive to consider the origin of genera (Stanley, 1977). During the Pleistocene, at least 25 genera of mammals evolved within Europe. It would seem reasonable to estimate that in order for a lineage to exhibit enough phyletic evolution that a new genus be recognized as having evolved within it, the lineage would have to be divided into at least eight or ten intergrading chronospecies. It is unlikely that fewer chronospecies transitions would produce the required degree of evolution because most intergrading chronospecies of mammals differ from each other only in size or other minor features of morphology. For an average phyletic lineage containing eight or ten chronospecies to produce one new genus of mammals during the Pleistocene, which lasted about 1.8 million years, an average chronospecies in the lineage would last only about 200,000 years. As noted above, however, hardly any Pleistocene chronospecies of mammals have lasted for such a short period of time. It may be that an occasional lineage underwent enough phyletic evolution to be regarded as having produced a new genus, but examples of this type must have been rare indeed. In Europe, at any given time during the Pleistocene, there was a diversity of only about 150 species, or in the order of only 100 continuous pathways of phyletic evolution through the Pleistocene. Given the great longevity of chronospecies, very few of these pathways could have produced phyletic transitions to new genera, certainly not enough to acount for more than a small fraction of the 25 or more new genera that evolved. This leaves us with the inference that most of the new genera arose through rapidly divergent speciation. (Certainly, some genera may have evolved by way of more than one speciation event.) The Pleistocene, with its advancing and retreating ice sheets, is hardly a time when one would expect phyletic evolution to have been unusually slow for mammals of the European region. If, in general, few genera of mammals arise by phyletic evolution, it seems unlikely that many families or orders do either. Thus, the great longevity of chronospecies here opposes the gradualistic model of evolution.

We have reviewed several kinds of evidence that seem to support the punctuational model of evolution. These include the evolutionary traits of "living fossil" taxa, the apparent correlation between rate of speciation and rate of large-scale evolution, and the longevity of species in geologic time. It should also be noted that we know of certain instances where speciation resulting in marked morphologic divergence has occurred very rapidly. Examples of such speciation in large lakes will be discussed in Chapter 12 (page 415). Still, readers are encouraged to evaluate the evidence independently. They may wish to consult the writings of workers who have expressed skepticism of the idea that most evolutionary change occurs in speciation. Among these writers are Hecht (1974) and Harper (1975, 1976).

RATES OF EXTINCTION ABOVE THE SPECIES LEVEL

We have already discussed rates of extinction of species, emphasizing the marked difference between ***phyletic extinction,*** sometimes called ***pseudoextinction,*** and extinction by termination of a lineage. We have also considered general causes of extinction. Sudden pulses of extinction will be considered under the heading "Patterns of Evolution and Extinction," later in this chapter, and also in Chapter 12.

It is important to understand that there are great problems in evaluating rates of extinction when using higher taxa as units of measure. Some of these were discussed near the beginning of the chapter. Of special significance is a problem not present in measuring rate of splitting (Box 11-A), namely, the problem that a higher taxon does not become extinct as a unit. It is made extinct only when all of its component species die out. The species are discrete entities, generally having distinct niches and often occupying discrete habitats. Only if a single extinguishing event happens to reach all

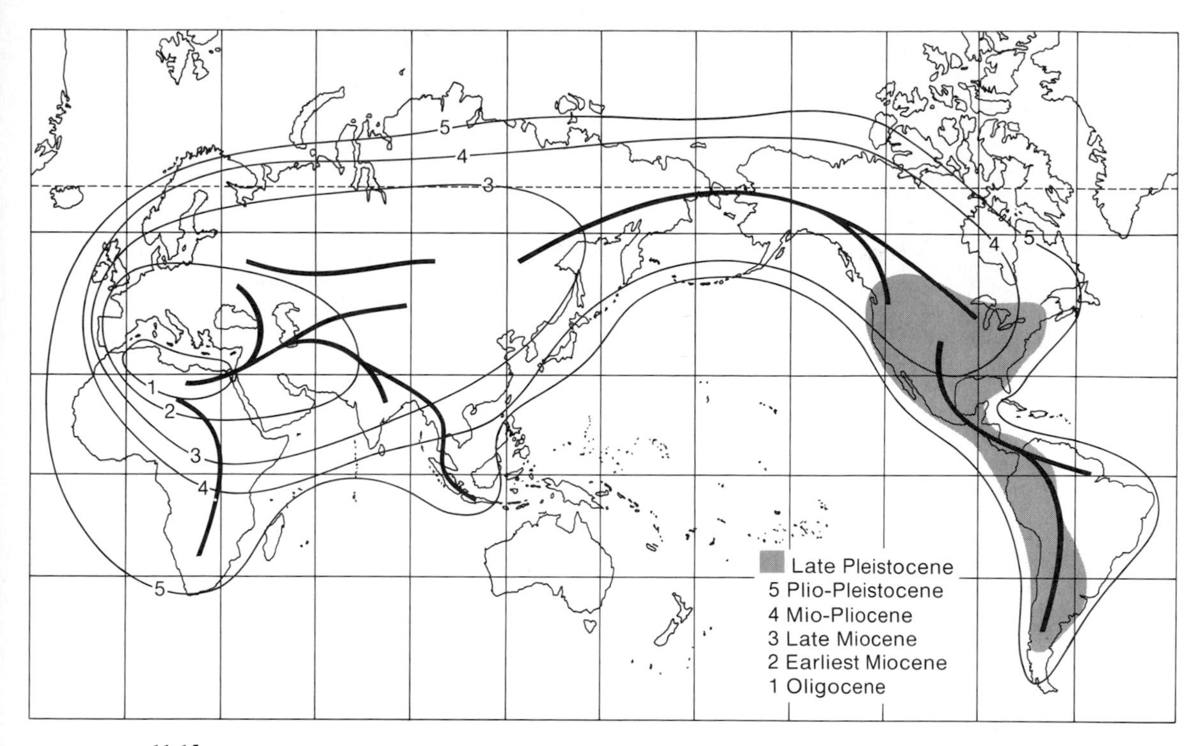

FIGURE 11-15
Changes in the geographic range of mastodonts throughout their evolutionary history. (From Simpson, 1944.)

of them with fatal results will they behave as a unit, and the evidence suggests that this seldom happens. The geographic aspect of extinction is shown in Figure 11-15. While the mastodonts originated in North Africa, they later spread throughout the world and the last species became extinct in the Americas.

Duration of a higher taxon will also tend to depend on the number of species it contains, and the average number differs greatly from taxon to taxon. Using raw data on stratigraphic ranges, Simpson (1944) found mean longevity for genera of bivalve molluscs to be 10 times that for genera of mammals (Figure 11-16). The data that he used are now out of date, but more recent estimates (Van Valen, 1973a) yield similar relative longevities. The long-accepted implication is that mammals have extinction rates 10 times higher, though in fact, their extinction rate at the species level is probably not this much higher, as we have already seen. As we noted at the start of this chapter, the disparity is widened at the genus level because genera of mammals contain fewer species. An average genus of living mammals contains only about 3 species, whereas an average genus of living bivalves contains more than 10.

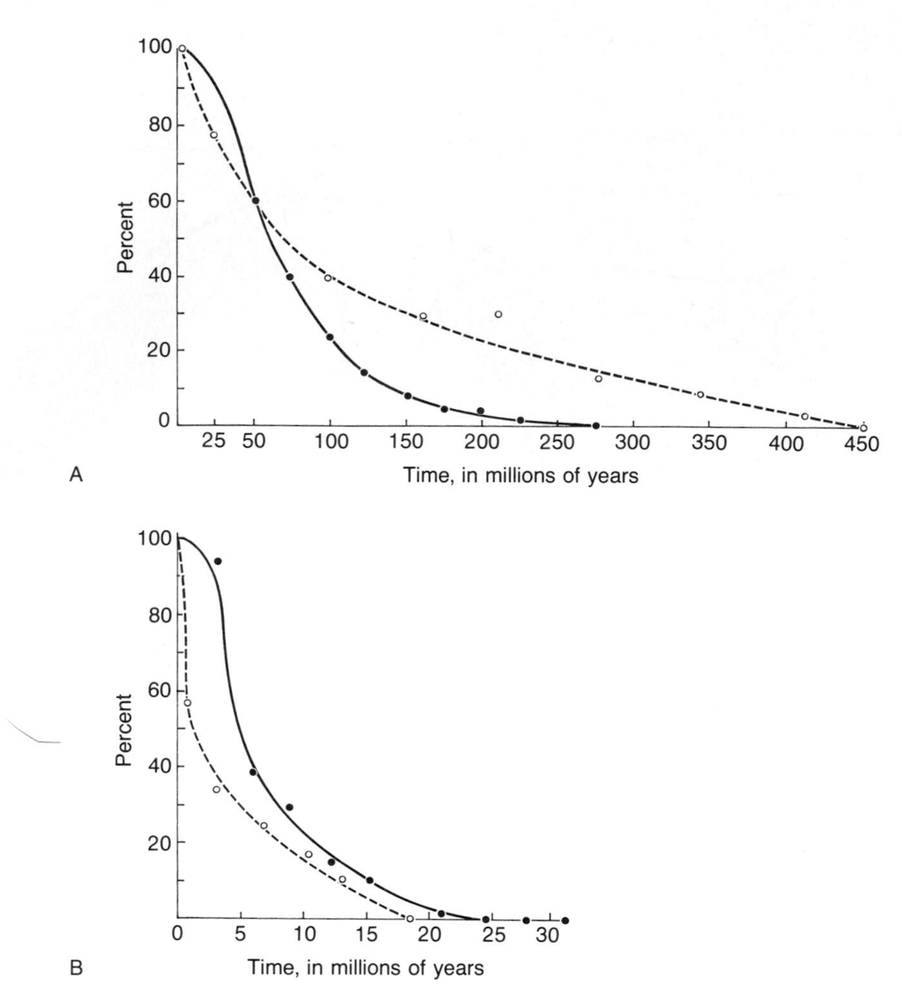

FIGURE 11-16
Survival times of living and extinct genera. The plots are cumulative (the points indicate percentages of genera whose survival times are equal to or greater than the times shown on the horizontal axis). Solid lines are for extinct genera and dashed lines, living genera. A: Bivalves (mean survival time for living plus extinct genera is 78.0 million years). B: Carnivorous land mammals (mean survival time for living plus extinct genera is 8.1 million years). (From Simpson, 1953.)

MASS EXTINCTION

Pulses of widespread extinction, often termed mass extinctions, are among the most interesting phenomena revealed by the fossil record. Such events tend to be world wide in extent, and we will discuss them in a geographic context in Chapter 12.

DIVERSITY CONTROLS

It should be evident that rates of splitting and termination of lineages determine standing diversity of species (diversity at any moment) within higher taxa. As we have seen, crowding of species often affects survival of species. Once a certain diversity is attained, crowding may become the main controlling factor. The subject of diversity is a very complex one that is tightly interwoven with the study of geographic distributions of environmental conditions and of taxa. For this reason, we will treat it in the final chapter, which covers biogeography.

The Nature of Trends

We can define a trend in evolution as a persistent change in a given direction. Many early evolutionists viewed directional change that was only crudely documented by the known fossil record as representing straight-line phyletic evolution. By "straight-line" we mean a rate of change that does not deviate greatly, or a plot of morphologic condition versus time that is linear. Early in this century the view that this kind of change prevailed in evolution became known as ***orthogenesis.*** Some extreme proponents of orthogenesis believed that evolutionary momentum of some sort actually carried trends to the point where they ceased to be adaptive and led to extinction. In effect, natural selection was rejected as the process of evolutionary change. In part, this view reflected the somewhat bizarre morphology of advanced species in certain groups. We now recognize that the structures in question were probably adaptive, and no mysterious driving force seems to have been required for their evolution. Some of these bizarre features are shown in Figure 11-17. The increasingly coiled shape of some Mesozoic oysters did not eventually prevent closure of shells, as some orthogenesists argued. It seems that curatorial negligence was the source of the problem: matrix had been left beneath the umbo (beak) of a particularly well-studied specimen (Gould, 1972). The enormous antlers of the Irish "elk" seem to have evolved with the function of serving for competitive display among males in the winning of mates (Gould, 1974). The enormous horn-like bony growth on the nose of advanced titanotheres seems to have evolved to protect the head and neck region from injury in butting combat (Stanley, 1974). The evolution of enormous canine teeth in sabertooth cats did not prevent biting or eating, as once thought. The teeth apparently served a stabbing function, and the lower jaw dropped down well out of the way when they were used.

Not only has the idea that these kinds of evolution were inadaptive been generally rejected, the view that they followed a linear path has also been refuted. Simpson (1951) showed that the classic sequence of change in the evolution of the modern horse did not follow a single phyletic path. Rather, it represented what we will call a phylogenetic trend—net change within a very complex phylogeny. The phylogenetic

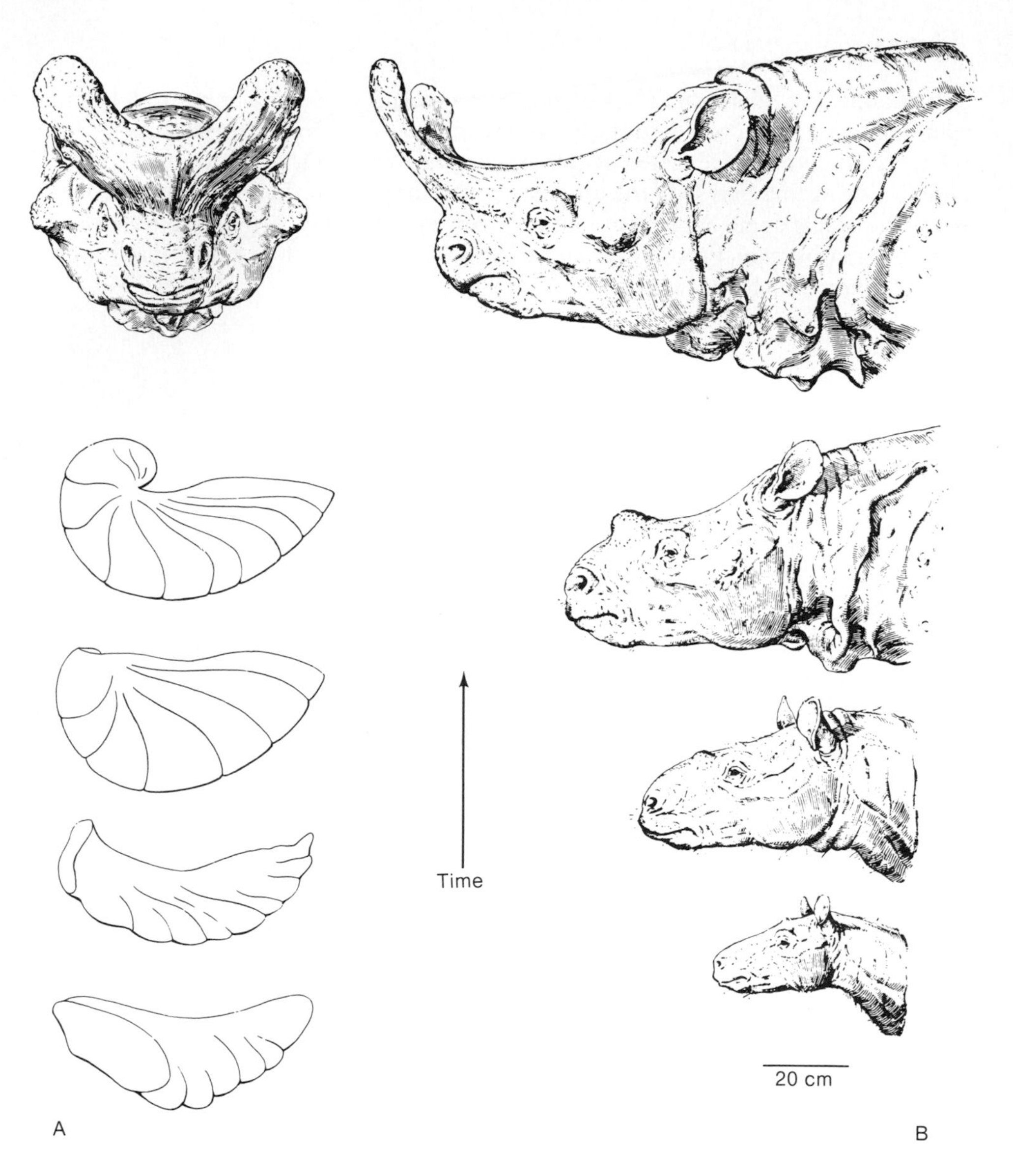

FIGURE 11-17
Creatures that in the past were thought to have evolved bizarre morphologic features inadaptively, by orthogenesis. A: Coiling of the Mesozoic oyster *Gryphaea.* B: Enlargement of the nasal "horn" in Cenozoic titanotheres. C: The enormous canine teeth of the Oligocene saber-toothed cat *Eusmilus.* D: Enormous antlers of the Pleistocene Irish "elk." (A and C from Simpson, 1944; B from Osborn, 1929; D from Romer, 1966 [copyright © 1966 by The University of Chicago], after Reynolds.)

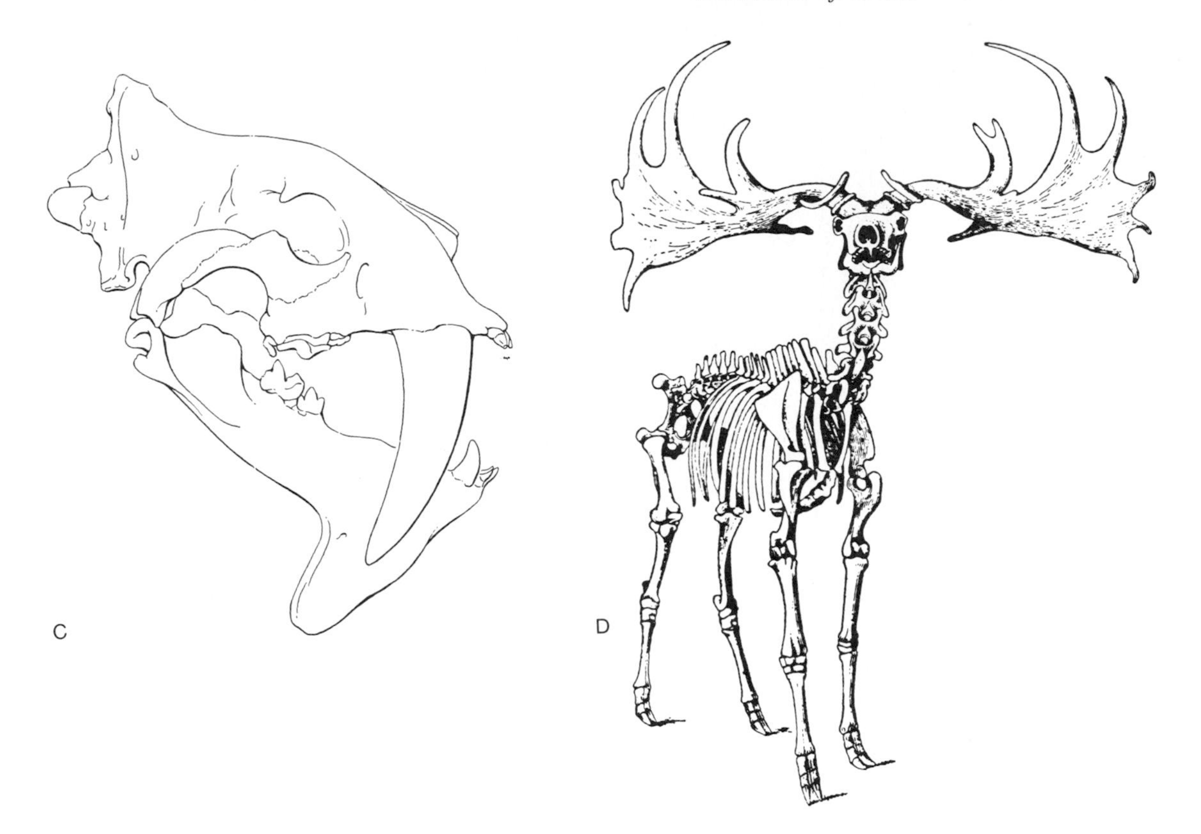

tree of horse evolution shows many reversals and intervals of stagnation. For example, although there was a net increase in size, certain lines of descent show diminution.

In our discussion of rates of evolution we introduced a dual concept of evolution that we may apply also to evolutionary trends. A ***phyletic trend*** results from evolution within a single lineage. Such a trend may produce arbitrarily delimited species-to-species transition. We may define a ***phylogenetic trend*** as a net change within a more complex, branching phylogeny. The historical modification of our view of changing size in horse evolution, described in the preceding paragraph, amounts to recognition that an alleged phyletic trend is instead a phylogenetic trend. We will now consider both kinds of trends in more detail.

PHYLETIC TRENDS

Phyletic evolution is difficult to study successfully. One problem, already considered is the difficulty of documenting a trend (page 308). Many species exhibit geographic gradients in morphology at single moments in time. Such gradients can

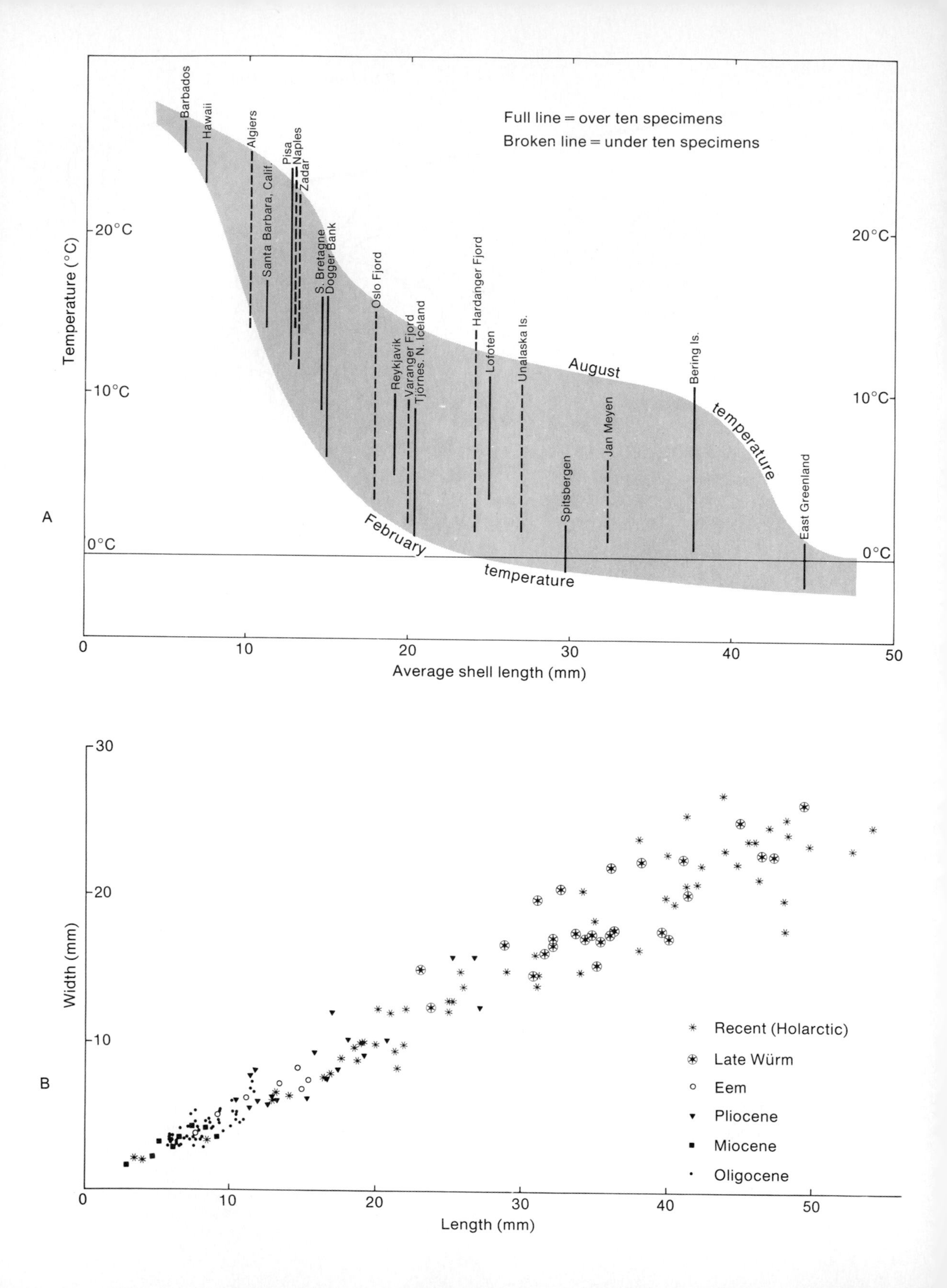

Full line = over ten specimens
Broken line = under ten specimens
Barbados
Hawaii
Algiers
Santa Barbara, Calif.
Pisa
Naples
Zadar
S. Bretagne
Dogger Bank
Oslo Fjord
Reykjavik
Varanger Fjord
Tjörnes. N. Iceland
Hardanger Fjord
Lofoten
Unalaska Is.
Spitsbergen
Jan Meyen
Bering Is.
East Greenland
August temperature
February temperature
Temperature (°C)
20°C
10°C
0°C
A
Average shell length (mm)
0
10
20
30
40
50
B
Width (mm)
Length (mm)
Recent (Holarctic)
Late Würm
Eem
Pliocene
Miocene
Oligocene

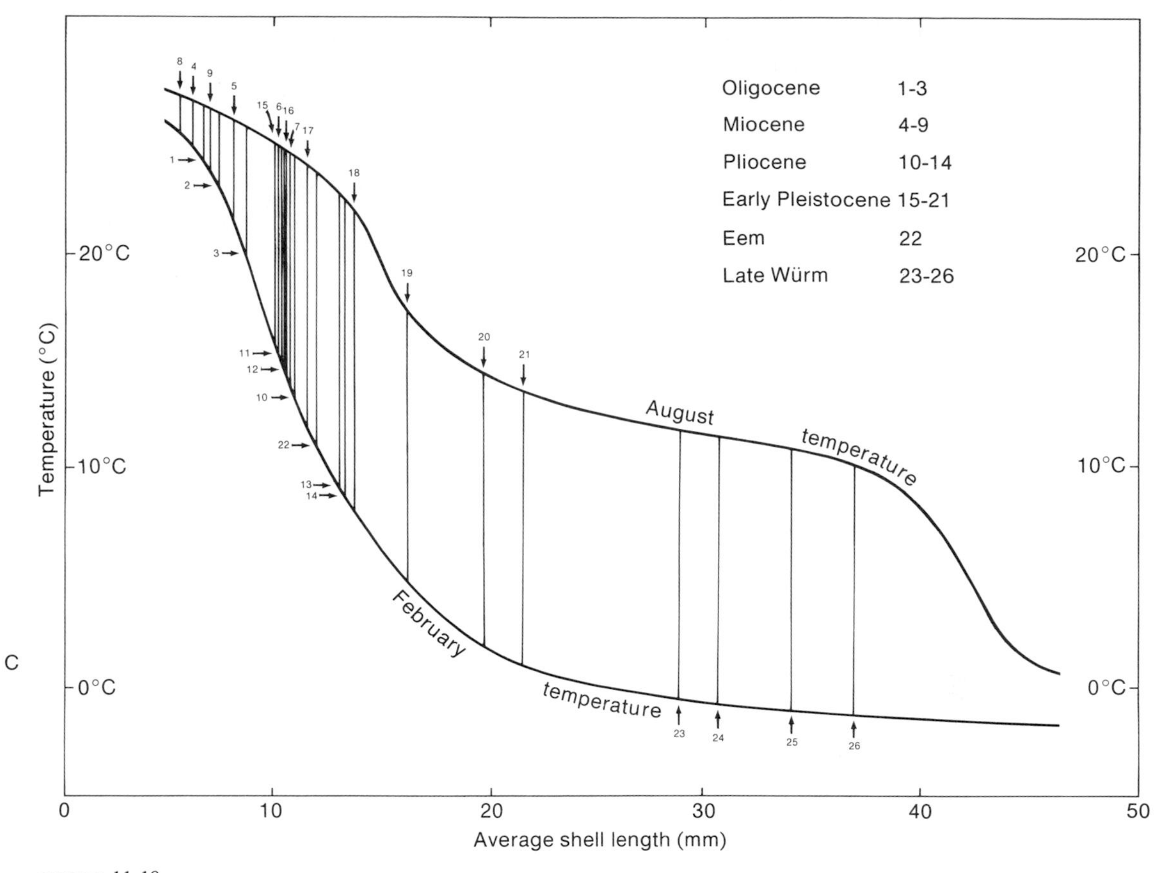

FIGURE 11-18
Variation in size of adult individuals of the bivalve species *Hiatella arctica.* A: Shell lengths for representatives living today in various temperature regimes. B: Sizes of European fossil specimens of various geological ages. C: Decrease in temperature since the mid-Cenozoic, inferred from size increase in *Hiatella arctica.* Numbers (1 through 26) serve as labels for the samples included. (From Strauch, 1968.)

lead to imitation of an evolutionary trend in any one locality if the controlling geographic condition shifts through time. As an example, the mean and maximum shell size of the clam *Hiatella arctica* in Central and Western Europe increases greatly throughout the long interval from Oligocene to Recent (Figure 11-18) (Strauch, 1968). From geologic data alone, we might conclude that the species evolved toward larger size. If we study living populations of the species throughout the world, however, we find that size varies consistently with temperature regime. The apparent evolutionary trend in Europe is only a local response to climatic cooling, which we know from other evidence took place during the general interval in question. The response may have been only phenotypic, but if genotypic, it was only a regional phenomenon.

Another problem in the study of phyletic trends is interpretation of adaptive significance. This problem arises in part from our lack of understanding of the functional morphology of the species displaying the change. Another contributing factor is our difficulty in reconstructing ancient environmental conditions. Some trends may reflect gradual adjustment to new environmental conditions that were imposed suddenly. Others may be guided by gradual environmental change. Still others may simply represent improved adaptation to unchanging conditions.

For many years apparent trends have been studied in the echinoid genus *Micraster* in the Cretaceous Chalk of England. Phyletic trends in *Micraster* were first recognized by the amateur paleontologist Rowe (1899). Later, Kermack (1954) seemed to confirm Rowe's basic observations biometrically, though rejecting some details of his account. More recently, Nichols (1959a, 1959b) analyzed the adaptive significance of the apparent morphologic trends in *Micraster.*

Echinocardium is an irregular sea urchin belonging to the group of echinoids that have altered their ancestral radial symmetry to bilateral symmetry. This change is associated with unidirectional movement of the animals on or within sediment. A typical burrowing species of the Recent Epoch is shown in Figure 11-19. The animal occupies an open burrow and moves forward by excavating sediment with movable spines, chiefly on the ventral surface. Connection with the overlying water is maintained through a vertical "funnel," formed by special respiratory tube feet. The animal is a deposit feeder that excavates organic-rich sediment with special tube feet that surround its mouth. Some food is also passed down to the mouth from the dorsal surface via the anterior groove (Figure 11-20). Spines and tube feet surrounding the anus excavate a posterior sanitary drain, where feces are left in the animal's wake.

The morphologies of three *Micraster* species and three Recent heart urchin species are shown in Figure 11-20. Spines are almost never preserved in attached position, but much information about their morphology and location can be inferred from their knob-like attachment sites, or tubercles, on the test surface. In nearly all heart urchin species, certain regions of the test bear bands of tubercles to which heavily ciliated spines are attached in life. There regions are termed "fascioles." Cilia of the inner fasciole are largely responsible for drawing water down the respiratory funnel. Cilia of the anal fasciole provide currents to pass feces away from the body and downward, where cilia of the subanal fasciole aid other body cilia in producing currents that carry waste into the sanitary drain.

Figure 11-21 shows the distribution of several recognized *Micraster* species of the Cretaceous Chalk in relation to time and inferred depth of burrowing in life. The species intergrade and are separated arbitrarily. The original species, *M. leski,* is thought to have been a shallow burrower. It is thought to have given way to a range of forms, including both shallow and deep burrowers, that are arbitrarily divided into two species, *M. corbovis,* and *M. cortestudinarium.* Only representatives of the latter

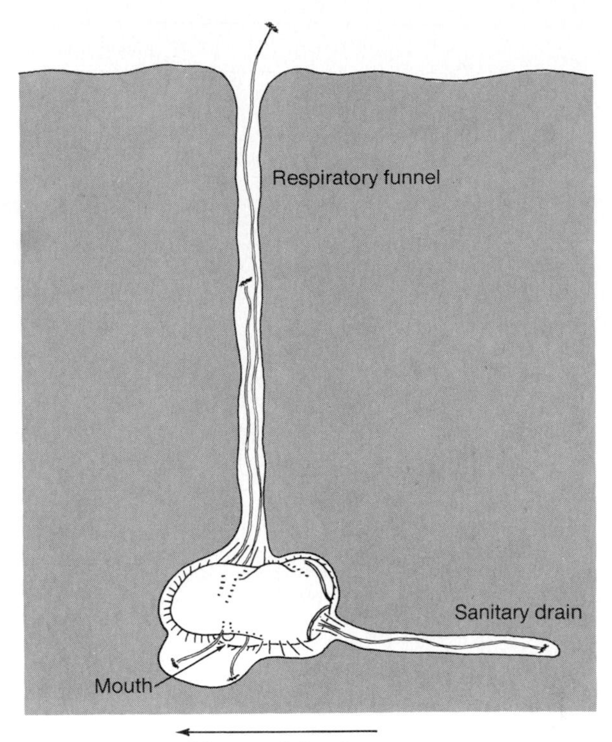

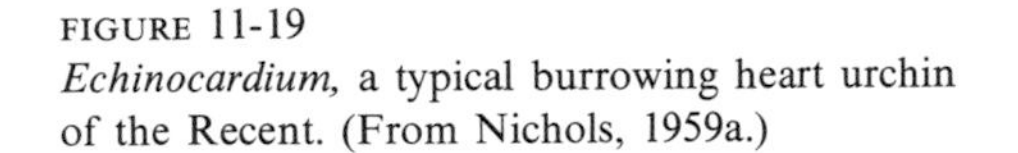

FIGURE 11-19
Echinocardium, a typical burrowing heart urchin of the Recent. (From Nichols, 1959a.)

species survived into the Senonian Age, when they are believed to have given rise to the lineage having the species name *M. coranguinum.* The shallow-burrowing species, *M. (Isomicraster) senonensis,* is thought to have entered the region of study in the early Senonian from a geographically separate region, probably having been isolated from the main stock for a substantial period of time. It had apparently not attained sufficient genetic isolation to prevent interbreeding with *M. coranguinum* (though paleontologists have designated the two as separate species); intermediate coexisting forms suggest that there was interbreeding between the two groups. This would indicate that phyletic evolution produced rather little genetic change during the interval in question.

Nichols inferred the life habits of the *Micraster* representatives from their adaptive morphology by homology with Recent forms. The supposed burrowing depths of the three *Micraster* species and the known burrowing depths of three Recent species are

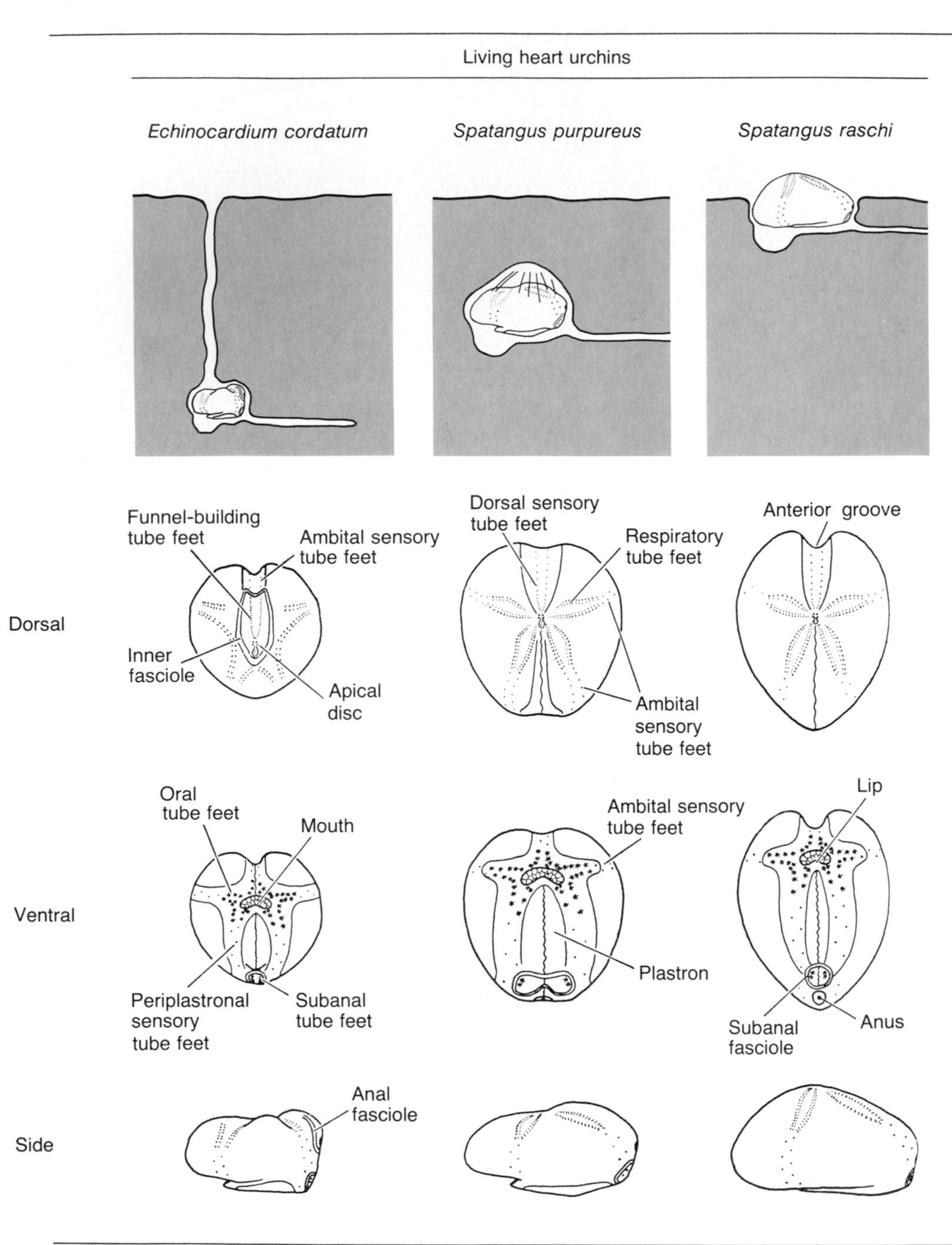

FIGURE 11-20
Morphology and habits of living and fossil heart urchins. (From Nichols, 1959b.)

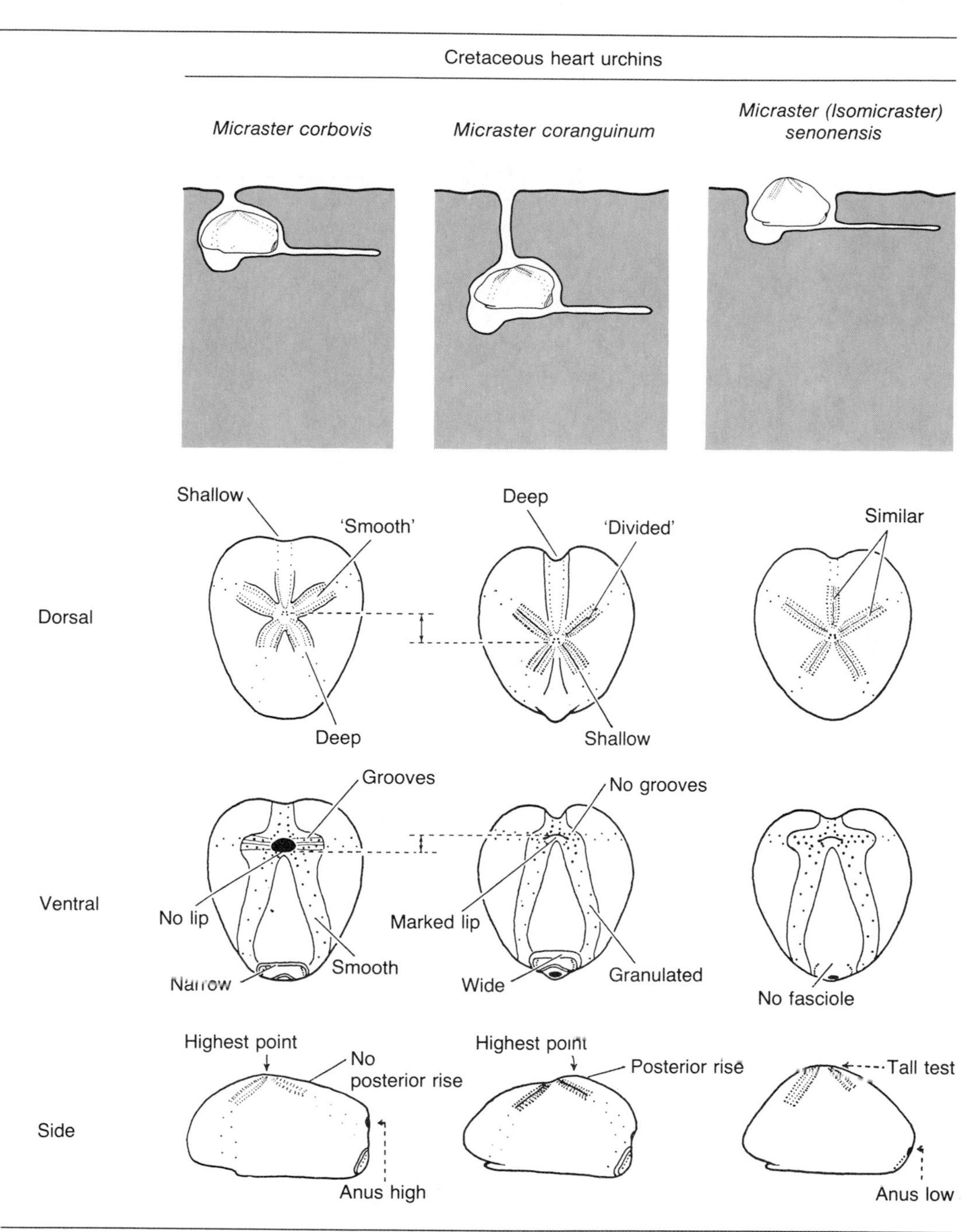
Cretaceous heart urchins
Micraster corbovis
Micraster coranguinum
Micraster (Isomicraster) senonensis
Shallow
'Smooth'
Deep
'Divided'
Similar
Dorsal
Deep
Shallow
Grooves
No grooves
Ventral
No lip
Marked lip
Smooth
Granulated
Narrow
Wide
No fasciole
Highest point
No posterior rise
Highest point
Posterior rise
Tall test
Side
Anus high
Anus low

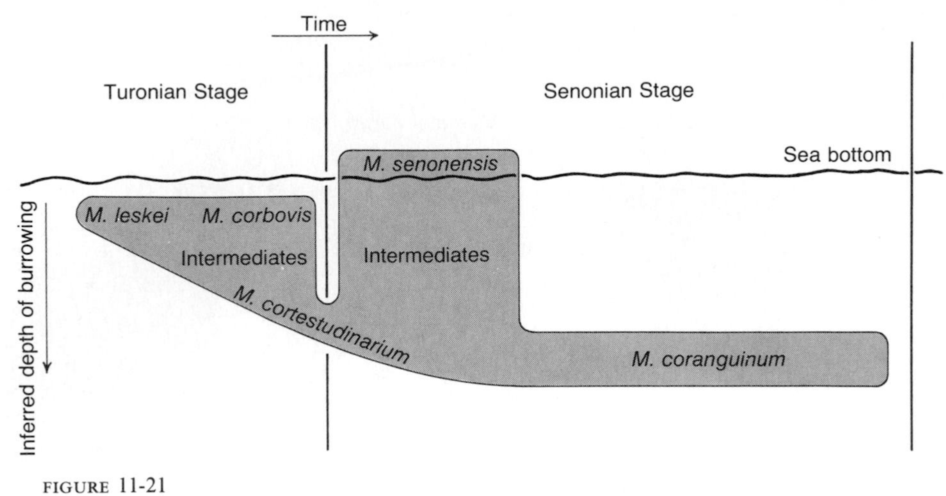

FIGURE 11-21
Life habit changes in the evolution of *Micraster* in the Cretaceous of the south of England. (From Nichols, 1959b.)

shown in Figure 11-20. Some of the adaptive features that were progressively developed in the *M. cortestudinarium-coranguinum* lineage, leading Nichols to recognize the trend toward deeper burrowing, are as follows (see Figure 11-20 for illustrations):

1. A broader test, with the broadest and highest region shifting posteriorly. This change apparently reflected the need for streamlining and elimination of the need for the dorsal surface to shed sediment, as it must for organisms that live only partly buried in sediment.
2. Increased ornamentation in the region of the respiratory tube feet, apparently to increase the area of ciliated surface for drawing strong water currents down through the long respiratory funnel.
3. Increase in the granularity of the ventral surface, apparently to increase the area of ciliated surface for movement of fine-grained particles posteriorly in locomotion.
4. Widening of the subanal fasciole, apparently for improved sanitation in the confines of a burrow. The test of *M. (Isomicraster) senonensis,* the alleged shallow burrower or surface "plower," lacks a subanal fasciole.

The absence of preserved burrows in the Cretaceous Chalk has led recent authors to question whether any species of *Micraster* burrowed as deeply as Nichols suggested (Bromley and Asgaard, 1975). According to their interpretation, even *M. coranguinum* may have lived near the surface of the sediment.

Moderate change of size is one of the simplest morphologic transformations that can be brought about through phyletic evolution. The Pleistocene geologic record, which displays clear alternations between glacial and interglacial conditions, also reveals corresponding fluctuations in the body sizes of mammals that may reflect phyletic evolution, though it is also possible that they simply represent phenotypic response to changing environmental conditions. Most of the fluctuations illustrate Bergmann's Rule, that animal size tends to be larger in cold climates than in warm climates. The reason that many mammal species, like the brown bear (Figure 11-22), exemplify this rule is that in these animals the ratio of body surface to volume declines as size increases. Because rate of heat loss increases in proportion to surface area and heat production increases in proportion to volume of body tissue, the larger the animal, the less rapidly it will lose heat. One Pleistocene species that displays the opposite relationship (smaller size in colder climates) is the woolly mammoth (Figure 11-23). It seems reasonable to guess that this tendency results from the large size of the species in general. Modern elephants have such a small surface-to-volume ratio that they require cooling adaptations, notably large, constantly flapping ears that emit excess heat. Pleistocene mammoths had small ears, probably because cold environmental temperatures increased their rate of heat loss. Presumably their surface-to-volume ratio still enabled them to conserve body heat more efficiently than smaller mammals could. The crucial factor may have been food supply. Modern elephants feed about 80 percent of the time. In large animals, the area used for chewing tends to be small relative to volume of body tissue to be fed. During glacial stages, food supplies may have been too scarce for mammoths to grow to sizes they normally attained during interglacial intervals. If so, the animals may have been phenotypically

FIGURE 11-22

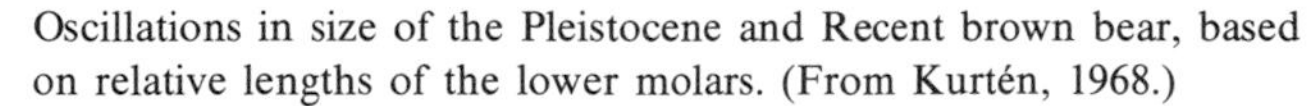

Oscillations in size of the Pleistocene and Recent brown bear, based on relative lengths of the lower molars. (From Kurtén, 1968.)

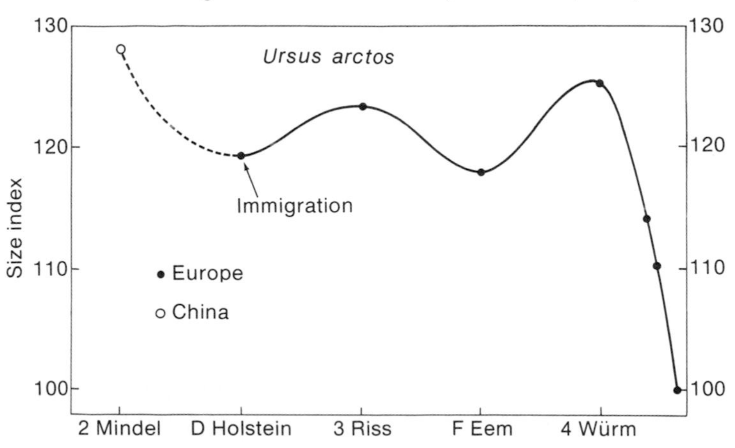

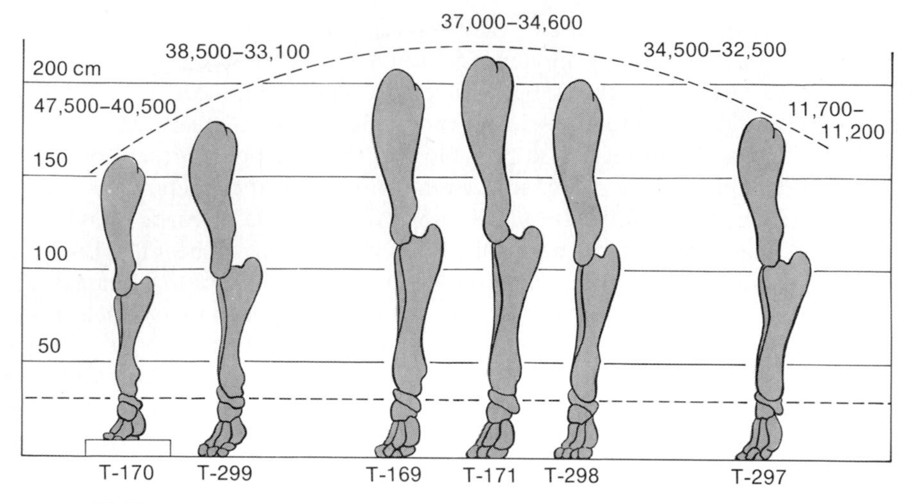

FIGURE 11-23
Sizes of the forelegs of the woolly mammoth, *Mammuthus primigenius,* from Siberia. Radiocarbon dates are above drawings of the legs. The three large legs in the center represent an interglacial interval. The three smallest legs represent glacial intervals. (From Kurtén, 1968.)

stunted, or selection pressure favoring small individuals may have produced an evolutionary decrease in size.

If the enormous horn-like structures of the titanotheres and antlers of the Irish "elk" (Figure 11-17) arose by phyletic evolution, then we have a likely functional explanation for one kind of phyletic trend. The structures were apparently advantageous to males in the winning of females. This kind of natural selection was termed ***sexual selection*** by Darwin. Sexual selection favors certain morphologic features or behavior patterns that tend directly to improve the chance of an individual to attract a mate. Such attributes may be of no value for survival of the individual or for success or survival of the species. Sometimes they are quite bizarre. On the other hand, sexual selection would seem to be a very effective source of change because it operates directly on reproductive success. Some males may sire many offspring, and others, none at all. It is possible that sexual selection sometimes accounts for more rapid phyletic rates than are normally produced by other forms of selection.

Another manner in which phyletic evolution may occasionally be accelerated is sometimes called the "bottleneck effect." It may happen that an environmental change causes a species to decline nearly to extinction. The small remaining population, confronted with uniform new conditions in a local area, may evolve rapidly to meet these conditions and re-expand into a large, successful species. This scenario is much like that of speciation through a peripheral isolate, except that the entire species becomes equivalent to an isolate. There is reason to question the general importance of this sort of mechanism, in that one might expect drastic decline of a species usually to lead to extinction. In the fossil record, it is almost impossible to distinguish such a mechanism from allopatric speciation.

Allometric relationships (page 62) can aid us in distinguishing between phyletic evolution and rapid change produced through splitting of lineages. A plot for a wide variety of species shows that brain weight tends to vary with body weight raised to the two-thirds power (Jerison, 1973). In other words, it increases roughly in proportion to body surfaces. (These surfaces increase with the square of linear dimensions, whereas volume increases with the cube.) As shown in Figure 11-24, a linear relationship is produced by a logarithmic plot, because

$$\text{Brain weight (or volume)} = b^{2/3},$$

or

$$\log b = \frac{2}{3}\log b + \log b,$$

where b, a constant, stands for "body weight." The power ⅔ becomes the slope of the plot, and b, the y-intercept. When individuals of a single species or two or more closely

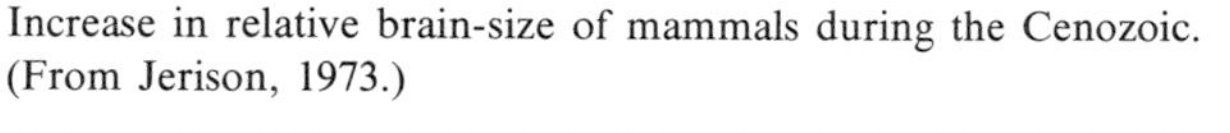

FIGURE 11-24
Increase in relative brain-size of mammals during the Cenozoic. (From Jerison, 1973.)

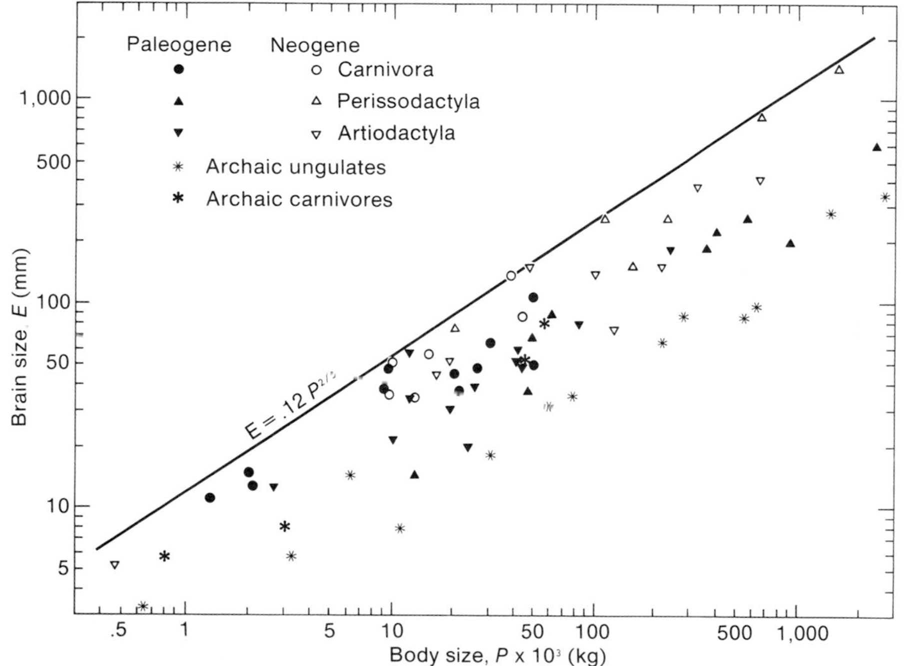

A

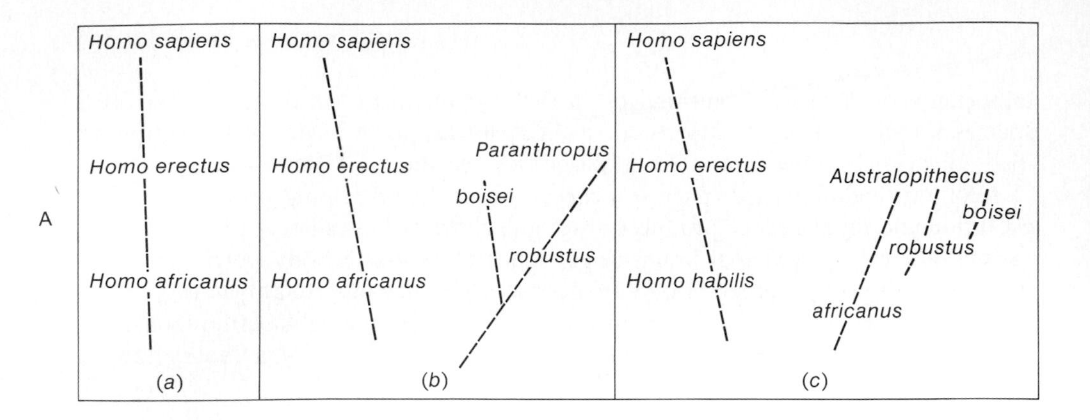

B

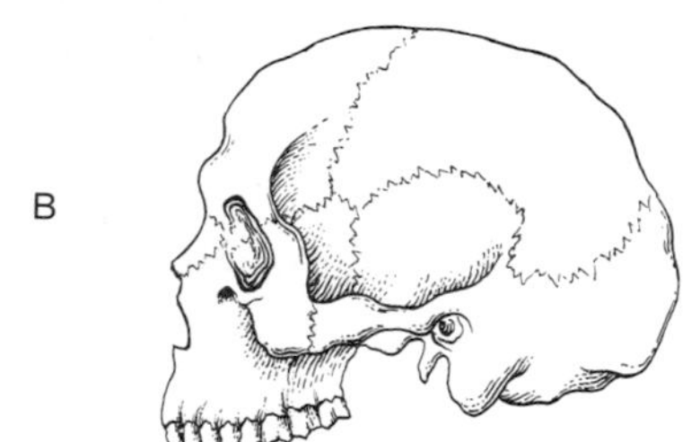

Homo sapiens

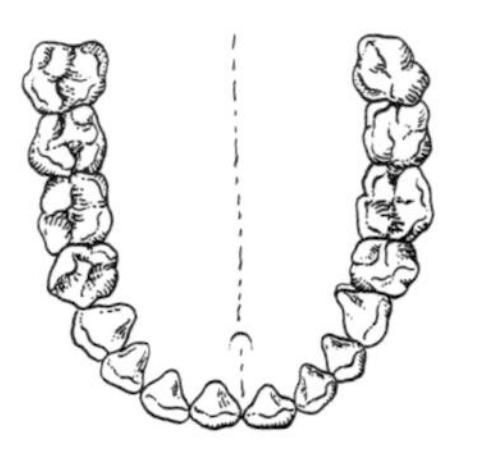

C

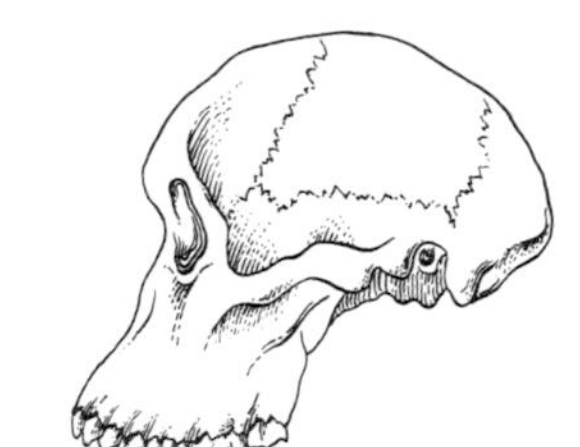

Australopithecus

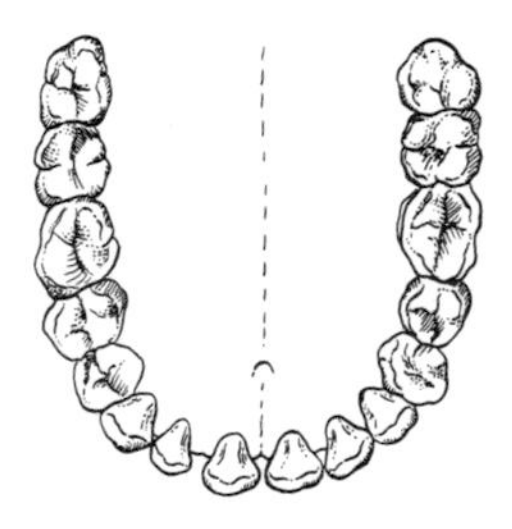

D

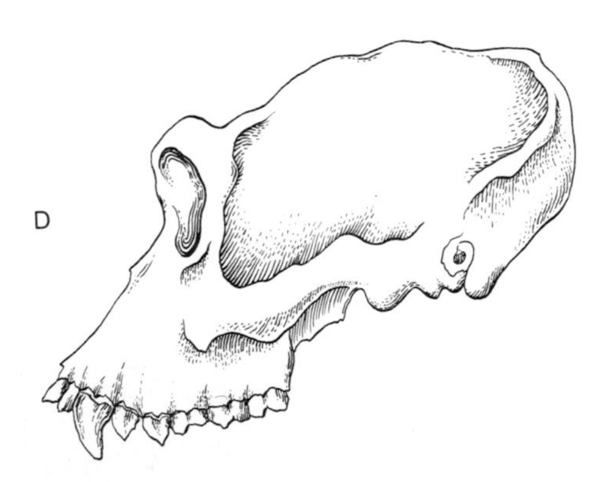

Gorilla

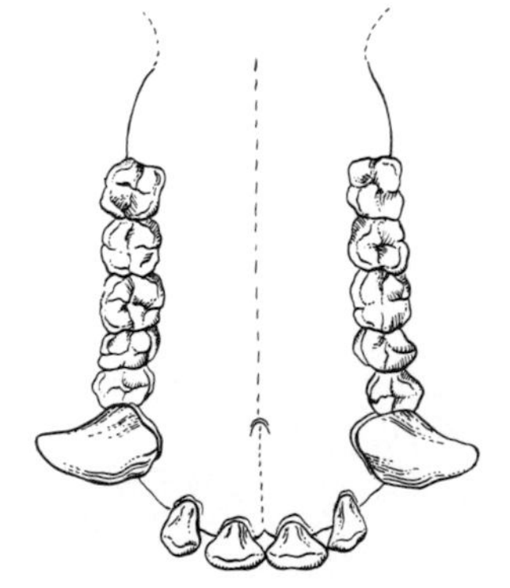

FIGURE 11-25
Allometry of brains and teeth in the Pongidae (Great Apes) and Hominidae. A: Possible phylogenies of late Cenozoic hominids. B–D: Skulls and palates and upper teeth of three taxa. E: Logarithmic plot showing how the volume of the brain case measures relative to body weight. F: Similar plot for tooth area. (A, E, and F from Pilbeam and Gould, 1974; skulls of B, C, and D from Le Gros Clark, 1966 [by permission of the Trustees of the British Museum (Natural History)], and teeth after Pilbeam, 1970.)

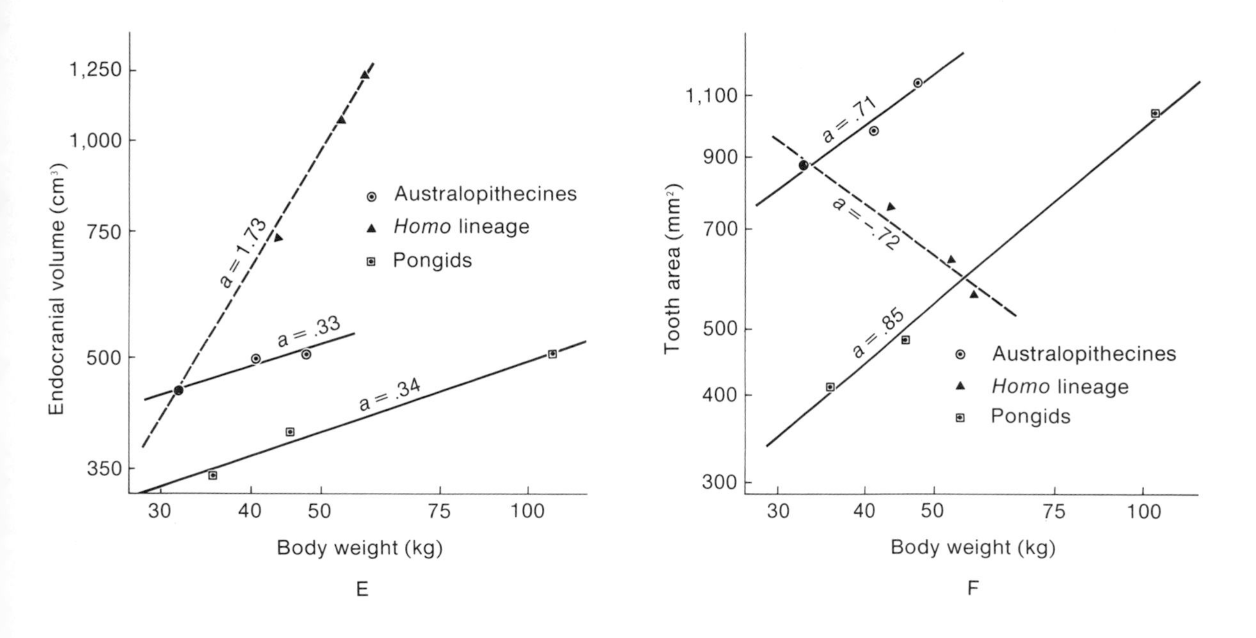

related species are compared, the slopes are much lower, ranging from 0.2 to 0.4. Apparently the slope of ⅔ observed for less closely related mammals reflects increase in number of neurons. The lower slope for similar animals is believed to reflect simple enlargement of old neurons without addition of new ones. Pilbeam and Gould (1974) examined the phylogeny of Plio-Pleistocene hominids, including modern man, in this light. Several alternative phylogenies for the species in question are shown in Figure 11-25. Pilbeam and Gould (1974) found that a logarithmic plot of brain volume against estimated body weight for three species of the genus *Homo* yielded a reasonably straight line, which also included the plot for an earlier form having the specific name *africanus.* Some people have considered this form to belong to the primitive genus *Australopithecus,* while others have assigned it to *Homo.* As it turns out, the point for this species falls at the intersection of the lines for the two genera.

The interpretation is that the species should be retained in the genus *Australopithecus.* The slope of the line for this genus is only 0.33, which suggests that points along the line represent closely related species. The genus *Homo* may have evolved from *Australopithecus africanus,* but *Homo* displayed an enormous allometric increase in brain size only after divergence from this species (slope: 1.73). The allometry of the surface area of cheek teeth shows the same kind of relationship, except that in *Homo* there was a relative decrease in this area (slope: −0.72). In *Australopithecus,* geometric similarity was maintained. If similarity had been maintained exactly, tooth area would have scaled with the ⅔ power of body weight. The measured value is 0.71, which is not very different from ⅔. Again, *A. africanus* falls at the intersection of the two curves, suggesting that it may represent the link between the two genera. Pilbeam and Gould (1974) suggested that species of *Australopithecus* were exclusively herbivorous and that the relative reduction of the cheek teeth in *Homo* accompanied the development of omnivorous habits. The brain size analysis raises the possibility that continuous selection pressures may have produced certain of the phyletic changes in the transition from *Australopithecus* to *Homo sapiens.* These selection pressures would need to have been far stronger than those that operated within *Australopithecus.* Whether most evolution within *Homo* was indeed phyletic is presently unknown, but the allometric considerations at least give us an idea of the amount of change that would have been required beyond the change expected as a natural product of size increase.

PHYLOGENETIC TRENDS

Phyletic trends are not easily recognized for many taxa, especially as we venture far back into the geologic record. Also, as we have already seen, most documented phyletic trends have not produced very rapid evolutionary change (Figures 11-3 through 11-6). It has been pointed out that a corollary of the punctuational model is that important long-term trends tend to be stochastic, or statistical in nature (Eldredge and Gould, 1972). In other words, they are phylogenetic trends (see page 335) rather then phyletic trends. This is true because if most evolutionary change occurs in discrete speciation events, some of these events will produce change in one direction and others, change in the opposite direction. If we accept the punctuational model, we can view large-scale evolution as being guided to a considerable degree by a process called ***species selection*** (Stanley, 1975). As shown in Table 11-1, species selection is analogous to natural selection but operates at a higher level of biological organization. The species, rather than the individual, is the unit of selection. The kinds of species that are favored are ones that speciate at high rates or last for long intervals of geologic time so that they tend to produce more total descendant species. As Mayr (1963) has noted, new species represent evolutionary experiments. Some possess new adaptations that turn out to be of widespread benefit. These species will tend to leave a

TABLE 11-1
Analogous Features of Natural Selection and Species Selection

Process	*Unit of selection*	*Source of variability*	*Type of selection*
Microevolution	Individual	Mutation/recombination	Natural selection A. Survival against death B. Rate of reproduction
Macroevolution	Species	Speciation	Species selection A. Survival against extinction B. Rate of speciation

SOURCE: Stanley, 1975.

disproportionately high number of descendant species. This may happen because the species survive a long time or because their small, geographically isolated populations have a higher-than-average rate of survival and evolution into new species. Other species display adaptations that serve them well for the particular local conditions where they originate, but that are ***inadaptive*** (have a negative value) or ***nonadaptive*** (have no value) under most other circumstances. These species will tend to leave few or no descendant species. There is a strong random element in speciation, in that we cannot predict where, when, or in what ecological settings new species will arise (Mayr, 1963). Acceptance of species selection as the dominant source of phylogenetic trends does not require a denial that natural selection is the source of most evolution. It simply requires, first, a belief that natural selection operates most intensely in the formation of new species, and, second, a belief that the set of new species forming in any interval of time will be determined by a variety of circumstances that are in part determined by chance. A hypothetical trend produced by species selection is shown in Figure 11-26. Only a small percentage of the net change (change in mean morphology) is produced by phyletic evolution within species.

It is important to recognize that species selection is not the only mechanism that can produce a large-scale trend within the framework of the punctuational model. A tendency for speciation events to shift morphology in a particular direction can also produce a net trend (Figure 11-27). Grant (1963) has emphasized this mechanism.

Large-scale phylogenetic trends are more easily explained in functional terms than are small-scale phyletic trends. Phyletic trends may reflect local environmental influences and short-term selection pressures that we cannot easily reconstruct. Furthermore, such trends often produce only minor morphologic change. Phylogenetic trends, in contrast, tend to produce major morphologic transitions that require less subtle interpretation. The evolution of modern horses entailed, among other things, phylogenetic trends toward fewer toes, a relatively longer neck, and more complex molars. These and other trends were in part related to the ecologic shift from browsing on leafy vegetation to grazing on grass (Simpson, 1951). Reduction in number of toes and enlargement of the middle toe to form a hoof improved

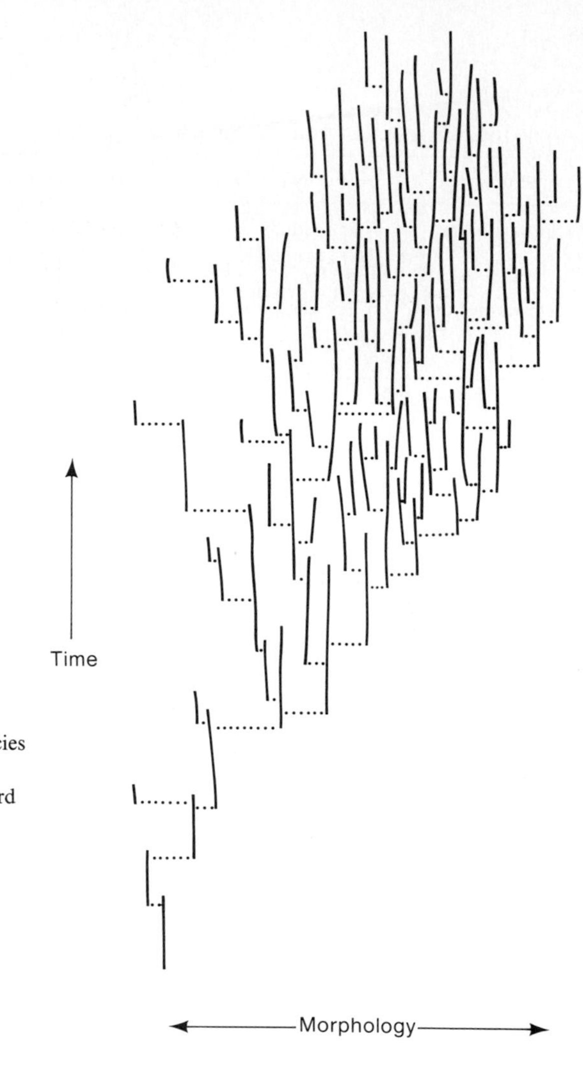

FIGURE 11-26
Hypothetical large-scale trend produced by species selection. Species, represented by upright bars, survive longer and speciate at higher rates toward the right. Speciation events are represented by dashed lines.

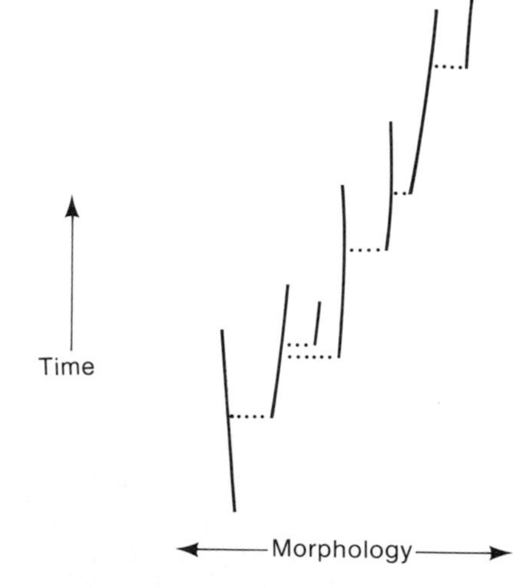

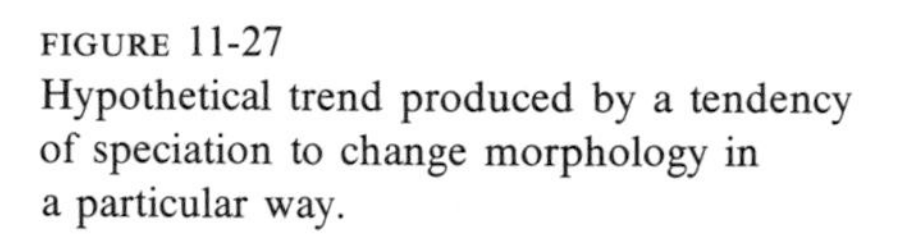
FIGURE 11-27
Hypothetical trend produced by a tendency of speciation to change morphology in a particular way.

long-distance running ability on the hard turf of grasslands. The longer neck facilitated grazing at ground level, and complication of the molars improved efficiency of chewing on siliceous grasses. As noted earlier, the horse family also underwent an increase in body size, which probably functioned to increase endurance and defense against predation.

The brain of the horse family also increased in size and became more intricately fissured (Figure 11-28). In discussing Figure 11-25 we noted that brains in different kinds of mammals tend to increase by the power of $\frac{2}{3}$ body size. Figure 11-25 also shows two other important things. First, relative mammalian brain size has generally increased since the early Cenozoic. For both early and late Cenozoic forms, the slope is about $\frac{2}{3}$, but it shifts upward through time. Second, carnivores at all times seem to have had larger average relative brain-sizes than ungulate herbivores, upon which

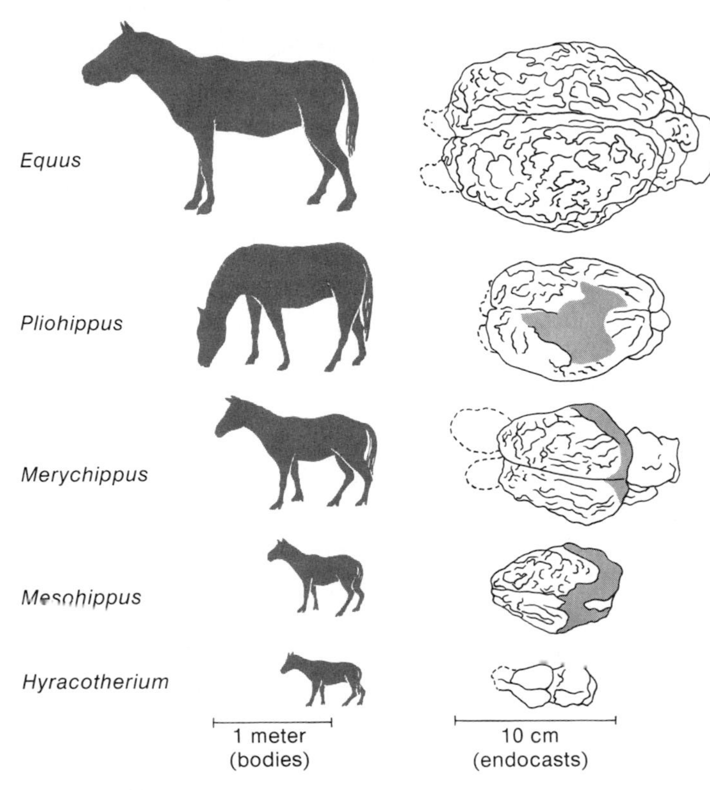

FIGURE 11-28
Increase in size and complexity of the horse brain from the Eocene to the present. The front of each brain is on the left. (From Jerison, 1973.)

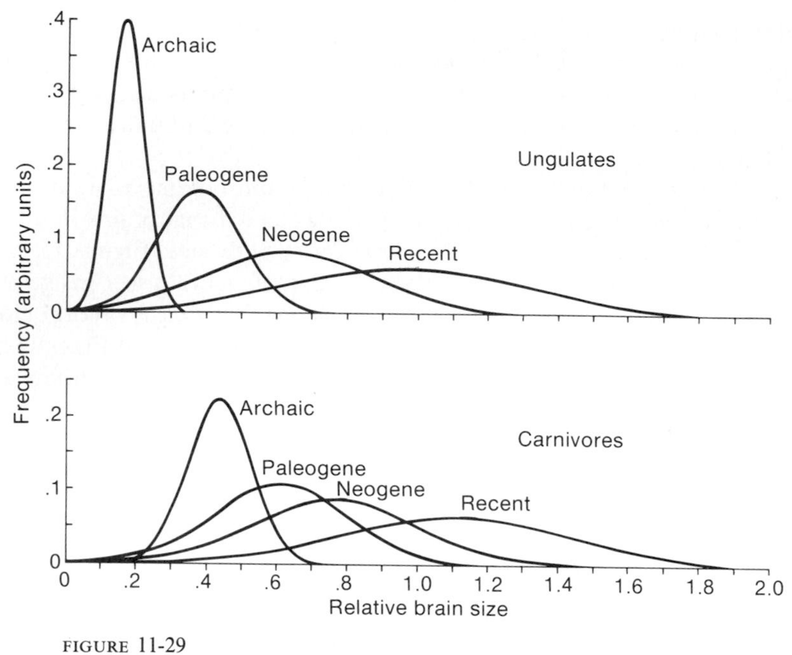

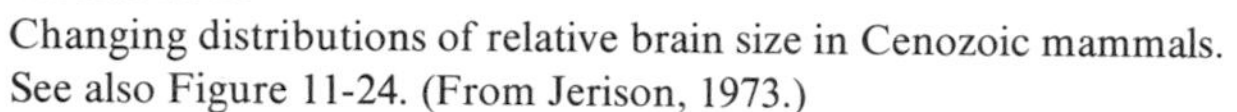

FIGURE 11-29
Changing distributions of relative brain size in Cenozoic mammals. See also Figure 11-24. (From Jerison, 1973.)

many of them feed. This relationship, depicted in Figure 11-29, suggests that high intelligence is more important for catching prey than for avoiding capture.

Many animal taxa exhibit long-term phylogenetic trends toward increased body size. This tendency is termed Cope's Rule. The reason seems to be that most groups arise at small size relative to their potential range of sizes. Major evolutionary transitions are unlikely at relatively large sizes for a given body "plan" because adaptive solutions to problems of large size usually lead to specialized morphological and physiological traits that represent evolutionary dead ends. It is difficult to imagine a modern elephant, for example, giving rise to a new order of mammals. On the other hand, many small members of most higher taxa remain less specialized and can more easily give rise to distinct new taxa. As in the horse example discussed previously, as diversification proceeds from small ancestors, there will be an average increase in mean and maximum size among all descendant species, and many minor trends toward smaller size are not likely to reverse the overall pattern. Three stages in the size diversification of rhynchonellid brachiopods are shown in Figure 11-30. Another factor that may tend to produce increase in size during diversification is the tendency of large taxa to become extinct at higher rates than small taxa (Van Valen, 1975). Higher extinction rates for large taxa may result from both small population size and a high degree of specialization. One result of this condition may be that after certain mass extinctions most surviving species will be small. Inevitably, radiation from these species will produce a net increase in size (Bakker, 1977).

Knowledge of the functional morphology of living species can be very useful in interpreting the adaptive significance of phylogenetic trends documented by the fossil record. In Chapter 8 we showed that living mussels of the family Mytilidae, which attach to the substratum by a byssus, can be divided into two life-habit groups. The endobyssate group consists of species living partly buried in soft substrata and the epibyssate group, of species living on the surface of the sediment (Figure 10-28). It was pointed out that biologists have viewed byssally attached bivalves as having evolved from burrowing clams. All modern burrowing clams seem to employ a byssus temporarily, in the post-larval stage of their existence. This prevents tiny clams that have just settled out of the plankton from being swept away by currents and killed. Yonge (1962) suggested that the byssus originated as a juvenile feature in primitive clams and was passed on to the adult stage by what is termed ***neoteny,*** or ***paedomorphosis.*** This amounts to evolutionary juvenilization. In the extreme form of the process, a juvenile form simply develops reproductive maturity and never passes through the full course of its previous ontogeny. This seems to have happened

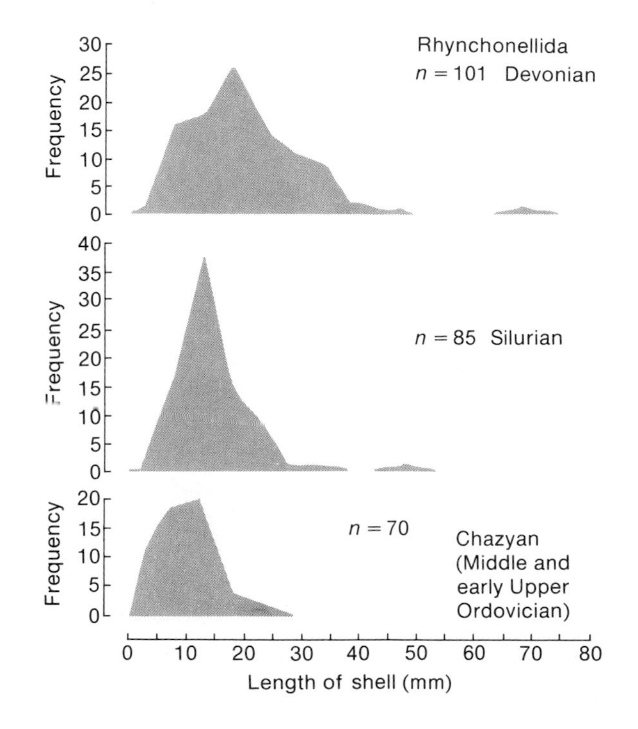

FIGURE 11-30
Distributions of shell lengths of species of rhynchonellid brachiopods early in the group's history. (From Stanley, 1973.)

numerous times in the evolution of the Bivalvia. Various tiny, byssally attached clams exist today, and some belong to the families otherwise characterized by large, free-living clams whose juveniles the tiny adult species resemble (Stanley, 1972). This process of juvenilization must entail the splitting of a lineage because only local populations, not entire species, would be likely to undergo such a sudden transformation. Examples are well known in the Amphibia. The Mexican axolotl, for example, lives its entire life as an aquatic form equivalent to a tadpole but can be induced to metamorphose into a terrestrial adult, as its ancestors obviously did, by injection of a certain hormone (Romer, 1959). We can view the process of sudden juvenilization, even in its simplest form, as producing a sudden, simple phylogenetic trend involving only one parent species and one divergent descendant species. Neoteny can also occur less dramatically, with one or a few juvenile features being gradually retained into the adult stage. This has apparently also happened repeatedly within the Bivalvia. The byssus seems to have been retained further and further into the adult stage in some groups without reduction of adult size or other evidence of juvenilization. The origin of the endobyssate condition by neoteny formed a polyphyletic pathway to the origin of epibyssate habits, which was not a neotenous transition. As discussed in Chapter 8, in the Mytilidae, all middle Paleozoic forms have the characteristic endobyssate shape (Figure 8-18). The epibyssate shape is unknown before the Mesozoic. Intermediate forms, named *Promytilus,* are found in the Permian. Endobyssate forms persisted during the phylogenetic trend toward epifaunal habits, and some survive today. In this sense, the pattern resembles that just described as typical of phylogenetic size increase within a major group: small species are maintained, while larger and larger species are added to the spectrum.

Neoteny is only one of a variety of evolutionary changes that can occur in the sequence of ontogeny. The study of such changes has a long history. Early in the nineteenth century, von Baer noted that some stages in the embryologic development of higher animals resemble those of lower animals. A human embryo, for example, passes through successive stages in which it bears superficial resemblance to the embryos of fish, amphibians, reptiles, and lower mammals. This pattern became known as the Biogenetic Law. Haeckel and other early evolutionists later expanded the idea into what became known as the Principle of Recapitulation. This "principle" asserted that higher animals pass through stages resembling successive *adult* stages in their evolutionary ancestry. This idea, usually expressed by the statement "ontogeny recapitulates phylogeny," is largely in error. Von Baer's original generalization is closer to the truth, in that the development of an advanced group frequently resembles the *development* of ancestral forms. Still, there is usually considerable alteration of developmental rates and sequences during evolution. For understanding such changes it is useful to think of evolution as a *sequence of ontogenies* (or partial ontogenies, depending on when reproduction takes place).

At any stage of ontogeny, evolution may operate in ways that have little direct effect on later stages. For example, marine invertebrate larvae commonly exhibit elaborate features that have evolved semi-independently of any changes in the adult form.

Evolution has also commonly produced profound changes in adults through alteration of the ontogenetic sequence through which their ancestors had developed. For example, the sequence in which two organs develop may be reversed, or development of one may be retarded to the point at which it becomes a mere vestige or disappears altogether. Evolutionary changes in the sequence of development of organs are referred to as ***heterochrony.*** The great variety of recognized patterns of heterochrony has given rise to a large and cumbersome vocabulary that will not be reviewed here; the interested reader is referred to an excellent summary by DeBeer (1958). We will, however, consider several examples.

During the late nineteenth and early twentieth centuries, many students of Mesozoic ammonites attempted to apply Haeckel's recapitulation theory to all ammonite species, believing that the course of ammonite evolution could thus be read from ontogenetic changes in shell ornamentation and suture patterns. In 1901, Pavlow invalidated the strict recapitulation concept by showing that in certain Jurassic lineages of ammonites new evolutionary features arose in the early stages of ontogeny; not until later in the evolutionary history of their respective lineages were these changes retained in the adult stages. In other words, ontogenetic development of the new features was retarded, relative to time of reproductive maturation.

Certain graptolite lineages exhibit heterochrony in colony formation (which is not a true ontogenetic sequence, because the colony is composed of a series of many individuals, each with its own ontogeny, formed by successive budding events.) Bulman (1933) noted that many structural changes that occurred during graptolite evolution made their appearance in the early stages of colony formation. Through subsequent evolution they were retained in the later stages of colony development.

Heterochrony may have played an important role in the origin of certain higher categories. Schindewolf and Cloud have suggested such a mechanism for the origin of modern hexacorals (Scleractinia) from the Paleozoic tetracorals (Rugosa) that apparently became extinct near the end of the Permian. The oldest known hexacorals are of Middle Triassic age. As shown in Figure 11-31, the septa (blade-like vertical partitions) in hexacorals are arranged radially in cycles of six or multiples of six. In tetracorals, however, septa are arranged in four quadrants in the adult. Nevertheless, the adult pattern in tetracorals developed through modifications after the emplacement of six primary septa. It is possible that by heterochrony the six primary tetracoral septa came to remain dominant to the adult stage, and tetracorals thus gave rise to hexacorals. Morphologies of an intermediate type are, in fact, shown by certain late Paleozoic tetracorals and early Mesozoic hexacorals. The origin of the hexacorals from the tetracorals is still uncertain, however; their skeletons differ from those of tetracorals not only in symmetry but in mineralogy, being aragonite rather than calcite. Hexacorals may have arisen separately from naked anemone-like ancestors.

We must recognize that phylogenetic trends can be produced by chance factors, in addition to the directional processes of species selection and non-random speciation. Raup and Gould (1974) have shown by computer simulation that random processes can produce distinct morphologic trends, especially in small taxonomic groups.

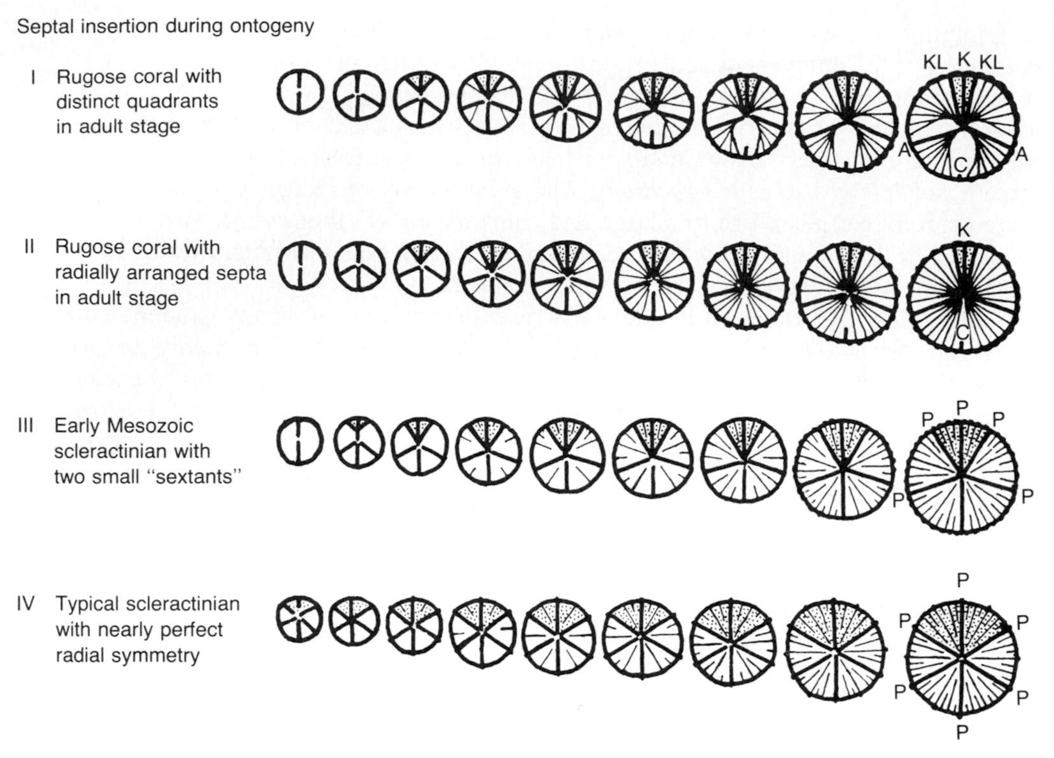

FIGURE 11-31
Diagram showing possible derivation of the Scleractinia (hexacorals) from the Rugosa (tetracorals). The drawings are of coral cups viewed from above. Ontogenetic patterns of septal insertion are sketched for a typical rugose coral (I), a typical scleractinian coral (IV), and corals that are somewhat intermediate in form (II and III). The six primary septa in all forms are represented by heavy lines. C: Cardinal septum. A: Alar septa. K: Counter septum, KL: Counter lateral septa (Rugosa), P: Primary septa (Scleractinia). The minor septa, which are emplaced between major septa, are shown as finer lines. The evolutionary transition from tetracorals to hexacorals is suggested to have occurred by insertion of minor septa in the stippled areas. (From Moore, Lalicker, and Fischer, 1952, copyright © 1952 by McGraw-Hill Book Company.)

Patterns of Evolution and Extinction

A pattern of evolution can be defined as a configuration of phylogeny. We normally do not apply the term "pattern" to events within a single lineage. As done earlier in this chapter, we speak instead of phyletic trends within single lineages. A phylogenetic trend, representing average or net change within a branched sequence of lineages, can be considered a kind of pattern. We have chosen, however, to discuss phylogenetic trends with phyletic trends and will confine the following discussion to other kinds of patterns.

ADAPTIVE RADIATION

We introduced the concept of adaptive radiation in our discussion of rates of evolution (page 307). Most evolutionary change seems to occur during bursts of speciation and proliferation of higher taxa that are termed adaptive radiations or episodes of explosive evolution. An adaptive radiation commonly occurs immediately following an evolutionary breakthrough, or adaptive innovation, and leads to the origin of what, in retrospect, may be recognized as a new higher category. An example is the origin of the large incisors of rodents, which permitted evolutionary access to a wide variety of niches in which food has been obtained by gnawing (Simpson, 1953). Rudwick (1970) attributed much of the success of the terrebratuloid brachiopods, the most abundant articulate brachiopods of the Cenozoic, to their evolution of transapical resorption (Figure 11-32). This process has allowed enlargement of the attachment organ (the pedicle) during growth. Pedicle-bearing brachiopods lacking the capacity for transapical resorption have not been able to retain strong attachment at large sizes.

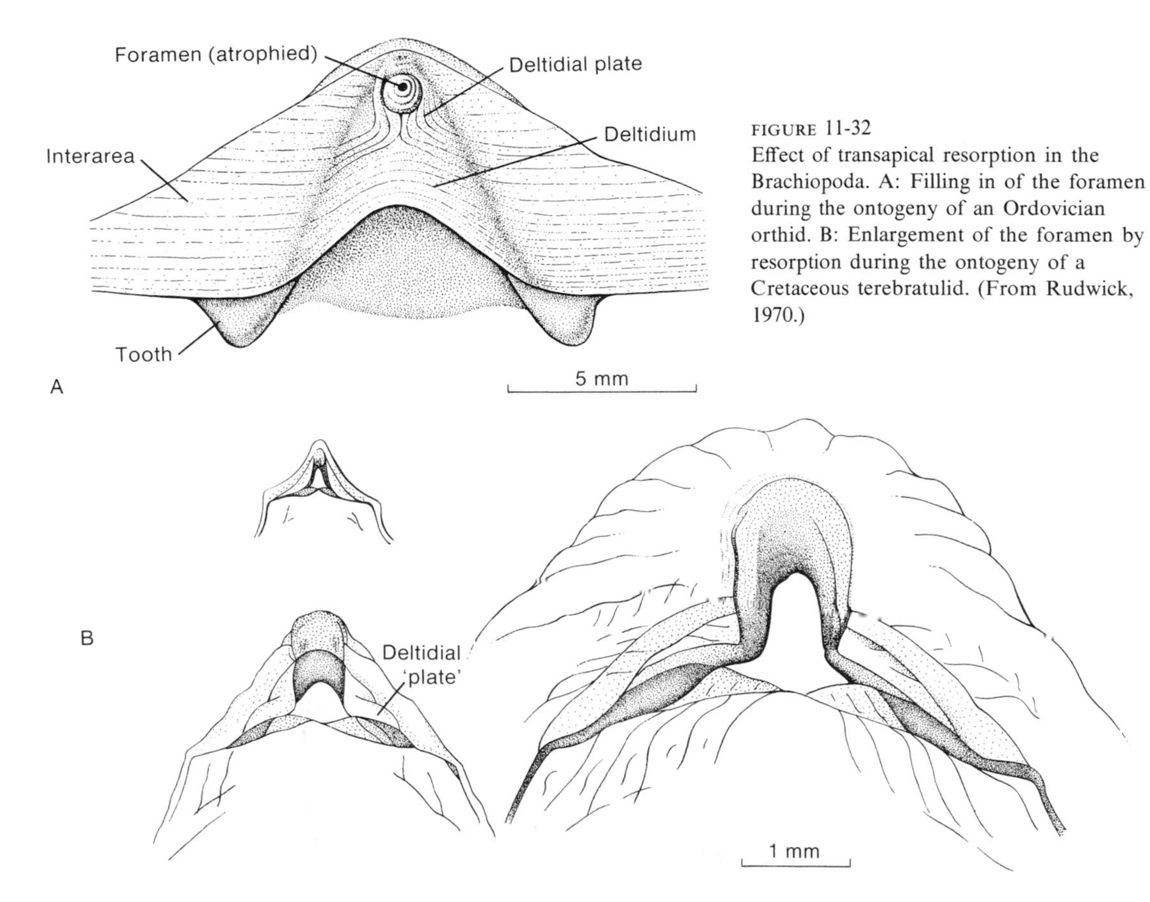

FIGURE 11-32
Effect of transapical resorption in the Brachiopoda. A: Filling in of the foramen during the ontogeny of an Ordovician orthid. B: Enlargement of the foramen by resorption during the ontogeny of a Cretaceous terebratulid. (From Rudwick, 1970.)

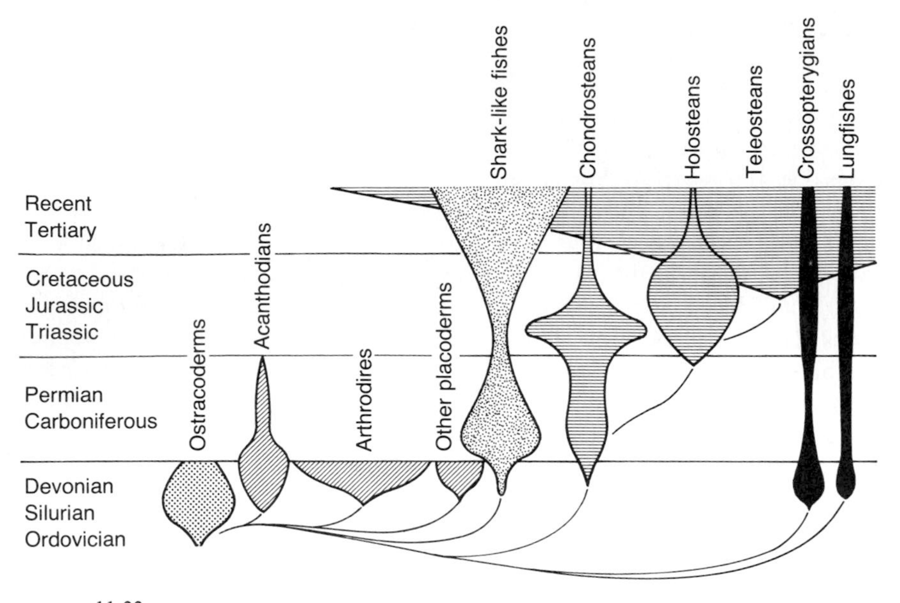

FIGURE 11-33
Stratigraphic ranges and relative abundances of major fish groups. (From Colbert, 1955.)

Many other modifications have led to the origin and diversification of major higher categories. Some of the most notable are: the shell-covered amniote egg of reptiles, which freed early reptiles from dependence on the aqueous environment for reproduction and permitted them to exploit late Paleozoic terrestrial habitats unavailable to their precursors, the amphibians; the pollen-and-seed reproductive mechanism of gymnosperm plants (pines, spruces, etc.) and angiosperm (flowering) plants, which freed these groups from dependence on moist surroundings for union of egg and sperm and enabled them to invade many terrestrial habitats inaccessible to their seedless predecessors; and the opposable thumb of primates, which—in conjunction with stereoscopic vision—gave early primates dexterity, the most important prerequisite for manipulation and tool making, which in turn facilitated the development of a large, highly-developed brain.

Occasionally in the geologic past, a group has made such an important adaptive breakthrough that it has won in competition with another group with similar life habits and habitat preferences, thus causing the other group's extinction. This phenomenon is termed ***ecologic displacement.*** Obviously, displacement can be documented only by temporal overlap in the histories of two groups and a decline of the group considered to be adaptively inferior. The geologic history of the fishes is marked by examples that seem to meet these conditions. The primitive ostracoderms, most of which were apparently scavengers or deposit feeders, declined in the Devonian during the ascendency of the placoderms (Figure 11-33). The placoderms were the first jawed vertebrates and their jaw mechanism was the adaptive innovation that led to their radiation in the Devonian, and perhaps to their displacement of many ostracoderms. The remaining ostracoderms and the placoderms both gave way in the late Paleozoic to shark-like fishes with cartilaginous skeletons and bony fishes. The chief advantage of the two new groups seems to have been their possession of an advanced and efficient set of fins, which made them far more adept swimmers than their predecessors.

The subsequent adaptive radiation of the bony fish took place in three major steps; sketches of specimens from the three taxonomic groups that were each dominant in turn are reproduced here as Figure 11-34. Each group displayed adaptive improvements over the preceding group. These improvements were achieved primarily through modification of the scales, jaws, tail, and fins, and through alteration of primitive lungs into a hydrostatic air bladder.

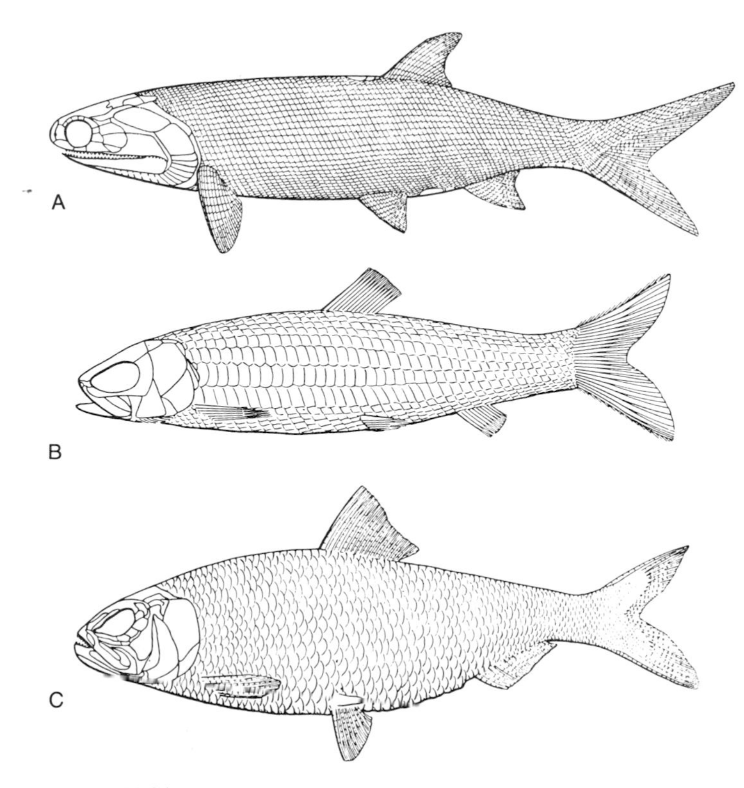

FIGURE 11-34
Representatives of the three major bony fish groups shown in Figure 11-33, which show stages in evolution toward thinner scales, a more symmetrical tail, forward movement of pelvic fins, and shortening of the jaws. A: *Palaeoniscus* (a Permian chondrostean). B: *Pholidophorus* (a Jurassic holostean). C: *Clupea* (a Cenozoic teleost). (From Colbert, 1955.)

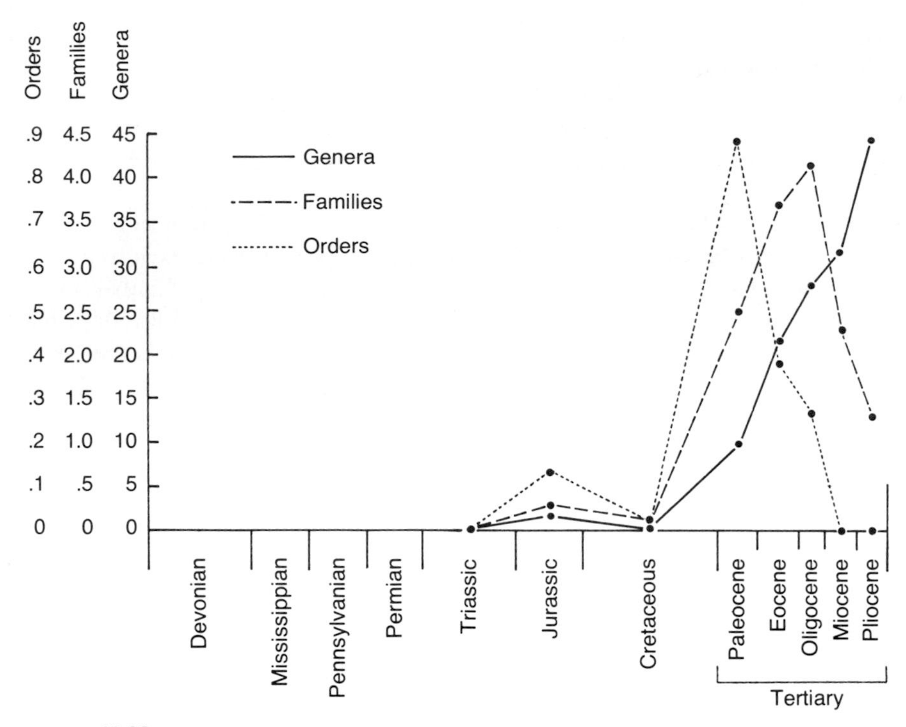

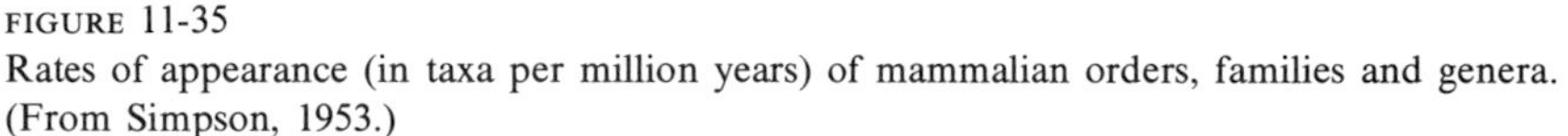
FIGURE 11-35
Rates of appearance (in taxa per million years) of mammalian orders, families and genera. (From Simpson, 1953.)

Adaptive radiations occur at various levels within phylogeny. We have already noted the hierarchical nature of phylogeny (Figure 11-9). This accounts for a characteristic pattern of adaptive radiation (Figure 11-35). The higher taxonomic categories comprised by a new group tend to appear relatively early and their rate of origin tends to decline as the radiation progresses. Lower taxonomic categories originate most rapidly later in the radiation. The curves that are compared in Figure 11-35 reflect a certain amount of bias from the fossil record. As discussed in Chapter 1, completeness of preservation increases toward the Recent. This means that the peaks of the curves of Figure 11-33 may lie further toward the Recent than they would if preservation were complete. Furthermore, as discussed in Chapter 1, stratigraphic ranges are better known for higher taxa than for lower taxa. The shift toward the Recent may therefore be greater for genera than for families (Raup, 1972). Thus, the peaks may be artificially spread out somewhat along the horizontal axis. Our knowledge of the fossil record of Cenozoic mammals is reasonably good, so this effect may not be great in Figure 11-35, but it should be borne in mind when plotting graphs of this type.

Some higher taxa appear suddenly in the known fossil record at surprisingly high diversities. Such occurrences have tended to lead to two kinds of speculation. One is that a long period of diversification, for some reason not documented by known fossil evidence, preceded the time at which the oldest recognized biotas existed. The other kind of speculation is that no earlier record has been found because diversification was very sudden, occurring in a brief interval and perhaps in restricted habitats or geographic areas. For many years the record of the angiosperms (flowering plants) presented this sort of problem. Angiosperm floras are present in substantial diversity in the Lower Cretaceous, but have not been unequivocally identified from Jurassic or Triassic deposits.

Darwin considered the early angiosperm record an "abominable mystery," and until recently it was regarded as such by most paleobotanists. Recently Doyle and Hickey (1976) have provided evidence that seems to solve the mystery. For one thing, they have found that many fossil leaves of the Cretaceous long thought to belong to living genera are actually distinct forms that happen to resemble leaves of living genera. More important, they have uncovered evidence, primarily in the Atlantic Coastal Plain of the United States, that angiosperms were undergoing their initial adaptive radiation in the Early Cretaceous. The interval of diversification that they have documented lasted only about 10 million years.

In the upper part of the Aptian Stage, Doyle and Hickey have found only simple kinds of leaves and pollen. In the succeeding Albian Stage, however, a much greater variety of morphologic types is found. What is especially striking is that increases in diversity and complexity are found in the fossil record of both pollen and leaves. The upper Aptian pollen is ***monosulcate,*** which means that it has only a single germination furrow. In the Albian, ***tricolpate*** pollen, which has three furrows, makes an appearance. Toward the upper part of the Aptian, tricolpate pollen diversifies in size, shape, and surface sculpture (Figure 11-36). The value of having three pores instead of one seems to have been that chances were increased that a grain landing on the surface of a stigma (female portion of a flower) would have a pore in contact with the stigma for fertilization. Surface sculpture seems to help grains to stick to pollenating insects. Other advanced shapes seem to have more subtle adaptive values, some of which relate to the strength of individual grains. The earliest leaves—those of the upper Aptian—are also archaic. Among other features, most have ***entire*** (smooth) margins, highly disorganized venation, and poor differentiation of the blade of the leaf from the ***petiole*** (stem). During the Albian (Figure 11-34) there was an increase in the diversity of nonentire (toothed and lobed) margins, an increase in the organization of venation, an increase in the number of species in which the blade was well differentiated from the petiole, and an increase in the variety of compound leaves (leaves clustered as in Figure 11-36,v and 11-36,y).

Doyle and Hickey favor the idea of Stebbins (1965) that the reproductive mechanism that characterizes angiosperms may have arisen in a (semiarid) ***mesic*** habitat, but the earliest angiosperms in their study (those in zone I of Figure 11-36) are found in stream deposits. They seem to have occupied ***riparian*** habitats, or habitats

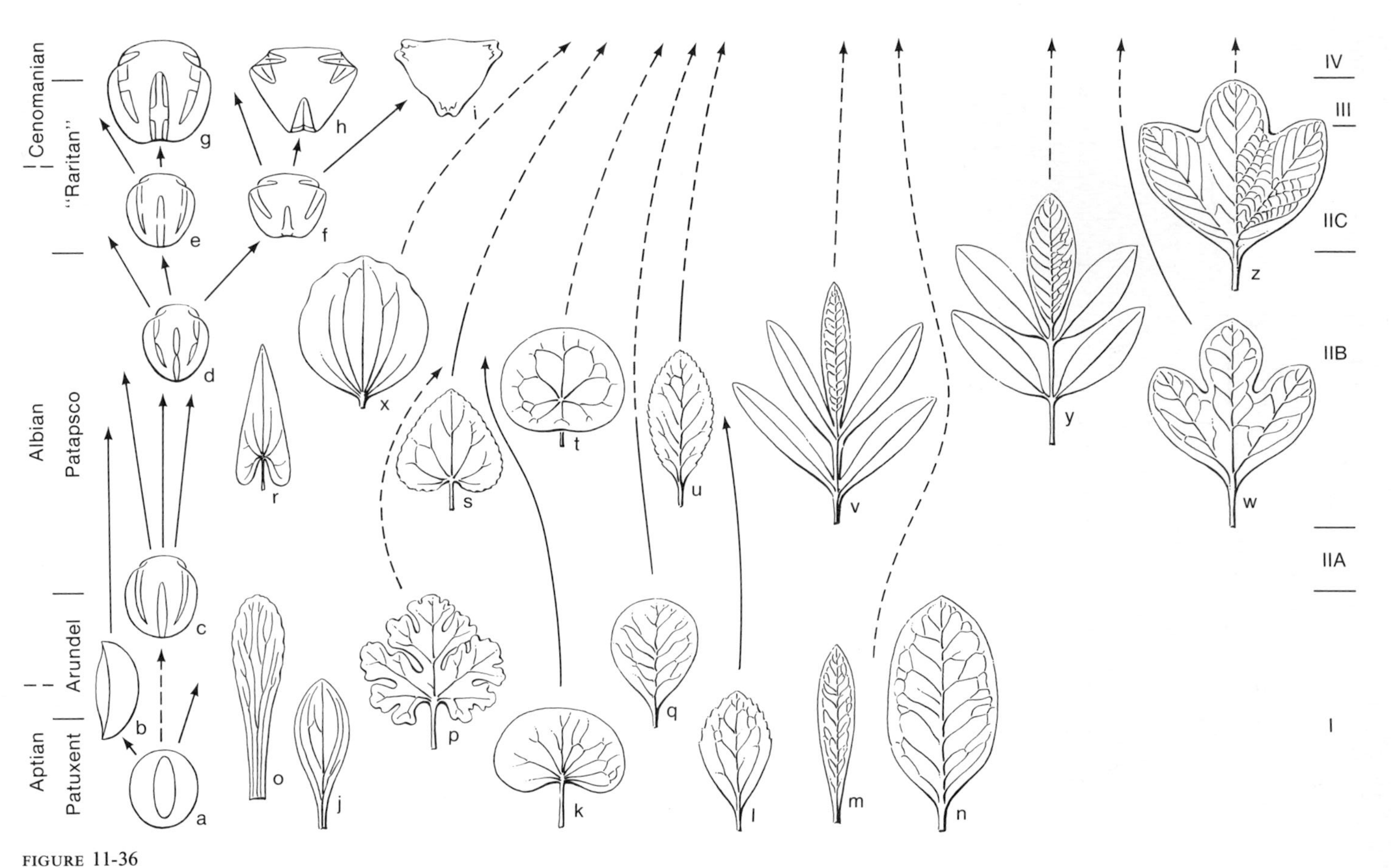

FIGURE 11-36
Increase in the diversity and complexity of pollen (a–i) and leaves (j–z) of early angiosperms in the Lower Cretaceous Potomac Group of the Atlantic Coastal Plain of the United States. (From Doyle and Hickey, 1976.)

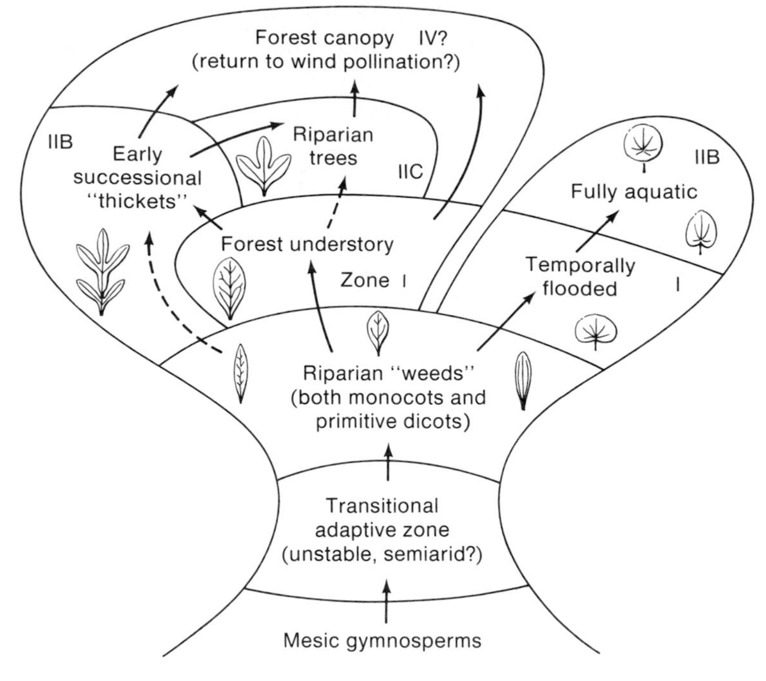

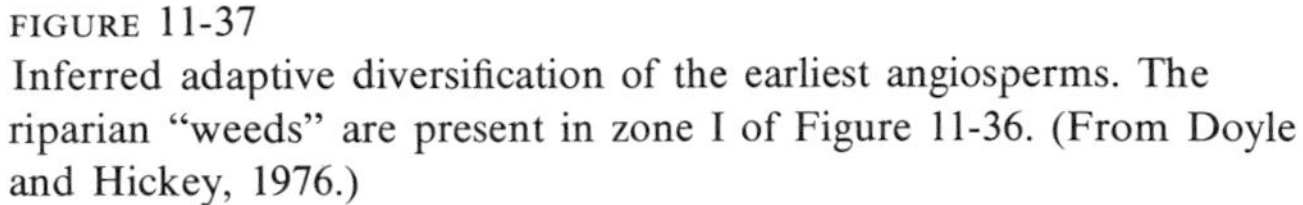
FIGURE 11-37
Inferred adaptive diversification of the earliest angiosperms. The riparian "weeds" are present in zone I of Figure 11-36. (From Doyle and Hickey, 1976.)

along the margins of streams (Figure 11-37). Doyle and Hickey believe that initially they were able to become established in a world dominated by gymnosperms (conifers and other "naked seed" plants) only by inhabiting such frequently disturbed habitats, where competition was weak (see page 280). Then, during the Albian, the angiosperms diversified remarkably, producing aquatic forms that had leaves like lily pads, and also low trees, followed by large trees that formed forest canopies. This extraordinary diversification seems to have occurred in little more than 10 million years.

EVOLUTIONARY CONVERGENCE AND PARALLELISM

In the course of evolution close morphologic similarity may arise between two unrelated groups as they take on similar life habits. This is termed ***adaptive convergence,*** and the similar taxa are referred to as ***homeomorphs.*** Some biologists reserve the term homeomorph for one of a pair of species whose members are indistinguishable to the untrained observer. Most paleontologists, however, apply the term to one of a pair of species whose members simply exhibit a strong resemblance.

Homeomorphy can also result from what is referred to as ***parallel evolution,*** or ***parallelism,*** in which two closely related stocks with minor morphologic differences undergo a series of similar evolutionary changes through time. Parallelism and convergence are distinguished on the basis of the amount of similarity between the ancestral groups giving rise to the similar forms. In convergence the ancestors are quite distinct, whereas in parallelism they are similar and the similarity of the evolutionary pathways followed results in part from the ancestral similarities (Figure 11-38). The terms parallelism and convergence have frequently been applied to crudely similar trends in only one or a few characters, rather than in whole organisms. Both can occur by either phyletic or phylogenetic change. In other words, not only may two lineages converge or evolve in parallel, but phylogenetic trends (Figure 11-26, 11-27) may also document net convergence or parallelism.

Similar taxa that have arisen by parallel or iterative evolution are often said to have obtained the same ***grade*** of evolution. This kind of resemblance must be distinguished from resemblance arising from close relationship within a ***clade*** (see page 149).

FIGURE 11-38
Ways in which evolving lineages may be related. The horizontal distance separating two species at any point is proportional to the differences between them relative to the differences between them at other points in their evolutionary histories. In heterochronous convergence, as in iterative evolution, similarities arise at different times. (Partly from Cloud, 1949.)

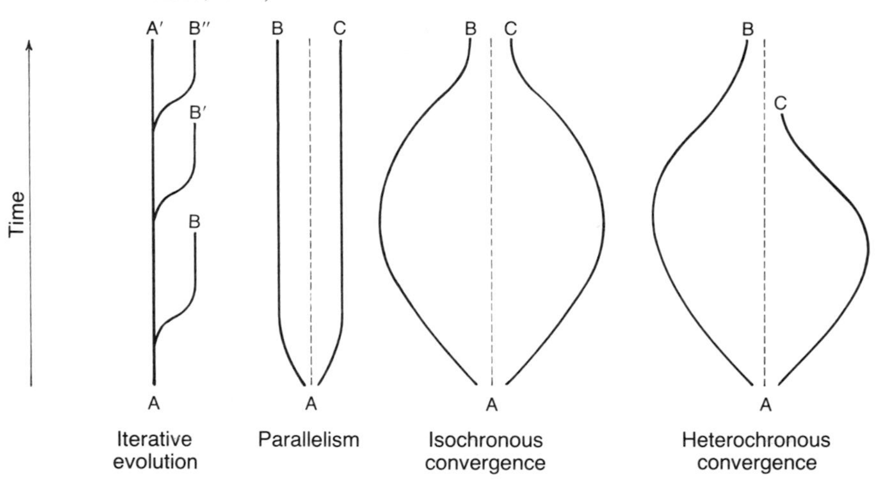

ECOLOGIC REPLACEMENT

Simpson (1944) introduced the term ***adaptive zone*** to describe a group of niches occupied or potentially occupied by a higher taxon or certain members of a higher taxon. Thus, we might speak of the adaptive zone of the Echinoidea, or the adaptive zone of burrowing echinoids. An episode of adaptive radiation represents the progressive occupation of an adaptive zone. Not only are adaptive breakthroughs required, but the group must have ecological access to the adaptive zone.

A new group of organisms cannot easily evolve to occupy niches already filled by well-adapted members of another group, even if the new group has the potential to produce more efficient and more diverse adaptations than the entrenched group. The earliest members of the new group may possess rudimentary features offering great evolutionary potential, but in their primitive state these features may be inferior to the advanced features of the entrenched group. A new group may, however, diversify in a region that is geographically separated from the one occupied by the group it is to displace, and then gain access to the region after attaining adaptive superiority. If the older group is geographically widespread, however, this is unlikely to happen. During the Mesozoic Era, diversification of the earliest mammals, which were small and unspecialized, was suppressed by the domination of terrestrial habitats throughout the world by dinosaurs. As is well known, following the mass extinction of dinosaurs in the Late Cretaceous, mammals have radiated to attain greater adaptive diversity than Mesozoic reptiles. The Cenozoic mammalian diversification (Figure 11-35) provides a striking illustration of the evolutionary pattern described earlier as being typical of adaptive radiation. As the ecologic successors of the dinosaurs, the mammals exemplify what is known as ***ecologic replacement.*** (This phrase does not imply competitive displacement, but most workers consider displacement (see page 358) to be a variety of replacement.)

In a special type of repetitive ecologic replacement, a basic parent stock has given rise to successive groups of higher taxa, each replacing the former. The resulting pattern is called ***iterative evolution.*** Iterative evolution was apparently characteristic of Mesozoic ammonite groups, although it is difficult to determine exact relationships between ancestral and descendant groups. Salfeld (1913) established the generalization that two slowly evolving suborders (the Phylloceratina and Lytoceratina) lived predominantly in deep-water regions and gave rise to successive groups of short-lived, shallow-water taxa that are grouped together as the Ammonitina. The Ammonitina is thus considered to be a polyphyletic suborder (Figure 11-39).

Cambrian trilobites are also known to have undergone iterative evolution. The best-documented example has been provided by Kaufmann (1933, 1935) and discussed by Simpson (1953). Kaufmann recognized four successive lineages of the genus *Olenus* in an Upper Cambrian rock sequence in Sweden. He divided each lineage into three species or subspecies (Figure 11-40). Three of the lineages exhibit continuous morphologic gradation. All four show certain morphologic trends, the

FIGURE 11-39
Iterative evolution of Jurassic ammonite families derived from the suborders Phylloceratina and Lytoceratina. The suffix -aceae denotes a superfamily; the suffix -idae denotes a family. (From Arkell, 1957.)

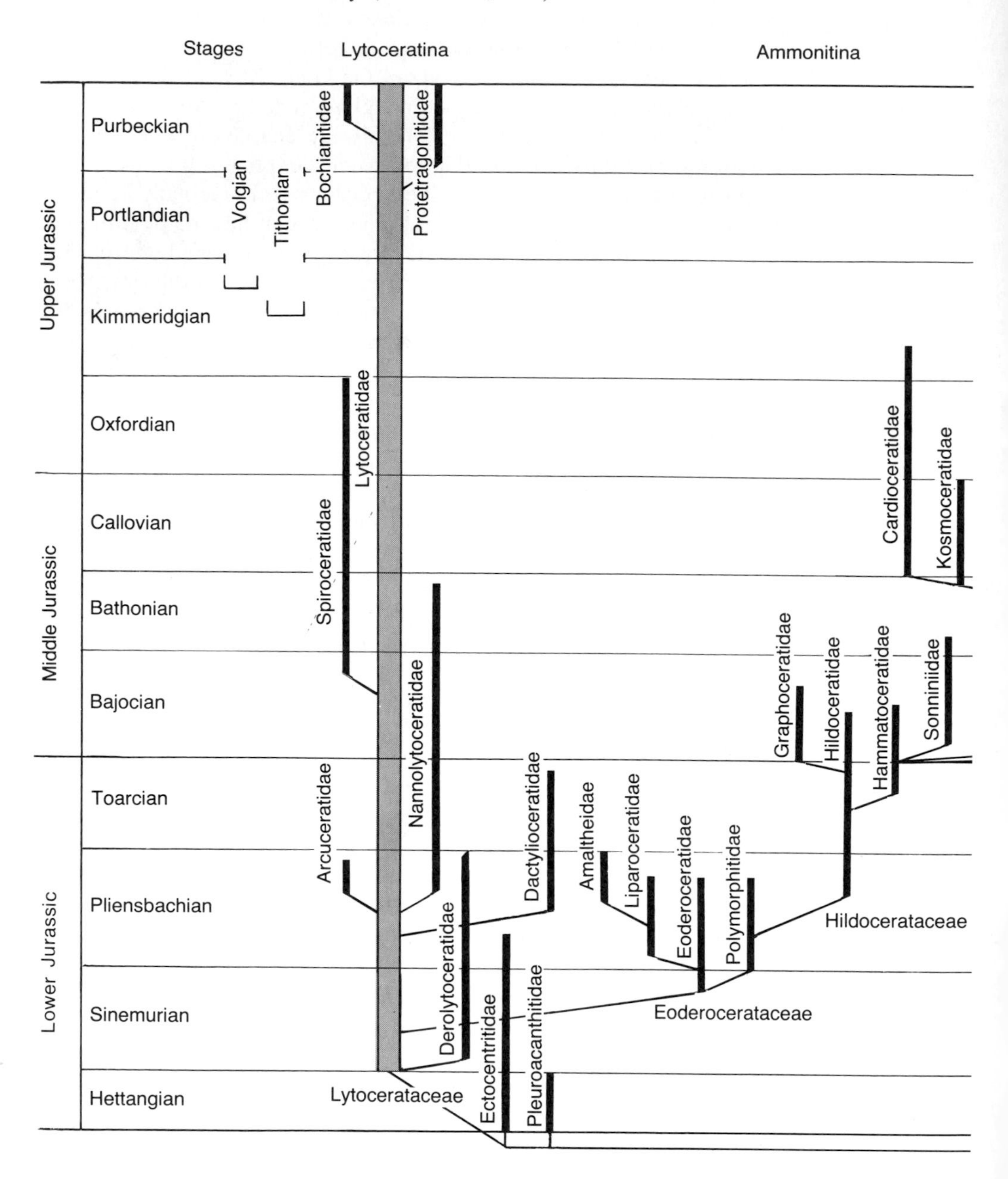

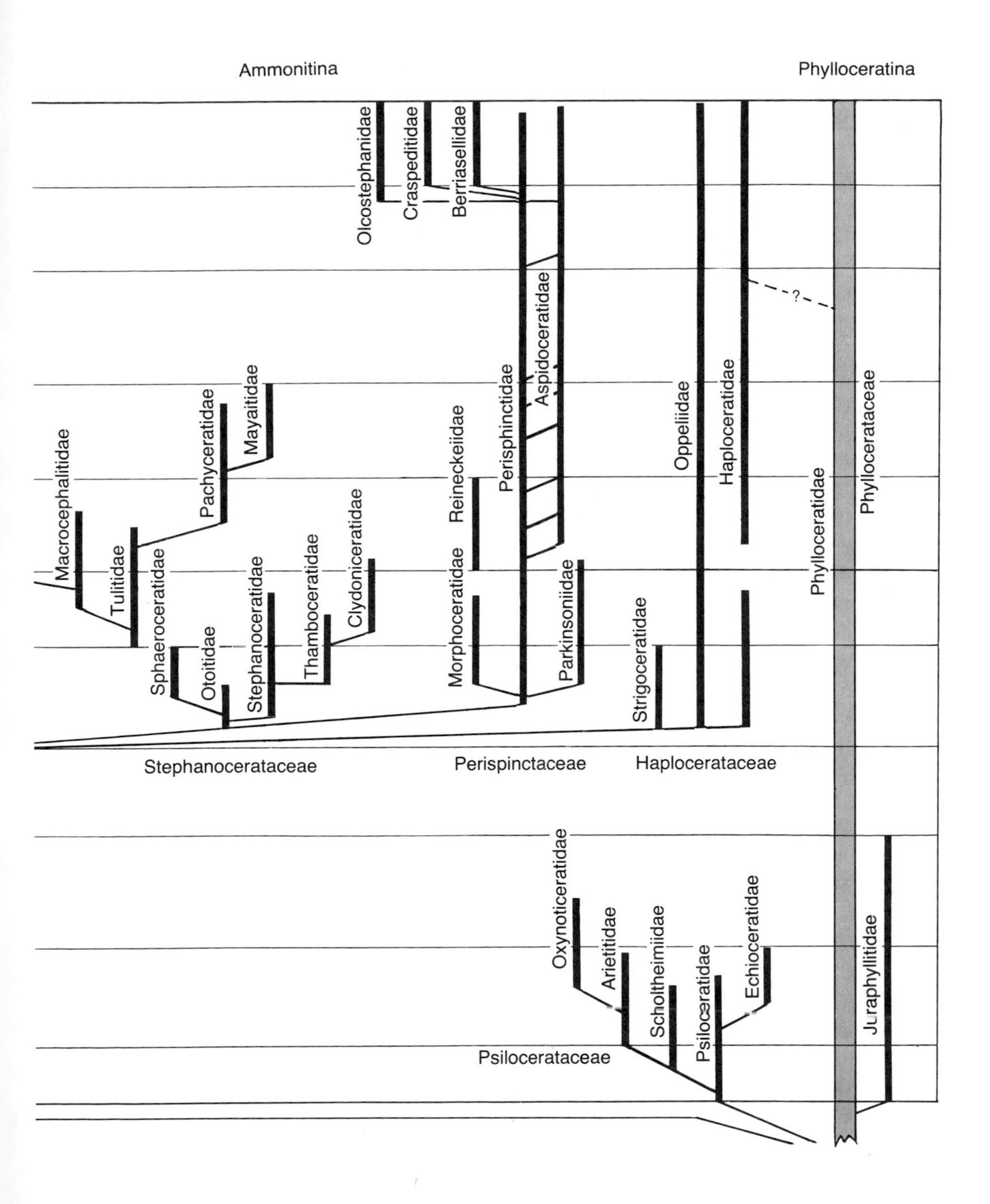

Ammonitina
Phylloceratina
Olcostephanidae
Craspeditidae
Berriasellidae
Aspidoceratidae
?
Macrocephalitidae
Tulitidae
Pachyceratidae
Mayaitidae
Sphaeroceratidae
Otoitidae
Stephanoceratidae
Thamboceratidae
Clydoniceratidae
Reineckeiidae
Morphoceratidae
Perisphinctidae
Parkinsoniidae
Strigoceratidae
Oppeliidae
Haploceratidae
Phylloceratidae
Phylloceratacea
Stephanocerataceae
Perispinctaceae
Haplocerataceae
Oxynoticeratidae
Arietitidae
Schoitheimiidae
Psiloceratidae
Echioceratidae
Juraphyllitidae
Psilocerataceae

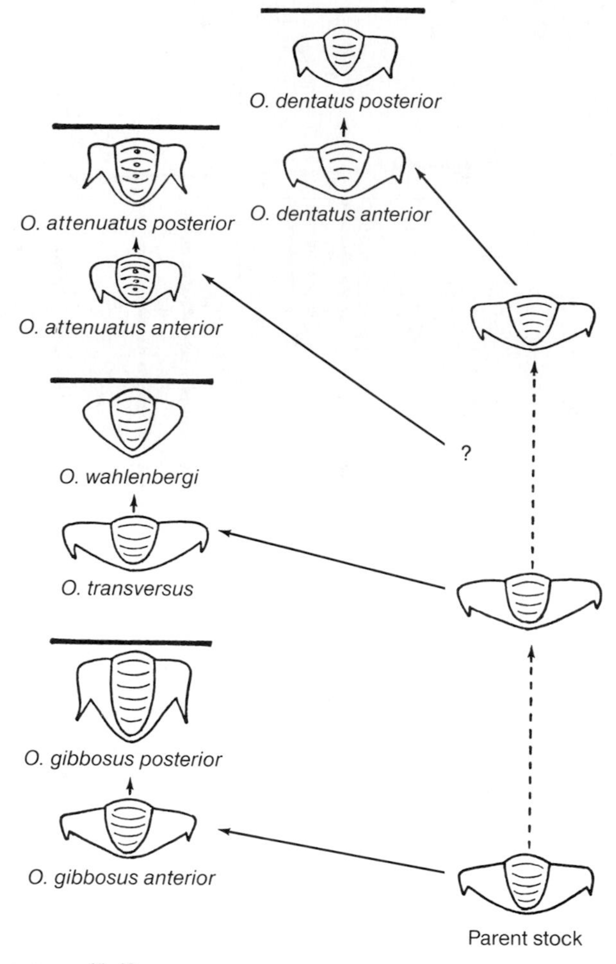

FIGURE 11-40
Iterative evolution in the Cambrian trilobite genus *Olenus,* represented by outlines of pygidia (tail regions). Notice that the four descendent lineages underwent similar morphologic changes. (From Simpson, 1963; after Kaufmann, 1933.)

most striking being relative elongation of the pygidium (tail region). Furthermore, the earliest representatives of lineages 1 and 2 are nearly identical. The four lineages are believed to have evolved in succession from a more slowly evolving, geographically removed parental stock. Each lineage invaded the region of study, followed the recognized evolutionary trends, and then became extinct, to be replaced by another invasion of taxa derived from the parent stock.

Cenozoic planktonic foraminiferans provide one of the most striking recognized examples of iterative evolution (Cifelli, 1969). These forms belong to the Globigerinacea, which arose in the Jurassic. This taxon nearly disappeared at the close of the Cretaceous (Figure 11-41). Only the globigerine type survived into the Cenozoic, but from it a variety of other types arose in the Paleocene and early in the Eocene. Another major extinction occurred in the Oligocene, and again only the globigerine type survived. In the early part of the Miocene there evolved from the surviving globigerine type an array of forms that very closely resembled those that had arisen from earlier globigerines following the Cretaceous-Paleocene extinctions. Before Cifelli recognized the iterative pattern, early Cenozoic and late Cenozoic members of each morphology were considered to represent a single line of descent and were placed in a single genus. We do not know for sure whether early and late representatives of each morphologic type occupied similar niches, but this kind of iterative evolution seems to illustrate that a given kind of organism has a limited evolutionary potential. Morphologic, genetic, and ecologic constraints seem to confine evolution to a few adaptive pathways. It should be evident that iterative evolution is similar to parallelism. In effect it is a kind of parallelism in which the groups evolving in parallel do so at different times.

Determining Phylogenetic Relationships

Phylogenetic relationships among taxonomic groups are determined from both biologic information and the fossil record, either separately or together. In studying extinct groups whose taxonomic relationships are uncertain, there has been a common tendency to attempt to establish affinities with phyla that have Recent representatives. This bias, often based on wishful thinking, has led to many erroneous conclusions. There has, however, been some success in determining relationships, as is exemplified by study of the extinct graptolites.

Most workers now accept the argument of Kozlowski (1966 and earlier studies) that the graptolites were hemichordates, closely resembling living pterobranchs (Figure 11-42). The detailed structure and chemical composition of the graptolite skeleton, as well as the pattern of colonial budding, all support such a relationship. Kozlowski's interpretations stem from his work with exceptionally well-preserved colonies etched from chert with hydrofluoric acid. Kozlowski imbedded free specimens in paraffin and cut serial sections with a microtome in the same way that biologists section Recent organisms. In contrast, most graptolite skeletons are found flattened along bedding planes of black shale. Kozlowski's success demonstrates the importance of seeking unusually well-preserved fossil material for taxonomic study. It also appears to represent a fruitful attempt to establish affinities between a problematic fossil group and a living group.

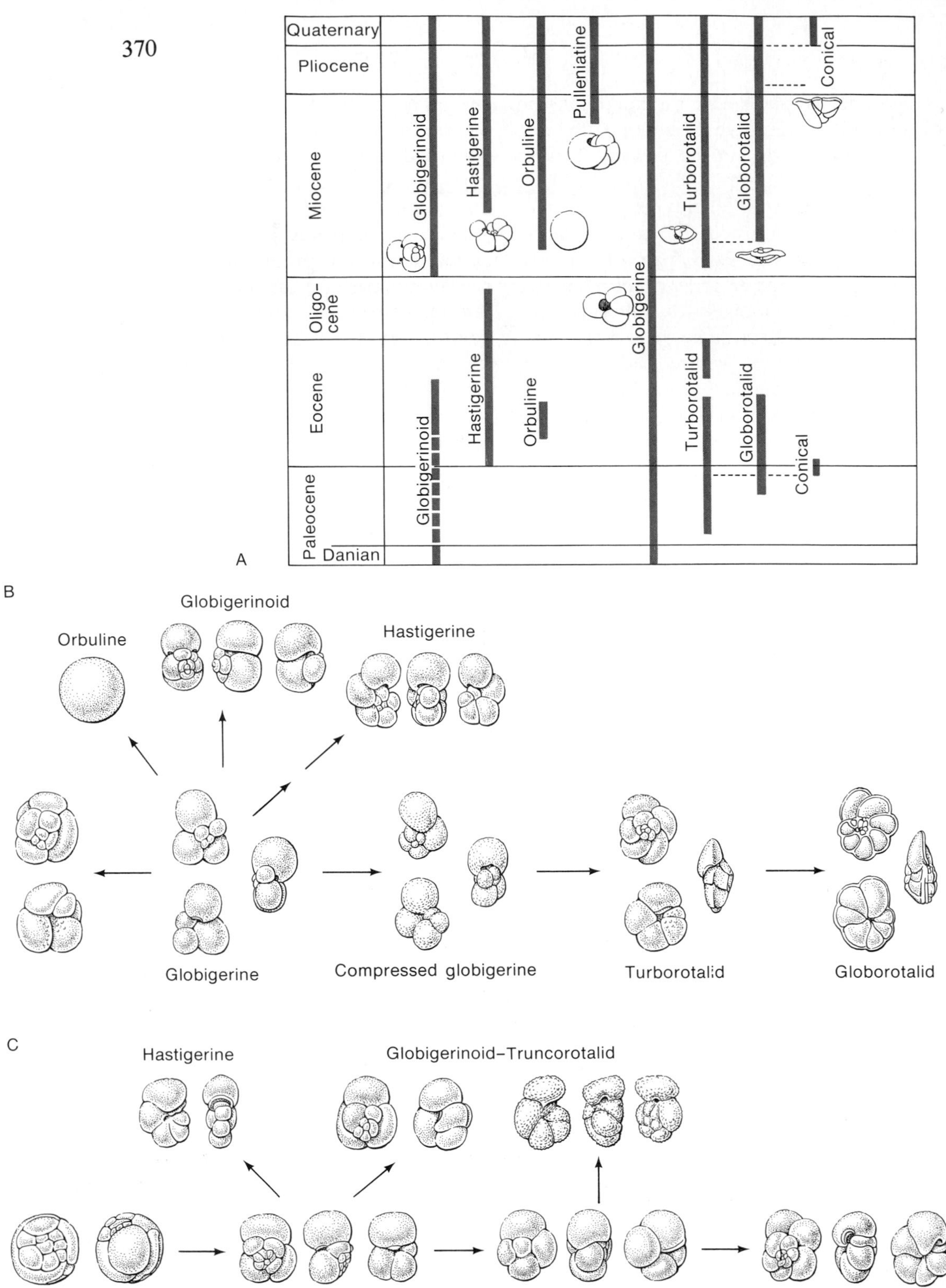

Quaternary
Pliocene
Miocene
Oligo-cene
Eocene
Paleocene
Danian
Globigerinoid
Hastigerine
Orbuline
Pulleniatine
Globigerine
Turborotalid
Globorotalid
Conical
A
B
Orbuline
Globigerinoid
Hastigerine
Globigerine
Compressed globigerine
Turborotalid
Globorotalid
C
Hastigerine
Globigerinoid–Truncorotalid
Orbuline
Globigerine
Compressed globigerine

FIGURE 11-41
Iterative evolution of planktonic foraminiferans. A: Stratigraphic ranges of particular morphologic types. B: Miocene adaptive radiation from the globigerine type. C: Similar early Cenozoic adaptive radiation from the globigerine type. (From Cifelli, 1969.)

Looking at the opposite side of the coin, there are certain living taxonomic groups whose affinities are understood largely through study of the fossil record. The subungulate mammals, for example, include modern-day conies (small rodent-like forms), elephants, and sea cows. It is highly unlikely that these three strikingly different groups would be lumped together were it not for their fossil records, which show basic similarities among their primitive Tertiary ancestors on the continent of Africa, where they apparently shared a common ancestry.

In general, it is much easier to establish phylogenies for major vertebrate groups than for major invertebrate and plant groups because all recognized classes and orders of the Vertebrata have originated since the Cambrian. Although all higher vascular plant taxa have apparently originated since the early Paleozoic, the fossil record of plants is less complete. Furthermore, fossil plant remains usually reveal less about whole-organism morphology than do vertebrate remains. Invertebrate animals fall into several phyla whose Late Precambrian and Cambrian origins are almost universally undocumented by the known fossil record. In some instances, however, we have a moderately good knowledge of the post-Paleozoic phylogenies within invertebrate phyla.

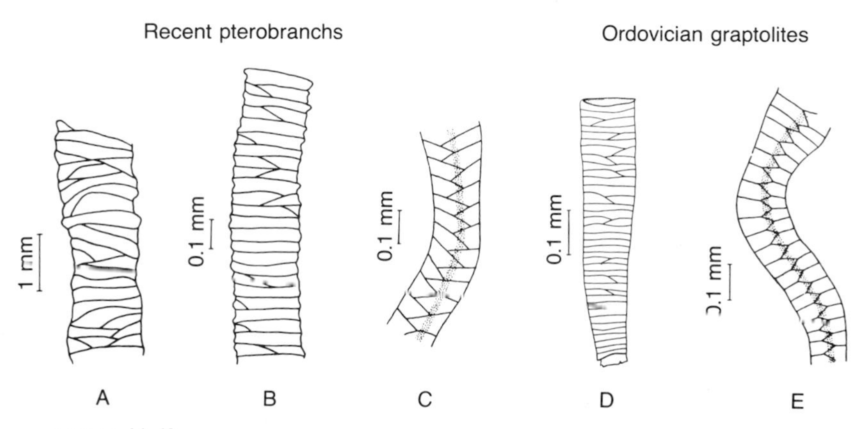

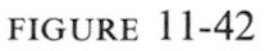

FIGURE 11-42
Morphologic comparison of pterobranchs and graptolites. A and B: Wall structure, consisting of "fusellar" bands, of *Cephalodiscus* and *Rhabdopleura.* D: Stolon (stippled) of *Rhabdopleura.* D and E: *Mastigograptus* and *Bulmanicrusta,* showing similar structures. (From Kozlowski, 1966.)

Unfortunately, the origins of most higher categories are shrouded in mystery; commonly new higher categories appear abruptly in the fossil record without evidence of transitional ancestral forms. Simpson (1953) has listed several reasons for this situation. Among them are the following:

1. Appearance of a new higher category has usually marked a major adaptive breakthrough, often accompanying inhabitation of previously unoccupied niches; evolution under such conditions has tended to be *very rapid.*
2. Any lineages of the ancestral group that were similar enough to enter into competition with the new group are likely to have been rapidly displaced.
3. Often, times of higher category appearance are represented by gaps in the geologic record. (In some instances, rapid evolutionary turnover and unconformities may have resulted from the same widespread environmental change.)
4. Change in habitat during the adaptive breakthrough has made discovery of certain transitional forms unlikely.
5. Major adaptive breakthroughs have commonly occurred in relatively small populations or taxonomic groups.
6. Transitions have commonly been made in taxa whose members were small relative to average size in both the ancestral and descendant higher categories.
7. Transitions have commonly taken place in restricted geographic areas, and possibly the same transitions occurred at different times in different areas.

The fossil record does occasionally provide what might be termed as a "missing link," a species that appears to represent a transitional stage between higher taxa. One such form is the dinosaur-like bird *Archaeopteryx,* of the Middle Jurassic, already discussed in other connections and pictured in Figure 1-11. *Archaeopteryx* possessed both dinosaur and avian characters. Its possession of feathers suggests that it was warm-blooded, like modern birds, but it also had large teeth, solid bones, and other skeletal features derived from dinosaurs.

A spectacular transition, described by Erben (1966), is shown by a graded morphologic series of cephalopods in the Lower Devonian Hunsrück Shale, of Germany. In the 1930's Schindewolf postulated the origin of the ammonoids from the bactritid nautiloids. His arguments stressed the similarity of the initial chambers of bactritid and early ammonoid shells. The siphuncle (a fleshy tube passing from the living chamber to all older chambers) of the bactritids was also marginal, like that of most ammonoid shells. The morphologic series reported by Erben apparently confirms Schindewolf's hypothesis. The series contains bactritids, ammonoids, and intermediate forms (Figure 11-43). The various taxa constituting the series have not been studied with respect to relative stratigraphic position, but their existence together in a local rock sequence strongly supports the idea of a bractritid origin for the ammonoids.

Just as the fossil record may provide transitional links between major taxa, so may discoveries of hitherto unknown Recent organisms. The deep-sea mollusc *Neopilina*

FIGURE 11-43
Adult shells of species from the Devonian Hunsrück Shale of Germany showing the apparent evolutionary sequence leading to the ammonoids. A–E: Bactritid nautiloids. F–L: Early ammonoids. (From Erben, 1966.)

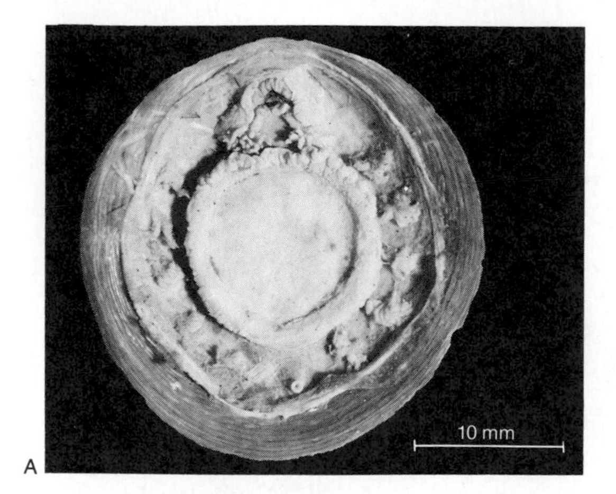
10 mm

A

10 mm

B

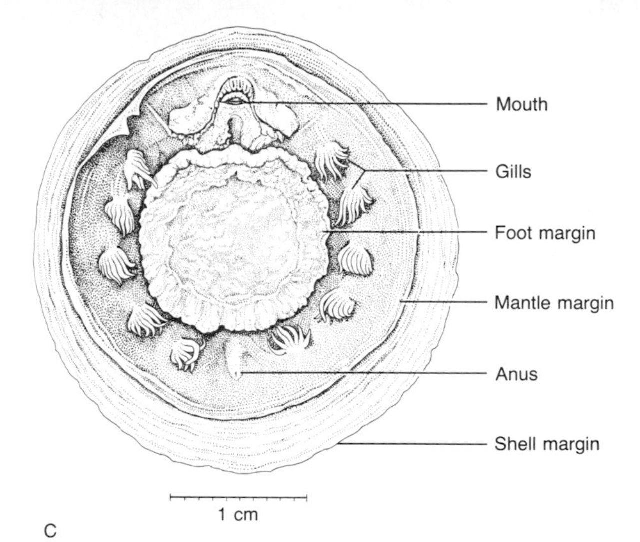
Mouth
Gills
Foot margin
Mantle margin
Anus
Shell margin
1 cm

C

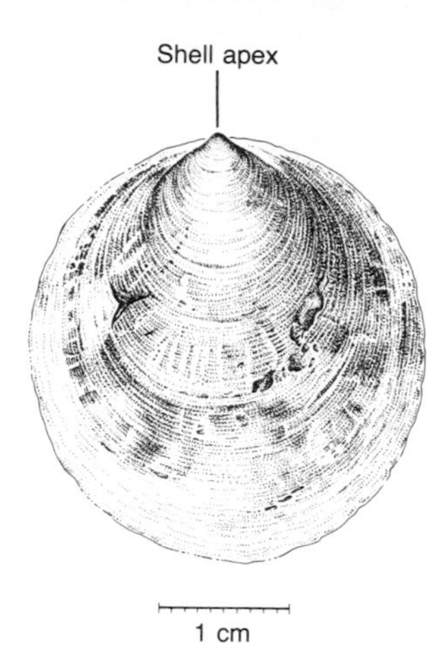
Shell apex
1 cm

D

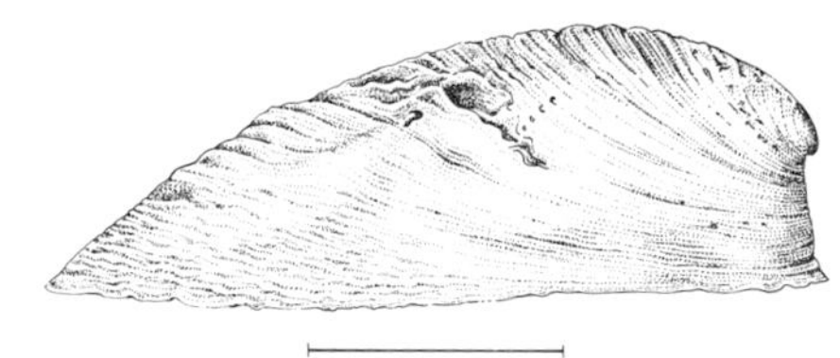

1 cm

E

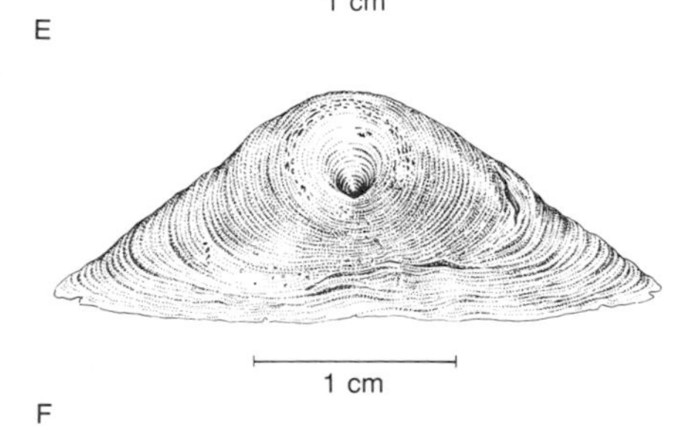
1 cm

F

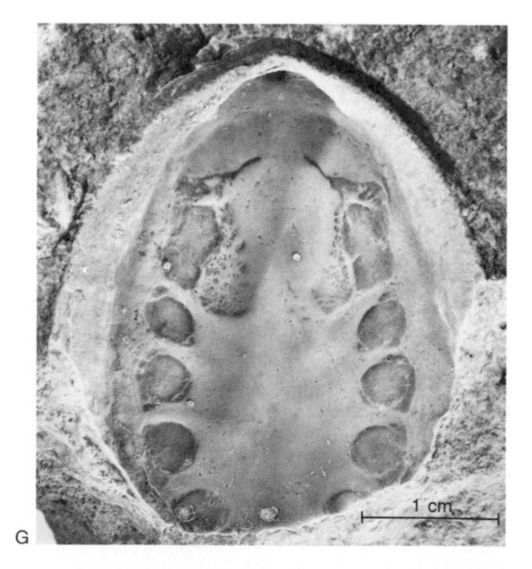
1 cm

G

FIGURE 11-44
Monoplacophora. A–F: The living species *Neopilina galatheae* Lemche. A and C: Ventral views. B and D: Dorsal views. E: Right lateral view. F: Anterior view. G: The Silurian fossil *Pilina unguis* (Lindstrom); shell interior, showing paired muscles by which the body and foot were attached. (A–F from Lemche and Wingstrand, 1959; G courtesy of the Swedish Museum of Natural History.)

provides a possible illustration. In 1952, workers of the Danish Galathea Expedition dredged members of this unusual genus from the ocean bottom off the Pacific Coast of Central America at a depth of about 3,500 meters. Lemche and Wingstrand (1959) have described the anatomy of the genus, which has many primitive characters. The most unusual feature is the general body plan, which displays paired series of pedal muscles, kidney-like structures, and gills (Figure 11-44). Other molluscan groups possess only one pair of each organ type, or multiple pairs that appear to have had a secondary (nonprimitive) origin.

During the two decades preceding the discovery of *Neopilina,* some paleontologists, including Wenz and Knight, had suggested that early Paleozoic fossils belonging to the molluscan class Monoplacophora (see Figure 11-44) were the cap-shaped ancestors of gastropods, and perhaps of other higher molluscan classes. Upon its discovery, *Neopilina* was recognized as a monoplacophoran, though fossil representatives of the group were known only from the lower and middle Paleozoic. *Neopilina* is a "living fossil," representing a group that has apparently found refuge for many millions of years in the deep sea.

The evolutionary position of the Monoplacophora, whose anatomy is known only through *Neopilina,* is still a matter of debate. One issue is whether the semisegmented body plan of the primitive Monoplacophora indicates that the Mollusca arose from segmented annelid worms. Molluscs and annelids have long been considered to be related because of similarities in their early ontogenetic development. Another question is whether *Neopilina*-like monoplacophorans did, indeed, give rise to higher molluscan classes. Most hypothetical models of the "ancestral mollusc" have portrayed an organism with single pairs of organs. If *Neopilina* is in the main line of early molluscan evolution, loss of the semisegmented structure that it possesses probably preceded the origin of higher mollusc classes. It has not been demonstrated conclusively that the group to which *Neopilina* belongs was ancestral to any of the other molluscan classes. Despite its uncertain phylogenetic position, *Neopilina* is one of the most significant living taxa to have been discovered in recent decades.

A comparable discovery made more recently was of a variety of sponges placed in a new class, the Sclerospongiae (Hartman and Goreau, 1970). These tropical reef animals tend to live in crevices or caves. Partly because of their cryptic occurrence, they were not discovered until the 1960's. Sclerosponges occur in both the Caribbean and Indo-Pacific regions, and there are several living species. Morphologic similarities show that the living sclerosponges are closely related to problematic fossil creatures

A
B
C
D

FIGURE 11-45
Comparison of the gross morphologies of modern sclerosponges and stromatoporoids. A: A sclerosponge colony of *Ceratoporella,* about 15 centimeters long, showing bumpy structures where astrorhizae are centered. B: Closeup view of astrorhizae of *Ceratoporella.* C: Devonian stromatoporoid from Iowa, showing the protuberances that are the sites of astrorhizae (×.63). D: Surface of a Devonian stromatoporoid from Petoskey, Michigan, showing astrorhizae (×2.28). (A and B provided by Willard Hartman.)

called stromatoporoids. Before the discovery of the living forms, stromatoporoids had been assigned by different workers to a variety of phyla. The most recent consensus was that they were coelenterates. One of the most striking skeletal resemblances between these organisms and modern sclerosponges is the presence of ***astrorhizae,*** which in living sclerosponges house radial canal systems where water flowing through the sponge is collected to exit from a central opening (Figure 11-45). These systems are fundamentally related to the basic body plan of sponges, and yet the functional relationship went unappreciated until the discovery of the living forms.

Most progress in the establishment of phylogenies is made through consideration of both fossil and Recent evidence. Used alone, either type of evidence may lead to misinterpretation. There is no need here to cite specific examples, but many unreasonable hypotheses linking taxonomic groups have been proposed by both paleontologists and neontologists whose approaches include data from only the fossil record or only the Recent.

Supplementary Reading

DeBeer, G. R. (1958) *Embryos and Ancestors.* London, Oxford University Press, 197 p. (A thorough review of heterochrony, with many examples.)

Hallam, A., ed. (1977) *Patterns of Evolution.* Amsterdam, Elsevier, 591 p, (A collection of evolutionary contributions by several paleontologists.)

Mayr, E. (1971) *Populations, Species, and Evolution.* Cambridge, Mass., Harvard University Press, 453 p. (An abridged and readable version of Mayr's 1963 book, emphasizing the species as the unit of large-scale evolution.)

Olson, E. C. (1965) *The Evolution of Life.* London, Weidenfeld and Nicolson, 300 p. (A discussion of the major aspects of evolution, especially those evident in the vertebrate fossil record.)

Rensch, B. (1960) *Evolution above the Species Level.* New York, Columbia University Press, 419 p. (A treatise on the evolution of higher taxa with references to many German articles.)

Schopf, T. J. M., ed. (1972) *Models in Paleobiology.* San Francisco, Freeman, Cooper, and Co. 250 p. (A collection of paleobiological papers, many of which are of an evolutionary nature.)

Simpson, G. G. (1953) *The Major Features of Evolution.* New York, Columbia University Press, 434 p. (A classic discussion of rates, trends, and patterns of evolution.)

Valentine, J. W. (1973) *Evolutionary Paleoecology of the Marine Biosphere.* Englewood Cliffs, N.J., Prentice-Hall, 511 p. (A general review of ecological and evolutionary theory as applied to marine life.)

CHAPTER 12

Biogeography

Biogeography is the study of the geographic distribution of plants and animals. It is sometimes considered to be a branch of ecology. The distinction between biogeography and ecology proper is quite arbitrary. It is a matter of scale, and there is no clear-cut division between a local pattern of distribution of the sort commonly studied in ecology and a pattern that is of geographic proportions.

Nowhere is the importance of geology to biology more evident than in the study of biogeography. This is because biogeography is basically a historical subject. First of all, the modern distribution of life on earth has arisen gradually over millions of years. In addition, climates have changed through time, and the geologic record offers evidence of this change. Finally, the physiography of the earth has changed through time. Of special importance here are the revolutionary new concepts of plate tectonics, which are based on convincing evidence that during Phanerozoic time, land masses have moved relative to one another through distances measured in thousands of kilometers. The geological concepts of plate tectonics have provided a new framework not only for study of the earth's crust, but also for study of the present-day distributions of plants and animals. As a consequence, many long-standing beliefs of biogeographers have been discredited in the past decade. It should be appreciated that geology generally contributes less to the study of local ecology than it contributes to the study of biogeographic distributions. Within a small habitat recent environmental history can only rarely be deciphered in sufficient detail to provide insight into the ecology of living biotas. The scale of biogeography, in both space and time, is more amenable to the application of geological data.

In this chapter, we will introduce basic concepts of biogeography and show how ancient geographic conditions are reconstructed and how living and fossil biotas can be related to these conditions. Our use of the title "Biogeography" for this chapter, instead of "Paleobiogeography" or some comparable word or phrase, reflects our view that biogeography is inherently so much a historical subject that it cannot be separated from studies of the geographic distribution of biotas that are entirely extinct.

The Earth's Climate

A basic understanding of present climatic conditions of the earth is essential to the study of biogeography. Ancient climatic conditions can only be reconstructed on the basis of what is known about Recent climates, and the history of climatic change represents part of the framework in which changing biogeographical patterns must be studied. The patterns of circulation of the atmosphere and oceans play a major role in determining climatic patterns. We will begin our discussion with a brief review of these patterns of circulation.

WINDS AND OCEAN CURRENTS

Every point on the surface of the earth receives the same number of hours of daylight in a year. This may at first seem surprising, but the explanation is actually quite simple. Near the equator daylength changes little from season to season. Near the poles, short winter days are balanced by long summer days. Here, then, we have an explanation for increasing seasonality toward the poles. On the other hand, because number of hours of daylight does not vary with latitude, we do not have an explanation for the temperature gradients from the equator to the poles. These gradients exist because sunlight strikes the earth at a lower angle near the poles than near the equator. Thus, a beam of light of a given cross-sectional area is spread over a larger area of the earth's surface near the poles, providing less heat per unit area.

Patterns of rainfall relate in part to the local distribution of temperatures. Let us consider first what happens at the equator. Heavy rainfall in tropical regions results from the rising of warm equatorial air, which upon cooling releases its moisture. To understand what else happens in the real atmosphere, we can simplify things. Imagine a nonrotating earth with a smooth, homogeneous surface and a sun circling it to spread warmth around its entire surface. Then ascent of air near the equator and descent of air near the poles would produce a rather simple atmospheric pattern (Figure 12-1).

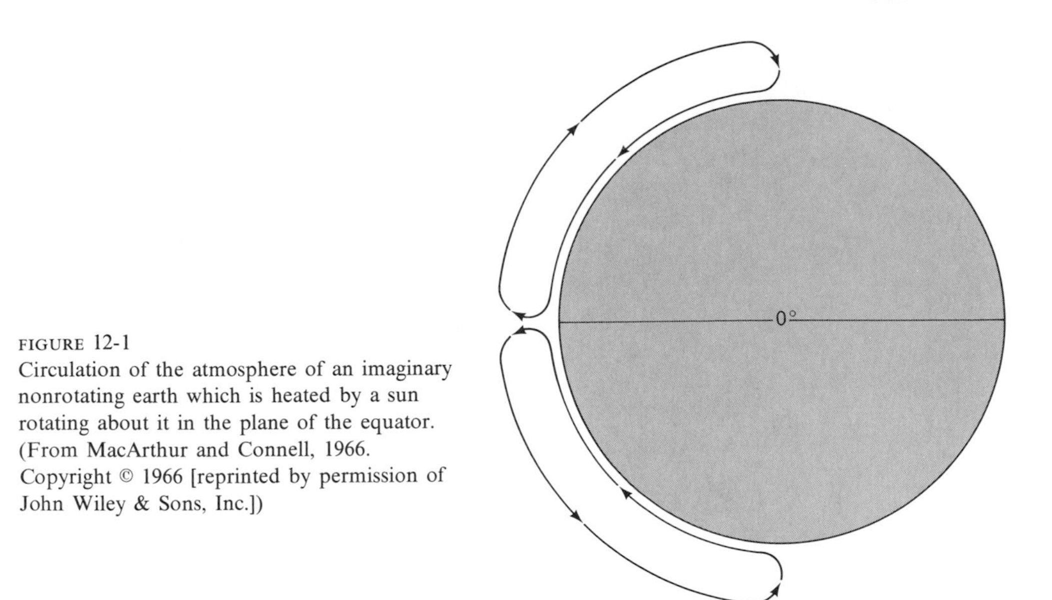

FIGURE 12-1
Circulation of the atmosphere of an imaginary nonrotating earth which is heated by a sun rotating about it in the plane of the equator. (From MacArthur and Connell, 1966. Copyright © 1966 [reprinted by permission of John Wiley & Sons, Inc.])

Not surprisingly, the more complex conditions of the real earth produce a more complex pattern of atmospheric motion. Not only does the atmosphere tend to be cooler near the poles, as the model just described would predict, but at any given latitude it tends to be cooler with increased elevation above the earth's surface, which is warmed by the sun. For this reason, air masses often drop their moisture on mountains, over which they naturally tend to rise. This commonly leaves a "rain shadow" of dry conditions on the leeward side of mountain ranges. Further complexity is added by the real earth's rotation, which has profound effects upon atmospheric circulation. The primary effect is production of the Coriolis Force. This force is most easily envisioned for a current of air or water passing from pole to equator. As a current flows southward from A (Figure 12-2) the earth rotates, so that instead of arriving at point B, the current moves to point B′. The current follows a curved path, as observed from a point on the earth's surface, because the surface speed of the earth increases from pole to equator. Just as currents in the Northern Hemisphere tend to curve toward the right, currents in the Southern Hemisphere tend to curve toward the left. The Coriolis Force in part determines the orientation of the trade winds (Figure 12-3). These winds are produced as air rushes toward the equator to replace the warm air that is rising there. The trade winds do not travel directly equatorward but, because of the Coriolis Force, curve toward the west. Air rising from the equator elevates the upper atmosphere so that at any given altitude there is more air pressing down near the equator than at higher latitudes. Therefore, the air rising at the equator spreads north and south. It is deflected by the Coriolis Force, and part of it

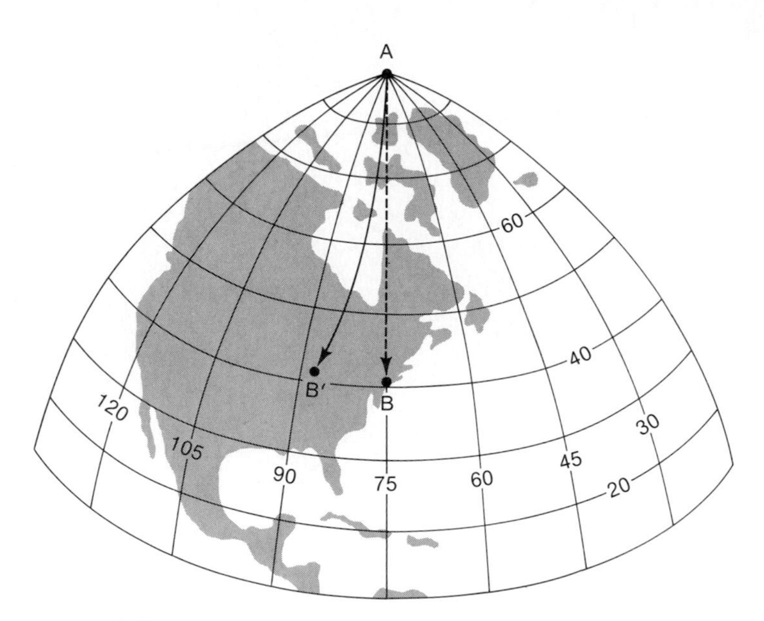

FIGURE 12-2
Effect of the Coriolis Force. An imaginary current flowing south from the North Pole follows the solid line to B′, rather than the dotted line to B, because of the earth's rotation. (From Strahler, 1971. Copyright © 1971 by Arthur N. Strahler [reprinted by permission of Harper & Row, Publishers, Inc.])

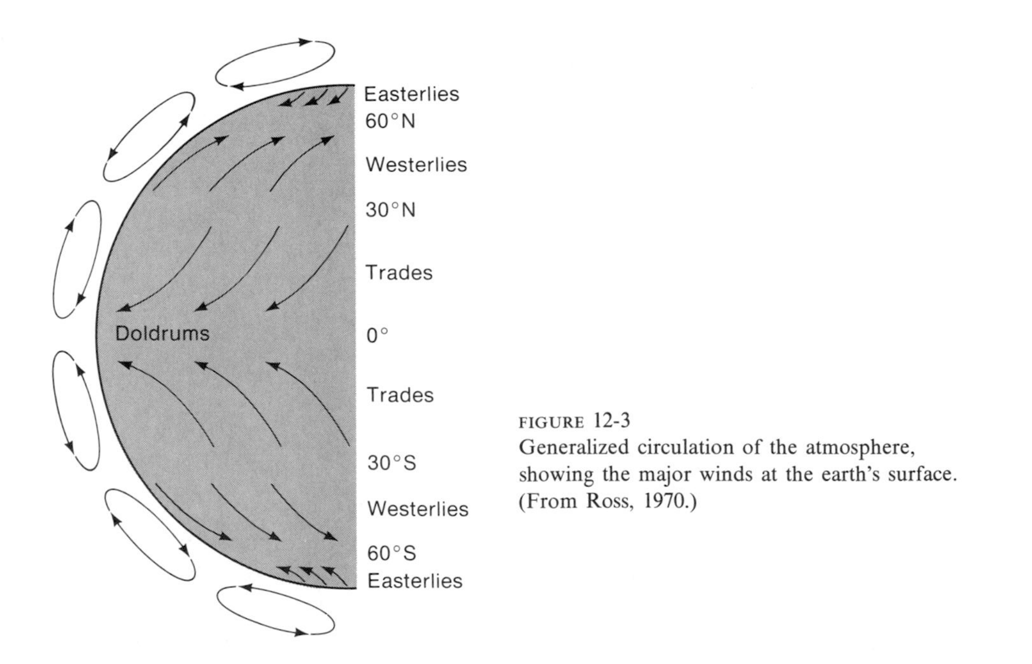

FIGURE 12-3
Generalized circulation of the atmosphere, showing the major winds at the earth's surface. (From Ross, 1970.)

descends at about 30° N or S latitude, to move back to the equator as trade winds. This completes a gyre, or circular path. As one might expect, some of the air descending at about 30° N or S latitude does not follow the equatorward path, but spreads toward the poles. The Coriolis Force turns the resulting winds, which are therefore known as westerlies. Air of the westerlies is heated again and rises at about 60° N latitude, becoming part of another gyre. In complementary fashion, air is cooled at the pole and descends toward the equator to form still another gyre. Because the winds of this gyre are deflected toward the west and therefore come from the east, they are known as easterlies. Air tends to move about the poles from west to east. It moves in the opposite direction at the equator, where the northern and southern trade winds meet.

Winds are of prime importance in determining the pattern of ocean circulation. Large-scale winds passing over the ocean move surface waters great distances (Figure 12-4). The trade winds produce north and south equatorial currents that flow westward. Water that these currents pile up on the eastern sides of continents returns eastward as an equatorial countercurrent. The trade winds carry the equatorial currents northward along the east side of continents and the westerlies turn their continuations eastward, back across ocean basins. A familiar current of this type is the

FIGURE 12-4
Major surface currents of the modern oceans. (After Ehrlich, Ehrlich, and Holdren, 1977. W. H. Freeman and Company. Copyright © 1977.)

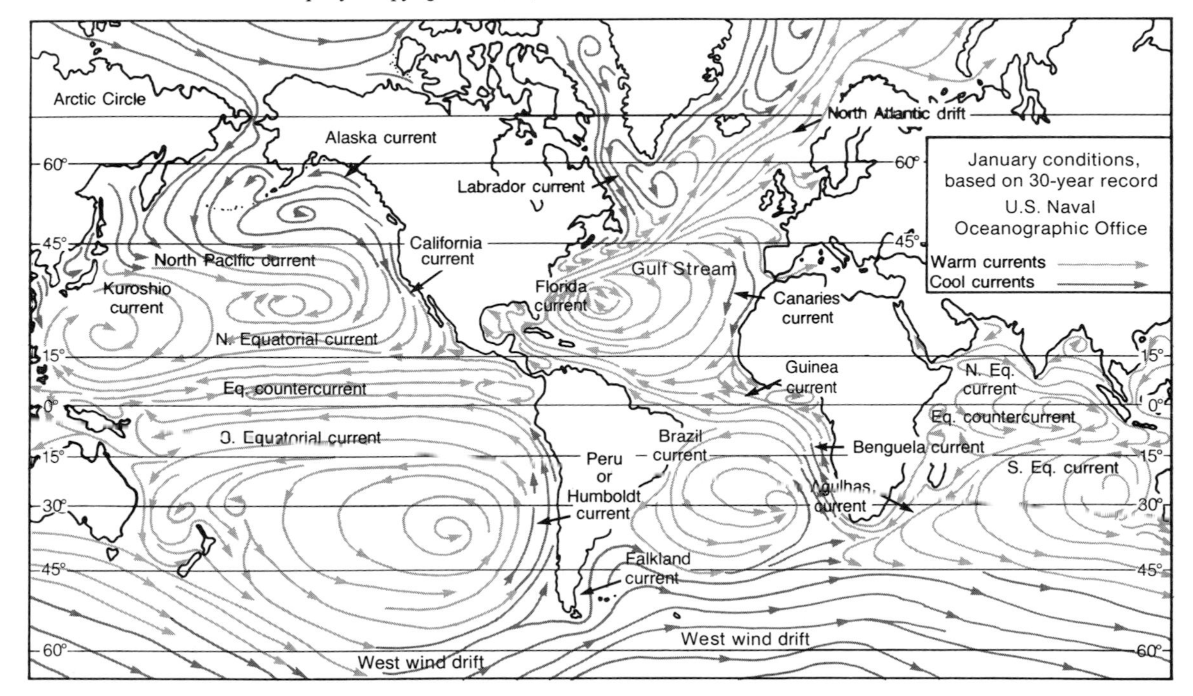

Gulf Stream, which carries warm seawater across the Atlantic. The Gulf Stream is largely responsible for the unusually equable climate of Great Britain. Although lying at the same latitude as northern Newfoundland, southwest England has a climate suitable for growing many types of palm trees. The gyres of a major ocean basin pass southward on the western sides of continents, carrying cold water back to the equatorial current. At its poleward extremity a gyre becomes part of the west wind drift. In the Southern Hemisphere, where west wind drifts of the various ocean basins are largely unobstructed by land masses, they join to produce a circular current around Antarctica (Figure 12-5).

FIGURE 12-5
Antarctic circumpolar current. (From Strahler, 1971, copyright © 1971 by Arthur N. Strahler [reprinted by permission of Harper & Row, Publishers, Inc.]; after Sverdrup, 1942.)

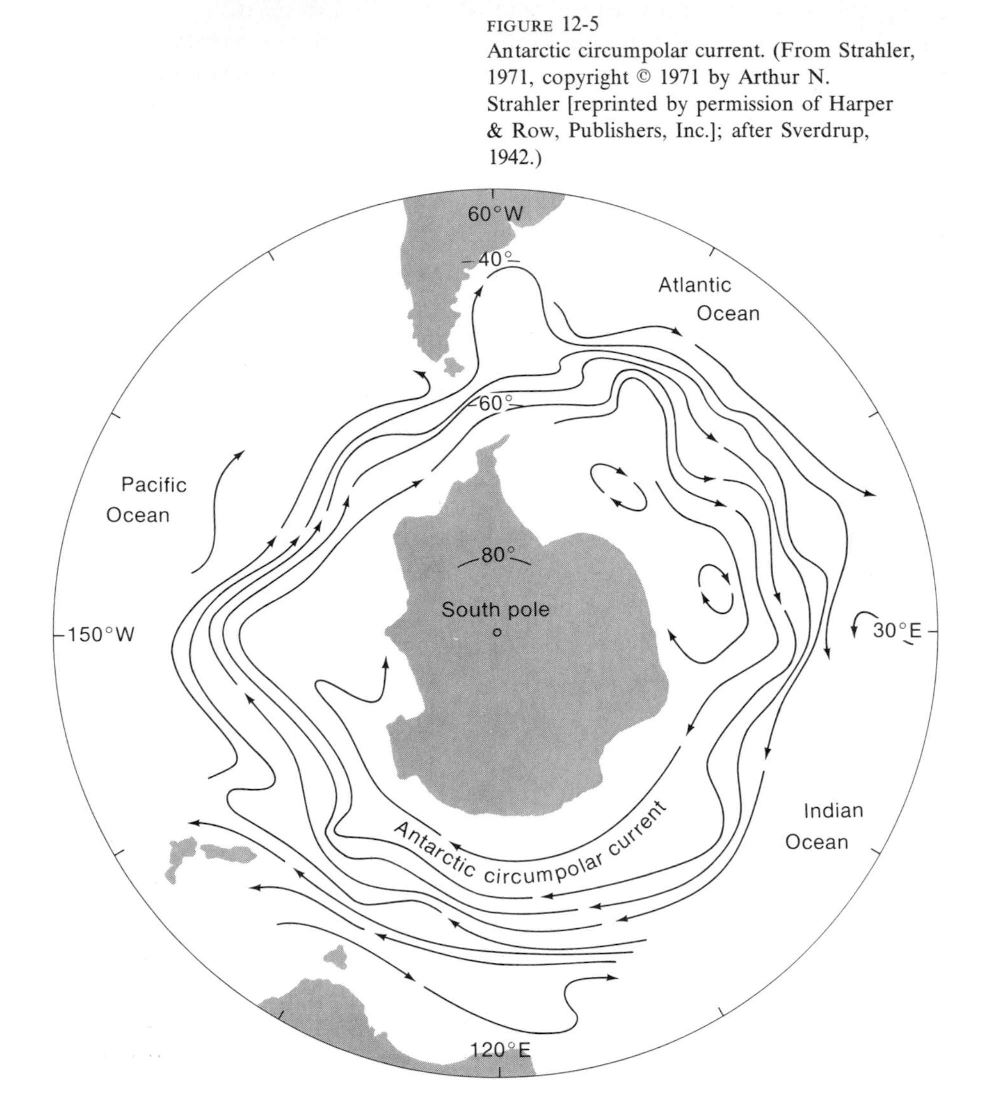

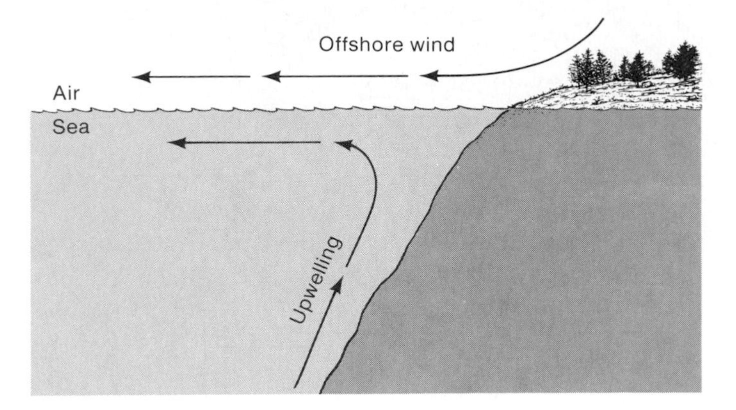

FIGURE 12-6
Diagrammatic illustration of upwelling produced by offshore winds along the coast of an ocean.

Major currents also flow at depths of a few hundred or thousand meters. Such currents remain poorly understood, and we will omit them from our discussion. The major currents of the deep sea are simpler and, for our purposes, more important to understand. Near both poles, cold water descends to the floor of the deep sea. Water sinking from Antarctica is particularly cold. It spreads northward along the seafloor past the equator and well into the Northern Hemisphere, displacing less dense, warmer water as it flows. Above it, slightly warmer water that has descended from the arctic regions surrounding the North Pole spreads into the Southern Hemisphere. This submergence of polar waters is what maintains near-freezing temperatures in the deep sea. It also supplies the deep sea with oxygen, without which most benthos could not survive.

Other kinds of vertical water movement are also of special significance. Locally, offshore winds and seaward movement of longshore currents produce upwelling of water along the continents (Figure 12-6). Upwelling is especially important where the continental slope is steep, because here deep water, rich in nutrients released by bacterial decay, rises to the surface. Passage of such water into the photic zone, produces high productivity and biomass at all levels of the food web. Commercial fishing is often most successful in areas of large-scale upwelling.

TYPES OF CLIMATE

Regional climates are intimately related to the large-scale current systems that we have briefly described. The distribution of regional climates, according to one classification, is shown in Figure 12-7. We have already explained that the wet equatorial belt (climate 1) exists because equatorial air is heated near the earth's

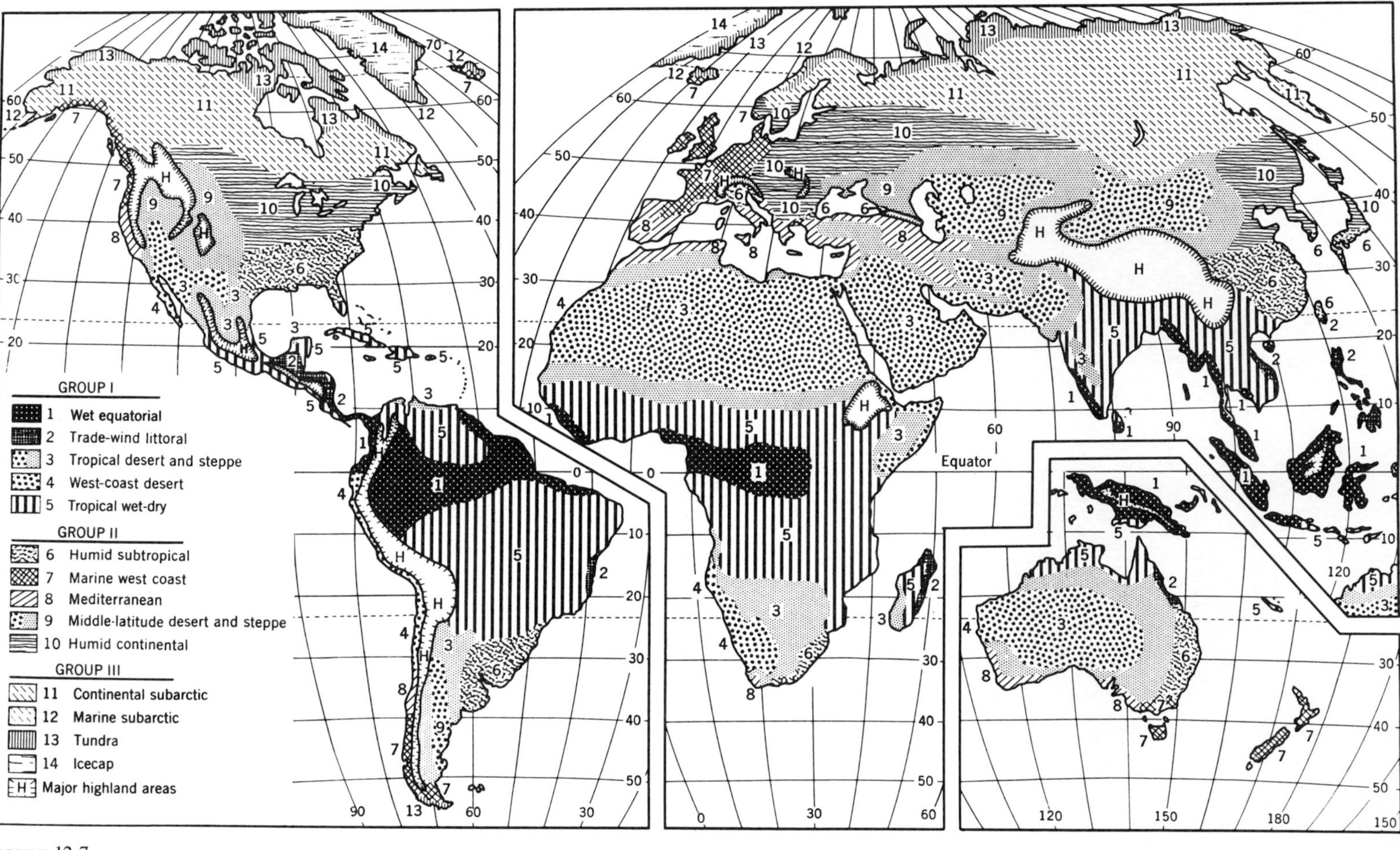

FIGURE 12-7
The distribution of climates in the modern world. (From Strahler, 1971. Copyright © 1971 by Arthur N. Strahler [reprinted by permission of Harper & Row, Publishers, Inc.])

surface, rises, and loses its moisture. Tropical deserts occur in belts 20° to 30° from the equator. Here cool air descends (Figure 12-3) and as it warms, picks up moisture rather than losing it. Other deserts and semiarid grasslands known as steppes (climate 4) are found at higher latitudes (30°–50°) in the interiors of continents where prevailing westerly winds have ascended upon encountering mountain ranges to the west and lost their moisture. The Great Basin of the United States is an area of this type. There are not sufficient expanses of mid-latitude continental areas in the Southern Hemisphere to produce such deserts. On eastern coasts of continents, between the wet equatorial belts and about 25 to 30° N or S latitude, narrow zones of high rainfall (climate 5) are found. These are areas where the moist trade winds rise over hills or mountains and are cooled as they sweep inland. The southeastern sections of North America and Asia and comparable sections of continents of the Southern Hemisphere (climate 6) are relatively warm and humid for this reason. These conditions result from the buildup of air that has risen from the equator and spilled poleward to produce high pressure centers. From these centers warm, moist air from oceanic areas spreads poleward over adjacent continental areas and loses its moisture as it cools. On the west coasts of continents at middle latitudes (35°–65°) moist conditions also prevail (climates 7 and 8). Here westerlies coming from the ocean often rise as they reach the land and drop their moisture. The southern parts of such regions are said to have Mediterranean climates, with moderate annual temperature ranges, wet winters, and dry summers.

The west coast of North America illustrates how ocean currents modify the effects of the westerlies. The currents in these regions flow toward the equator (Figure 12-4). Having circled around from the tropics, the ocean currents are warmer than the land in areas like British Columbia. Hence westerlies drop an enormous amount of rain throughout the year. Farther south, the land becomes warmer and the ocean current becomes cooler because, as it curves gradually westward, it pulls water away from the coast and this is replaced by upwelling cold water. The ocean currents flow on south as far as southern California: there nonmountainous terrain is warmer year-round than adjacent water, and westerlies pick up moisture as they pass inland to produce arid conditions. Northern California is in an intermediate climate: in summer the land is warmer than the water, so that there is little rain; in winter the situation is reversed, and there is continual rain.

Biotic Distributions

BIOMES AND PROVINCES

Not surprisingly, the worldwide distribution of certain types of terrestrial vegetation tends to reflect the distribution of climatic conditions (compare Figure 12-7 with Figure 12-8). This relationship suggests that fossil plants should be useful indicators of ancient climatic conditions, and indeed they are. And to a considerable degree, the distribution of land animals follows the distribution of plant associations.

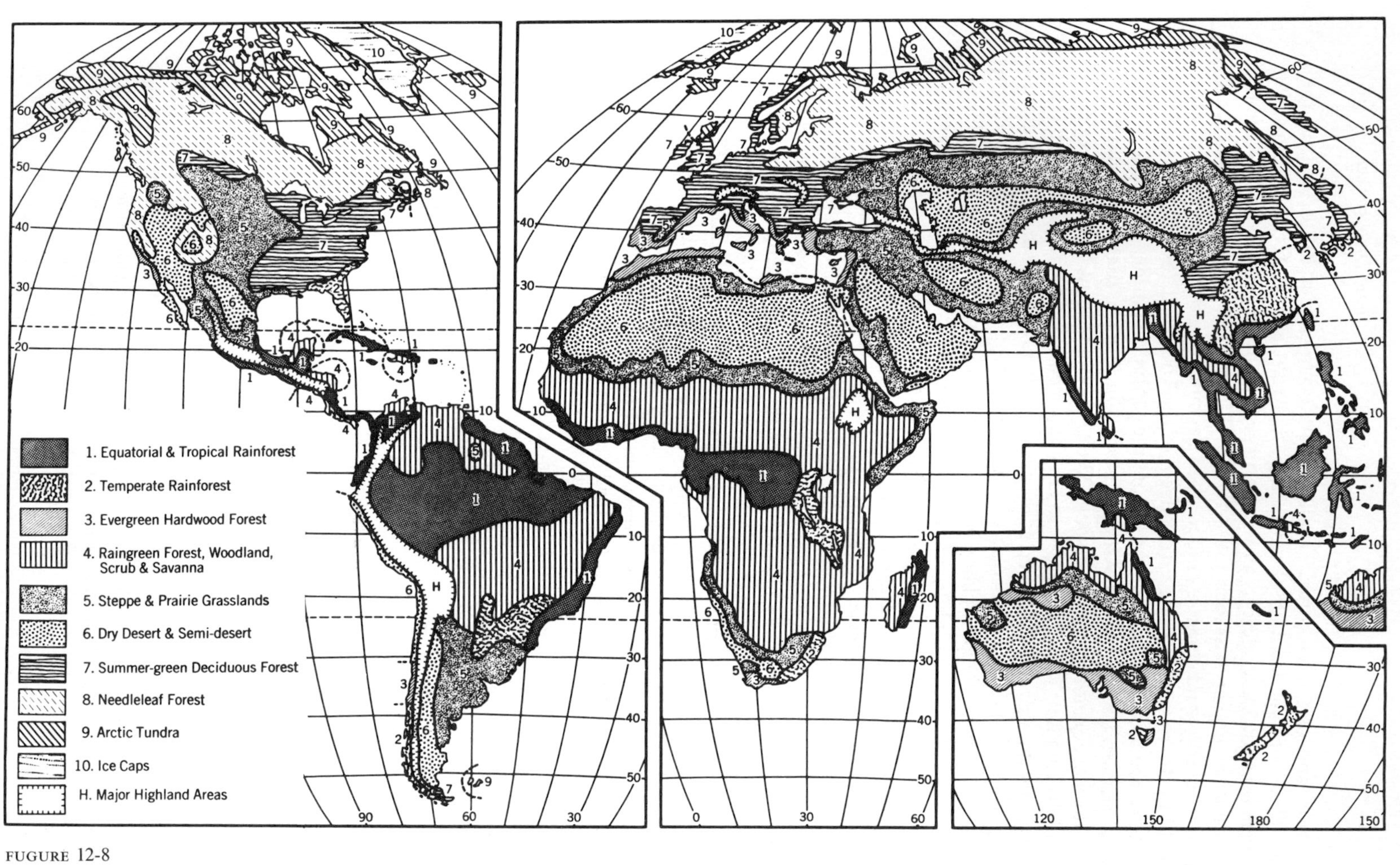

FUGURE 12-8
The general distribution of vegetation in the modern world. (From Strahler, 1971. Copyright © 1971 by Arthur N. Strahler [reprinted by permission of Harper & Row, Publishers, Inc.])

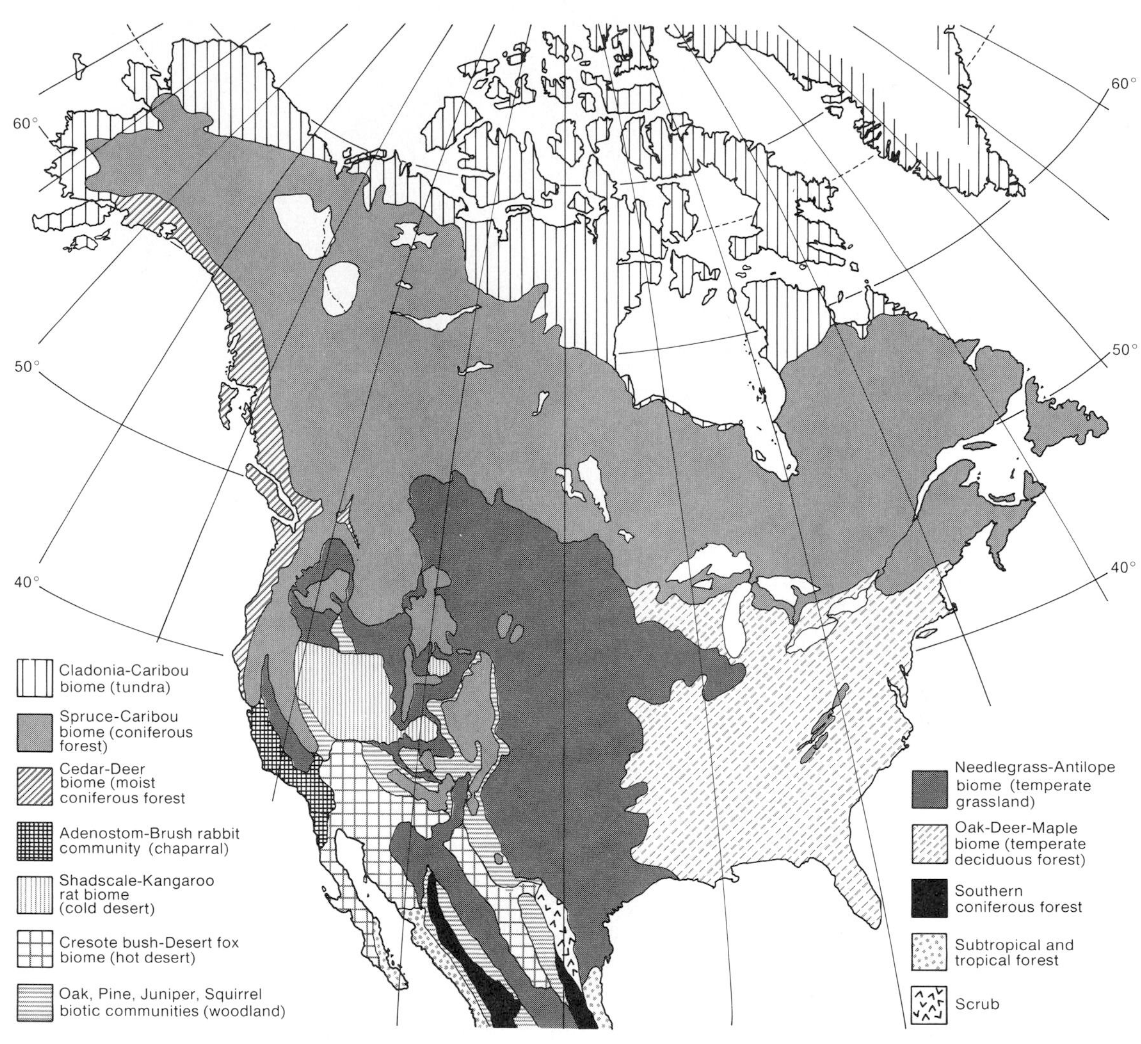

FIGURE 12-9
Distribution of biomes in North America. (From Shelford, 1963. The University of Illinois Press. Copyright © 1963.)

Broadly homogeneous biotic associations occurring over large areas of land have become known as ***biomes.*** North American biomes recognized by Shelford are shown in Figure 12-9. Within each biome various smaller communities can also be recognized, depending on the scale of analysis. It should be understood that even though particular classes of climate and vegetation occur on widely separated land masses (Figures 12-7 and 12-8), particular biomes do not. This is because there exist

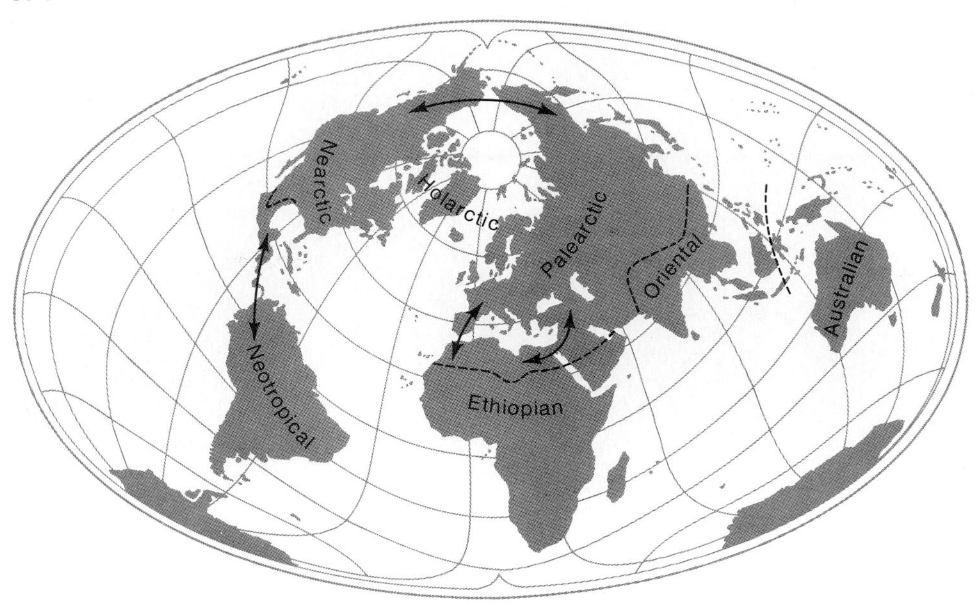

FIGURE 12-10
Distribution of modern zoogeographic regions of the terrestrial realm. Arrows indicate intercontinental similarities. (From Colbert, 1971.)

barriers to the dispersal of various kinds of organisms. Certain taxa that arise and diversify in one region may be confined to that region, while other taxa arise to assume similar roles in other areas.

In some instances, the distribution of living associations of species makes little sense in terms of present-day geographic conditions. Here we begin to see the importance of the historical aspect of biogeography. We must trace geographic conditions, including climatic features, into the past and learn what changes have taken place if we are to understand how modern biotas have attained their present distributions.

Zoogeographic regions of the modern terrestrial realm are shown in Figure 12-10. The divisions shown came to be recognized largely on the basis of the distributions of living animals, with little knowledge of past histories. Much the same classification is used for terrestrial floras. All classifications of this type are subjective. Their categories intergrade.

The marine realm presents a distinct set of problems. Pelagic species tend to be widely distributed. Some benthic species are also widely distributed, but others are limited by barriers of various sorts. Barriers are especially evident for some nearshore species, but other nearshore species have planktonic larvae which, as we have already seen (Figure 9-3), can sometimes travel for thousands of miles. On page 205 we discussed biogeographic provinces. In Figure 9-4 we presented alternative classifications of biotic provinces for shallow water benthic marine life of the Atlantic coast of North America. It was pointed out that many boundaries are placed at physiographic barriers to water movement.

LATITUDINAL GRADIENTS IN TAXONOMIC DIVERSITY

One of the most conspicuous phenomena of biogeography is the tendency for particular taxa to display increases in species diversity toward the equator. The result, of course, is an increase in average species diversity of communities in the same direction. Examples of latitudinal gradients in diversity are displayed in Figure 12-11. It should be borne in mind that there are exceptions. A few taxa have their centers of distribution in very cold regions and decline in specific diversity toward the equator. Comparison of diversities is not simple, and, in fact, can never be accomplished with complete consistency. The size of an area sampled and the variety of habitats included will affect the number of species recorded.

The origin of diversity gradients has been widely debated. An important point to bear in mind is that some hypotheses that have been advanced to explain the gradients are based on the assumption that environments tend to be saturated with species. The corollary of this assumption is that saturation levels decrease toward the poles. Other hypotheses are not based on an assumption of environmental saturation, but view regional diversity as a product of the history of a region.

One suggestion is that most higher taxa arise in the tropics and only gradually invade colder regions, because of the severity of living conditions that colder regions offer. Another is that rates of speciation increase toward the tropics. Another is that, though environments of high latitudes may potentially support as many species as environments of tropical regions, diversification at high latitudes is more frequently interrupted by large-scale environmental catastrophies so that the potential carrying capacity is never approached (Figure 12-12). None of these ideas consider saturation of the environment to be a major factor. An idea that does invoke the concept of saturation is that taxonomic diversity increases toward the equator because the physical environment increases in complexity toward the equator.

Actually, many of the arguments presented to explain high diversity in the tropics only offer explanations for why a certain basic form of increased diversity is compounded in the tropics. For example, it may be reasonable to claim that more niches are potentially available for animals in the tropics because there are more plant species to provide food and refuge there, but then we must explain why there are more plant species. Similarly, it can be argued that there are more predators in the tropics than in high latitudes, allowing for more prey species to coexist (page 279), but then we must explain why there are more predators in the tropics. It seems apparent that latitudinal gradients in diversity reflect a variety of interrelated factors that are not yet fully understood.

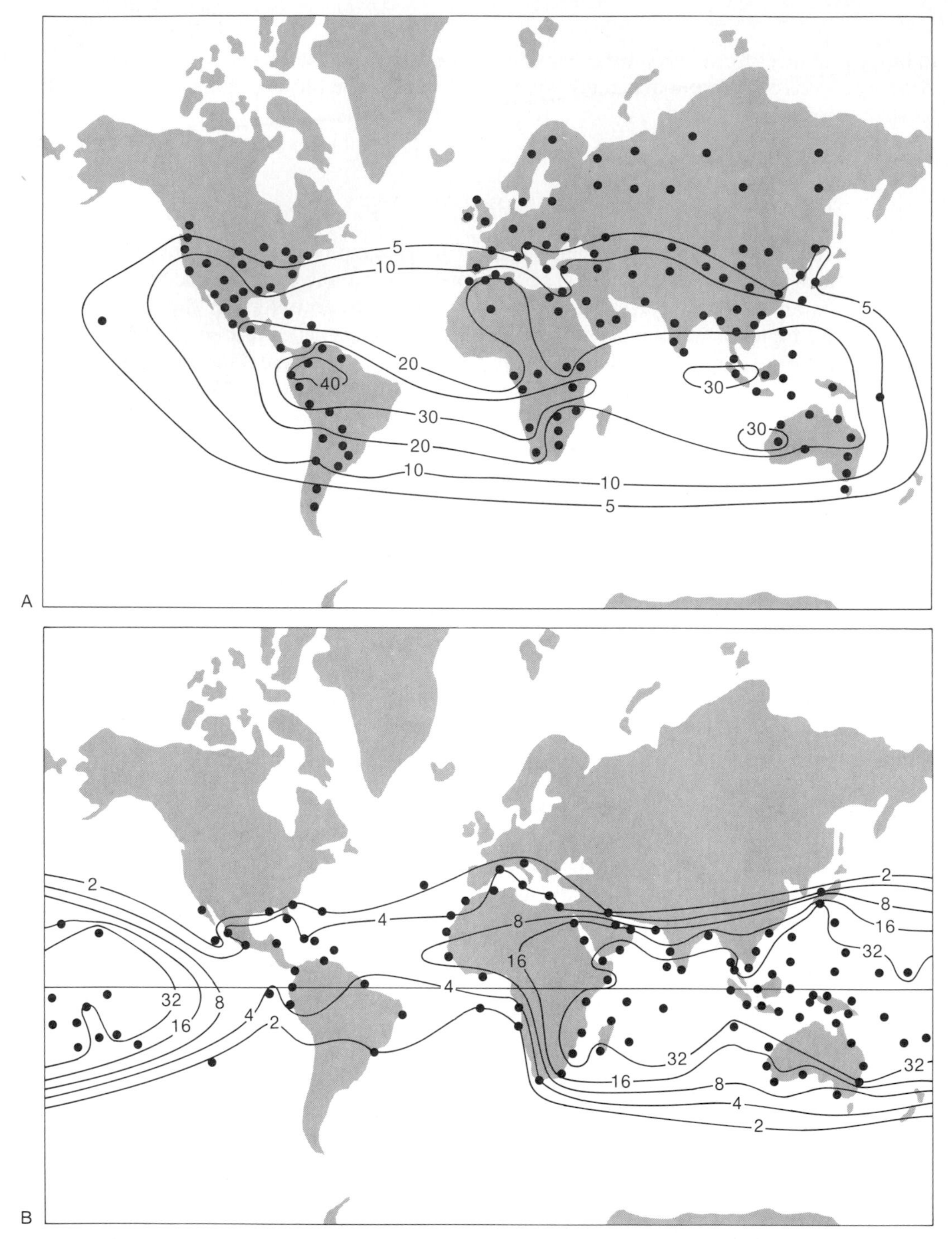

FIGURE 12-11
Diversity gradients. A: Contours showing diversity (number of genera) of Recent lizards throughout the world. Dots represent localities for which data were compiled. B: The same kind of plot for diversity of cypraeid gastropods. (From Stehli, 1966.)

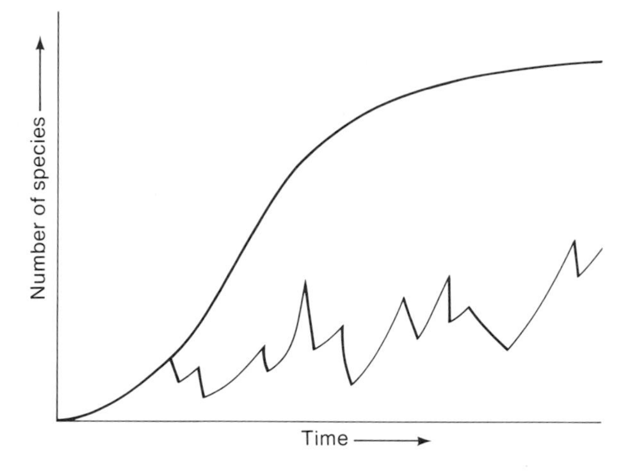

FIGURE 12-12
Hypothetical diversification of a taxon at high latitudes. The lower, jagged curve represents the idea that diversification is frequently interrupted by catastrophic extinctions. The upper, smooth curve represents the hypothetical course of diversification for the same taxon suffering no catastrophic extinctions. (Modified from Fischer, 1960.)

Climates of the Past

The recentness of the earth's emergence from the last Pleistocene glacial episode has left even many laymen with a strong impression of the great variability of climate in geologic time. Still, it is not easy to evaluate climatic changes that occurred before the Pleistocene. We have at our disposal a variety of tools for reconstructing ancient climatic conditions, but they provide us with only a partial picture. The following sections will introduce some of these tools.

BOTANICAL EVIDENCE

Plants, especially angiosperms (flowering plants), have been particularly useful in the reconstruction of Cretaceous and Cenozoic climatic conditions. We noted earlier the close correspondence in the modern world between type of vegetation and climate (Figures 12-7 and 12-8). Certain living genera and families that have both restricted climatic occurrences and long geologic histories represent useful "paleothermometers." The palms are a good example. Long ago it was also noticed that certain features of leaf morphology are useful as climatic indicators regardless of taxonomic occurrence. The utility of leaf characters has been summarized by Dorf (1970). The most useful feature appears to be the nature of the leaf margin. ***Nonentire*** (toothed, indented, or lobed) margins are most common in cold climates. ***Entire*** (smooth) margins are most common in the tropics (Table 12-1). Also more common in the tropics are large leaves, thick leaves, and leaves with ***pinnate*** (feather-like) venation or vein arrangements, as opposed to ***palmate*** (radiating) venation. ***Compound*** leaves—leaves in which several leaflets grow as a cluster—are more common in the tropics

TABLE 12-1
Percentages of Leaves with Entire Margins in Floras Living in Various Climatic Zones

Climate and location	*Trees*	*Shrubs*
COLD TEMPERATE		
East-central North America	10	37
Rocky Mountains	0	40
Central Russia	0	36
England	3	37
WARM TEMPERATE		
Southeastern United States	36	54
Los Angeles	18	58
Spain	13	59
East-central China	38	52
SUBTROPICAL AND TROPICAL		
Florida Keys	84	83
West Indies	88	71
Hong Kong	73	71
Brazil	90	87

SOURCE: Bailey and Sinnott, 1915; 1916.

(Table 12-2), but the range of variation within climatic zones is rather large, which makes this feature less useful than percentage of venation. Finally, many tropical species have an elongated lobe, called a drip-point, at the tip of the leaf. This structure serves to draw water from the surface of the leaf for rapid drying in the tropics, where rains fall with great frequency.

When climatic belts shift, floras tend to shift with them. In some cases, of course, extinction may occur instead. This may happen if the shifting of climates is too abrupt or severe for successful floral migration or if the required climatic conditions are

TABLE 12-2
Percentages of Compound Leaves Found in Australian Rain Forests in Various Climatic Zones

Climatic zone	*Percent*
Cool temperate	0
Warm temperate	10–30
Subtopical	30–40
Tropical	15–50

SOURCE: Webb, 1959.

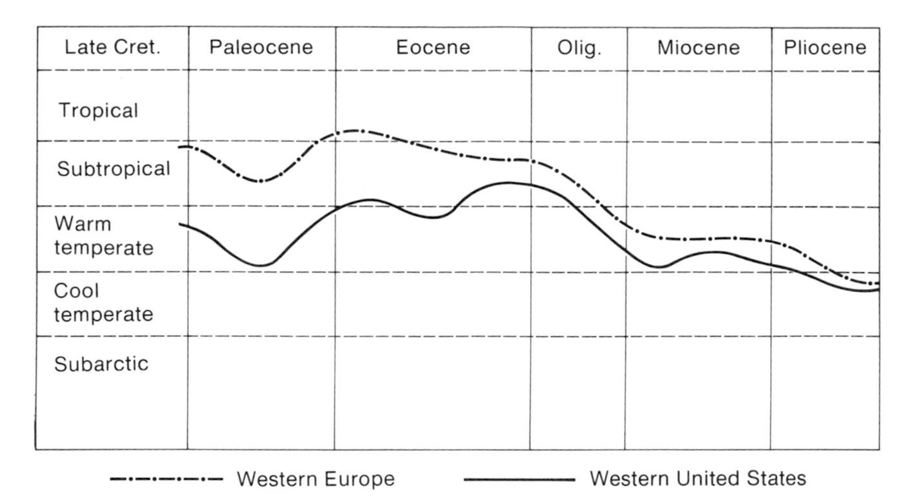

FIGURE 12-13
Inferred climatic regimes of western Europe and the western United States during the Cenozoic. (From Dorf, 1964. Copyright © 1964 [reprinted by permission of John Wiley & Sons, Inc.])

eliminated altogether. The waxing and waning of temperature within continents, however, can often be read from the fossil record of leaves. Evaluations of past climates are continually revised. Curves for Western Europe and the Western United States published by Dorf in 1964 are shown in Figure 12-13. A curve produced by Wolfe and Hopkins (1967) provides a more detailed picture of conditions in the Oligocene and Miocene (Figure 12-14). The discrepancy for the Eocene between this curve and that of Dorf remains unresolved. One thing is clear: climates have undergone a net cooling since the early Cenozoic.

Even for Triassic and Jurassic deposits, which predate the origin of angiosperms, floras are employed in the reconstruction of ancient climates. Of value here are certain ferns belonging to modern taxa and cycads. Cycads are largely restricted to tropical and subtropical areas today (Figure 12-15) and probably lived in similar conditions in the past. Whereas annual growth rings are conspicuous in the trunks of trees of temperate latitudes, they are weakly developed or absent in humid tropical regions; some Paleozoic plants seem to have existed before the evolution of the physiological mechanisms by which growth rings are produced, but rings have been found in others. Chaloner and Creber (1973) have provided a useful evaluation of the potential of such rings in paleoclimatic reconstruction.

There is an important aspect of floral distribution that complicates efforts to interpret regional climates of the past. This is the tendency of floras at high elevations near the equator to resemble lowland floras of higher latitudes. One of the most

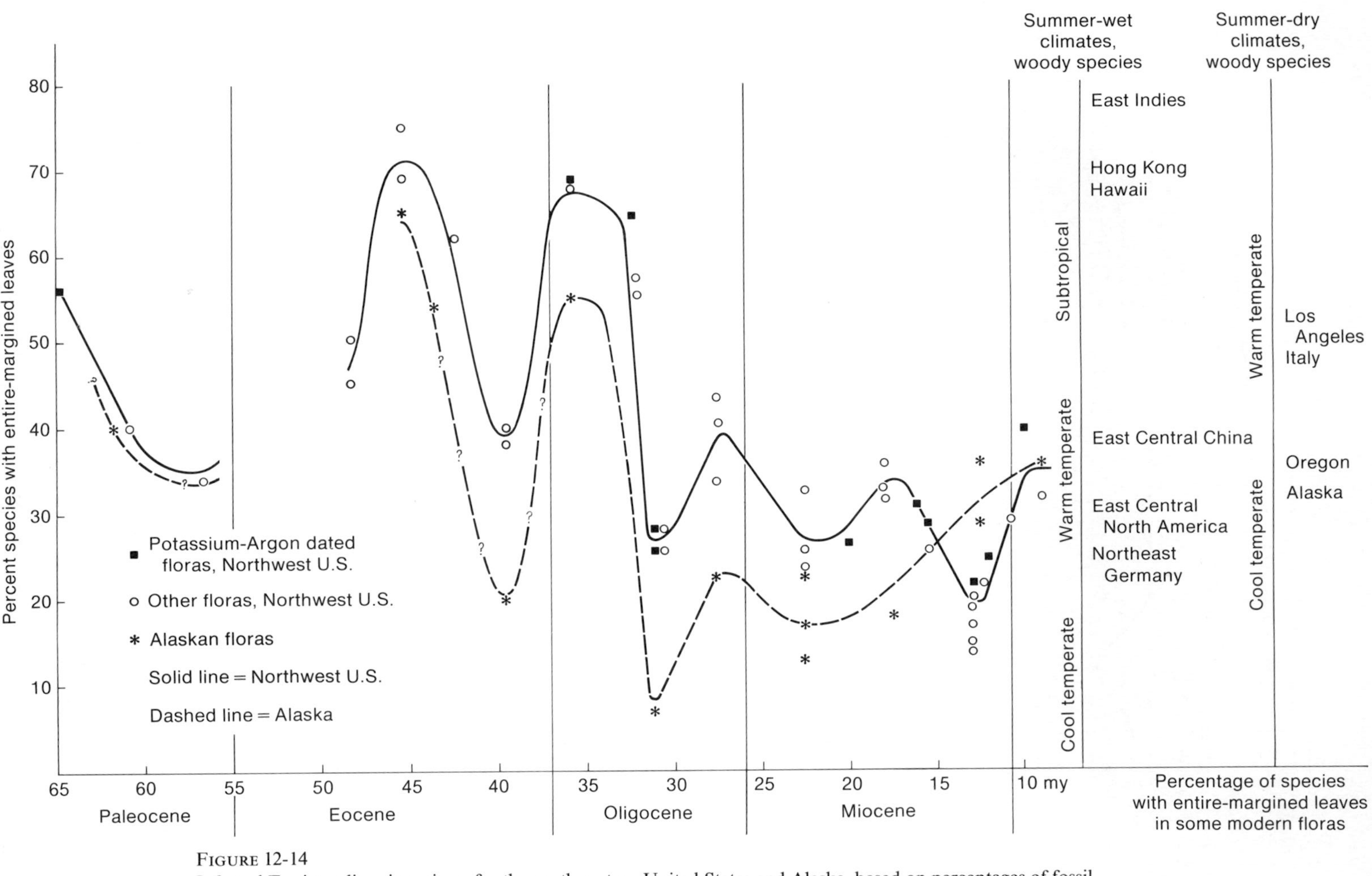

FIGURE 12-14
Inferred Tertiary climatic regimes for the northwestern United States and Alaska, based on percentages of fossil species having entire leaf margins. (Modified from Wolfe and Hopkins, 1967, with new data supplied by J. A. Wolfe.)

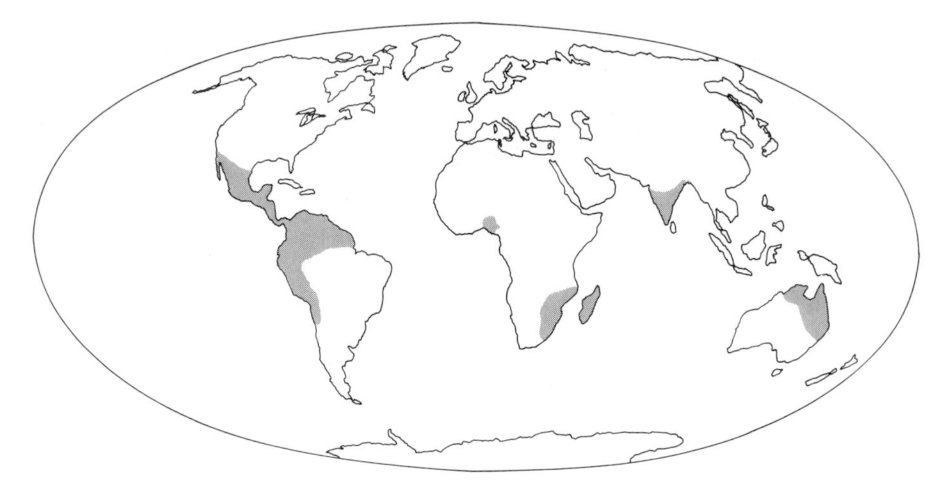

FIGURE 12-15
The global distribution of living cycads. (From Cox, Healey, and Moore, 1973.)

obvious examples is the increase in the percentage of conifers at low latitudes and increased elevation. Fortunately, the occurrence of most depositional environments in lowland areas reduced the degree of confusion that the effects of elevation contribute to the fossil record.

Not only leaves, but also pollen and spores, which can often be identified to the level of genus, are of great value as paleoclimatic indicators. Climatic interpretation of the Quaternary (Pleistocene and Recent) is based largely on study of these reproductive structures, which frequently resist decay and tend to occur abundantly in ancient sediments. The study of fossil pollen and spores is termed ***palynology.*** Of course, these structures can be transported by wind and water, but studies of the distribution of modern pollen and spores have shown that this distribution often faithfully reflects the general geographic distribution of modern floras (Davis, 1967). In Figure 12-16, the present locations of major vegetational zones are compared with the distributions of major zones during the Wisconsin, the last glacial stage of the Pleistocene. Pollen data show that during the glacial age there was a dramatic southward shift of isotherms (lines of equal maximum, minimum, or average temperature measured for some time of the year). It should be appreciated that large areas now forming part of the continental shelf of North America were emergent during the Wisconsin and undoubtedly supported terrestrial floras extending seaward from the zones shown in Figure 12-16,B. In general, the Pleistocene floras successfully shifted northward and southward with isotherms rather than going extinct. In fact, during the Pleistocene there seem to have been hardly any extinctions of major components of North American terrestrial floras.

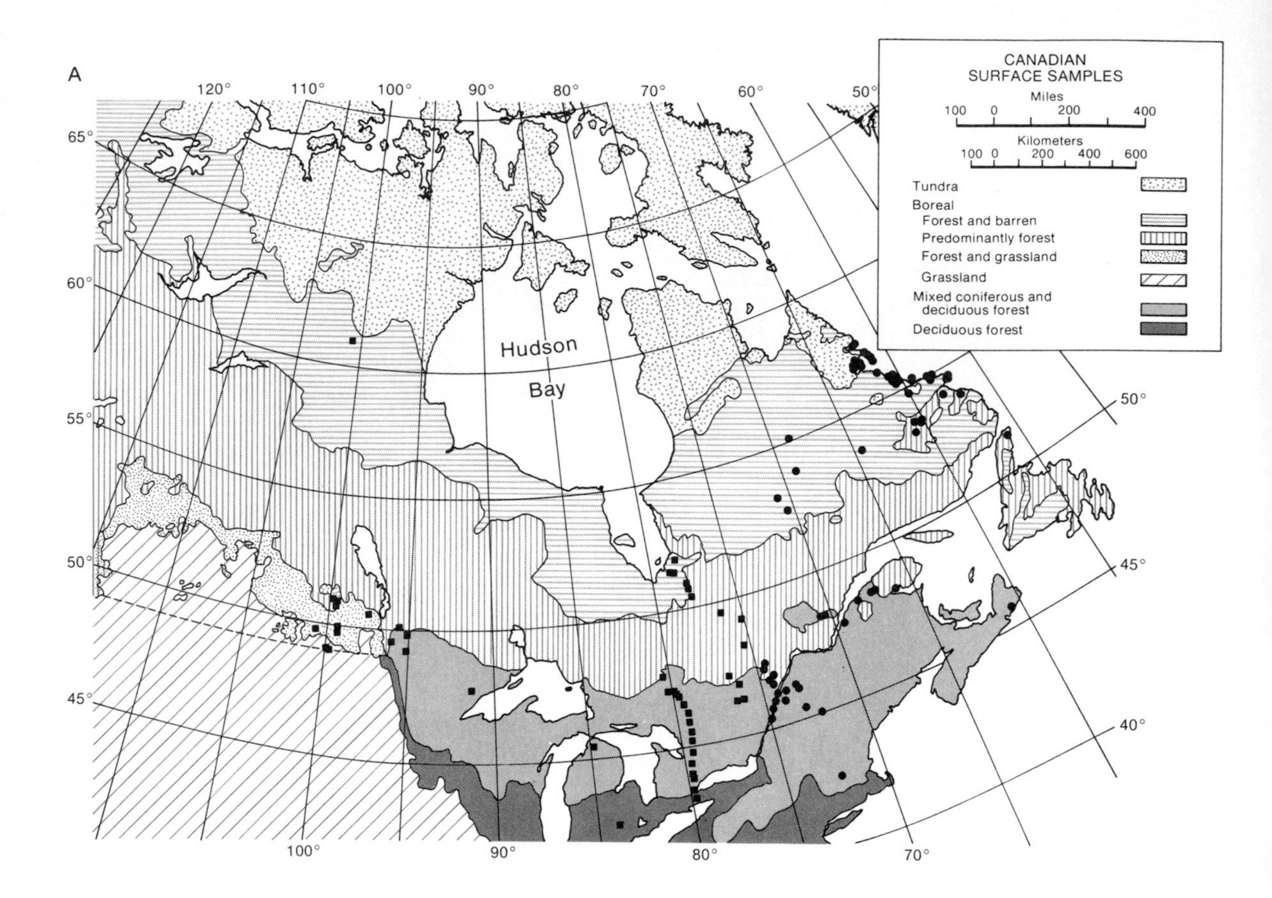

Certain groups of terrestrial animals are also of value in the reconstruction of past climates. For the most part these are taxa like reindeer and musk oxen that seem always to have occupied certain narrow climatic zones.

SEDIMENTARY EVIDENCE

Sedimentology has produced much of the paleoclimatic framework in which fossil biotas are studied. Some of the evidence here comes from deposition on land. For example, a particular suite of sedimentary features characterizes glaciated topography (Flint, 1975). Glacial sediments are especially easy to recognize if associated with other diagnostic features like grooves or striations gouged into bedrock by the movement of boulders frozen into the bases of flowing ice sheets. Not only is the gross fabric and composition of glacial "till" quite characteristic, but it has been discovered

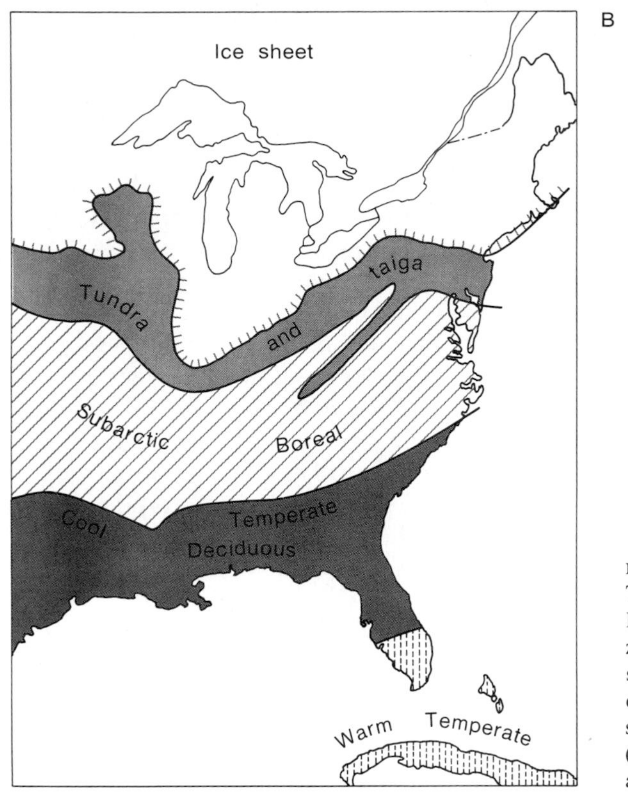

FIGURE 12-16
The use of paleobotanical data to evaluate Pleistocene climates. A: Major vegetational zones of present-day northern North America, showing sites from which pollen data were evaluated by Davis (1967). B: Position farther south of vegetational zones during the last (Wisconsin) glacial stage. (From Dorf, 1959, after Martin, 1958.)

by electron microscopy that stresses beneath glaciers produce a particular kind of surface texture on grains forming the till (Krinsley and Donahue, 1968). Types of soil representing certain climatic conditions are also commonly recognizable in the stratigraphic record. One of these is laterite, a brownish, nodular soil in which relatively insoluble oxides and hydroxides of iron, aluminum, and manganese are concentrated. Laterite forms in moist tropical climates, where heavy rainfall removes relatively soluble salts. In certain warm, semiarid climates, on the other hand, evaporation produces characteristic nodular, layered deposits of calcium carbonate known as caliche. Sandstones formed by lithification of desert dunes are among the most easily recognized sedimentary units that tell us a great deal about climate. Dunes consist of large-scale, cross-stratified clean sands. The orientation of the cross-stratification often reveals the direction of prevailing winds. As discussed earlier in the chapter, global circulation tends to restrict desert conditions to certain parts of the globe. Thus the location of ancient desert dunes and the prevailing direction of the winds that form them can usually be compared to regional conditions predicted from the distribution of land masses and oceans at the time when the dunes formed.

On the average, sediments deposited in aquatic settings are less diagnostic of climates than are terrestrial sediments. Perhaps the most notable exceptions are the evaporites. These are salt deposits of various compositions produced by the evaporation of isolated or semi-isolated bodies of water that may originally have been either fresh or marine. Evaporites, such as halite, anhydrite, and gypsum, tend to form in warm, arid climates. They are sometimes associated with dune deposits, as in the Permian System of the southwestern United States. Limestone and dolomite, the most common sedimentary carbonate rocks of shallow marine settings, accumulate largely in the tropics. In part, this reflects the fact that algae and marine invertebrates secrete calcium carbonate skeletons at much higher average rates here than in colder regions, and the remains of these skeletons, in the form of microscopic needles or larger particles, comprise the bulk of modern carbonate sediments. In addition, carbonate is precipitated more rapidly by bacterial and inorganic processes in the tropics. Still, carbonate sediments do accumulate today in cooler waters where the production of skeletal creatures is unusually high, as in the Irish Sea, so we must be careful in using carbonate rocks as indicators of tropical conditions.

DATA FROM MARINE LIFE

Marine invertebrates are also of some value as indicators of temperature regimes in ancient seas. Perhaps the most useful living group in this regard are the reef-building, or hermatypic, corals, which we discussed in Chapter 10. These colonial animals, which it will be recalled harbor symbiotic algae in their tissues, form reefs only within about 30° of the equator, where the water seldom falls below 18°C. Hermatypic corals survive at slightly higher latitudes, but are unable to flourish sufficiently to build reefs (Figure 12-17). It seems apparent that similar temperature limits have obtained since the Triassic, when hermatypic scleractinian corals arose. Most species of bivalve molluscs and certain other organisms living in waters far enough from the equator to undergo marked seasonal temperature changes exhibit growth banding in their shells. The molluscan shell consists of calcium carbonate elements embedded in an organic matrix, much like bricks embedded in mortar. During the winter, secretion of calcium carbonate is slowed down, and the resulting concentration of organic matrix appears as a thin, darkish band, sometimes referred to as a growth line or ring. Growth rings generally indicate nontropical conditions, but care must be taken in drawing conclusions from a small sample of shells because similar bands, called interruption rings, commonly result from slow growth of an animal after it is physically disturbed. Gastropod molluscs tend to have thicker, more highly ornamented shells in the tropics than in cool regions (Graus, 1974). Vermeij (1976) has related this kind of morphologic gradient in several groups of snails to claw size and strength in predaceous crabs increasing toward the equator.

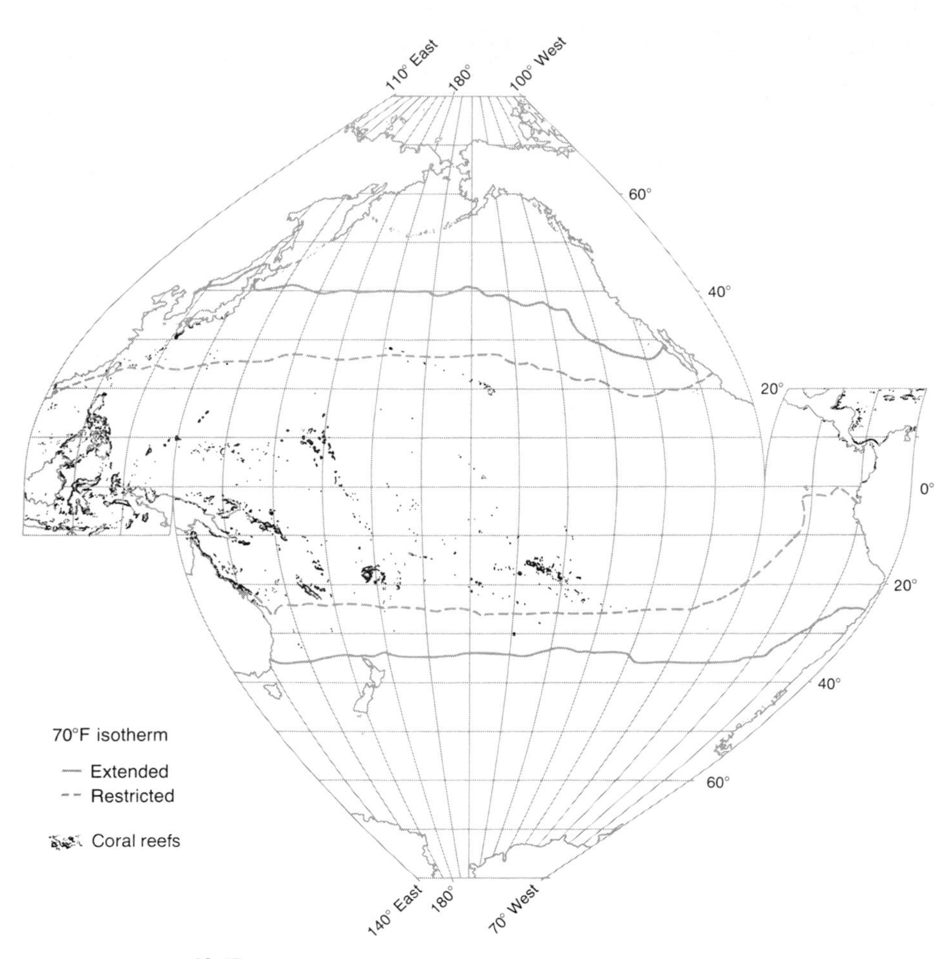

FIGURE 12-17
Latitudinal limits of living coral reefs (dashed lines) and hermatypic corals (solid lines) in the Pacific Ocean. (After Wells, 1957.)

The record of fossil plankton in sediments of the open ocean provides essential information for the reconstruction of Cenozoic climatic conditions. In particular this record sheds light on the complex pattern of waxing and waning of global temperatures that accompanied the buildup and retreat of Pleistocene ice sheets. Rates of evolution and extinction are far too low to be of help here, either in providing a detailed chronology or in offering evidence about temperature changes. In fact, the

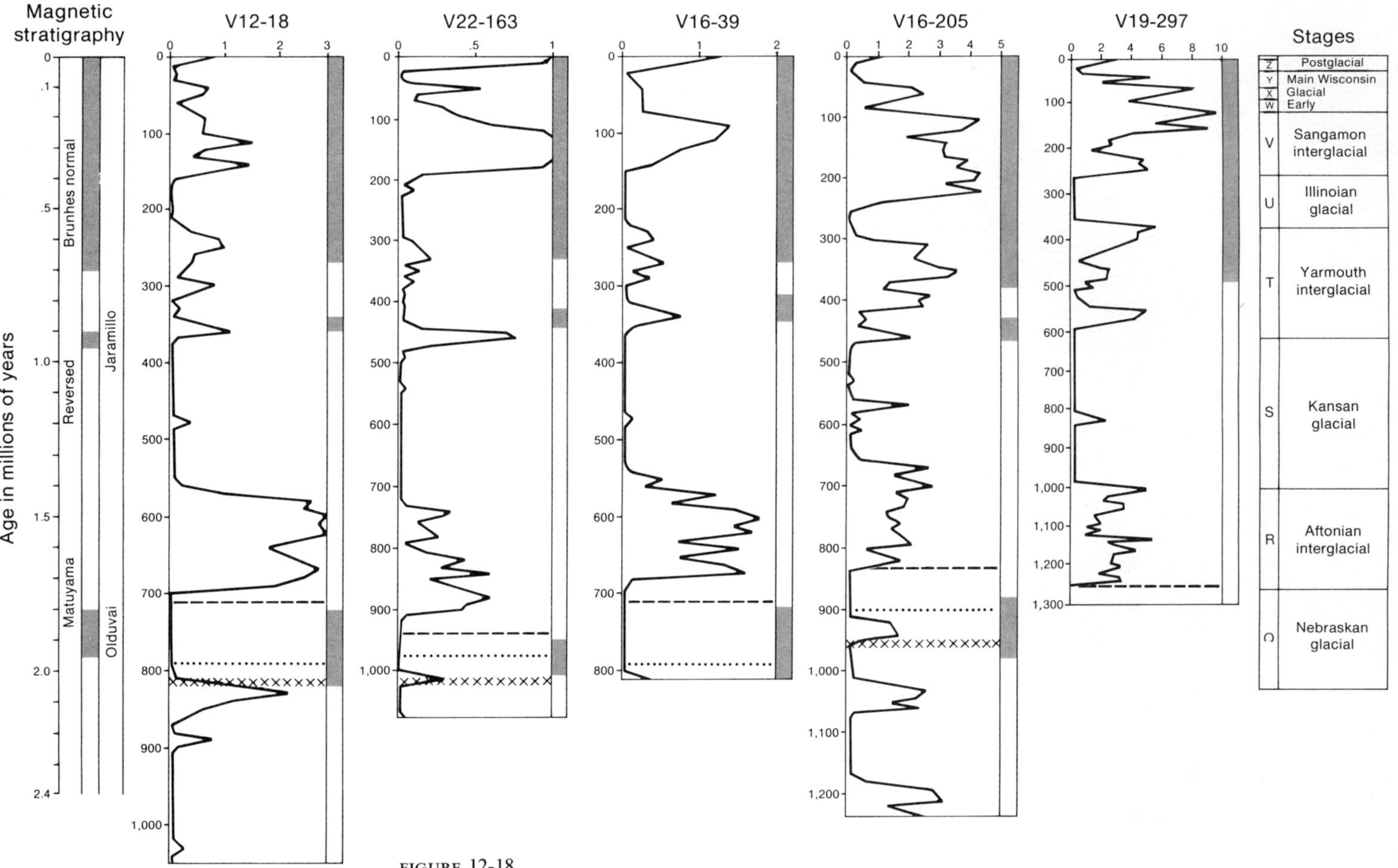

FIGURE 12-18
Ratios of numbers of specimens of *Globorotalia menardii* to total numbers of foraminiferans in five cores (V12-18 through V19-297) of deep-sea sediments from the Atlantic Ocean are shown across the top of each V-column. Scales along the left sides of the ratio plots are in centimeters. Horizontal lines of x's, dots, and dashes are biostratigraphic markers. The scales of the five cores have been adjusted to show correlation of climatic zones, which can be compared to the glacial and interglacial stages shown at the right. (From Ericson and Wollin, 1968. Copyright © 1968 by the American Association for the Advancement of Science.)

lack of appreciable evolutionary change turns out to be useful because observed effects of temperature on certain living species can be projected backward into the fossil record of these species to reveal temperature changes in the earth's climate. One effect of temperature discussed in Chapter 9 is on the coiling direction of planktonic foraminiferans that rain down upon the floor of the deep sea (Figure 9-8). Another effect is upon the geographic occurrence of species. Of particular importance is the planktonic foraminiferan species *Globorotalia menardii,* which is found in equatorial regions. The abundance of tests (shells) of this species relative to those of other foraminiferan species varies in cores of deep-sea sediment in a way that obviously reflects the Pleistocene history of temperature change in the surface water of the overlying ocean (Ericson and Wollin, 1968). During glacial episodes, the relative abundance of the species is very low (Figure 12-18). During interglacial episodes, its relative abundance is high, apparently varying somewhat wtih minor changes in climate. Fortunately, an accurate chronology is provided by the sedimentary record of reversals of the polarity of the earth's magnetic field. Sediments commonly contain magnetic minerals that, when deposited, become aligned with the earth's magnetic field. The polarity of the earth's magnetic field has been reversed many times in the past few million years. Because reversals affect sediments throughout the world, they provide a valuable means of correlation. A variety of radiometric dating techniques has provided an approximate absolute chronology for the scale of magnetic reversals.

Varying ratios between the oxygen isotopes O^{18} and O^{16} in foraminiferan tests provide an independent method for distinguishing glacial and interglacial intervals. Tests secreted by foraminiferans generally have the same isotopic composition as local seawater. Warm seawater is generally enriched in the heavier isotope, O^{18}, for two reasons. First, warm water evaporates more rapidly than cold water and O^{16}, the lighter isotope, is preferentially lost during evaporation. Second, because they accumulate through the precipitation of evaporated water, polar icecaps are enriched in O^{16}. Returning to the sea, their melt waters further decrease the O^{18}/O^{16} ratio of seawater at high latitudes. The use of oxygen isotope ratios to recognize temperature fluctuations in the Pleistocene has been somewhat controversial; in some ways the results have not matched the results of other lines of evidence. One of the problems is that temporal changes in O^{18}/O^{16} ratios seem not to reflect only temperatures of local surface waters: during glacial episodes, the locking up of vast amounts of O^{16}-enriched water in continental ice sheets seems to have elevated the O^{18}/O^{16} ratio of seawater throughout the world.

A group of workers has recently formed a multidisciplinary project known as CLIMAP, in which long-term climatic changes are being investigated. Using meteorologic, oceanographic, and geologic data, this group has reconstructed for a time 18,000 years ago the extent of large ice sheets and the distribution of sea surface temperatures, climatic conditions, and zones of terrestrial vegetation. The fossil record of ocean plankton has been of great importance here in the estimation of sea surface temperatures. How much further into the past such detailed reconstructions can be undertaken remains to be seen.

Determining Geographic Ranges of Extinct Taxa

Assuming that we can reconstruct climatic conditions to some degree for past intervals of time, how are biogeographic ranges of extinct taxa reconstructed? Most ranges are pieced together directly from fragmentary fossil data, but sometimes indirect methods are required. For example, Brasier (1975) has analyzed the biogeographic history of sea grass through study of the distribution of particular kinds of Foraminifera that are associated with the grass. The grass itself decomposes readily and is seldom preserved, but the record of Foraminifera is exceptionally good.

In general, post-mortem transport is less likely to mislead the paleontologist when he studies biogeography than when he studies paleoecology. Nonetheless, caution must be exercised in dealing with certain groups. Shells of the Recent cephalopod genus *Nautilus* often float for some time after death. The recorded distribution of dead shells is much broader than the distribution of the living animals (Figure 12-19). This observation should serve as a caution to those who undertake biogeographic studies of extinct nautiloids and ammonoids. Even many mobile structures, however, generally do not travel far enough to cause great problems in fossil interpretation. We have already noted that the record of Recent pollen gives quite a faithful picture of the large-scale geographic distribution of living floras (Figure 12-16).

We have discussed the potential of the planktonic larvae of many marine invertebrate species to travel great distances (Figure 9-3). This kind of dispersal tends to offset the limited mobility of adult animals. The waters of the ocean provide a much better medium for dispersal of living creatures, whether larval or adult, than does the gaseous, nonnutritious atmosphere. Nonetheless, airborne creatures that have overcome the problems imposed by the harshness of the atmosphere can travel great distances. Spores, small seeds, and other resistant reproductive structures are commonly adapted to aid the dispersal of plants, protists, and a few simple animals. Frequently an evolutionary trade-off exists between the value of extending reproductive energy to produce many zygotes or offspring for long distance dispersal or a few specially protected ones that will individually face lower risks of early death.

Barriers to Dispersal

Clearly the limits of the geographic ranges of species reflect barriers to dispersal. In some instances, as we have seen in examining climates, these barriers are not linear features, but vast tracts of territory that are inhospitable to a species. Such tracts may constitute most regions of the globe not occupied by a species. A species of tropical plankton, for example, may occupy all regions of the world suitable for its growth but never spread poleward into nontropical climates because of inherent physiological limitations. In other instances, barriers are narrow bands of environment that act like fences, preventing species from immigrating to regions suitable for their colonization. It is barriers of this latter type that we will now discuss.

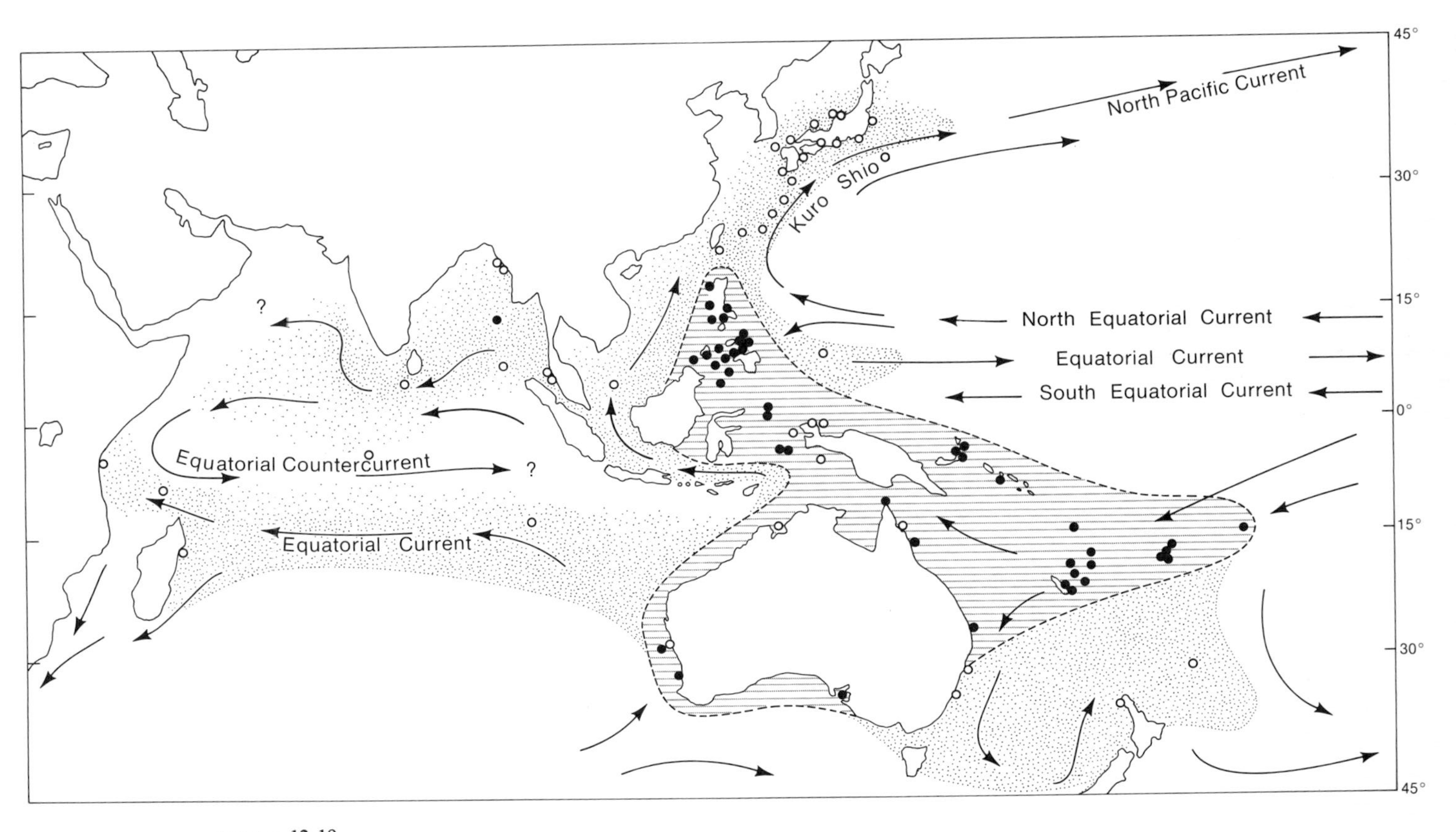

FIGURE 12-19
Map of the distribution of modern *Nautilus* shells. Black dots represent localities where live animals have been found, and horizontally-ruled area shows the probable biogeographic range. Open circles represent localities where drifted shells have been reported, and stippling represents the approximate post-mortem distribution. Current speeds in the Indian Ocean indicate that some specimens have probably drifted for years. (From House, 1973. Copied with the permission of the Palaeontological Association.)

THE NATURE OF BARRIERS

In a number of contributions published in the 1940's and early 1950's, George Gaylord Simpson laid the groundwork for modern investigations of biogeographic barriers. He recognized that barriers are not absolute. Routes for the dispersal of species may be largely unobstructed; these he termed ***corridors.*** Migration along corridors is generally rapid. A good example is the Bering Land Bridge, which for much of the Tertiary Period permitted passage of mammals back and forth between Asia and North America, via Alaska. At the other extreme are dispersal routes across largely hostile territories that make migration events rare. Under these conditions chance plays a major role. For example, one cannot predict if or when a few small mammals or terrestrial invertebrates will be rafted across a narrow ocean on a log, or a few insects will be blown over a mountain pass in a storm, or a stray bird will carry a few plant seeds across a desert in its digestive tract. Simpson termed pathways of this sort ***sweepstakes routes,*** in recognition of the importance of chance factors. He labelled intermediate kinds of routes ***filters.*** An example would be a chain of islands separated by narrow channels that represent partial barriers against the passage of terrestrial forms. In biogeography, the concept of an island has been expanded well beyond the literal meaning of the word. In terms of the dispersal of their biotas, lakes and mountain peaks, for example, are "islands" on the land.

Simpson and others have quantified the comparison of geographically separated biotas using various indices of comparison. The index favored by Simpson was discussed on page 212. In Table 12-3 this index is employed to compare living mammal faunas of Oregon and New York at various taxonomic levels. The relatively high values of percentages at all levels are characteristic of faunas connected by a corridor. Numerous other quantitative measures of biotic similarity are available. The interested reader can consult a review by Sneath and McKenzie (1973).

TABLE 12-3
A Comparison of the Living Mammal Faunas of Oregon and New York (Percentages)

	Common to both States (x)	*Peculiar to one or the other* $(100 - x)$	*New York fauna occurring also in Oregon* $(100C/N_1)$	*Oregon fauna occurring also in New York* $(100C/N_2)$
Orders	83	17	100	83
Families	70	30	74	93
Genera	44	56	56	68
Species	19	81	25	46

SOURCE: McKenna, 1973, after Simpson.

The late Cenozoic history of the Americas illustrates what a dramatic change can be effected when a sweepstakes route turns into a corridor. Throughout most of the Cenozoic, South America, like Australia, was a huge island. There was no land connection with North America, and only a sweepstakes route existed across what is now the Isthmus of Panama. The Cenozoic history of South American mammal faunas has been reviewed by Patterson and Pascual (1972). Throughout most of the Cenozoic it harbored a largely ***endemic fauna*** (one not found elsewhere). This archaic fauna diversified from a few primitive kinds of animals that in the late Mesozoic and early Cenozoic made their way to the South American land mass when it was connected to other continents. The isolation of South America early in the Cenozoic will be discussed in the following section. From the time of that isolation until the Pliocene, when the Isthmus of Panama formed, the huge island received very few immigrants. The most notable were the caviomorph rodents and the playrrhine primates (monkeys). Both arrived by sweepstakes routes under circumstances that are not well understood. The two groups may have been rafted in on floating logs.

In the Late Pliocene the tectonic uplift of the Isthmus of Panama initiated a dramatic biogeographic change. The land bridge thus formed became a corridor for faunal interchange between North and South America. The traditional view has been that widespread extinction of primitive South American taxa ensued. Patterson and Pascual (1972) have challenged this view. Traditionally it has been thought that because most of the southern forms evolved on a single, isolated continent they underwent relatively slow evolution; being inferior competitors, they suffered widespread extinction during the initial mixing of faunas. Patterson and Pascual, however, conclude that it was only South American predators, which were all marsupials, that were wiped out. Marsupials are commonly thought of as being less highly evolved than placentals. Nonetheless, there now seems to be no evidence that either marsupial or placental herbivores suffered abnormally high rates of extinction following the invasion by North American species. Patterson and Pascual suggest that some persisting herbivores had niches that overlapped those of invading species very little, and that others developed narrower niches in the face of competition. Furthermore, various South American herbivores made their way north into Central and North America. Of these the oppossum, which is really an omnivore rather than a strict herbivore, has spread the farthest. Its increased success in very recent times has clearly resulted from man's alteration of the environment, especially his elimination of predators. It is perhaps no surprise that in the dramatic faunal interchange between the Americas competitive displacement was most significant among predators. These animals are typically less extensively preyed upon themselves than are herbivores, which means that their populations may more commonly be limited by food supply. This condition would be expected to make them vulnerable to competitive displacement (page 358).

Ironically, while the uplift of Panama formed a corridor for land animals, it simultaneously produced a biogeographic barrier in the marine realm, severing the previous marine corridor between the Atlantic and Pacific Oceans. As a result, many so-called "species pairs" now exist, with one member of a pair found on each side. It is debatable whether the Atlantic and Pacific representatives of pre-existing species are now really separate species or are only given separate names because they now occur in separate ocean basins. This is a problem that concerns many scientists because the natural situation may soon be disrupted. The building of a larger Panama Canal will permit a degree of faunal mixing and interbreeding between the two oceans.

The idea of ***centers of dispersal*** arose early in the history of biogeography. According to this idea, major taxa arise and undergo their initial period of diversification in particular geographic regions. As they diversify, their species spread to other regions, but the diversity of species tends to decline away from the area of origin, or center of dispersal. This idea has commonly been applied to latitudinal gradients in species diversity, with regions that are now tropical and subtropical being viewed as major centers of origin and dispersal. One problem with this latitudinal application is that, as we have seen, many regions now characterized by cool climates had much warmer climates in the mid-Cenozoic (Figures 12-13 and 12-14). Even if groups ancestral to modern European taxa, for example, were adapted to warm climates, this does not necessarily mean they originated near the equator.

Hermatypic corals of the Cenozoic seem to provide an example of the concept of centers of dispersal (Stehli and Wells, 1971). In both the Atlantic Ocean and the Indo-Pacific, the average age of coral genera is highest in a particular region and decreases toward the periphery of the total area occupied (Figure 12-20,A). Furthermore, generic diversity follows a gradient trending in the opposite direction (Figures 12-20,B and 12-21). In other words, it appears that new genera (and presumably new species) arise at higher rates in a central region near the equator than they do near the periphery of the overall distribution. Many genera found in peripheral regions also occur near the center, and it seems likely that most of these arose in central areas and spread outward. It is also possible that genera living near the periphery are exceptionally hardy. This condition and their widespread geographic occurrence would also increase their longevity and perhaps also contribute to the pattern of Figures 12-20,B and 12-21.

ISLAND BIOGEOGRAPHY

In recent years there has arisen a rather extensive body of biological theory that is partly summarized in a book by MacArthur and Wilson (1967) entitled *The Theory of Island Biogeography.* Basically, the theory considers the manner in which species appear and disappear on islands, the rates at which they appear and disappear, and the compositions of island biotas that result. As noted above, an isolated habitat of any sort can, for theoretical purposes, be considered an island. A lake is an "island" in a "sea" of land. A mound-like submarine patch reef is an "island" in a "sea" of sand. Thus the theory of island biogeography has widespread application.

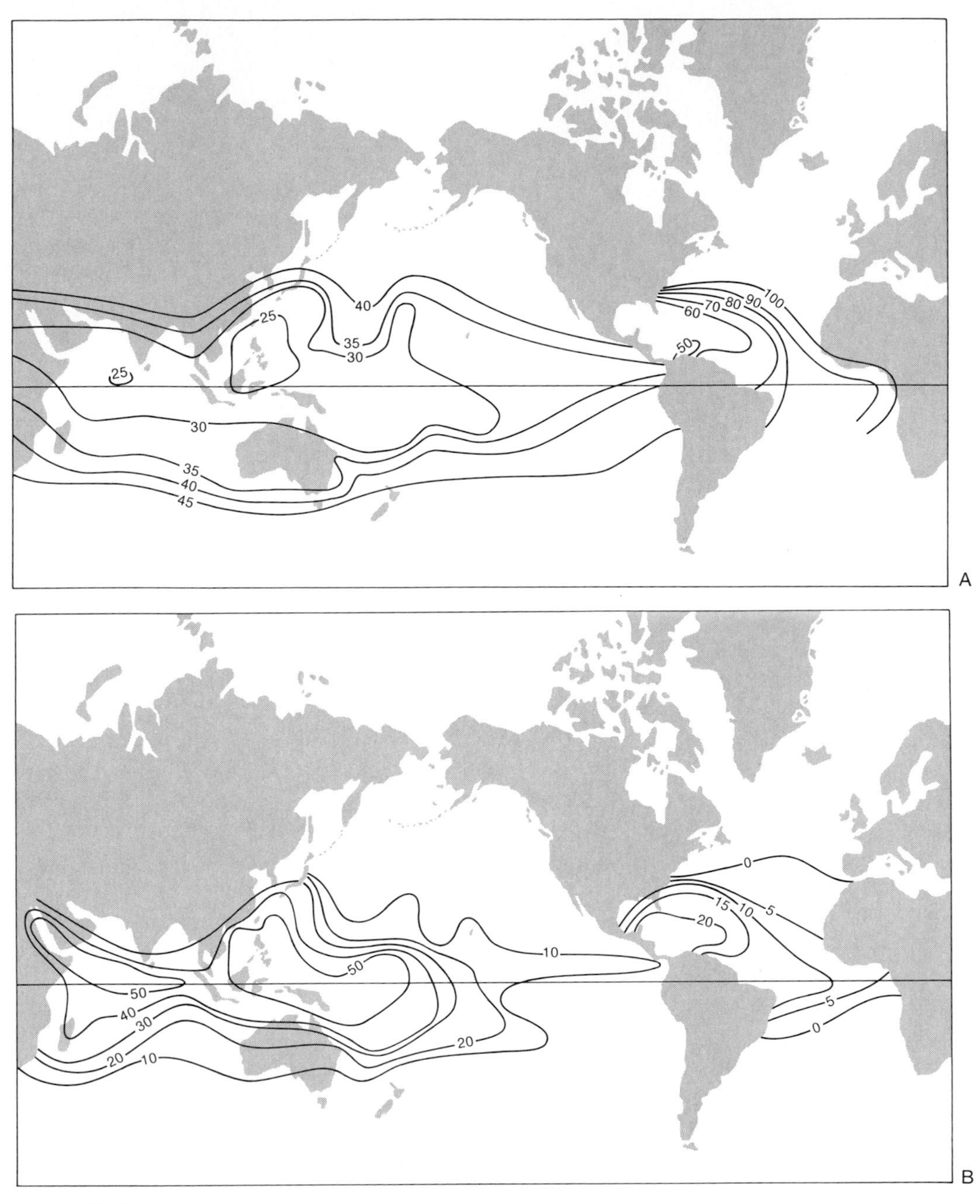

FIGURE 12-20
Geographic distribution of age and diversity for living hermatypic corals. A: Contours showing average geologic age of genera, in millions of years. B: Contours showing diversity. (From Stehli and Wells, 1971.)

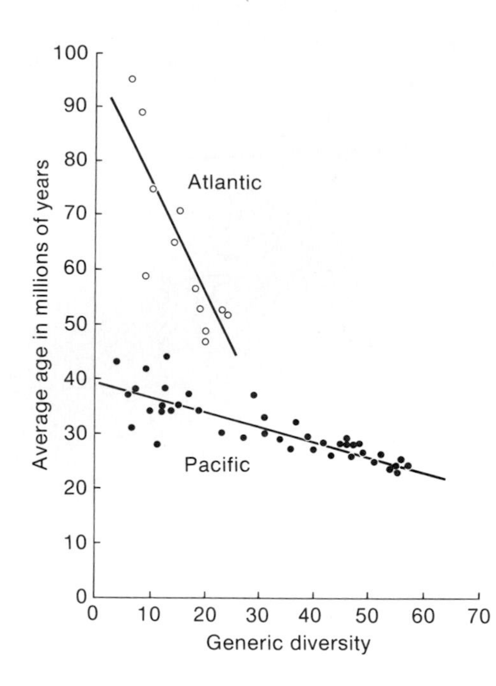

FIGURE 12-21
Inverse relationships between average age of genera and generic diversity for hermatypic coral faunas of the Atlantic and Pacific Oceans. (From Stehli and Wells, 1971.)

One important observation about island biotas is that their diversity tends to increase by some power of island area:

$$S = CA^z,$$

where S is species number, C is a constant, A is island area, and z is a number less than 1, generally ranging from 0.20 to 0.35. Note that this is the same kind of power function as the one commonly used to describe allometric growth (page 62). Because of the power function relationship, logarithmic plots of S versus A tend to be linear. An example is given in Figure 12-22. We will return to this relationship after considering the rudiments of a model presented by MacArthur and Wilson to account for the number of species that occur on islands.

In part, our presentation comes from the simplified version of Wilson and Bossert (1971). It can be predicted that as a new island is colonized and fills up with species, the number of species arriving per unit time should drop. For simplicity, we will assume that the decline in rate of immigration (λ_S) is linear (Figure 12-23,A). If we let P stand for the total number of potentially colonizing species, then the rate of immigration will be 0 when P species are on the island. As species colonize the island, we would also expect the number of species on the island that are exterminated per unit time to increase. (We use the term "exterminate" to indicate local rather than complete extinction of a species.) Again, for simplicity, we will assume that the total rate of extermination (μ_S) increases linearly with S. Because the number of species becoming extinct will be proportional to the number on the island, the percentage of species becoming extinct per unit time will remain constant. At some value of S, which we will label $\hat{S}$, values of λ_S and μ_S will be equal. We can predict the value of $\hat{S}$ if we know λ_S and μ_S. We do this as follows. The *total* immigration rate (λ_S) will be equal to the average rate of immigration *per species* (λ_A) multiplied by the number of species not yet on the island. If there are P species that can potentially colonize the island, then at any stage of colonization total rate of immigration will be equal to λ_A $(P - S)$. As mentioned above, percentage of species exterminated per unit time, or rate of extermination *per species,* which we can label μ_A, will be constant. Then for any stage of colonization, the *total* rate of extermination will be $\mu_A \cdot S$. The rate at which S increases will equal the total rate of immigration minus the total rate of extermination. At equilibrium ($S = \hat{S}$) this rate will be zero:

$$\lambda_A(P - \hat{S}) - \mu_A\hat{S} = 0.$$

Solving for $\hat{S}$,

$$\hat{S} = \frac{\lambda_A P}{\lambda_A + \mu_A}.$$

FIGURE 12-22
The species-area curve for West Indian amphibians and reptiles. (From MacArthur and Wilson, 1967. Copyright © 1967 by Princeton University Press and reprinted by permission.)

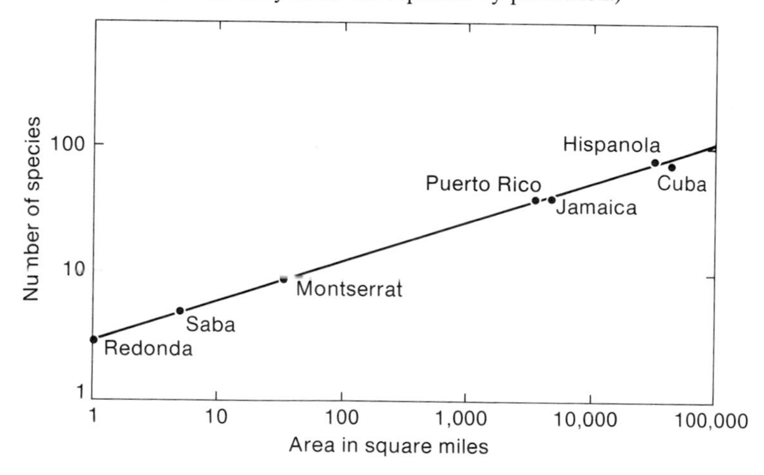

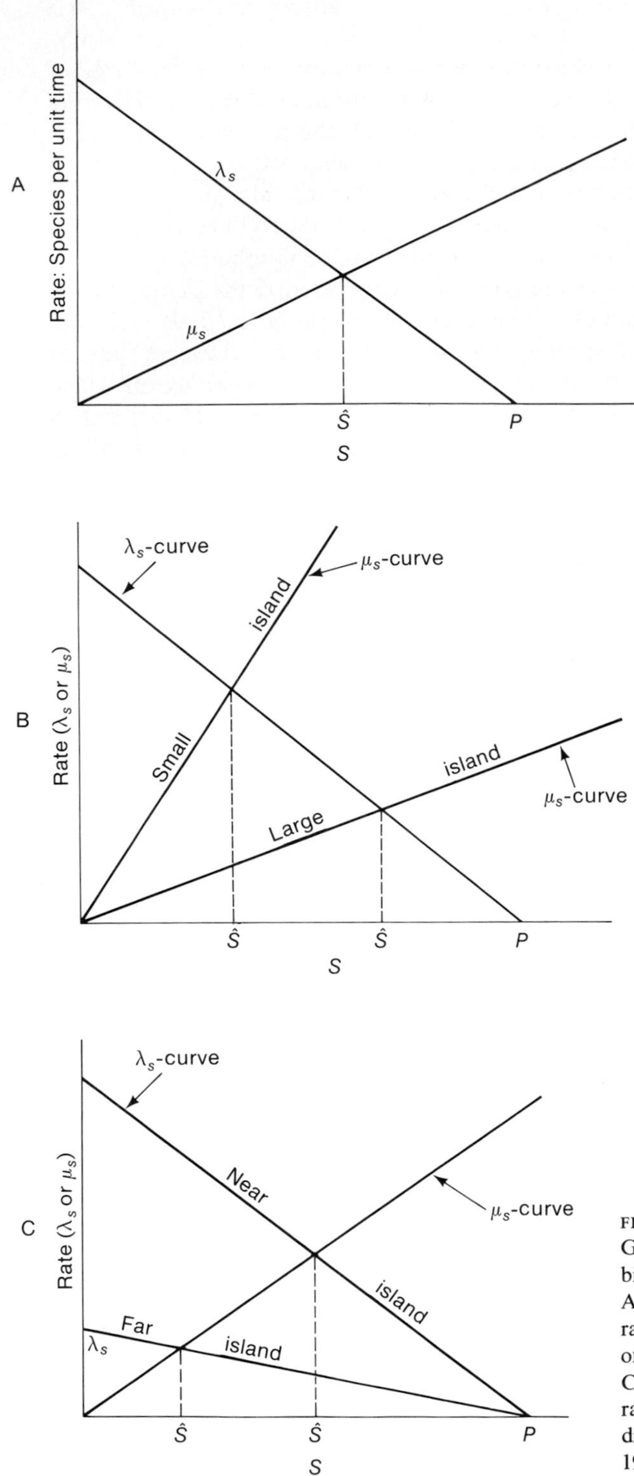

FIGURE 12-23
Graphs illustrating the theory of island biogeography. A: The basic model. At $\hat{S}$ extinction rate equals immigration rate. B: The effect of island area on rate of extinction and equilibrium diversity. C: The effect of distance from source on rate of immigration and equilibrium diversity. (From Wilson and Bossert, 1971.)

In other words, if we can measure the percentage of extermination and immigration per unit time and we know the total number of species that can potentially colonize the island, we can estimate the equilibrium diversity for the island.

Of course, our presentation has been somewhat simplistic. MacArthur and Wilson (1967) and others have undertaken more refined analyses and carried them in directions not considered here. In assuming that the immigration curve should be linear, we have, for example, ignored the effect of crowding of species. We have also ignored the possibility of speciation occurring within the island. Using this simplistic model, however, we can examine some additional factors. For example, we can return to the empirical observation that diversity on real islands increases with island area (Figure 12-22). This relationship has led to the conclusion that rate of extinction on islands is inversely related to island area. Crowding and reduced population sizes probably increase rates of extermination on small islands. Thus, as shown in Figure 12-23,B, number of species at equilibrium should increase with island size. Another factor not considered here is that variety of habitats also tends to increase with island size, permitting a greater variety of species to colonize large islands than small islands.

The degree of isolation of an island also tends to affect diversity at equilibrium because the distance of an island from other colonized areas will have an effect on rate of colonization (Figure 12-23,C). In fact, diversity of species on islands is often observed to decrease toward the extremities of archipelagos (Figure 12-24,A). In effect, archipelagos are filter routes of dispersal (page 406). The importance of accessibility also shows up when we compare the effects of area on species diversity for islands and for nonisolated sample areas within single land masses. In Figure 12-24,B, the slope of the cluster of points for islands is steeper than the slope for sampled areas of New Guinea. The lower slope for areas of New Guinea reflects a lower value of z in the power function that relates diversity to area. Because numerous species have access to small, nonisolated land areas, many that are not especially well adapted to the habitats of these areas will nonetheless be found there also, as transients or temporary inhabitants. Furthermore, within continents most of the increase in diversity that is observed when the sample area is increased reflects the fact that larger land areas tend to contain larger varieties of habitats. The same kind of argument applies to areas of continental shelf. Certainly the limited variety of habitats in the relatively small Caribbean area, for example, would seem to be partly responsible for the limited diversity of living reef corals and molluscs in this region compared to the diversity of these groups in the Indo-Pacific region.

EVOLUTION OF TAXA OCCUPYING ISLAND-LIKE HABITATS

Many families, genera, and even species that occupy small oceanic islands also occupy large land masses. Taxa of this type will not tend to display adaptations that have evolved specifically for island life. In contrast, taxa that inhabit lakes, having no

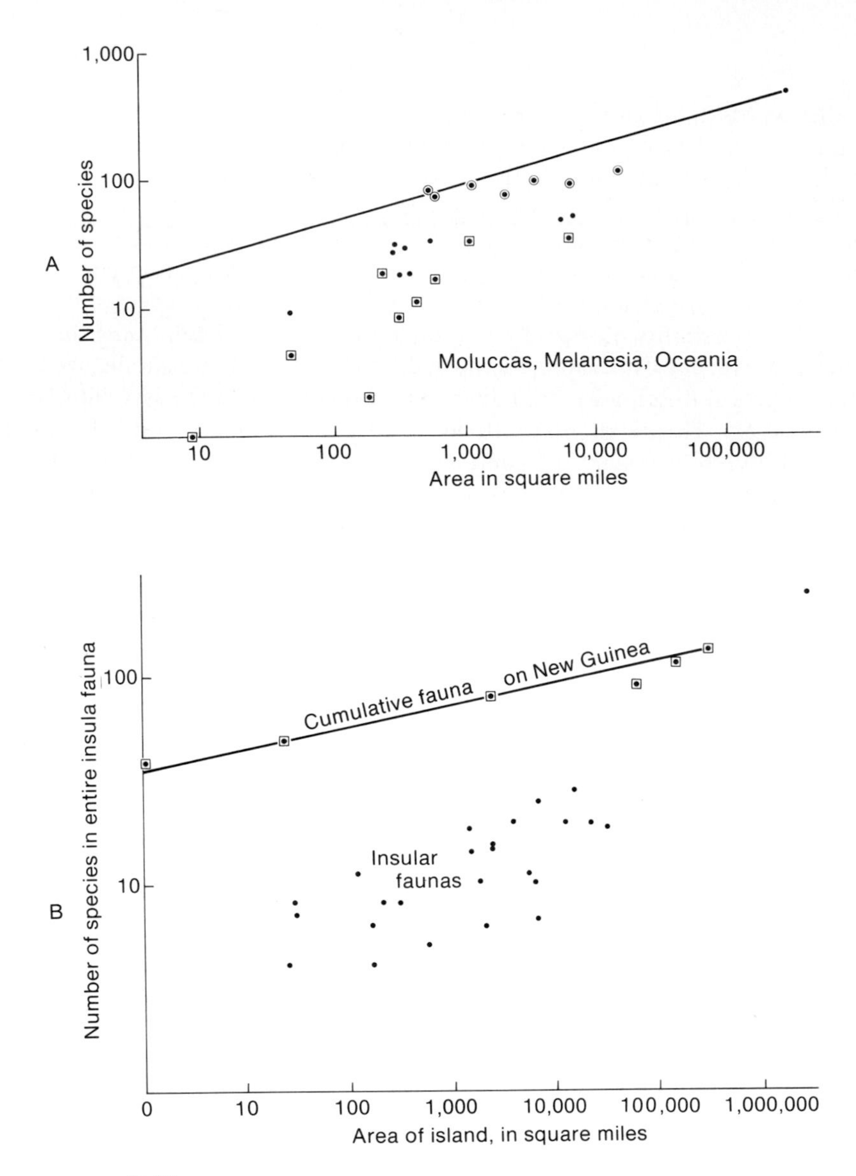

FIGURE 12-24
Effects of distance from source and island area on species diversity.
A: Species-area curves for Pacific islands. "Near" islands (less than 500 miles from New Guinea) are enclosed in circles. "Far" islands (more than 2,000 miles from New Guinea) are enclosed in squares. Islands at intermediate distances are unenclosed. B: Comparison of species-area plots for the Moluccas and the Melanesian islands (solid dots) and plots for expanding sample areas of the larger land mass of New Guinea (dots within squares). (From MacArthur and Wilson, 1967. Copyright © 1967 by Princeton University Press and reprinted by permission.)

freshwater bodies the size of continents to occupy, show distinct adaptations to the small "islands" of water that they can occupy. One might think that dispersal of lake dwellers would be slow and ineffective because of the extensive terrestrial barriers that exist between lakes. However, because they are restricted to such a discontinuous habitat, most successful groups of lake dwellers possess special mechanisms for dispersal. Indeed, many lake-dwelling species have enormous geographic ranges, and circumglobal ranges are common. Many lake dwellers of small size enter dormant resting stages when a habitat dries up. They can then be transported by wind. Others are transported by mammals or birds, sometimes in the form of eggs. They travel in fecal material, on feet or feathers, or in mouths (from which they later happen to escape).

Some living species of terrestrial organisms that have poor mechanisms for dispersal across hostile territory occupy mountaintop "islands" as a result of late Cenozoic climatic change. During the Pleistocene many cold-adapted species of the Northern Hemisphere spread southward ahead of advancing ice sheets (Figure 12-16). When the glaciers retreated and climatic conditions of North America and Europe ameliorated, there were many regions in which cold climates remained only on mountaintops. Today on these mountaintops we find disjunct populations, sometimes termed ***relicts*** of species that otherwise today occur much farther north.

Some biotas are relicts in time rather than in space. These consist of taxa that have survived in restricted areas long after the disappearance of other similar taxa that were more widely distributed. Relict biotas of this type are found in islandlike environments for other reasons as well. The most important of these is that such environments often provide primitive biotas with effective isolation from advanced competitors and predators that, if present, would contribute toward their extinction. We will discuss examples of island-like refugia in the following section.

A feature of large lakes and islands that contrasts sharply with the patterns described above, but is of special interest in macroevolution, is their tendency to support quite remarkable local adaptive radiations. The most publicized of these radiations are those that have occurred within large African lakes, such as Lake Tanganyika (Brooks, 1950). Here, for example, 25 of the 44 molluscan genera are endemic. Only two of these appear to be relicts. The rest must have arisen within the lake. In particular, many of the gastropod genera are quite distinctive. Lake Tanganyika apparently arose some time in the Pliocene, or between 2 and 5 million years ago, which makes the rate of adaptive radiation of gastropods within it quite remarkable. Just as remarkable has been the adaptive radiation of the cichlid fishes, of which there are 34 endemic genera. The cichlids have undergone comparable adaptive radiations in various other large lakes throughout the world. Liem (1973) has attributed the tendency of certain kinds of cichlids to speciate to the pharyngeal jaw mechanism, which is highly adaptable and "versatile" in evolution. Another group of fishes that has been prone to dramatic adaptive radiation in lakes is the Cyprinidae. In Lake Lanao of the Philippines, which is believed to be only about 10,000 years old, it appears that from a single invading cyprinid species there have already arisen in the lake 13 species, 5 of which are referred to 4 new genera. Some workers have argued that such speciation in large lakes must be sympatric, but most have accepted the idea that it occurs by geographic isolation, at least on a very small scale. Exactly why rapid, large-scale speciation is characteristic of some large lakes is much debated. It seems evident that the absence of other species preying upon or competing with early inhabitants is an important factor. In other words, the initial emptiness of the habitat may give isolated populations a good chance of becoming established as new species.

Plate Tectonics and Physiographic Change

We have now introduced enough of the biological concepts of biogeography to shift our focus into the past. As mentioned at the start of this chapter, two kinds of geological evidence about past conditions are particularly important to the study of biogeography: evidence about climate and evidence about the disposition of land masses. It should also be clear from our earlier discussion that these two subjects are intimately related.

Today at high latitudes climates are clearly cooler than they have been during much of the Phanerozoic: living tropical biotas and present-day tropical sedimentary deposits are restricted to rather narrow equatorial belts. In addition, more land area is exposed above sea level today than has been exposed during most of the Phanerozoic. Only about 71 percent of the earth's surface is now covered by water. We mentioned in Chapter 10 that few epicontinental seas exist that can serve as models of the ancient ones in which much of our continental sedimentary record accumulated. Exposure of large continental areas today is partly responsible for the cool climates that exist today at high latitudes. This is because the ***albedo*** of the earth's surface is affected by the nature of the surface; the albedo is the percentage of the sun's radiant energy that is reflected back from the earth's surface. For areas covered by ice or snow, albedos range from about 45 percent to 95 percent. For continental soil, grasslands, and forests they are lower, ranging from about 5 percent to 30 percent. For oceans, they tend to be even lower, ranging from about 6 percent to 10 percent. Thus we can readily see that the exposure above sea level of large land areas today at high latitudes in the Northern Hemisphere must in part account for the relative coolness of Eurasian and North American climates.

Scientists still debate the origin of the polar ice caps that have advanced and retreated during the Pleistocene. Their growth seems to generate further growth, because a land surface blanketed with reflective ice has decreased albedo, which results in further cooling. Similarly, the melting back of ice sheets decreases the albedo, which results in further melting. These relationships suggest that the sizes of ice sheets may be inherently unstable. What relationship there may be between continental glaciation and movements of continents remains uncertain. What does seem clear is that continental movements relate to degree of emergence of land areas. In this and other ways, such movements affect the distribution of climatic conditions and of species.

THE NEW PLATE TECTONICS FRAMEWORK

For years most geologists considered the movement of continents to be largely up and down (perpendicular to the earth's surface). Since the early 1960's, however, the emergence of the conceptual scheme termed ***plate tectonics*** has convinced most geologists that lateral movement of continents is far more extensive than up-and-down movement and that the two kinds of movement are genetically related.

We will provide only a sketchy account of plate tectonics here. More detailed treatments are provided in numerous other geological textbooks. The fundamental idea, developed by Hess (1962) and more recent authors, is that an outer shell of the earth, which includes the continents, is divided into a series of plates that move relative to one another. The plates comprise the lithosphere, which includes continents of low density that float upon the asthenosphere, which consists of material of higher density. Just how the plates move is not fully understood, but the prevailing idea is that convection cells positioned beneath the plates drag them along. Lithospheric plates are produced at spreading centers like the mid-Atlantic ridge, where dense rocks well up from below and flow laterally (Figure 12-25). As plate

FIGURE 12-25
Diagrammatic cross-section of the ridge system known as the Mid-Atlantic Rift. Lithospheric plates move laterally away from the rift (straight arrows), and plate material is added along the rift (curved arrows). The lithosphere ranges in thickness from about 65 to 145 kilometers. (From Dietz, 1972. Copyright © 1972 by Scientific American, Inc. All rights reserved.)

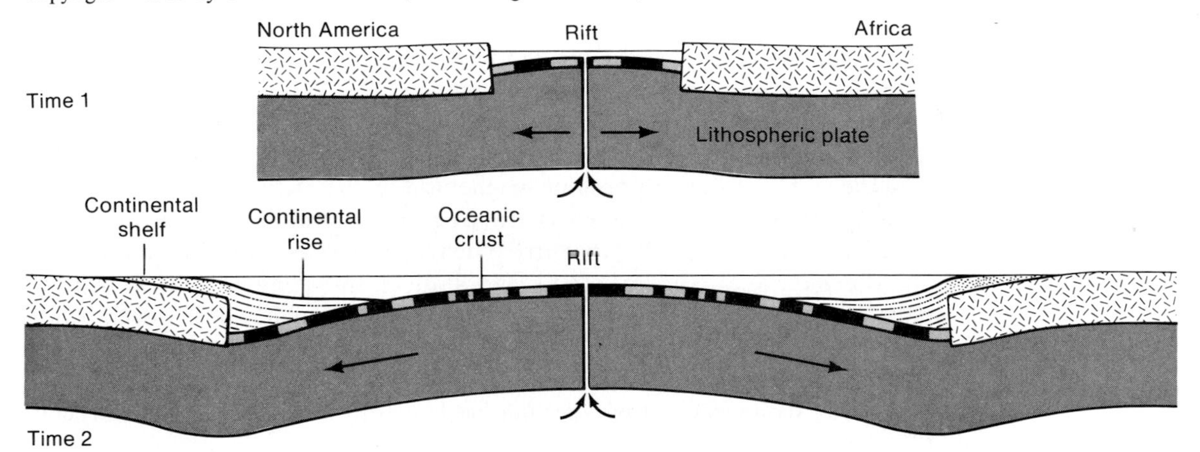

material moves away from a spreading center, or ***ridge system,*** it cools and its upper surface becomes sea floor. Typically the edge of a plate opposite the part forming at a ridge system passes diagonally downward into the earth's mantle along what is termed a ***subduction zone*** (Figure 12-26,A). Some of the plate material becomes part of the mantle and some of it (less dense material), after melting, is expelled from volcanoes. The subduction zone is demarcated as a trench along the sea floor. Where lava emerges on the side of the trench opposite the plate being subducted, a chain of volcanoes forms what is known as an ***island arc.*** Here also are located the foci of earthquakes that occur periodically, when pent up stresses on plates cause movement. It happens that at the present time the Pacific Ocean is encircled by subduction zones (Figure 12-27). Plates slide past each other along transform faults at boundaries that are neither spreading centers nor subduction zones. Continents of lower density than the rest of the lithosphere simply ride along as passengers on moving plates. When a continent on a plate that is being subducted reaches the subduction zone, however, its margin tends to be deformed into a mountain chain. This happens because the low density of the continent tends to prevent it from being subducted. Often the orientation of the subduction zone is then reversed, so that the plate whose edge lacks a continent becomes the subducted plate. When two continents meet at a subduction zone, an enormous mountain range, such as the Himalayas, is formed (Figures 12-26 and 12-27). The resistance of the continental masses to subduction tends to lead to formation of a new subduction zone at some other location where there is less resistance. Ridge systems, as well as subduction zones, shift their positions through time. When a new ridge system forms beneath a continent, the continent is split in two. The Red Sea represents an incipient ocean formed by relatively recent continental splitting that has separated the Arabian and Turkish plates (Figure 12-27).

The most obvious kinds of evidence that continents have broken apart and moved are the closely matching configurations and equidistant positions of continental margins on opposite sides of mid-oceanic ridges. The seemingly puzzle-like fit that would be had if the continents on opposite sides of the Atlantic were shoved together is, of course, the most widely cited example. However, it was not this but paleomagnetic evidence that formed the first compelling argument in favor of the basic model of plate tectonics. Vine and Matthews (1963) reasoned that if seafloor spreading really occurs, zones of positive and negative rock magnetism should be found on opposite sides of ridge systems. These were expected because as liquid rock cools, some of its components become magnetized by the earth's magnetic field, and the polarity of this field has been reversed at frequent intervals in the geologic past. As predicted, mirror image banding patterns of magnetism were found on opposite sides of mid-oceanic ridges (Figure 12-28), and seafloor spreading was confirmed. An absolute chronology of the stratigraphic sequence of reversal has been developed by radiometric dating of the magnetized rocks. Thus, rates of movement of plates and positions of continents can be determined for relatively recent geologic intervals. Magnetism of sediments and of volcanics has been of great utility in reconstructing

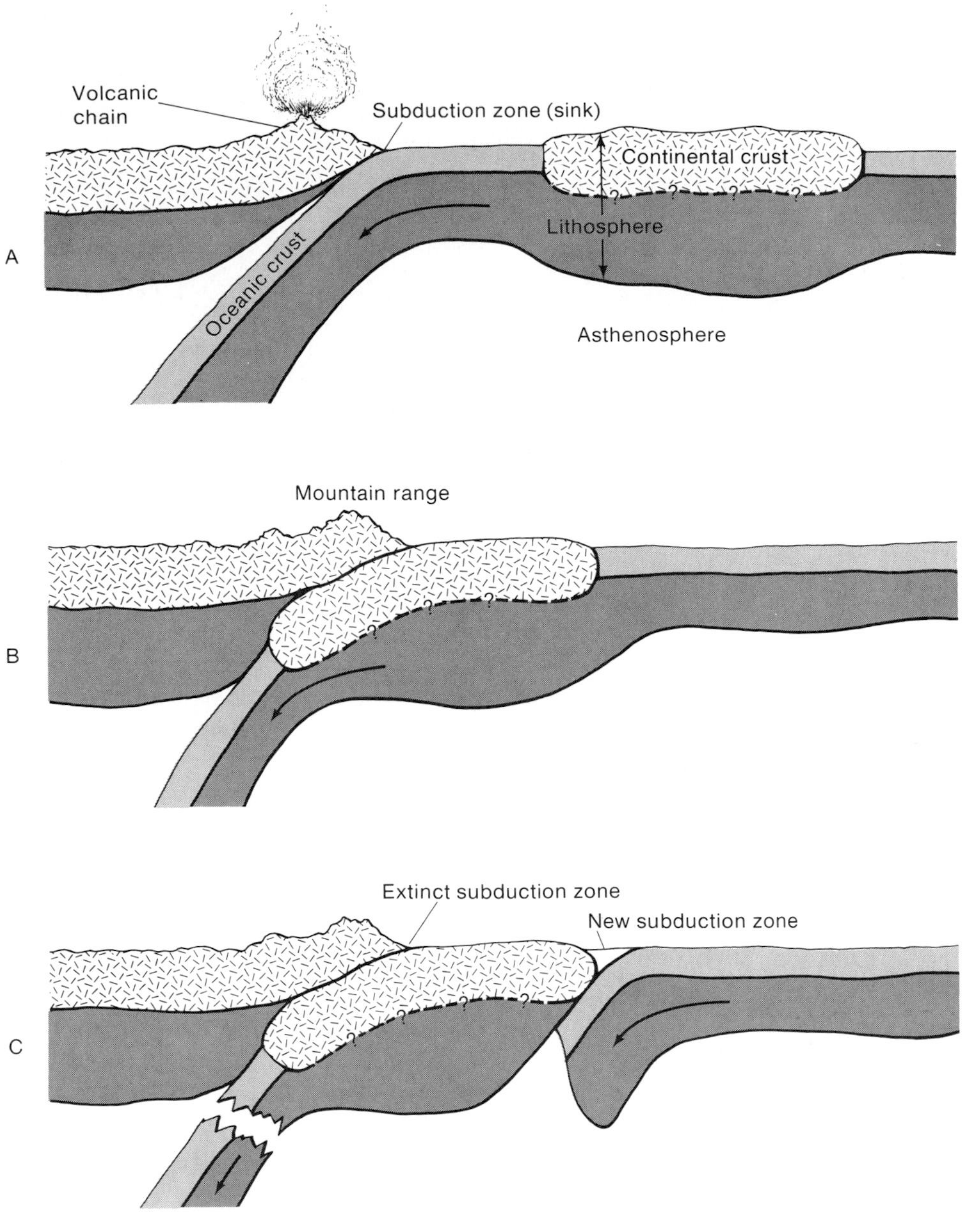

FIGURE 12-26
Subduction zones. A: A zone in which oceanic crust is subducted beneath a continent. B: A zone in which one continent is being subducted beneath another, elevating a mountain chain. C: Formation of a new subduction zone after the condition illustrated in B has caused extinction of the old one. (From Dewey, 1972. Copyright © 1972 by Scientific American, Inc. All rights reserved.)

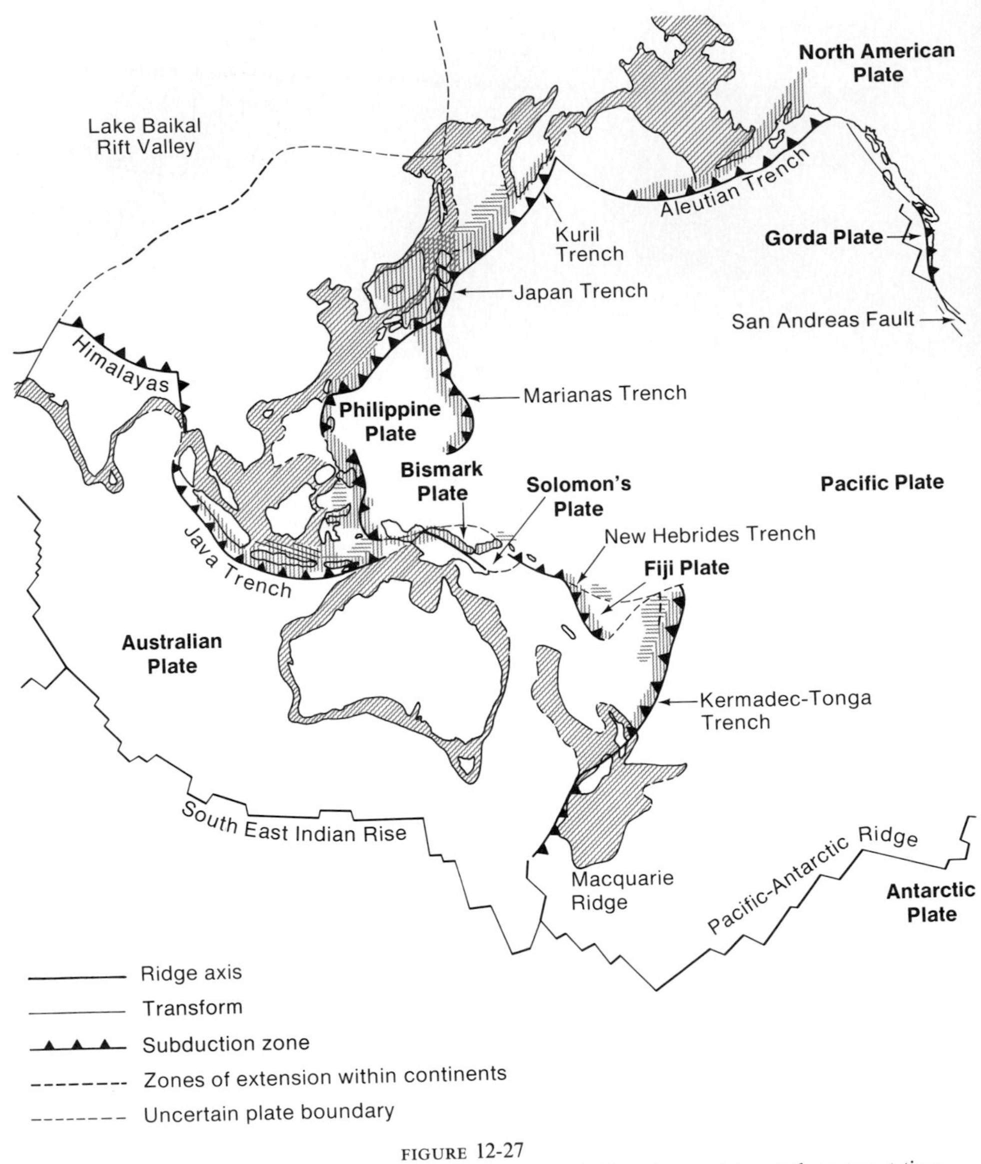

FIGURE 12-27
The arrangement of lithospheric plates at the present time. (From Dewey, 1972. Copyright © 1972 by Scientific American, Inc. All rights reserved.)

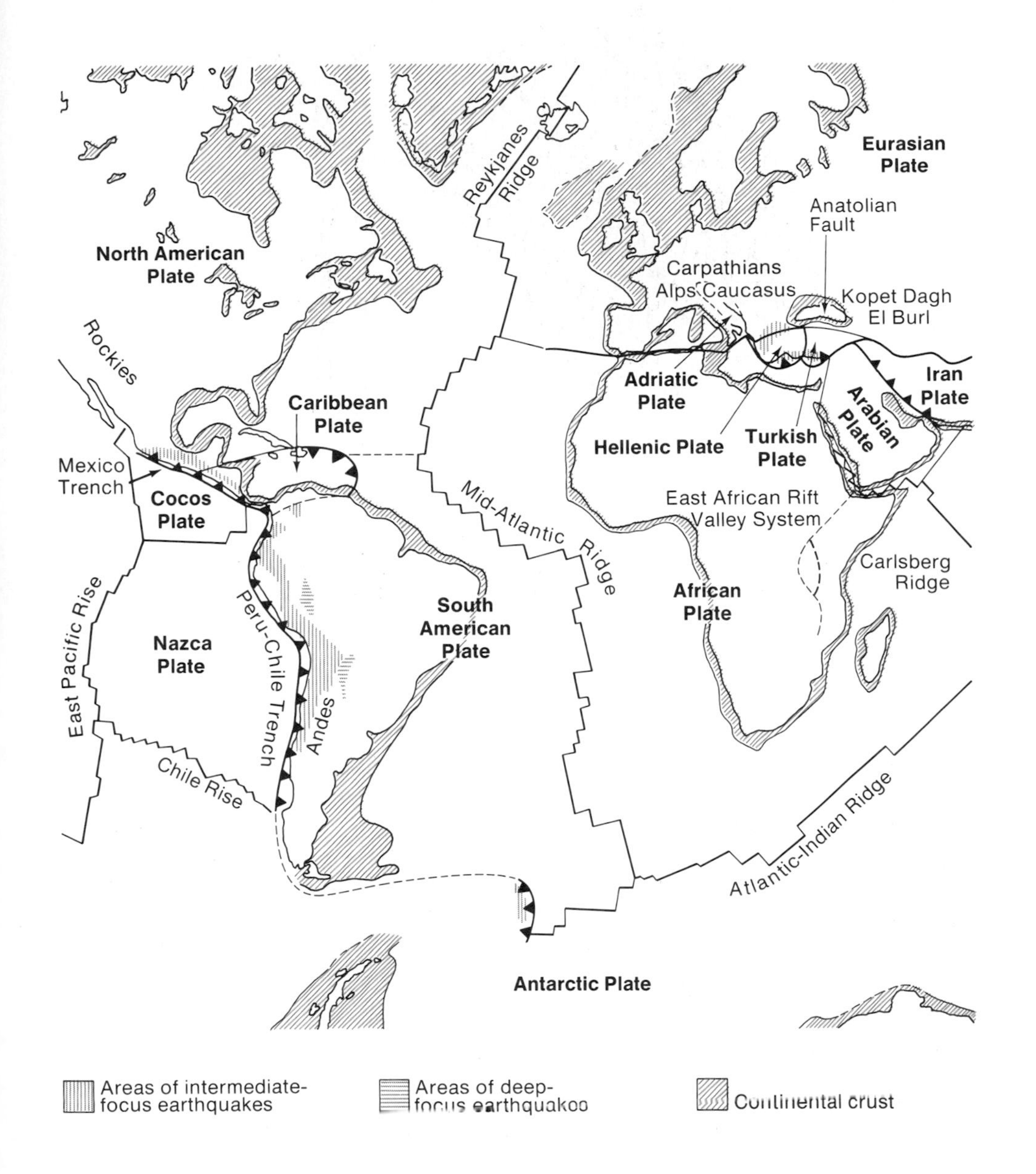

Eurasian Plate
Reykjanes Ridge
Anatolian Fault
North American Plate
Carpathians
Alps
Caucasus
Kopet Dagh
El Burl
Rockies
Adriatic Plate
Iran Plate
Caribbean Plate
Arabian Plate
Hellenic Plate
Turkish Plate
Mexico Trench
Cocos Plate
East African Rift Valley System
Mid-Atlantic Ridge
Carlsberg Ridge
African Plate
East Pacific Rise
Peru-Chile Trench
South American Plate
Nazca Plate
Andes
Chile Rise
Atlantic-Indian Ridge
Antarctic Plate
Areas of intermediate-focus earthquakes
Areas of deep-focus earthquakes
Continental crust

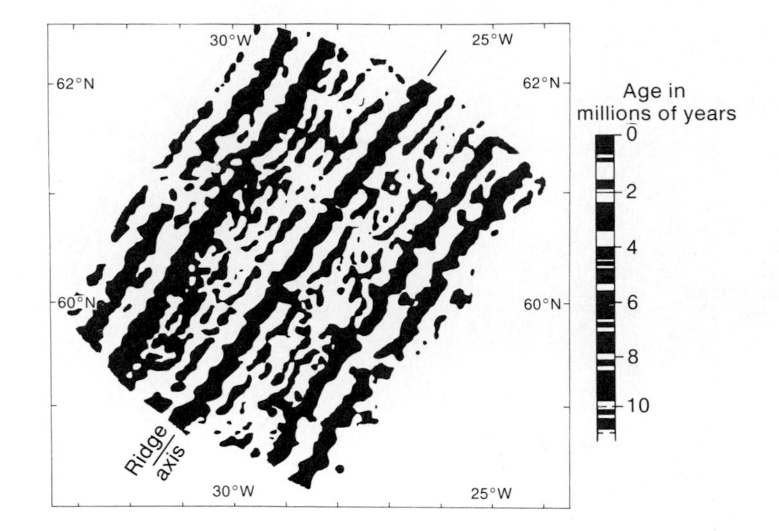

FIGURE 12-28
Mirror image striping of sea-floor magnetism on opposite sides of the Mid-Atlantic Rift southwest of Iceland. Black stripes represent "positive" magnetism (the same as is induced by the present magnetic field) and white stripes represent "negative" magnetism. (From Heirtzler, LePichon, and Baron, 1966.)

the past orientations of continents also by quite a different method. Rocks can, in effect, be used as compasses because their magnetic fields, which are "frozen" in place at the time of formation, are aligned with the earth's magnetic field. Thus, the orientations of rock units at the time of their formation can be determined within certain limits of accuracy. This procedure allows past orientations of continents to be reconstructed. Of use in reconstructing the movements of continents are "hot spots," or plumes, which are fixed regions of the mantle from which lava erupts periodically. As a plate passes over a hot spot, volcanoes are periodically pushed up through the plate to form a volcanic chain, which then traces the direction of movement. Using these methods and other kinds of geological evidence, including data from paleontology, geologists are piecing together the Phanerozoic history of continental movement. This history is now fairly well established for the interval from the late Paleozoic to the Recent. It is upon this interval that we will focus. The study of plate tectonics is so new that biogeography and paleoclimatology have been integrated with it to only a limited degree. Still, a number of fascinating relationships have been uncovered.

The idea that continents have undergone large-scale relative movement was first given comprehensive treatment by Alfred Wegener in the early part of this century.

The evidence that he presented, like much geologic evidence, was circumstantial. It included data on the distribution of both living and fossil organisms. There ensued a debate that, until the early 1960's, left most geologists believing that the continents had remained in fixed positions since early in earth history. One of the chief obstacles to the acceptance of ***continental drift,*** as Wegener's scheme became known, was the discovery of crustal rock lying everywhere beneath the oceans and between continents. Until the ideas of plate tectonics were popularized, most geologists ignored the idea that the entire crust of the earth might be moving, as suggested by Holmes (1931). Instead they simply accepted the argument of geophysicists that forces great enough to plow continents through oceanic crust could not possibly be developed within the earth.

BIOLOGIC AND PALEOBIOLGIC DATA

We can now see that there is a wealth of geologic evidence favoring continental movement. The tendency to rely on the more quantitative analysis of geophysics was probably one factor that caused many geologists, including paleontologists, to de-emphasize evidence of continental movements for half a century. However, some paleontologic evidence was interpreted as being contradictory to the idea of continental movements. One problem was that the history of plate movement envisioned by Wegener, though otherwise similar to that now in vogue, required most of the separation of continents now on opposite sides of the Atlantic to have taken place during the Cenozoic. This undoubtedly influenced Simpson (1940a) to oppose the idea of continental drift on the basis of evidence of the distribution of modern mammals. In fact, we know that continents of the Northern Hemisphere were near enough to their present positions by Paleocene time that the subsequent history of mammalian distribution in many areas was controlled primarily by factors other than continental movement. Furthermore, in the early days of the controversy no single worker was able to evaluate more than a small fraction of the total data available. Anyone attempting to establish an overview of the question was almost forced to include, without evaluation, conclusions reached by other workers. Some of these conclusions were less well thought out than others. A general flaw of the early work was that advocates of continental movement could not successfully refute negative evidence, evidence emphasized by opponents: often it was claimed that if some sort of postulated continental movement had occurred, then a particular living or fossil biota should resemble some other biota, and it was sometimes forgotten that if no similarity was found, there might be an explanation unrelated to continental positions. Therefore, weak negative evidence was permitted to offset certain kinds of strong positive evidence. Even today, paleontologists who have adopted the modern scheme of plate tectonics have difficulty explaining certain data within its framework. Another early problem was that it was impossible to assess the likelihood of dispersal by filter routes and sweepstakes routes. Not knowing if or when any two land masses had moved relative to one another, scientists were unable to develop standards for comparison.

Both in Wegener's time and today, the most striking stratigraphic evidence for continental movement is found in the so-called Gondwana strata of India, South America, Southern Africa, Antarctica, and Australia. These strata, ranging in age from Carboniferous to Triassic, are to varying degrees arranged in similar sequences on the various continents and contain many similar fossils (Figure 12-29). There is unquestionable evidence of glaciation in these strata, so it is not surprising that the Carboniferous floras are depauperate in comparison to floras of comparable age elsewhere in the world. What is remarkable is that there is such a strong resemblance of fossil floras among the now widely separated areas. Two genera of seed ferns, *Glossopteris* and *Gangamopteris*, are particularly well represented. As an example of the similarity of the floras, India and Antarctica have in common about 20 species of *Glossopteris*. Here, however, we come up against the kind of uncertainty that prevented many early workers from favoring continental drift. Some workers argued

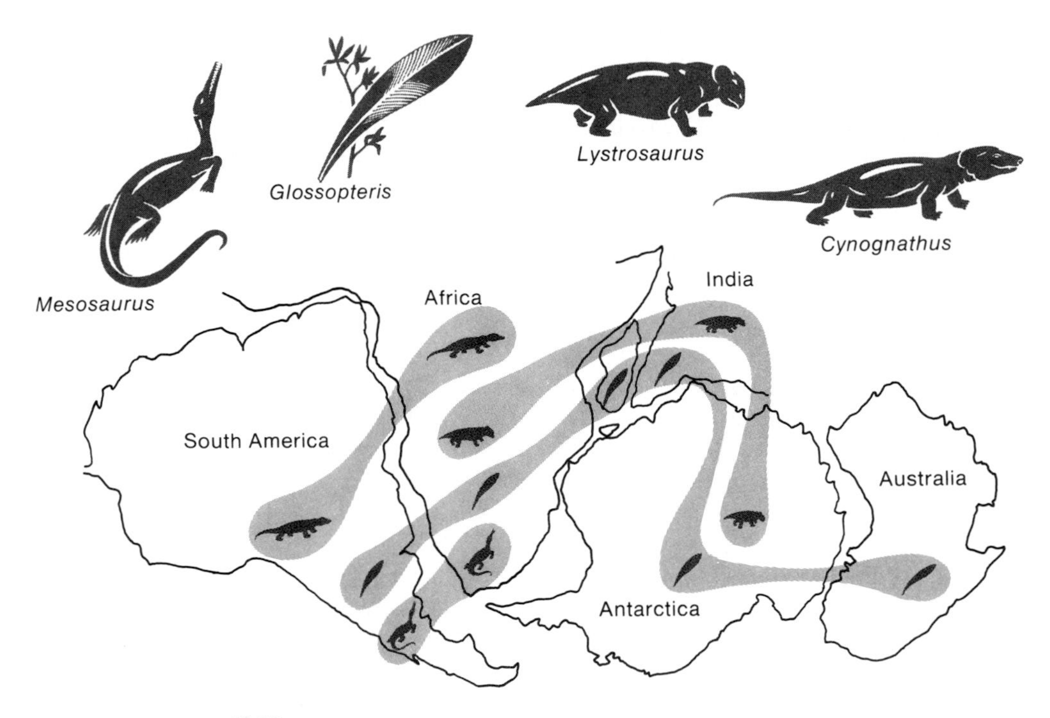

FIGURE 12-29
Reconstruction of Gondwanaland, showing the recorded distribution of key Permian and Lower Triassic genera. (From Colbert, 1973. Copyright © 1973 by Edwin H. Colbert [reprinted by permission of the publishers, E. P. Dutton and the Hutchinson Publishing Group Ltd.])

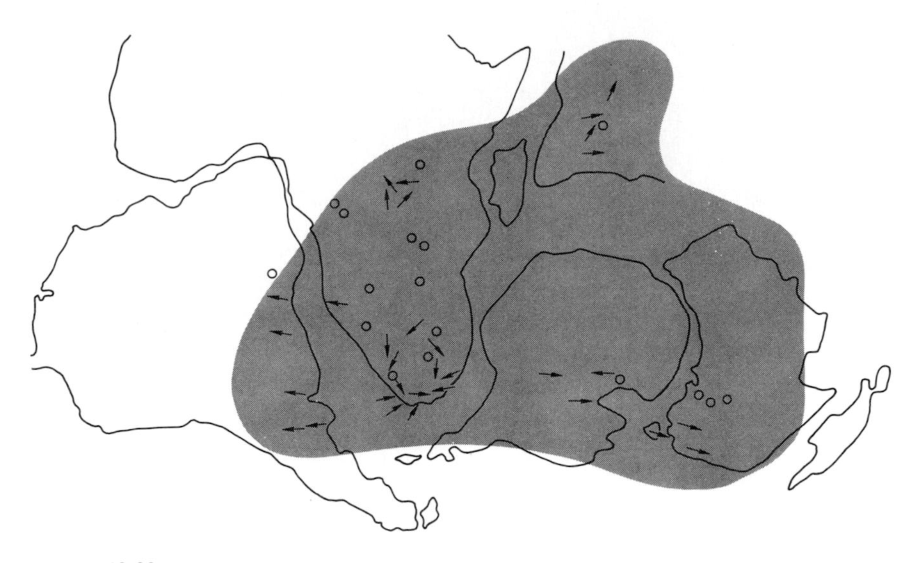

FIGURE 12-30
Evidence of Permo-Carboniferous glaciation in Gondwanaland. Arrows indicate directions of glacial striations on rocks. Circles indicate the presence of rocks considered to be tillites. Gray area shows the possible extent of the ancient ice cap. (From Colbert, 1973. Copyright © 1973 by Edwin H. Colbert [reprinted by permission of the publishers, E. P. Dutton and the Hutchinson Publishing Group Ltd.])

that the seeds of *Glossopteris* might have blown across thousands of miles of ocean. Similarly, though the reconstructed jigsaw-puzzle fit of land masses into the hypothetical continent ***Gondwanaland*** was indeed remarkable, it was regarded by many as possibly fortuitous. Even the evidence that all five land masses were glaciated during part of the Permo-Carboniferous (Figure 12-30) failed to convince many workers of the existence of Gondwanaland. Like the pattern of occurrence of seed ferns, the presence of the small aquatic reptile *Mesosaurus,* which lived during the Early Permian in South America and southern Africa (Figure 12-29), was cited for years as demonstration of a former proximity of these two regions, but skeptics claimed that even though it was a small, freshwater animal, it might somehow have crossed the Atlantic Ocean. Until the advent of plate tectonics, it was only in some areas where geologists knew Gondwana deposits intimately that continental drift remained generally popular. The foremost of its proponents was a South African, A. L. DuToit.

In 1969 a fossil discovery was made in Antarctica that provided dramatic, positive evidence in favor of continental drift. Had the balance of geological opinion not been tipped earlier, this discovery might well have been decisive: the discovery, followed by similar discoveries in subsequent years, was of the distinctive skull of a Lower Permian reptile *Lystrosaurus* (Figure 12-31). This genus is a dominant member of Lower

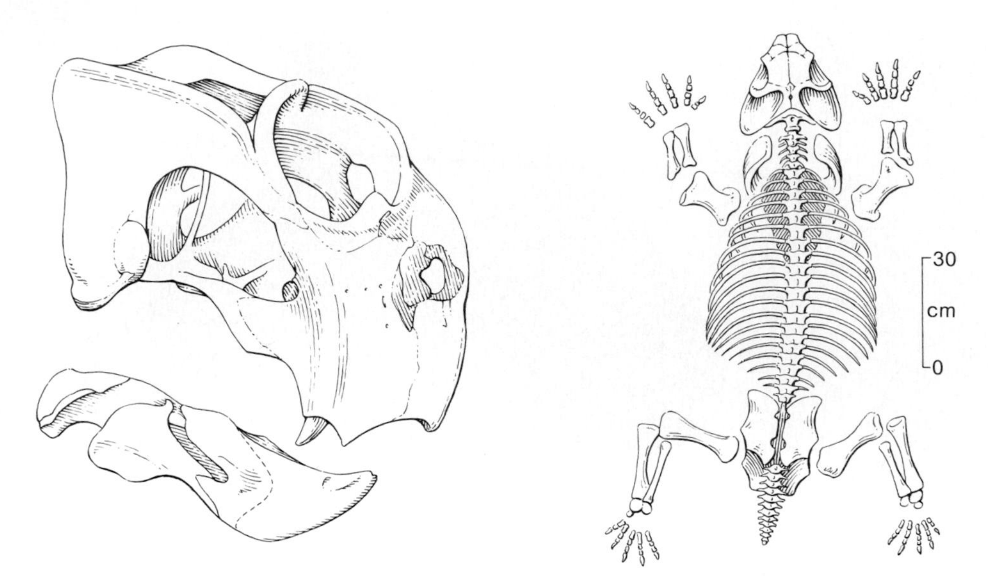

FIGURE 12-31
Skull and complete skeleton of *Lystrosaurus* from South Africa. (From Colbert, 1971.)

Permian deposits in South Africa and India. In fact, the Antarctic specimens are nearly identical to those of a particular South African species and occur as part of a fauna that is nearly identical to the South African fauna that includes *Lystrosaurus* (Elliott et al., 1970). Here is evidence that there was not some sort of filter between South Africa and Antarctica in the Permian, but a continental connection—a corridor. Many members of the similar faunas were fully terrestrial. There is simply no way that they could all by chance have made their way across thousands of miles of deep ocean. The known distributions of *Lystrosaurus* and some of the other Gondwana taxa we have mentioned are summarized in Figure 12-29.

If, partly through historical accident, paleontology did not initially convert the geological profession to continental drift, the study of fossils is at least modifying the view of plate tectonic movements pieced together from other lines of evidence. Currently, the Paleozoic history of continental movements is quite uncertain. What is clear is that, in the late Paleozoic, a huge supercontinent called ***Pangea*** was a conglomeration of all of the separate continents in existence today. The continent seems first to have split parallel to the equator, but we remain unsure of the degree of division that had occurred by the end of the Permian. One interpretation is shown in Figure 12-32,A. Pole positions, figured from paleomagnetic data, are at ends of a line through "A." Much of Gondwanaland was at rather high latitudes. As indicated by the

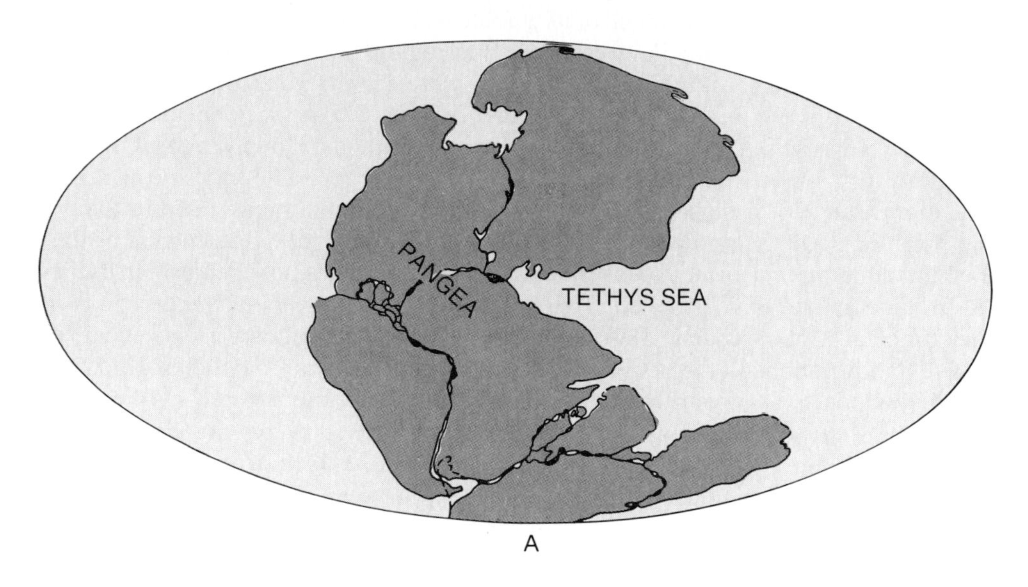

FIGURE 12-32
The breakup of Pangea and Gondwanaland. A: Reconstruction of the supercontinent Pangea as of the end of the Permian. B: Fragmentation of land masses as of the end of the Triassic. Arrows indicate continental movement. New ocean floors are shown with light tint. Solid lines indicate centers of spreading; finely broken line is a subduction zone. (After Dietz and Holden, 1970.)

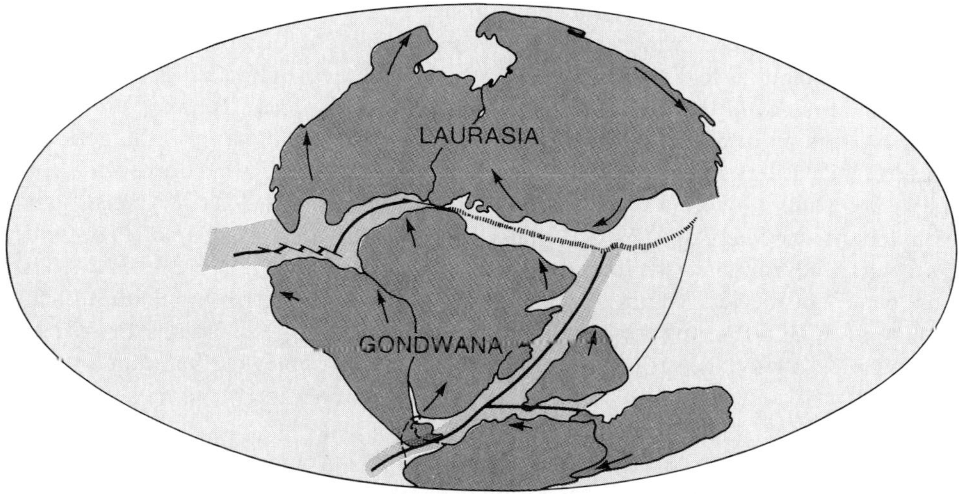

proximity of glaciation (Figure 12-30) and the presence of rings in tree trunks (Challoner and Creber, 1973), much of its climate was temperate. Bakker (1975) noted also that large fin-backed reptiles were limited to regions that were equatorial at this time, which seems to confirm the idea that they were ectothermic (cold blooded).

According to Dietz and Holden (1970), the Tethyan Sea was extended westward during the Triassic, as Pangea split into two large continents, Gondwanaland, which we have already discussed, and ***Laurasia,*** to the north (Figure 12-32,B). At the same time, Gondwanaland began to break apart into some of the continents we know today. The Atlantic Ocean had not yet begun to form. One interesting phenomenon of the breakup of southern continents was the splitting off of what is now peninsular India. This triangular land mass travelled north and slightly east on a plate bounded on two sides by centers of spreading. Here is one place where paleontologic data seem to require modification of the sequence of events constructed from other kinds of geologic evidence. Dietz and Holden (1970) believe India was isolated by the Late Triassic and remained so during northwest movement through the end of the Cretaceous (Figure 12-33,A). The problem here is that Jurassic and Cretaceous dinosaur faunas of the Indian peninsula appear to have required land connections with the rest of the world during much of the Mesozoic, probably by way of Africa. Another puzzle posed by fossil data concerns the position of eastern China during the Permo-Triassic (Colbert, 1971). As suggested in Figure 12-32, the disposition at that time of what is now Southeast Asia remains uncertain. There is no evidence that a section of what is now China formed part of Gondwanaland, yet a *Lystrosaurus* fauna occurs in China. Colbert (1971) suggested that migration may have occurred around the Tethys via the Iberian Peninsula. He compared the apparently disjunct distribution of this fauna with that of living alligators in China and America. It appears that in middle Tertiary times, alligators spread between the two continents by the way of a northern land bridge between Asia and Alaksa.

The history of plate movements that is now generally accepted has suggested possible explanations for many previously puzzling patterns of biotic distribution. We will consider some of the more interesting here. In the Southern Hemisphere one of the most striking distributions of a plant genus is that of *Nothofagus,* the genus of southern beech trees (Figure 12-34,A). This genus appears to have evolved from *Fagus,* the genus of beeches of the Northern Hemisphere. Melville (1973) constructed a phylogeny for living species of *Nothofagus,* using a large number of taxonomic characters, and estimated the times at which species originated (Figure 12-34,B). The most primitive species, *N. alessandri,* occurs in Chile. The interesting feature of the phylogeny is that the present geographic distribution of species seems to bear no relationship to phylogeny. Melville suggests that the somewhat random present distribution of closely related species has been brought about by the movements of continents.

To accommodate the effects of plate movements, McKenna (1973) has added two modes of dispersal to Simpson's corridors, sweepstakes routes, and filter routes. The

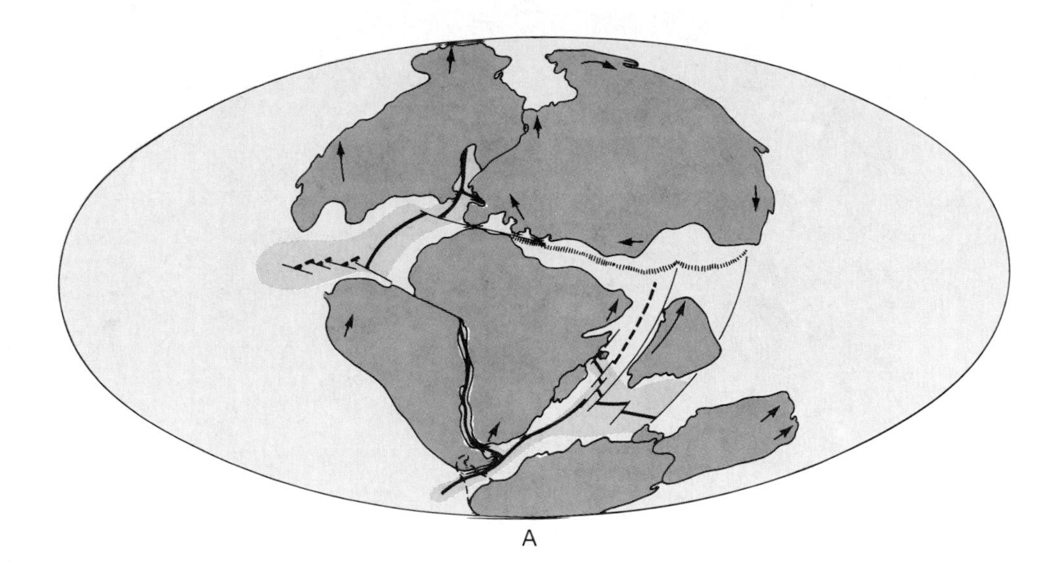

FIGURE 12-33
Reconstructed deposition of continental land masses as of the Late Jurassic (A) and the end of the Cretaceous (B). (After Dietz and Holden, 1970.)

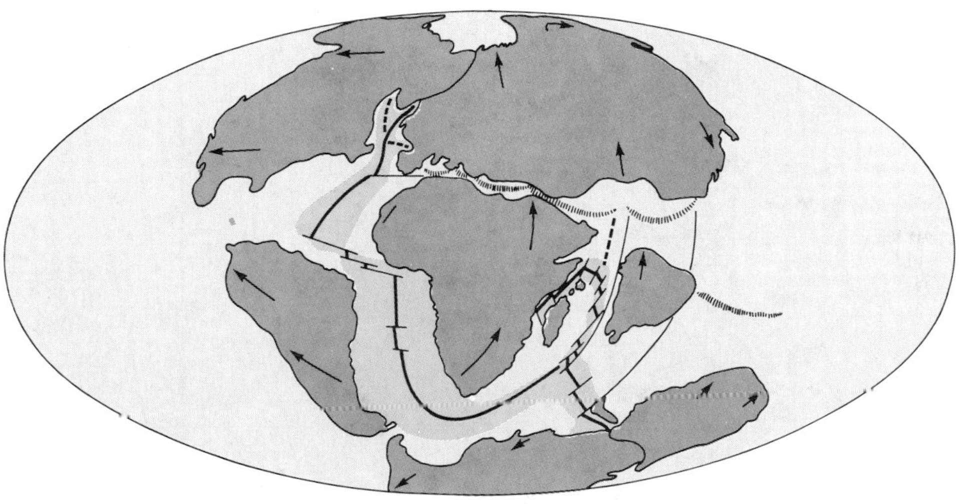

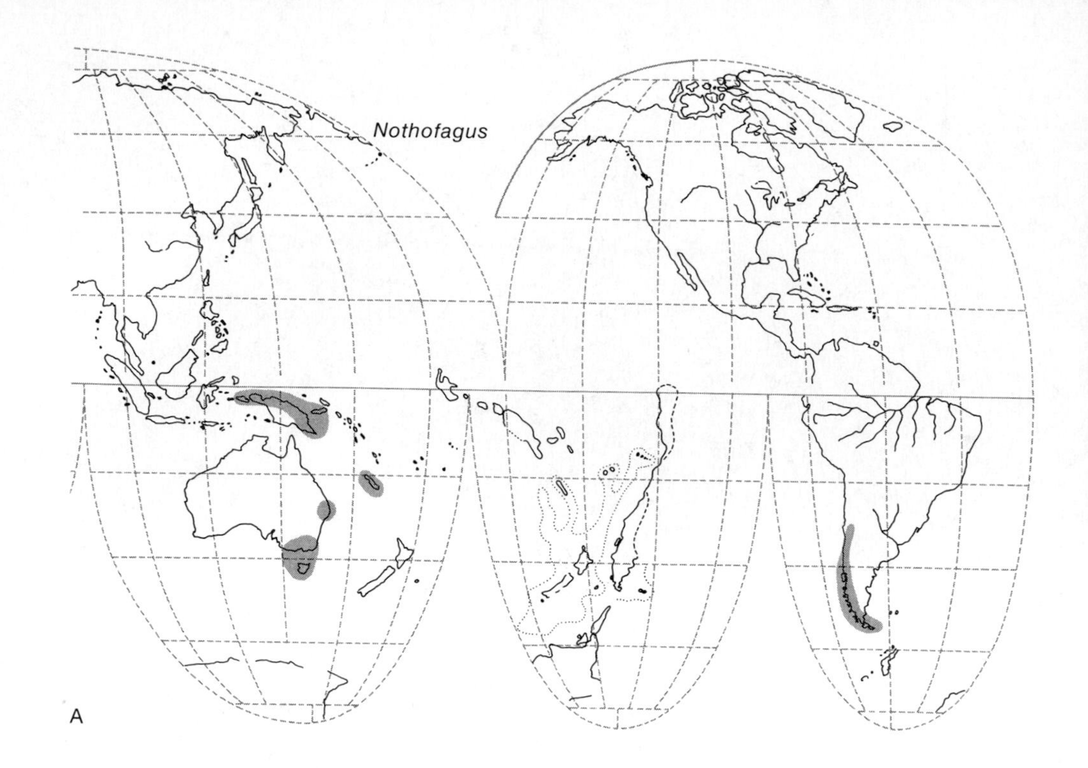
Nothofagus
A

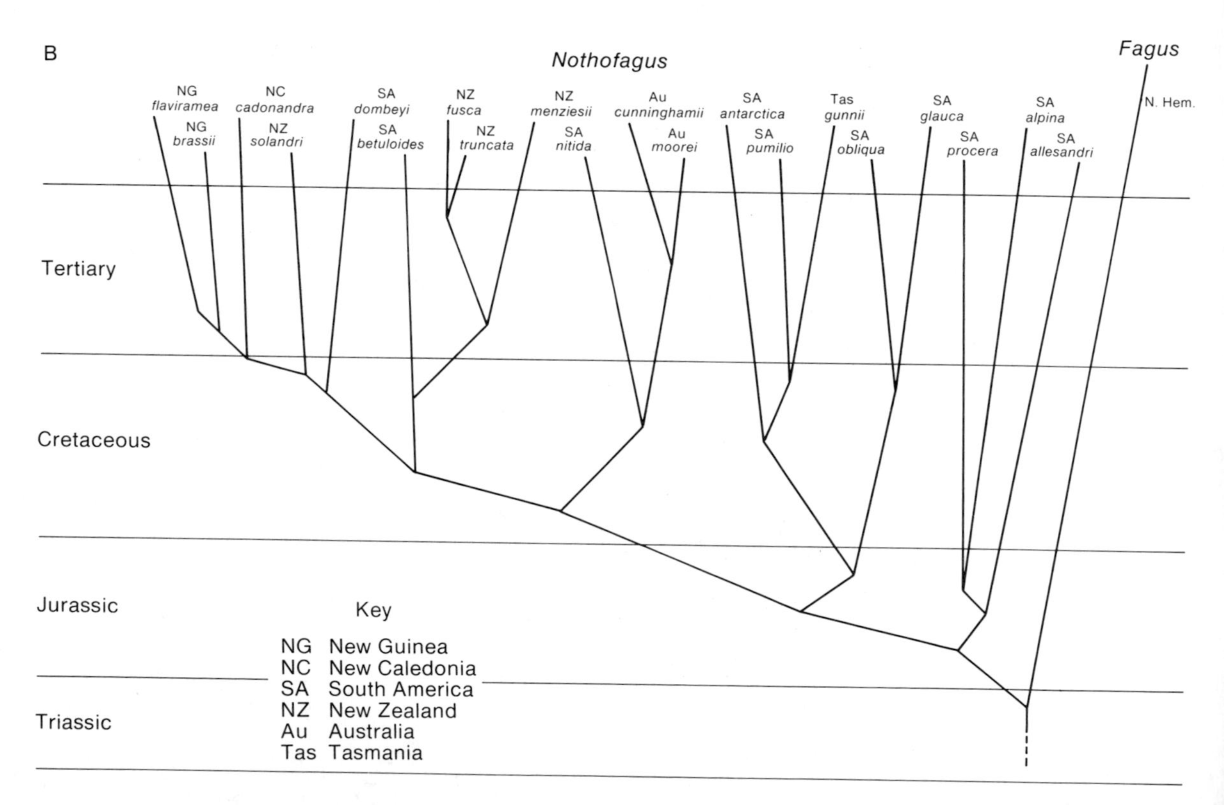
B
Nothofagus
Fagus
NG flaviramea
NG brassii
NC cadonandra
NZ solandri
SA dombeyi
SA betuloides
NZ fusca
NZ truncata
NZ menziesii
SA nitida
Au cunninghamii
Au moorei
SA antarctica
SA pumilio
Tas gunnii
SA obliqua
SA glauca
SA procera
SA alpina
SA allesandri
N. Hem.
Tertiary
Cretaceous
Jurassic
Triassic
Key
NG New Guinea
NC New Caledonia
SA South America
NZ New Zealand
Au Australia
Tas Tasmania

FIGURE 12-34
The modern distribution of the genus *Nothofagus* (A) and the inferred phylogeny of the genus (B), showing geographic distributions of species. (From Melville, 1973.)

first is simply rafting of already fossilized creatures to a new land mass by passage aboard a floating plate. McKenna termed this the ***viking funeral ship*** mode of transport. The fossil continent of peninsular India, for example, was beached on the southern shore of Asia after a tremendous voyage (Figure 12-33,B). The second is ***Noah's Ark*** transport. The mechanism here is really the same, but the organisms are transported while still alive. An apparent example is dispersal of the genus *Podocarpus.* Florin (1963) observed that from the Carboniferous through the Eocene, every genus of conifers (gymnospermous trees) was restricted either to the Laurasian Region (Eurasia and North America) or to land masses that originally formed part of Gondwanaland. During the Tertiary, however, this division ended. For example, the previously southern genus *Podocarpus* spread into Central America and Southeast Asia. This expansion was apparently brought about by the northward rafting of South America and Australia.

Just as rafting has extended the distributions of some taxa, it has preserved other taxa from extinction by isolating them from predators and superior competitors. Exile on Australia has permitted certain primitive creatures like the platypus and echidna to survive. For many years scientists debated where the famous marsupial fauna of Australia came from. South America, as we have already discussed, harbors several marsupial groups today and supported flourishing marsupial faunas in the past. In the past, some workers believed that South America was the ancestral area for the Australian forms, but other workers favored a northern origin and entry by way of Indonesia. The acceptance of plate tectonics has settled this debate in favor of a South American ancestry. A general biogeographic scheme of phylogeny for the marsupials is shown in Figure 12-35. In latest Cretaceous, or Paleocene, times, the ancestral group of South American marsupials arrived from North America. By at least the Oligocene, when the known fossil record of Australian marsupials begins, the journey to Australia via Antarctica (Figure 12-36) had been made. The precise time at which Australia split from Antarctica is unknown, but the time of splitting and the marsupial invasion of Australia probably preceded the Oligocene. It remains a mystery why no placental mammals made their way to Australia from South America with the marsupials. In any event, the marsupials diversified in Australia and have survived there in isolation ever since.

Certain other taxa of the Australian region are also relicts produced by plate tectonic isolation. New Zealand was apparently split off from Australia and contiguous land masses in the Late Cretaceous. On it, a number of primitive taxa were preserved. Raven and Axelrod (1972) point out that the present lowland flora of New Zealand is similar to that of Gondwanaland of 80 million years ago! In addition, the primitive frog *Leiopelma* and the primitive reptile *Sphenodon* (the tuatara) survive

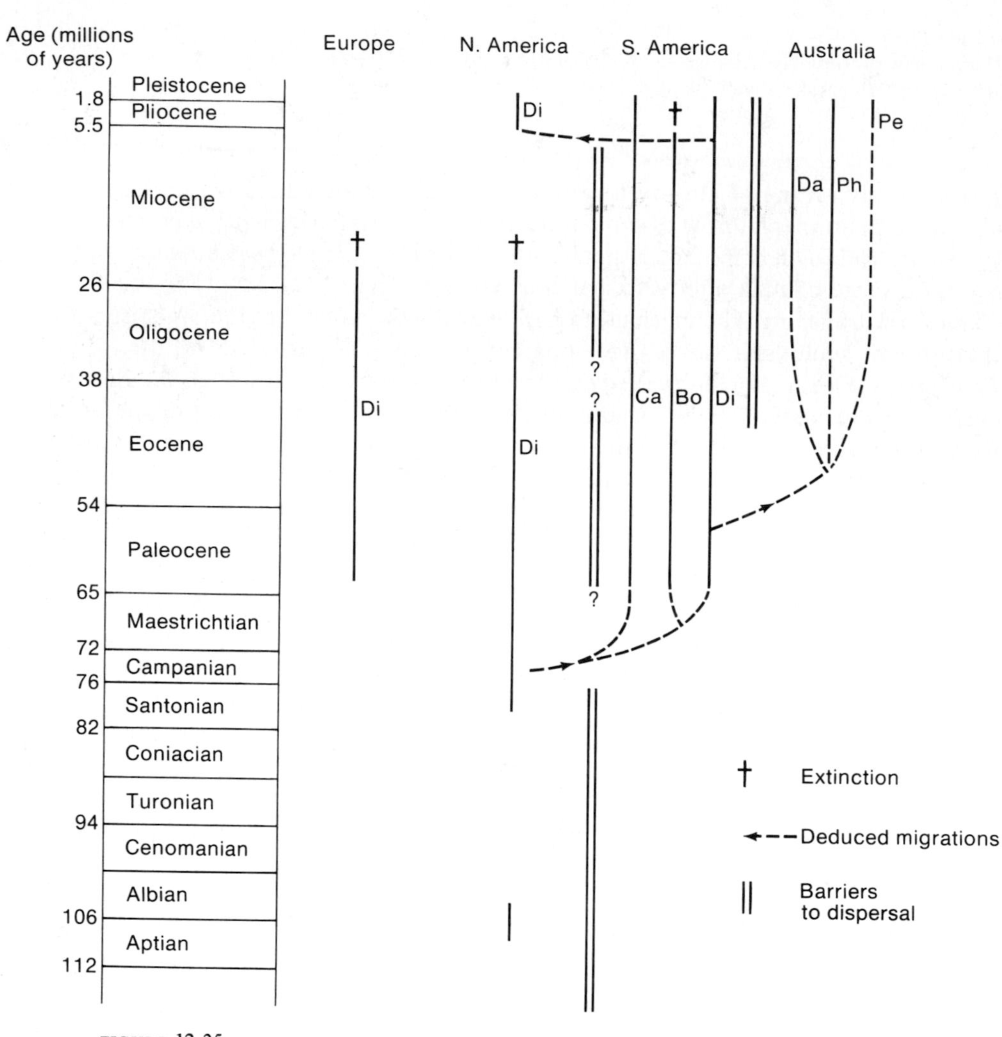

FIGURE 12-35
Geographic phylogeny of the marsupials. Di—Didelphoidea; Ca—Caenolestoidea; Bo—Borhyaenoidea; Da—Dasyuroidea; Ph—Phalangeroidea; Pe—Perameloidea. (From Jardine and McKenzie, 1972.)

there as relicts. The giant moa birds did too, until wiped out by man. Raven and Axelrod note that the remarkably primitive present-day flora of New Caledonia is also a relict, preserved by the early separation of that small island. This flora resembles the Late Cretaceous flora of eastern Australia.

One of the most significant biogeographic contributions of plate tectonics may be in providing an explanation for the existence of Wallace's Line (Figure 12-37). This narrow band is one of the most remarkable lines of biogeographic demarcation in the

FIGURE 12-36
Proximity of Australia (bottom center of map), Antarctica (lower left), and South America (right of center) in the Campanian (Late Cretaceous). (From Jardine and McKenzie, 1972.)

FIGURE 12-37
The location of Wallace's Line. Numbers of genera are indicated for the various islands. (Modified from Simpson, 1961a.)

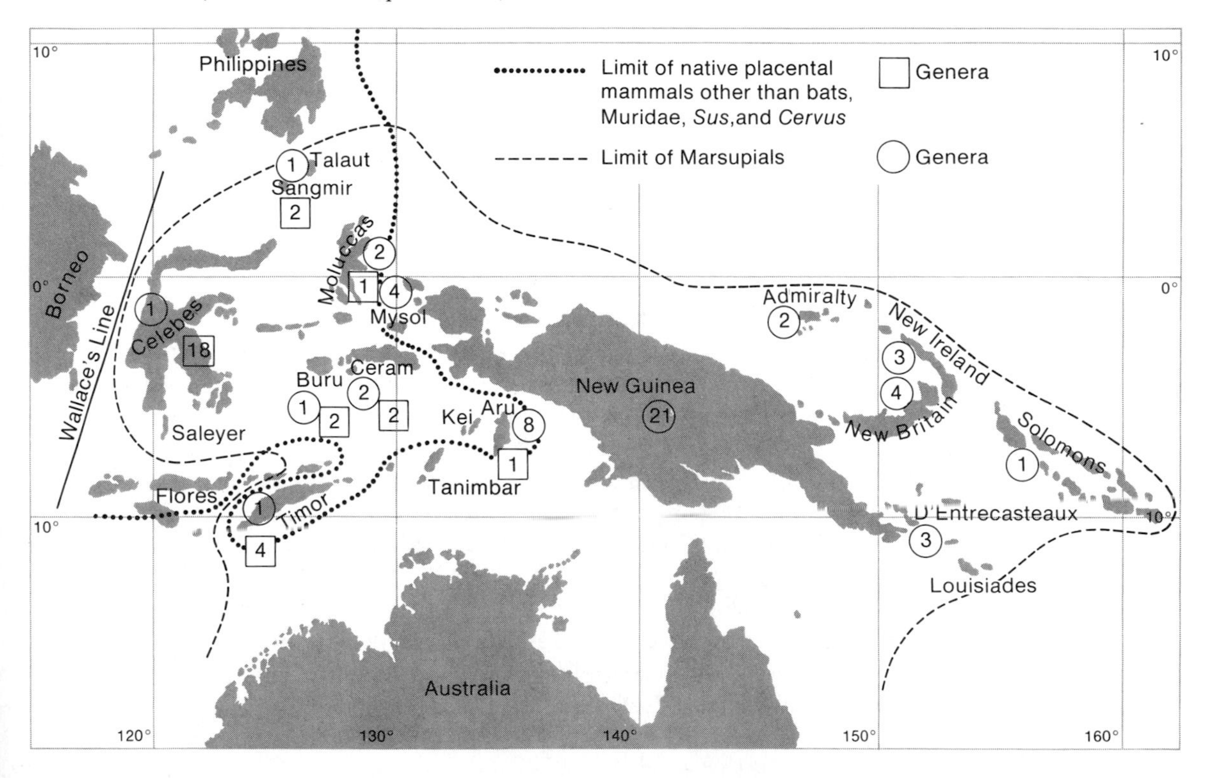

world. It was discovered in the last century by A. R. Wallace, a cofounder with Darwin of the concept of natural selection. Raven (1935) provided a summary of the taxa whose ranges are demarcated by this line. Of the diverse mammalian taxa to the west, only a few species of shrews, two monkeys, a deer, a pig, and a porcupine have crossed Wallace's Line to Celebes. Only one Australian marsupial has crossed in the opposite direction. Why the line is located exactly where it is and why certain plants and animals have crossed it when others have not are not fully understood and relate both to chance factors and to local tectonics, submarine physiography, and fluctuations in sea level. In any event, plate tectonic reconstructions suggest that the biotas were juxtaposed in the Miocene by collision of the northward-moving Australian plate with the Asian plate (Figure 12-27). Mixing of faunal elements has occurred slowly because of the great distances between islands.

The biogeographic effects of plate tectonics are evident even on a worldwide scale. The distribution of post-Paleozoic terrestrial quadrupeds, for example, reflects the history of fragmentation of Gondwanaland and Laurasia. Kurtén (1967) pointed out that the many orders of modern mammals have arisen in the Cenozoic, after land masses were highly fragmented. On the other hand, there were few orders of dinosaurs, and dinosaurs radiated early in the Mesozoic, when major land areas were generally interconnected (Figure 12-32,B). Much convergent evolution has occurred within the Mammalia, with different taxa occupying similar adaptive zones on different continents. The convergence between the isolated marsupials of Australia and placental mammals of other continents is particularly striking. For example, Australia harbors marsupial species that bear strong superficial resemblances to placental moles, flying squirrels, and anteaters.

The northward migration of Australia illustrates the consequence of plate movement across latitudes, namely the subjection of living taxa riding on a plate to changing climatic conditions. Raven and Axelrod (1972) point out that many kinds of plants and animals endemic to Australia were saved from extinction by the equatorward migration of the continent during the Cenozoic, when the world climate was undergoing a net cooling trend. At the same time, as the continent entered the dry zone between 20° and 30° South of the equator (Figure 12-7) much of it became arid. Recall that at these latitudes air that has risen near the equator, cooled, and lost its moisture descends and picks up moisture as it warms up (Figure 12-3). In Australia during the latter part of the Cenozoic, there has been an increase in the diversity and abundance of vegetation adapted to arid and semiarid conditions. As the continent continues to move north, savannah and then rain forest will spread over the central portion of the continent, which is now arid. The southern region, now bathed in a Mediterranean climate (page 387), will become arid and much of its biota will become extinct.

The separation of the Americas from Europe and Asia and the formation of the Atlantic Ocean serve as a convenient example of the evolutionary divergence of biotas on opposite sides of large plates being rifted apart. This final rifting began in the Triassic, and we will discuss its effects a bit later. First we will consider an earlier cycle of rifting and reuniting of the continents. In the Late Precambrian, rifting within a large supercontinent separated what is now Europe from what is now North America. By the Early Ordovician, the predecessor of the Atlantic Ocean was apparently rather wide (Figure 12-38). It is therefore no surprise that Williams (1973), using the index of faunal similarity presented on page 212, found faunas of Europe and North America to be relatively dissimilar. Shortly thereafter, the direction of plate movement was reversed. A subduction zone formed within the ocean, moving the continents back together (Figure 12-38). This seems to account for Williams's observation that British brachiopod faunas in the middle of the Caradocian (lowest stage of the Upper Ordovician) are quite similar to faunas of the same age in Maine and Newfoundland.

The evidence that Africa and South America formed a single land mass in the late Paleozoic offers an explanation for another paleontologic phenomenon that would otherwise remain a mystery. This is the occurrence of an unusual fauna of bivalve mollusks in the Estrada Nova Formation of the Paraná Basin, which lies near the eastern coastline of South America (Figure 12-39). The fauna, analyzed by Runnegar and Newell (1971), is so unusual that its exact age remains undetermined. Probably the fauna dates from late in the Permian, but it may date from very early in the Triassic. Many of the taxa belong to the Megadesmidae, a family that occurred primarily in Gondwanaland. The fauna occupied an inland position, in what was apparently a largely isolated body of water. The water may have been brackish or even fresh, but this remains uncertain. At any rate, the bivalve fauna includes 14 distinctive genera, and only one of these is known to occur outside the Paraná Basin. It seems that a few species initially colonized the basin from the open sea, possibly by way of a seaway to the south (Figure 12-39). It is not certain how long it took for the colonizers to diversify into the 14 genera and more than 20 species that are now recognized, but it would seem to have been a very short time. Runnegar and Newell (1971) compared the rapid adaptive radiation of bivalves in the Paraná Basin to a similar radiation within the Pontian Sea of the Pliocene, an inland body of brackish water that was the predecessor of the present-day Caspian Sea. During only a very short time, certainly less than 5 million years, a dramatic adaptive radiation occurred in the Pontian Sea. Well over 30 genera belonging to 5 subfamilies seem to have diverged from the single genus *Cerastoderma,* which includes the present-day European cockle. Two things are particularly striking about the evolution of bivalves in both the Paraná and Pontian Basins. One is that speciation was much more rapid than in the open ocean. The other is that an enormous variety of new morphologic features appeared. A large percentage of the speciation events produced dramatically divergent new forms that are recognizable as new genera. These episodes of diversification are comparable to those that occurred in certain large, extant lakes (page 415). While some workers have argued that sympatric speciation explains the diversification in both basins, others believe that speciation has resulted from microgeographic isolation of small populations.

The final rifting that divided Laurasia along what is now the mid-Atlantic ridge apparently commenced in the Triassic (Figure 12-32). Even the amount of divergence

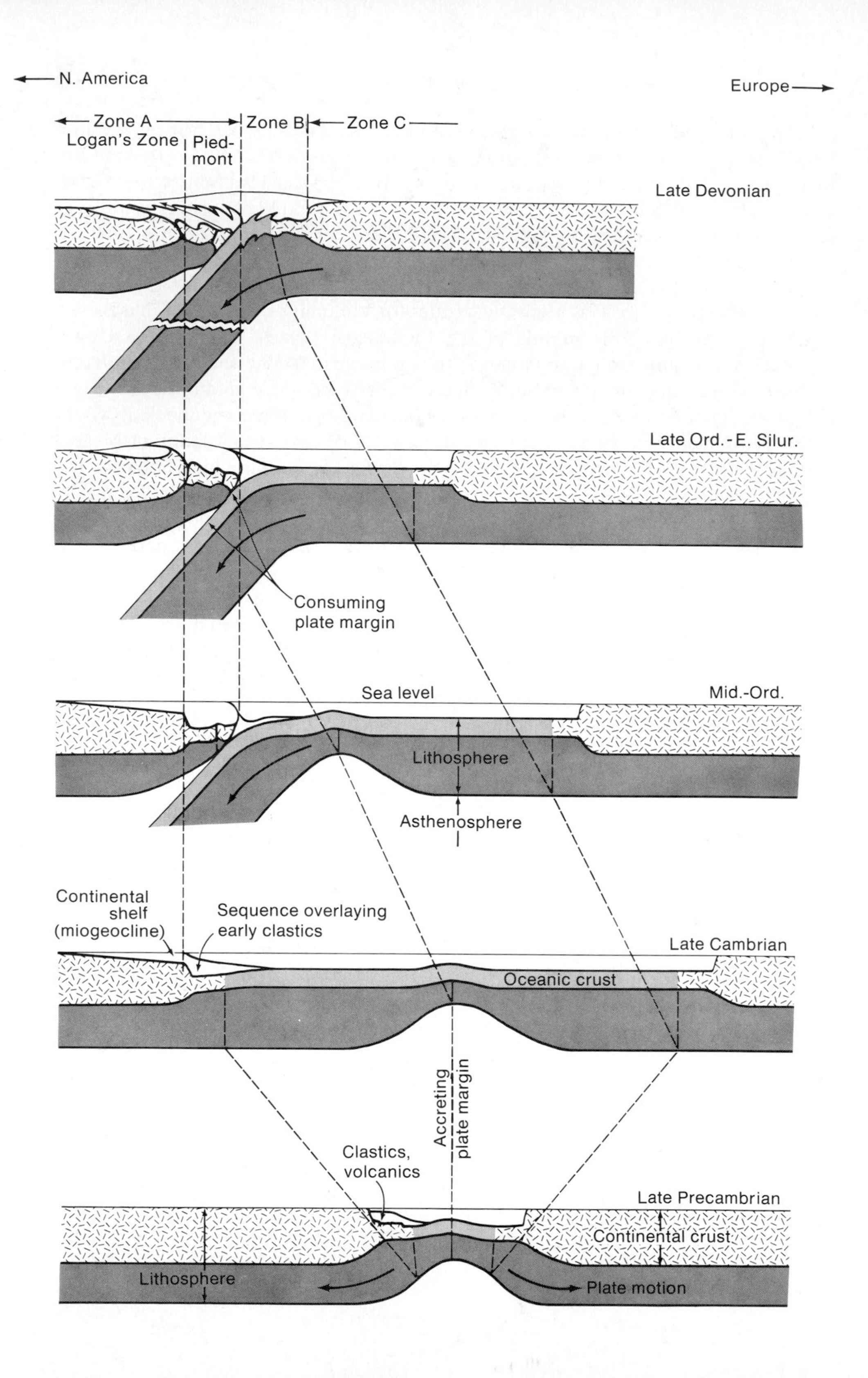

N. America
Europe
Zone A
Zone B
Zone C
Logan's Zone
Pied-
mont
Late Devonian
Late Ord.-E. Silur.
Consuming
plate margin
Sea level
Mid.-Ord.
Lithosphere
Asthenosphere
Continental
shelf
(miogeocline)
Sequence overlaying
early clastics
Late Cambrian
Oceanic crust
Accreting
plate margin
Clastics,
volcanics
Late Precambrian
Continental crust
Lithosphere
Plate motion

FIGURE 12-38
Rifting-apart of the European and North American land masses in the Late Precambrian and early in the Paleozoic, and reunification of the continents in the middle of the Paleozoic after the formation of a new subduction zone early in the Ordovician. (From Bird and Dewey, 1970, reprinted with permission.)

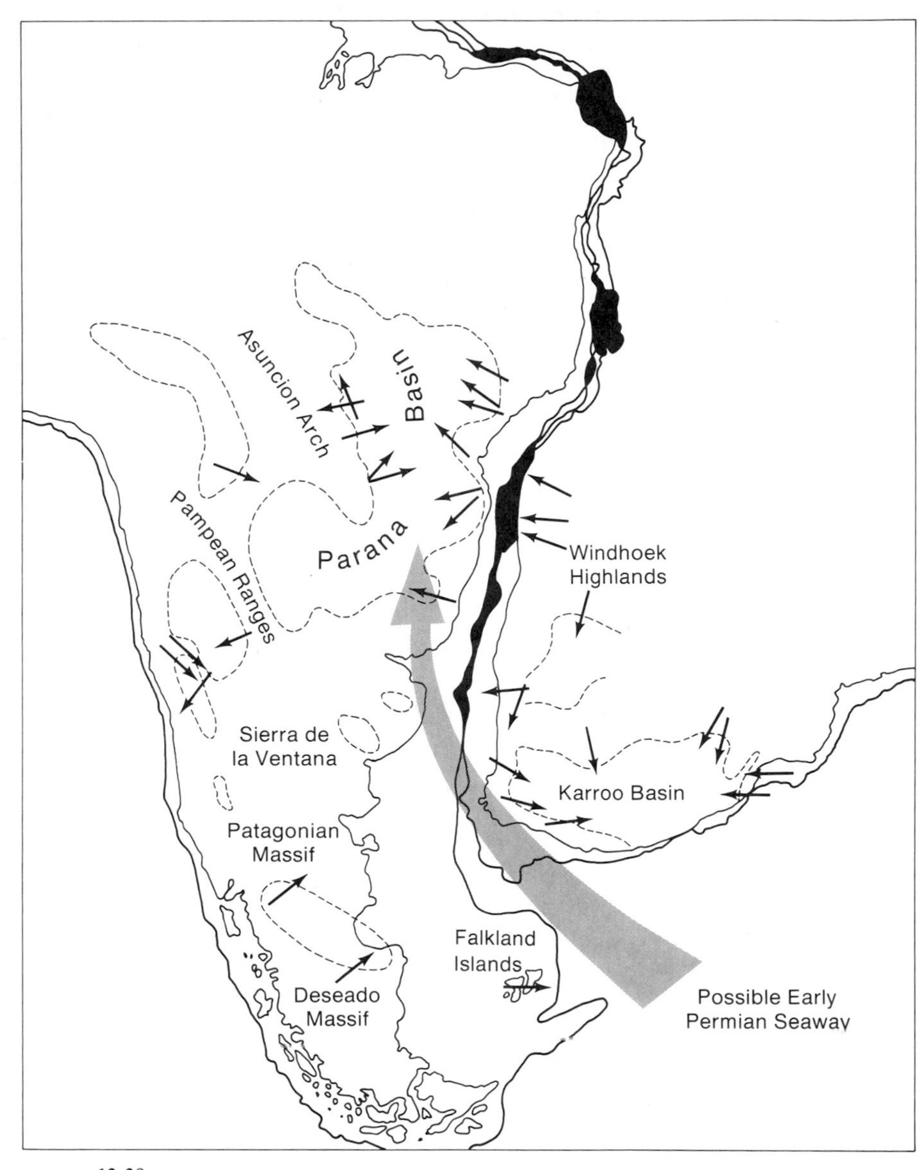

FIGURE 12-39
Paleogeography of the Paraná Basin in the Early Permian. Late Paleozoic marine and continental sedimentary deposits are indicated by dashed lines. Arrows indicate direction of sediment transport. (After Runnegar and Newell, 1971.)

that this rifting has caused since the Cretaceous has had a measurable effect on the dissimilarity of faunas on opposite sides of the Atlantic. Kauffman (1973) estimated that the percentage of endemic genera among Caribbean bivalves has risen to a present value of about 20 percent from a value of about 5 percent in the Coniacean (middle Late Cretaceous, about 85 million years ago). The increase in endemism reflects not only a weakening connection with the eastern Atlantic, but also a less effective westward connection with the Pacific. However, Coates (1973) showed that until well into the Cretaceous there was no Caribbean province of scleractinian corals: generic composition of Caribbean faunas was not distinct from that of the Indo-Mediterranean region. By about the beginning of the Late Cretaceous, it appears that coral larvae were unable to cross the expanding Atlantic, and during the Late Cretaceous, the percentage of endemism increased notably.

It has been suggested that species-area equations of the sort used in the study of island biogeography (page 410) may be of use in considering the biotic effects of plate tectonic fragmentation and coalescing of continents (Flessa and Imbrie, 1973). If, for example, we were to assume that in the equation on page 410 the value of z is 0.26, then if a continent of area A_1 is divided into three smaller continents of equal size (A_2), the number of species (S_2), on any of the newer continents might be expected to decline to 75 percent of the number of the parent continents (S_1):

$$S_2/S_1 = (A_2/A_1)^2 = (1/3)^{0.26} = 0.75.$$

There are various problems in undertaking this kind of analysis. One is that we have few measured values of z for large continents. Furthermore, the distribution of habitats on a continent will influence the effect of fragmentation. For example, consider what will happen if a continent that is part savanna and part tropical rainforest is rifted apart more or less along the boundary between the two major habitats. Suppose that the rifting carries the fragment dominated by rain forest a bit farther toward the equator, eliminating what little savanna bordered its rifted side. Suppose that a few fringes of rain forest on the other new continent are eliminated in the same way. Then overall biotic diversity (diversity of the total continental area considered) will remain more or less constant. Similarly, diversity within the rainforest and diversity within the savanna will remain more or less constant.

Despite the kind of problem outlined in the preceding paragraph, quantitative treatment of relationships between area and diversity may turn out to shed light on some evolutionary problems.

Mass Extinction

During Phanerozoic time, episodes of relatively sudden large-scale extinction have decimated significant fractions of the world's biota. These episodes are usually called mass extinctions. What is striking about them is that they tend to affect all or most representatives of certain major taxa almost simultaneously. This is not to say that the extinctions are instantaneous. Often fossil evidence indicates that extinctions proceed for several million years. Extinction is a geographic phenomenon (Figure 11-15). Mass extinctions, in particular, must be viewed in a biogeographic framework because they are generally worldwide in scale.

A variety of hypotheses—some ingenious and some preposterous—have been proposed to explain mass extinctions. Some of these hypotheses are meant to apply only to a single episode of mass extinction and others, to several or all mass extinctions. Rather than reviewing all kinds of hypotheses that have been presented over the years, we will focus upon certain biogeographic considerations that seem now to represent fruitful avenues for research.

The two most famous mass extinctions are those that occurred at the ends of the Paleozoic and Mesozoic Eras. Extinction late in the Permian had an especially profound effect upon marine invertebrates. Brachiopods and crinoids underwent great reductions in diversity then and, among other groups, fusulinid foraminiferans, tetracorals, and trilobites (already on the wane) became entirely extinct. The ammonoids came so close to extinction that only one or two genera seem to have made the transition into the Mesozoic Era. The Late Cretaceous extinction eliminated the dinosaurs, as well as marine groups like the ammonites, belemnites, and rudist bivalves. Mass extinctions have also occurred at the ends of several other geologic periods. This close correlation is in itself remarkable when one considers the generally haphazard way in which the geologic systems came to be recognized (page 220). It would be hard to deny that important biotic and stratigraphic changes caught the eye of pioneering geologists of the past century: these abrupt changes, most of which first became apparent within sedimentary rocks on the small land area of Great Britain, are repeated in varying ways and to varying degrees throughout the world.

One important aspect of many transitions between geologic periods, including the transition between the Permian and Triassic and the Cretaceous and Tertiary, is that in many parts of the world they are represented by stratigraphic gaps. These gaps represent episodes of widespread withdrawal of shallow seas from continental margins. While some regressions of this type have coincided roughly with transitions between geologic periods, others have not. Sloss (1963) evaluated the general record of transgressions and regressions in North America (Figure 12-40) and applied the name ***sequences*** to bodies of rock that have formed between major regressions. Newell (1967) observed that the simple fact that continents throughout the world are today emergent suggests there are worldwide controls of sea level. Of course, land movements can offset the effects of worldwide controls in certain regions. Nevertheless, for every million-year interval of Phanerozoic history there is some continental record of shallow-water deposition. The tendency of some mass extinctions to occur at times when seas withdrew from continents led early workers to propose that destruction of shallow water environments and crowding of species might have been responsible for wholesale extinction. Newell (1967) has provided evaluations of the early evidence.

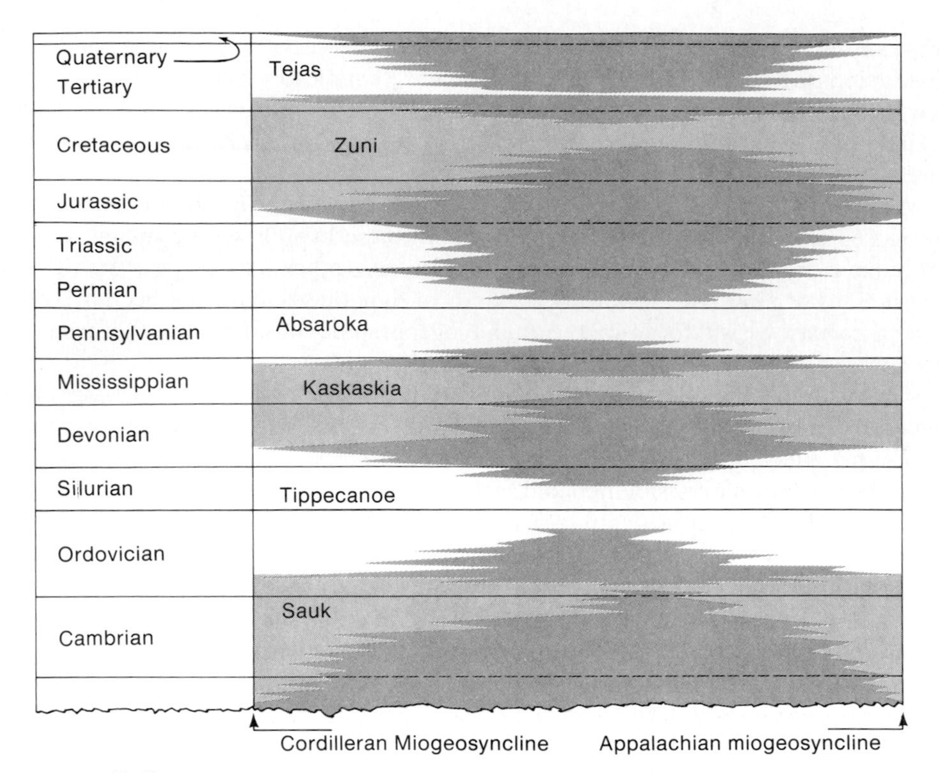

FIGURE 12-40
Sequences (Sauk through Tejas) of the North American continent. Alternating episodes of deposition are indicated by white and stippled areas. Black areas represent nondepositional gaps. (From Sloss, 1963.)

Before examining the ways in which regressions might be related to extinction, we will consider the way in which large-scale transgressions and regressions may be brought about; plate tectonics provides a possible mechanism. Shortly after the generally accepted new scheme of plate tectonic movement was proposed, various authors suggested that when seafloor spreading is occurring, the flow of heat and new crustal material elevates mid-oceanic ridge systems, which must then displace substantial volumes of water onto the continents. Here, then, we have a potential mechanism for worldwide transgression. On the other hand, at times when little spreading is occurring, previously elevated seafloors should subside, draining water from the continents. The forces that drive plate movements are not yet understood. For this reason we do not understand why spreading or subduction begins or ends in particular areas. It does seem clear that subduction between two continents is generally terminated when the continents collide (Figure 12-26). It may be that in the late Paleozoic the coalescing of all continental masses into the supercontinent Pangea (Figure 12-32,A) terminated many spreading systems, leading to subsidence of many areas of the sea floor and to worldwide regression. Hays and Pitman (1973) estimated the volume of the midoceanic ridge systems for various times in the Cretaceous and Tertiary. They found a crude agreement between the area of land that would have been covered by water, based on their estimates of ridge volume, and the area that geologic evidence indicates was, in fact, covered by water (Figure 12-41). There remain uncertainties about the degree of displacement caused by active ridge systems. What was hoped would be a simple scheme has been complicated by the observation that some spreading ridge systems are of relatively low relief.

At the turn of the century Emil Haug noted a relationship that Johnson (1971) documented in considerable detail. This relationship is a coincidence in time between ***orogenic*** (mountain building) movements and marine transgressions (Figure 12-42). Johnson (1973) named it the ***Haug Effect.*** The Haug Effect remains to be fully explained, but both mountain building and worldwide transgression seem to have occurred during times of active seafloor spreading. Intervals of relative quiescence seem to be intervals of regression and continental emergence.

How are mass extinctions related to the large-scale movements of land and ocean that we have been discussing? The answer is complex and we do not have a definite answer. Valentine and Moores (1970) suggested several possible relationships, one of which is that the widespread transgressions should serve to moderate climates. The reason is that water tends to retain heat longer than land. This is why at high latitudes areas adjacent to oceans or large lakes tend to remain warmer in winter than areas farther inland. Hays and Pitman (1973) suggested that worldwide regression played a

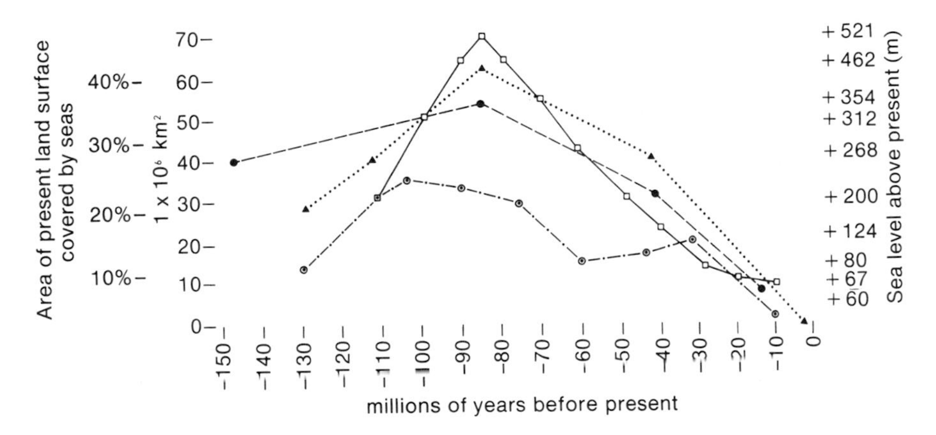

FIGURE 12-41
Area of present land areas covered by former seas, and the heights of these seas plotted against age. Maximum transgression occurred in the Late Cretaceous. Squares show the curve calculated by Hays and Pitman (1973) and other symbols show curves that these authors calculated from the analyses of other workers.

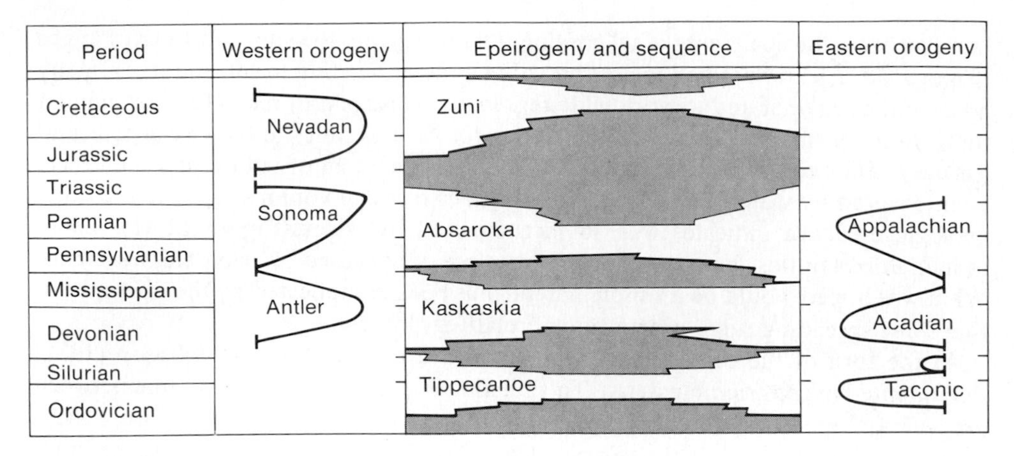

FIGURE 12-42
The coincidence in time of sequence deposition (see Figure 12-40) and orogenies in North America. (From Johnson, 1971.)

major role in the mass extinction at the end of the Cretaceous. Earlier in the Late Cretaceous broad epicontinental seas inundated the land, providing north-south corridors along which heat from the equator spread poleward. Using analogies with the modern world and evidence of past climates, Gordon (1973) reconstructed global ocean currents for Maastrictian time, just before regression (Figure 12-43). Then, as throughout the Mesozoic, there existed a Tethyan Seaway of warm water stretching around the equator. Currents created by the Coriolis Force must have prevailed in large oceanic areas of the Northern and Southern Hemispheres. Trade winds would have produced a more continuous westward equatorial current than exists today, when we have no Tethyan Seaway. Many Cretaceous taxa were confined to the tropical Tethyan belt. It seems reasonable to suggest that lowering of the seas not only cooled climates by exposing more land but also cooled low-latitude climates by breaking up the Tethyan current, from which warm water had been transported poleward. This current must have been further obstructed at about this time by the movement of Africa and India close to Eurasia. Clearly, latitudinal gradients in temperature steepened at this time, and seasonality of climates increased in regions not far from the equator. There may have been harmful biological effects on land as well as in the oceans.

Analyzing tetrapod animals from the Permian through the Cretaceous, Bakker (1977) has recognized seven successive dynasties, or complexes of communities. Each dynasty arose and diversified rapidly, only to be terminated by the sudden, nearly synchronous extinction of many of its major families. Each mass extinction was followed by the adaptive radiation of new taxa, which formed a new dynasty. Certain habitat groupings of Mesozoic families of aquatic and terrestrial tetrapods are shown

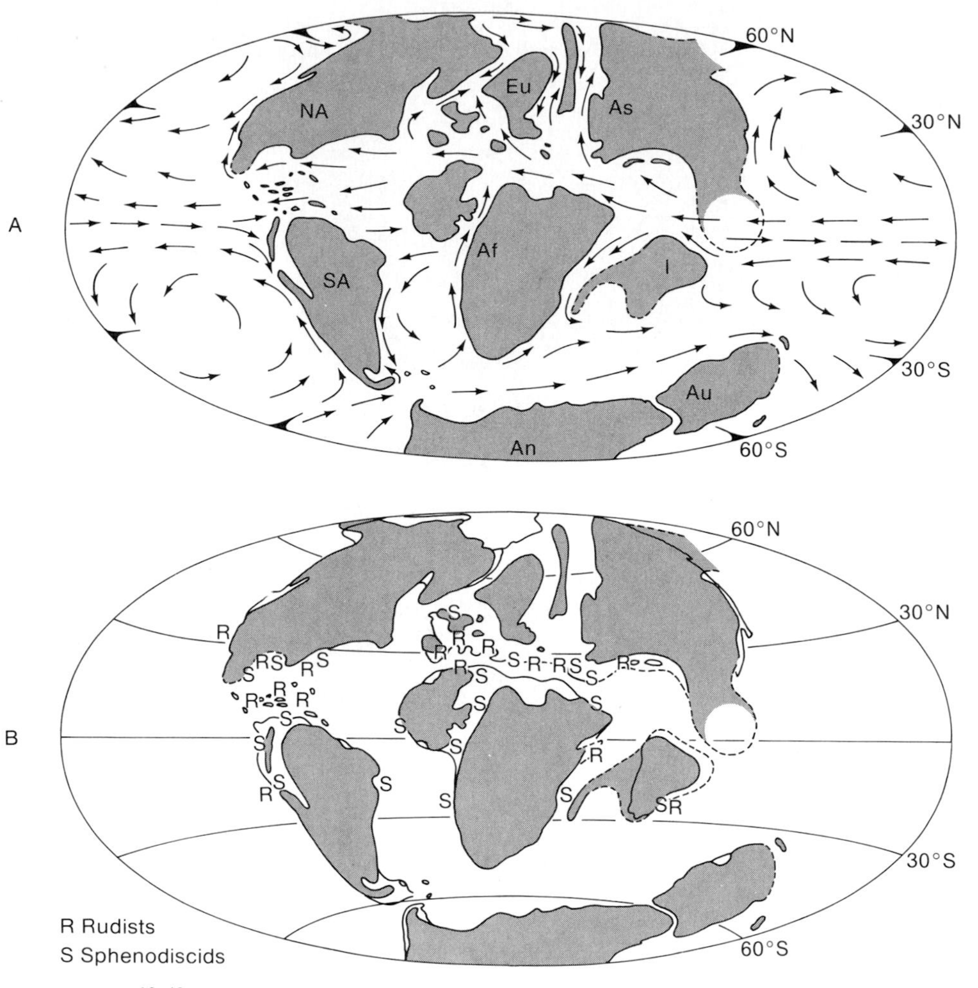

FIGURE 12-43
The Tethyan Seaway of the Late Cretaceous. A: Distribution of land masses, seaways, and major ocean currents for the Maastrichtian (latest Cretaceous). B: Restriction of rudist bivalves (see page 284) and sphenodiscid ammonoids to the equatorial Tethyan Seaway. (From Gordon, 1973. Copyright © 1973 by The University of Chicago.)

in Figure 12-44. At the ends of the three Mesozoic periods, mass extinctions are apparent for marine forms and large terrestrial herbivores. Termination of dynasties seems to have coincided neither with change in climate nor with change in plant life. Bakker has shown that neither small reptiles (those weighing less than about 10 kilograms) nor freshwater aquatic forms (Figure 12-44) suffered mass extinctions. He

200
150
100
50
10⁶ yr bp
Triassic
Jurassic
Cretaceous
Cenozoic
S A L C N R H S P T B B C O K T B V H B A A C T C S C M P E O M
Stages
Marine aquatics
Large freshwater aquatics
Large terrestrial herbivores
A
B
C
D
E
F
G
H
I
J
K
L
M
N
O
P
Q
R
S
T
U
V
W
X

FIGURE 12-44
Ranges of Mesozoic families of tetrapods. Stages of each Mesozoic Period are indicated by their first letters. Each continuous bar represents a family. Narrow segments of bars represent intervals in which families are present but very rare. Note that marine aquatic groups and large terrestrial herbivores, but not large freshwater aquatic taxa, underwent dramatic mass extinctions at the end of the periods. (From Bakker, 1977.)

has also concluded that mass extinction results not so much from increase in rate of extinction at the species level as from decrease in rate of speciation. It seems that at times of mass extinction too few new species arose from old species dying out at normal rates to be replaced. Bakker has suggested that this phenomenon in the history of large terrestrial mammals may relate to the Haug Effect described above. His idea is that at times of transgression and orogeny, there existed diverse terrestrial habitats, and rates of geographic speciation were high. When mountain building ceased, vast homogeneous lowlands were left and, though large tetrapods could inhabit them, the broad biogeographic ranges of these large species within the lowlands made speciation rare. Speciation could occur at high rates only in heterogeneous areas of high topographic relief. Small species, requiring smaller barriers for isolation in the speciation process, speciated at high rates in the monotonous lowlands and were unaffected. Also unaffected were freshwater taxa like alligators and crocodiles, which occupied wetlands that retained geographic heterogeneity in times of regression.

If correct, the argument developed by Bakker must relate in some way to mass extinctions in the marine realm, which are becoming increasingly well documented through study of deep-sea cores from the recent expeditions of JOIDES, the Joint Oceanographic Institutions for Deep Earth Sampling. The advantage of studying marine mass extinctions in such cores is that a largely uninterrupted record is available. It is important to be able to observe the fossil record spanning a mass extinction, and yet because most mass extinctions seem to coincide with regressions, only poor fossil records are available for shallow-water benthos. In fact, in the past this condition led to the reasonable argument that if continental unconformities produced by regressions represented long enough intervals of time, "mass extinctions" might be nothing but illusions: perhaps seemingly abrupt biotic changes were not abrupt at all. The record of the deep sea, however, is showing that this is not the case. Not only is the deep-sea record relatively complete, but it represents a pelagic system that is more or less cosmopolitan. In general, the content of deep-sea cores reflects neither localized biotic conditions nor localized tectonic events. The diversity of pelagic taxa that leave a deep-sea record has waxed and waned, in much the way that tetrapod dynasties have on land. Pulses of radiation and extinction of marine vertebrates are apparent in Figure 12-45. These are based on fossil data from present-day continents. Similar data for ammonites reveal almost total extinction, followed by reradiation at the ends of the Permian and Triassic Periods. In Chapter 11 we noted Cifelli's discovery of the near-extinction and reradiation of planktonic foraminiferans, not only at the end of

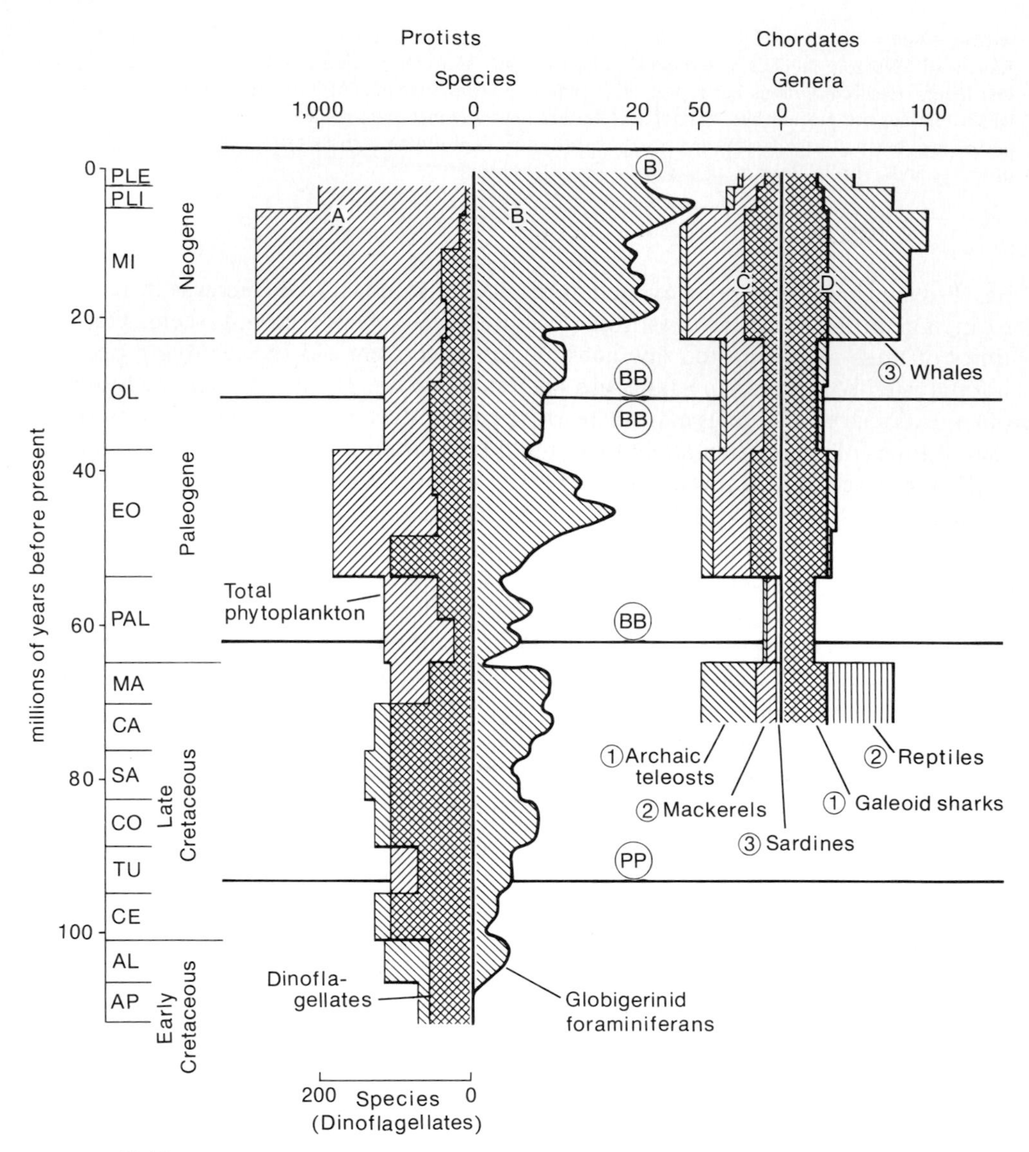

FIGURE 12-45
Changes in the diversity of pelagic taxa since the Early Cretaceous. The letters BB and PP indicate the abundant occurrence of unusual species of phytoplankton that spread throughout oceanic areas during intervals when the marine ecosystem seems to have broken down. From Fischer and Arthur (1977).

the Cretaceous but also in the Oligocene (page 369). This history of changing diversity is depicted in Figure 12-45, along with trends in diversity for other taxa from deep-sea cores and continental deposits, as analyzed by Fischer and Arthur (1977). While the picture is not simple, pulses of diversification and extinction are evident. Major mass extinctions occurred at the end of the Cretaceous and in the Oligocene. A less dramatic extinction, not documented in Figure 12-45, occurred in the mid-Cretaceous.

Fischer and Arthur (1977) suggest that between mass extinctions there evolved diverse communities in which large predators occupied the top positions in complex pelagic food webs. These communities were perhaps analogous to the terrestrial dynasties recognized by Bakker. At such times, there seems to have been a weak latitudinal temperature gradient in surface waters, and temperatures in the deep sea were rather high. Bottom waters were stagnant, as shown by the widespread deposition of black, anoxic sediments. An apparent increase in temperature gradients during intervals of mass extinction produced polar submergence of cold waters. Just as such waters do today (page 385), these cold waters flowed toward the equator at great depths. The deep sea was cold, but relatively well oxidized. Another fascinating discovery is that during each interval of mass extinction a particular kind of phytoplankton bloomed throughout much of the oceanic realm (Figure 12-45). One phytoplankter of this kind occurs today only sparsely in nearshore habitats. It seems to be an opportunistic form, waiting aside, under pressure of competition, to inherit the seas when the pelagic ecosystem breaks down once again. The observant reader will perhaps have noticed that Figure 12-45 offers an unfavorable forecast. Today we seem to be entering an episode of mass extinction. One aspect of this trend, the good circulation of waters in the deep sea, was mentioned above. Although perhaps likely, it is by no means certain that the pelagic mass extinctions will be explained by factors related to plate tectonics.

Some workers invoke crowding to explain mass extinction of benthic life during worldwide marine regression. Their contention is that competition and predation are intensified as species are concentrated into small areas of continental shelf. The idea has long been appealing, perhaps because it is potentially a simple and direct explanation, but it presents some difficulties. One difficulty is that as sea level falls, not only do areas of continental shelf shrink but shallow-water areas must increase around conical oceanic islands as well, and today the most diverse shallow-water faunas occur in the island-dotted areas of the Indo-Pacific region. Furthermore, though the kinds of species-area curve used for islands may possibly apply to continental shelf area, we have no information as to how large the value of z may be. In other words, it is uncertain whether area is very important on this scale.

Clearly, we are beginning to understand mass extinction. Many time-honored observations, such as the observation that mass extinctions tend to occur during widespread marine regressions, are likely to find a place in popular explanations of the future. Some suggestions, such as the proposal that magnetic reversals trigger extinction, seem to be discredited by recently discovered evidence: the polarity of the earth's magnetic field is simply reversed too often, without apparent biotic effects, to account for major mass extinctions (Figure 12-18). Whether changes in the partial pressure of oxygen in the earth's atmosphere may have played a role (Tappan, 1968; McAlester, 1970) is not at all certain. The same can be said for climatic fluctuations resulting from changes in solar radiation. One conclusion that we will always be safe in drawing is that the causes of mass extinction are not simple; simple correlations can be misleading, for they do not identify cause-and-effect relationships, and some apparent causes may only be biproducts of true causes. These biproducts may, however, lead us to the true causes. Furthermore, we must bear in mind the fact that the basic agents of mass extinction can operate in diverse ways. The variety of effects of plate tectonics illustrates this point: mass extinction on land and in the ocean may have similar general causes but quite different agents. Clearly the new principles of plate tectonics seem to offer some of the most promising avenues for future analysis of the intriguing problems of mass extinction.

Supplementary Reading

Cox, C. B., Healey, I. N., and Moore, P. D. (1973) *Biogeography: An Ecological and Evolutionary Approach.* Oxford, Blackwell, 184 p. (A brief general introduction.)

Hallam, A. (1973a) *A Revolution in the Earth Sciences.* Oxford, Oxford University Press, 127 p. (A summary and historical analysis of the emergence of plate tectonics.)

Hallam, A. (1973b) *Atlas of Palaeobiogeography.* Amsterdam, Elsevier, 531 p. (Presentation and interpretation by specialists of data on the geographic distribution of extinct organisms.)

Hughes, N. F., ed. (1973) *Organisms and Continents through Time.* Glasgow. Palaeontological Association. Special Paper No. 12, 334 pp. (A series of analyses by specialists of the geographic distributions of certain fossil taxa in the context of plate tectonics.)

MacArthur, R. H., and Wilson, E. O. (1967) *The Theory of Island Biogeography.* Princeton, N.J., Princeton University Press, 203 p. (A highly original analysis of species diversity on islands and in island-like habitats.)

MacArthur, R. H. (1972) *Geographical Ecology.* New York, Harper & Row, 269 pp. (An advanced account of the geographic distribution and abundance of organisms.)

Middlemiss, F. A., Rawson, P. F., and Newell, G., eds. (1969) *Faunal Provinces in Space and Time.* Liverpool, Seel House Press, 236 p. (A volume comparable to the one edited by Hughes that is listed above.)

Bibliography

Ager, D. V. (1963) *Principles of Paleoecology.* New York, McGraw-Hill, 371 p.

Alexander, R. M. (1968) *Animal Mechanics.* Seattle, University of Washington Press, 346 p.

Alexander, R. R. (1974) Morphologic adaptations of the bivalve *Anadara* from the Pliocene of the Kettleman Hills, California. *Jour. Paleont.,* **48**:633–650.

American Commission on Stratigraphic Nomenclature. (1961) Code of stratigraphic nomenclature. *Bull. Amer. Assoc. Petrol. Geol.,* **45**:645–665.

Andrews, H. E., Brower, J. C., Gould, S. J., and Reyment, R. A. (1974) Growth and variation in *Eurypterus remipes* DeKay. *Bull. Geol. Inst. Univ. Uppsala,* n.s. **4**:81–114.

Arkell, W. J. (1956) *Jurassic Geology of the World.* London, Oliver and Boyd, 806 p.

Arkell, W. J. (1957) Introduction to Mesozoic Ammonoidea. *In* Moore, R. C., ed. *Treatise on Invertebrate Paleontology.* Boulder, Colo., Geological Society of America and University Press of Kansas (joint publication), Part L, Mollusca 4, p. 81–129.

Bacescu, M. (1963) Contribution à la biocoenologie de la Mer Noire. L'étage périozoique et le faciès dreissenifère, leure charactéristiques. *Proc. Verb. Réun. C.I.E.S.M.,* **17**:107–122.

Bailey, I. W., and Sinnott, E. W. (1915) A botanical index of Cretaceous and Tertiary climates. *Science,* **41**:831–834.

Bailey, I. W., and Sinnott, E. W. (1916) The climatic distribution of certain types of angiosperm leaves. *Amer. Jour. Botany,* **3**:24–39.

Bakker, R. T. (1975) Dinosaur renaissance. *Scient. Amer.* **232** (4):p. 58–78.

Bakker, R. T. (1977) Cycles of diversity and extinction: A plate tectonic/topographic model. *In* Hallam, A., ed. *Patterns of Evolution.* Amsterdam, Elsevier, p. 431–478.

Bambach, R. K. (1973) Tectonic deformation of composite-mold fossil Bivalvia (Mollusca). *Amer. J. Sci.,* **273-A**:409–430.

Bell, W. C., Kay, M., Murray, G. E., Wheeler, H. E., and Wilson, J. A. (1961) Note 25—Geochronologic and chronostratigraphic units. *Bull. Amer. Assoc. Petrol. Geol.,* **45**:666–670.

Berry, W. B. N. (1968) *Growth of a Prehistoric Time Scale.* San Francisco, W. H. Freeman and Company, 158 p.

Bird, J. M., and Dewey, J. F. (1970) Lithosphere plate-continental margin tectonics and the evolution of the Appalachian orogen. *Geol. Soc. Amer. Bull.,* **81**:1031–1060.

Blackwelder, R. E. (1967) *Taxonomy: A Text and Reference Book.* New York, John Wiley & Sons, 698 p.

Blatt, H., and Jones, R. L. (1975) Proportions of exposed igneous, metamorphic, and sedimentary rocks. *Geol. Soc. Amer. Bull.,* **86**:1085–1088.

Boardman, R. S., Cheetham, A. H., and Oliver, W. A., Jr., eds. (1973) *Animal Colonies: Development and Function through time.* Stroudsburg, Penn. Dowden, Hutchinson, & Ross, Inc., 603 p.

Boardman, R. S., and McKinney, F. K. (1976) Skeletal architecture and preserved organs of four-sided zooids in convergent genera of Paleozoic Trepostomata (Bryozoa). *Jour. Paleont.,* **50**:25–78.

Bonner, J. T. (1952) *Morphogenesis.* Princeton, Princeton University Press, 296 p.

Boucot, A. J., Brace, W., and deMar, R. (1958) Distribution of brachiopod and pelecypod shells by currents. *Jour. Sedim. Petrol.,* **28**:321–332.

Bowen, Z. P., McAlester, A. L., and Rhoads, D. C. (1974) Marine benthic communities of the Sonyea Group (Upper Devonian) of New York. *Lethaia,* **7**:93–120.

Bramwell, C. D., and Whitfield, G. F. (1974) Biomechanics of *Pteranodon. Philos. Trans. Roy. Soc. London,* **B.267**:503–581.

Brasier, M. D. (1975) An outline history of seagrass communities. *Palaeontology,* **18**:681–702.

Bridges, P. H. (1975) The transgression of a hard substrate shelf: the Llandovery (Lower Silurian) of the Welsh Borderland. *Jour. Sedim. Petrol.,* **45**:79–94.

Bromley, R. G., and Asgaard, U. (1975) Sediment structures produced by a spatangoid echinoid: a problem of preservation. *Bull. Geol. Soc. Denmark,* **24**:261–281.

Brooks, J. C. (1950) Speciation in ancient lakes. *Quart. Rev. Biol.,* **25**:131–176.

Brown, C. A. (1960) *Palynological Techniques.* Baton Rouge, La., C. A. Brown, 188 p.

Bulman, O. M. B. (1933) Programme-evolution in the graptolites. *Biol. Rev.,* **8**:311–334.

Callomon, J. H. (1963) Sexual dimorphism in Jurassic ammonites. *Trans. Leicester Lit. Philos. Soc.,* **57**:21–56.

Camp, C. L., and Hanna, G. D. (1937) *Methods in Paleontology.* Berkeley, University of California Press, 153 p.

Campbell, K. S. W. (1957) A Lower Carboniferous brachipod–coral fauna from New South Wales. *Jour. Paleont.,* **31**:34–98.

Carpenter, F. M. (1953) The geological history and evolution of insects. *Amer. Scientist,* **41**:256–270.

Chaloner, W. G., and Creber, G. T. (1973) Growth rings in fossil woods as evidence of past climates. *In* Tarling, D. H., and Runcorn, S. K., eds. *Implications of Continental Drift to the Earth Sciences,* New York, Academic Press, v. 1, p. 425–437.

Chave, K. E. (1964) Skeletal durability and preservation. *In* Imbrie, J., and Newell, N. D., eds. *Approaches to Paleoecology.* New York, John Wiley & Sons, p. 377–387.

Cifelli, R. (1969) Radiation of Cenozoic planktonic Foraminifera. *System. Zool.,* **18**:154–168.

Clarke, G. L. (1965) *Elements of Ecology.* New York, John Wiley & Sons, 560 p.

Clarkson, E. N. K. (1966) Schizochroal eyes and vision of some Silurian acastid trilobites. *Palaentology,* **9**:1–29.

Clarkson, E. N. K. (1973a) The eyes of *Asaphus raniceps* Dalman (Trilobita). *Palaeontology,* **16**:425–444.

Clarkson, E. N. K. (1973b) Morphology and evolution of the eye in Upper Cambrian Olenidae (Trilobita). *Palaeontology,* **16**:735–764.

Clarkson, E. N. K., and Levi-Setti, R. (1975) Trilobite eyes and the optics of Descartes and Huygens. *Nature,* **254**:663–667.

Cloud, P. E. (1949) Some problems and patterns of evolution exemplified by fossil invertebrates. *Evolution,* **2**:322–350.

Coates, A. G. (1973) Cretaceous Tethyan coral–rudist biogeography in relation to the evolution of the Atlantic Ocean. *Palaeont. Assoc. Special Papers in Palaeont.,* **12**:169–174.

Colbert, E. H. (1948) Evolution of the horned dinosaurs. *Evolution,* **2**:145–163.

Colbert, E. H. (1955) *Evolution of the Vertebrates.* New York, John Wiley & Sons, 479 p.

Colbert, E. H. (1971) Tetrapods and continents. *Quart. Rev. Biol.,* **46**:250–269.

Colbert, E. H. (1973) *Wandering Lands and Animals.* New York, E. P. Dutton, 323 p.

Cooper, G. A., and Grant, R. E. (1976) Permian brachiopods of West Texas, IV. *Smithsonian Contrib. to Paleobiol.,* No. 21, p. 1923–2607.

Corliss, J. O. (1959) Comments on the systematics and phylogeny of the Protozoa. *System. Zool.,* **8**:169–190.

Cox, C. B., Healey, I. N., and Moore, P. D. (1973) *Biogeography: An Ecological and Evolutionary Approach.* Oxford, Blackwell, 184 p.

Craig, G. Y., and Hallam, A. (1963) Size-frequency and growth-ring analyses of *Mytilus edulis* and *Cardium edule,* and their paleoecological significance. *Palaeontology,* **6**:731–750.

Crompton, A. W. (1963) On the lower jaw of *Diarthrognathus* and the origin of the mammalian lower jaw. *Proc. Zool. Soc. London,* **140**:697–753.

Dacqué, E. (1921) *Vergleichende biologische Formenkunde der Fossilen niederen Tiere.* Berlin, Gebrüder Bornträger, 777 p.

Davis, M. B. (1967) Late-glacial climate in northern United States: A comparison of New England and the Great Lakes Region. *In* Cushing, E. J., and Wright, H. E., eds. *Quaternary Paleoecology.* New Haven, Yale University Press, p. 11–43.

Dayton, P. K. (1971) Competition, disturbance, and community organization: the provision and subsequent utilization of space in a rocky intertidal community. *Ecol. Monographs,* **41**:351–389.

Dayton, P. K., and Hessler, R. R. (1972) Role of biological disturbance in maintaining diversity in the deep sea. *Deep-Sea Res.,* **19**:199–208.

DeBeer, G. R. (1958) *Embryos and Ancestors.* London, Oxford University Press, 197 p.

Dewey, J. F. (1972) Plate tectonics. *Scient. Amer.,* **226** (5):56–68.

Dietz, R. S. (1972) Geosynclines, mountains and continent building. *Scient. Amer.,* **226**(3):30–38.

Dietz, R. S., and Holden, J. C. (1970) Reconstruction of Pangea: Breakup and dispersion of continents, Permian to present. *Jour. Geophys. Res.,* **75**:4939–4956.

Dobzhansky, T. (1951) *Genetics and the Origin of Species.* New York, Columbia University Press, 364 p.

Dobzhansky, T. (1970) *Genetics of the Evolutionary Process.* New York, Columbia University Press, 505 p.

Donovan, D. T. (1966) *Stratigraphy.* Chicago, Rand McNally, 199 p.

Dorf, E. (1959) Climatic changes of the past and present. *Contr. Mus. Paleontol. Univ. Mich.,* **13**:181–210.

Dorf, E. (1964) The use of fossil plants in palaeoclimatic interpretations. *In* Nairn, A. E. M., ed. *Problems in Palaeoclimatology,* London, Wiley-Interscience, p. 13–31.

Dorf, E. (1970) Paleobotanical evidence of Mesozoic and Cenozoic climatic changes *Proc. North American Paleontological Convention,* **1**:323–347.

Doyle, J. A., and Hickey, L. J. (1976) Pollen and leaves from the mid-Cretaceous Potomac Group and their bearing on early angiosperm evolution. *In* Beck, C. B., ed. *Origin and Early Evolution of Angiosperms.* New York, Columbia University Press, p. 139–206.

Dunbar, C. O., and Rodgers, J. (1957) *Principles of Stratigraphy.* New York, John Wiley & Sons, 356 p.

Durham, J. W. (1966) Classification. *In* Moore, R. C., ed. *Treatise on Invertebrate Paleontology.* Boulder, Colo., Geological Society of America and University Press of Kansas (joint publication), Part U, Echinodermata 3, p. 270–295.

Durham, J. W. (1967) The incompleteness of our knowledge of the fossil record. *Jour. Paleont.,* **41**:559–565.

Easton, W. H. (1960) *Invertebrate Paleontology.* New York, Harper & Brothers, 701 p.

Ehrlich, P. R., Ehrlich, A. H., and Holdren, J. P. (1977) *Ecoscience: Population, Resources, Environment.* San Francisco, W. H. Freeman and Company, 1051 p.

Eicher, D. L. (1976) *Geologic Time.* Englewood Cliffs, N.J., Prentice-Hall, 150 p.

Eldredge, N. (1971) The allopatric model and phylogeny in Paleozoic invertebrates. *Evolution,* **25**:156–167.

Eldredge, N., and Gould, S. J. (1972) Punctuated equilibria: an alternative to phyletic gradualism. *In* Schopf, T. J. M., ed. *Models in Paleobiology.* San Francisco, Freeman, Cooper and Company, p. 82–115.

Elliott, D. H., Colbert, E. H., Breed, W. J., Jensen, J. A., and Powell, J. S. (1970) Triassic tetrapods from Antarctica: evidence of continental drift. *Science,* **169**:1197–1201.

Elton, C. S. (1958) *The Ecology of Invasions by Animals and Plants.* London, Methuen, 181 p.

Eltringham, S. K. (1971) *Life in Mud and Sand.* New York, Crane, Russak, 218 p.

Emlen, J. M. (1973) *Ecology: An Evolutionary Approach.* Reading, Mass., Addison-Wesley, 493 p.

Erben, H. K. (1966) Über den Ursprung der Ammonoidea. *Biol. Rev.,* **41**:641–658.

Ericson, D. B., Ewing, M., and Wollin, G. (1963) Pliocene-Pleistocene boundary in deep-sea sediments. *Science,* **139**:727–737.

Ericson, D. B., and Wollin, G. (1968) Pleistocene climates and chronology in deep-sea sediments. *Science,* **162**:1227–1234.

Finks, R. M. (1960) Late Paleozoic sponge faunas of the Texas region. *Bull. Amer. Mus. Nat. Hist.,* **120**:1–160.

Fischer, A. G. (1960) Latitudinal variations in organic diversity. *Evolution,* **14**:64–81.

Fischer, A. G., and Arthur, M. A. (1977) Secular variations in the pelagic realm. Tulsa, Okla., Society of Economic Paleontologists and Mineralogists, Spec. Publication 25, p. 19–50.

Fisher, W. L., Rodda, P. U., and Dietrich, J. W. (1964) Evolution of the *Athleta petrosa* stock (Eocene, Gastropoda) of Texas. *Univ. of Texas Publ. 6413,* 117 p.

Fisk, H. N., McFarlan, E., Kolb, C. R., and Wilbert, L. J. (1954) Sedimentary framework of the modern Mississippi Delta. *Jour. Sedim. Petrol.,* **24**:76–99.

Flessa, K. W., and Imbrie, J. (1973) Evolutionary pulsations: Evidence from Phanerozoic diversity patterns. *In* Tarling, D. H., and Runcorn, S. K., eds. *Implications of Continental Drift to the Earth Sciences.* New York, Academic Press, v. 1, p. 247–285.

Flint, R. F. (1975) Features other than diamicts as evidence of ancient glaciations. *In* Wright, A. E., and Moseley, F., eds. *Ice Ages: Ancient and Modern.* Liverpool, Seel House Press, p. 121–136.

Florin, R. (1963) The distribution of conifer and taxad genera in time and space. *Acta Horti Bergiani,* **20**:121–312.

Forcier, L. K. (1975) Reproductive strategies and the co-occurrence of climax tree species. *Science,* **189**:808–810.

Gernant, R. E., and Kessling, R. V. (1966) Foraminiferal paleoecology and paleoenvironmental reconstruction of the Oligocene Middle Frio in Chambers County, Texas. *Trans. Gulf Coast Assoc. Geol. Soc.,* **16**:131–158.

Gingerich, P. D. (1974) Stratigraphic record of Early Eocene *Hyopsodus* and the geometry of mammalian phylogeny. *Nature,* **248**:107–109.

Ginsburg, R. H. (1957) Early diagenesis and lithification of shallow-water carbonate sediments in south Florida. *In* LeBlanc, R. J., and Breeding, J. C., eds. *Regional Aspects of Carbonate Deposition.* Tulsa, Okla., Society of Economic Paleontologists and Mineralogists. Special Publication 5, p. 80–100.

Gordon, W. A. (1973) Marine life and ocean surface currents in the Cretaceous. *Jour. Geol.,* **81**:269–284.

Gould, S. J. (1966) Allometry and size in ontogeny and phylogeny. *Biol. Rev.,* **41**:587–640.

Gould, S. J. (1967) Factor-analytic study of the Pelycosauria. *Evolution,* **21**:385–401.

Gould, S. J. (1970) Evolutionary paleontology and the science of form. *Earth-Science Rev.,* **6**:77–119.

Gould, S. J. (1974) The origin and function of "bizarre" structures: antler size and skull size in the "Irish Elk," *Megaloceros giganteus. Evolution,* **28**:191–220.

Grant, R. E. (1966) Spine arrangment and life habits of the productoid brachiopod *Waagenoconcha. Jour. Paleont.,* **40**:1063–1069.

Grant, R. E. (1968) Structural adaptation in two Permian brachiopod genera, Salt Range, West Pakistan. *Jour. Paleont.,* **42**:1–32.

Grant, R. E. (1972) The lophophore and feeding mechanism of the Productidina (Brachiopoda). *Jour. Paleont.,* **46**:213–249.

Grant, R. E. (1975) Methods and conclusions in functional analysis: A reply. *Lethaia,* **8**:31–33.

Grant, V. (1963) *The Origin of Adaptation.* New York, Columbia University Press, 606 p.

Grassle, J. F., and Sanders, H. L. (1973) Life histories and the role of disturbance. *Deep-Sea Res.,* **20**:643–659.

Graus, R. R. (1974) Latitudinal trends in the shell characteristics of marine gastropods. *Lethaia,* **7**:303–314.

Gregor, C. B. (1970) Denudation of the continents. *Nature,* **228**:273–275.

Hallam, A. (1969) Faunal realms and facies in the Jurassic. *Palaeontology,* **12**:1–18.

Hallam, A. (1973a) *A Revolution in the Earth Sciences.* Oxford, Oxford University Press, 127 p.

Hallam, A. (1973b) *Atlas of Palaeobiogeography.* Amsterdam, Elsevier, 531 p.

Hallam, A., ed. (1977) *Patterns of Evolution.* Amsterdam, Elsevier, 591 p.

Harland, W. B., et al. (1967) *The Fossil Record.* London, Geological Society of London, 827 p.

Harland, W. B., Smith, A. G., and Wilkock, B. (1964) *The Phanerozoic Time Scale.* London, Geological Society of London, 458 p.

Harper, C. W. (1975) Origin of species in geologic time: Alternatives to the Eldredge-Gould model. *Science,* **190**:47–48.

Harper, C. W. (1976) Reply. *Science,* **192**:269.

Hartman, W. D., and Goreau, T. F. (1970) Jamaican coralline sponges: Their morphology, ecology and fossil relatives. *Symp. Zool. Soc. London,* **25**:205–243.

Hayami, I., and Ozawa, T. (1975) Evolutionary models of lineage-zones. *Lethaia,* **8**:1–14.

Hays, J. D., and Pitman, W. C. (1973) Lithospheric plate motion, sea level changes and climatic and ecologic consequences. *Nature,* **246**:18–22.

Heaslip, W. G. (1968) Cenozoic evolution of the alticostate venericards in Gulf and East Coastal North America. *Palaeontographica Americana,* **6**:55–135.

Hecht, M. K. (1974) Morphological transformation, the fossil record, and the mechanisms of evolution: a debate. Part I. The statement and the critique. *Evol. Biol.,* **7**:295–303.

Hecker, R. F. (1965) *Introduction to Paleoecology.* New York, Elsevier Scientific Publications, 166 p.

Hedberg, H. D. (1951) Nature of time-stratigraphic units and geologic-time units. *Bull. Amer. Assoc. Petrol. Geol.,* **35**:1077–1081.

Hedgpeth, J. W., ed. (1957) *Treatise on Marine Ecology and Paleoecology, 1, Ecology.* Boulder, Colo., Geological Society of America. Memoir 67, 1296 p.

Heirtzler, J. R., LePichon, X., and Baron, J. G. (1966) Marine magnetic anomalies over the Reykjanes ridge. *Deep-Sea Res.,* **13**:427–443.

Hennig, W. (1966) *Phylogenetic Systematics.* Urbana, University of Illinois Press, 263 p.

Hertel, H. (1966) *Structure–Form–Movement.* New York, Reinhold Pub. Co., 251 p.

Hess, H. H. (1962) History of ocean basins. *In* Engel, A. E. et al. *Petrologic Studies: A Volume in Honor of A. F. Buddington.* Boulder, Colo., Geological Society of America, p. 599–620.

Hessler, R. R., and Sanders, H. L. (1967) Faunal diversity in the deep-sea. *Deep-Sea Res.,* **14**:65–78.

Heywood, V. H., and McNeil, J., eds. (1964) *Phenetic and Phylogenetic Classification.* London, Systematics Association. Publication 6, 164 p.

Holmes, A. (1931) Radioactivity and earth movements. *Trans. Geol. Soc. Glasgow,* **18**:559–606.

House, M. R. (1973) An analysis of Devonian goniatite distributions. *Palaeont. Assoc. Special Papers in Palaeont.,* **12**:305–317.

Hudson, J. D. (1963) The recognition of salinity-controlled mollusc assemblages in the Great Estuarine Series (Middle Jurassic) of the Inner Hebrides. *Palaeontology,* **6**:318–326.

Hudson, J. D. (1968) The microstructure and mineralogy of the shell of a Jurassic mytilid (Bivalvia). *Palaeontology,* **11**:163–182.

Hughes, N. F., ed. (1973) *Organisms and Continents Through Time.* Glasgow, Palaeontological Association. Special Paper No. 12, 334 p.

Hunt, A. S. (1967) Growth, variation, and instar development of an agnostid trilobite. *Jour. Paleont.,* **41**:203–208.

Hunt, O. D. (1925) The food of the bottom fauna of the Plymouth fishing grounds. *Jour. Marine Biol. Assoc. United Kingdom,* **13**:560–599.

Huxley, J. S. (1932) *Problems of Relative Growth.* New York, Dial Press, 276 p.

Imbrie, J. (1956) Biometrical methods in the study of invertebrate fossils. *Bull. Amer. Mus. Nat. Hist.,* **108**:211–252.

Imbrie, J., and Newell, N. D. (1964) *Approaches to Paleoecology.* New York, John Wiley & Sons, 432 p.

Israelsky, M. (1949) Oscillation chart. *Bull. Amer. Assoc. Petrol. Geol.,* **33**:92–98.

Jackson, J. B. C. (1972) The ecology of the Molluscs of *Thalassia* communities, Jamaica, West Indies, II. Molluscan population variability along an environmental stress gradient. *Mar. Biol.,* **14**:304–337.

Jackson, J. B. C. (1973) The ecology of molluscs of *Thalassia* communities, Jamaica, West Indies, I. Distribution, environmental physiology, and ecology of common shallow-water species. *Bull. Mar. Sci.,* **23**:313–350.

Jackson, J. B. C., and Buss, L. (1975) Allelopathy and spatial competition among coral reef invertebrates. *Proc. Nat. Acad. Sci. U.S.,* **72**:5160–5163.

Jackson, R. T. (1912) Phylogeny of the Echini with a revision of Paleozoic species. *Boston Soc. Nat. Hist., Mem.* **7**:443 p.

Jardine, N., and McKenzie, D. (1972) Continental drift and the dispersal and evolution of organisms. *Nature,* **235**:20–24.

Jeletsky, J. A. (1965) Is it possible to quantify biochronological correlation? *Jour. Paleont.,* **39**:135–140.

Jerison, H. J. (1973) *Evolution of the Brain and Intelligence.* New York, Academic Press, 482 p.

Johnson, J. G. (1971) Timing and coordination of orogenic, epeirogenic, and eustatic events. *Geol. Soc. Amer. Bull.,* **82**:3263–3298.

Johnson, J. G. (1972) Antler effect equals Haug effect. *Geol. Soc. Amer. Bull.,* **83**:2497–2498.

Johnson, R. G. (1959) Spatial distribution of *Phoronopsis viridis* Hilton. *Science,* **129**:1221.

Johnson, R. G. (1960) Models and methods for the analysis of the mode of formation of fossil assemblages. *Geol. Soc. Amer. Bull.,* **71**:1075–1086.

Jurva, R. (1952) Seas. *In* Societas Geographica Fenniae, Helsinki, Finland, eds. *A General Handbook on the Geography of Finland. Fennia,* **72**:136–160.

Kaesler, R. L. (1967) Numerical taxonomy in invertebrate paleontology. *In* Teichert, C., and Yochelson, E. L., eds. *Essays in Paleontology and Stratigraphy.* Lawrence, Kans., University Press of Kansas, p. 63–81.

Kaesler, R. L. (1970) Numerical taxonomy in paleontology: classification, ordination, and reconstruction of phylogenies. *Proc. N. A. Paleontol. Conv., Part B,* p. 84–100.

Kaesler, R. L., and Waters, J. A. (1972) Fourier analysis of the ostracode margin. *Geol. Soc. Amer. Bull.,* **83**:1169–1178.

Kauffman, E. G. (1973) Cretaceous Bivalvia. *In* Hallam, A., ed. *Atlas of Palaeobiogeography.* Amsterdam, Elsevier, p. 353–383.

Kauffman, E. G., and Sohl, N. F. (1974) Structure and evolution of Antillean Cretaceous rudist framework. *Verhandl. Naturf. Ges. Basel,* **84**:399–467.

Kaufmann, R. (1933) Variationsstatistische Untersuchungen uber die "Artabwandlung" und "Artumbildung" an der oberkambrischen Trilobitengattung *Olenus* Dalm. *Abh. Geol. Pal. Inst. Univ. Greifswald,* **10**:1–54.

Kaufmann, R. (1935) Exakt-statistische Biostratigraphie der *Olenus*-Arten von Sudöland. *Geol. Foren. Stockholm Förhandl.,* **1935**:19–28.

Kermack, K. A. (1954) A biometrical study of *Micraster coranguinum* and *M. (Isomicraster) senonensis. Philos. Trans. Roy. Soc. London,* **B, 237**:375–428.

King, P. B. (1948) Geology of the Southern Guadalupe Mountains, Texas. *U.S. Geol. Surv. Prof. Paper* 215, 183 p.

Kinne, O. (1970–1975) *Marine Ecology.* New York, John Wiley & Sons, 2 vols (5 parts).

Koch, D. L., and Strimple, H. L. (1968) A new Upper Devonian cystoid attached to a discontinuity surface. *Iowa Geological Survey Report of Investigations* 5, 49 p.

Koch, G. S., Jr., and Link, R. F. (1971) *Statistical Analysis of Geological Data,* vol. II. New York, John Wiley & Sons, 438 p.

Kozlowski, R. (1966) On the structure and relationship of graptolites. *Jour. Paleont.,* **40**:489–501.

Kraft, P. (1932) Die Mikrophotographie mit infraroten Strahlen. *Internat. Cong. of Photog. 8, Dresden, 1931, Berichte,* p. 341–345.

Krinsley, D. H., and Donahue, J. (1968) Environmental interpretation of sand grain surface textures by electron microscopy. *Geol. Soc. Amer. Bull.,* **79**:743–748.

Krumbein, W. C., and Sloss, L. L. (1963) *Stratigraphy and Sedimentation,* 2nd Ed. San Francisco, W. H. Freeman and Company, 600 p.

Kummel, B., and Raup, D. M., eds. (1965) *Handbook of Paleontological Techniques.* San Francisco, W. H. Freeman and Company, 852 p.

Kurtén, B. (1964) Population structure in paleoecology. *In* Imbrie, J., and Newell, N. D., eds. *Approaches to Paleoecology.* New York, John Wiley & Sons, p. 91–106.

Kurtén, B. (1967) Continental drift and the paleogeography of reptiles and mammals. *Comment. Biol. Soc. Scient. Fennica,* **31**:1–8.

Kurtén, B. (1968) *Pleistocene Mammals of Europe.* Chicago, Aldine Pub. Co. London, Weidenfeld and Nicolson, 317 p.

Ladd, H. S., ed. (1957) *Treatise on Marine Ecology and Paleoecology, 2, Paleoecology.* Boulder, Colo., Geological Society of America. Memoir 67, 1077 p.

Lang, J. (1971) Interspecific aggression by scleractinian corals. 1. The rediscovery of *Scolymia cubensis* (Milne Edwards & Haime). *Bull. Mar. Sci.,* **21**:952–959.

Laporte, L. F. (1968) *Ancient Environments.* Englewood Cliffs, N.J., Prentice-Hall, 116 p.

Lawrence, D. R. (1968) Taphonomy and information losses in fossil communities. *Geol. Soc. Amer. Bull.,* **79**:1315-1330.

Lawson, D. A. (1975) Pterosaur from the latest Cretaceous of West Texas: discovery of the largest flying creature. *Science,* **187**:947–948.

Le Gros Clark, W. E., ed. (1945) *Essays on Growth and Form Presented to D'Arcy Wentworth Thompson.* Oxford, The Clarendon Press, 408 p.

Le Gros Clark, W. E., (1966) *History of the Primates.* Chicago, University of Chicago Press, 127 p.

Leich, H. (1965) Eine neue Lebensspur von *Mesolimulus walchi* und ihre Deutung. *Aufschluss,* **1**:5–7.

Levinton, J. S., and Bambach, R. K. (1975) A comparative study of Silurian and Recent deposit-feeding bivalve communities. *Paleobiology,* **1**:97–124.

Levi-Setti, R. (1975) *Trilobites: A Photographic Atlas.* Chicago, University of Chicago Press. 213 p.

Lemche, H., and Wingstrand, K. G. (1959) The anatomy of *Neopilina galatheae* Lemche, 1957 (*Mollusca, Tryblidiacea*). *Galathea Rept.* (Copenhagen), **3**:1–71.

Leon, R. (1933) Ultraviolettes Licht endeckt Versteinerungen auf "leeren" Platten. Ein Pantpod im Jura-Kalk. *Senckenberg Naturf. Gesell., Natur. Mus.,* **63**:361–364.

Liem, K. F. (1973) Evolutionary strategies and morphological innovations: cichlid pharyngeal jaws. *System. Zool.,* **22**:425–441.

Linck, O. (1954) Die Muschelkalk-Seelilie *Encrinus liliiformis. Aus der Heimat,* **62**:225–235.

Lowenstam, H. A. (1950) Niagaran reefs of the Great Lakes area. *Jour. Geol.,* **58**:430–487.

McAlester, A. L. (1970) Animal extinctions, oxygen consumption, and atmospheric history. *Jour. Paleont.,* **44**:405–409.

MacArthur, R. H. (1972) *Geographical Ecology.* New York, Harper & Row, 269 p.

MacArthur, R. H., and Connell, J. H. (1966) *The Biology of Populations.* New York, John Wiley & Sons, 200 p.

MacArthur, R. H., and Wilson, E. O. (1967) *The Theory of Island Biogeography.* Princeton, N. J., Princeton University Press, 203 p.

McKenna, M. C. (1973)Sweepstakes, filters, corridors, Noah's Arks, and beached Viking funeral ships in palaeogeography. *In* Tarling, D. H. and Runcorn, S. K., *Implications of Continental Drift to the Earth Sciences.* New York, Academic Press, v. 1, p. 295–308.

McLean, J. D., Jr. (1959 to date) *Manual of Micropaleontological Techniques.* Alexandria, Va., McLean Paleontology Laboratory.

Macurda, D. B., ed. (1968) *Paleobiological Aspects of Growth and Development, a Symposium.* Menlo Park, Calif., Paleontological Society. Memoir 2, 119 p.

Makurath, J. H. (1977) Marine faunal assemblages in the Silurian-Devonian Keyser Limestone of the central Appalachians. *Lethaia,* **10**:235–256.

Martin, P. S. (1958) Pleistocene ecology and biogeography of North America: zoogeography. *Amer. Assoc. Adv. Sci. Publ.,* **51**:375–420.

Martin-Kaye, P. (1951) Sorting of lamellibranch valves on beaches in Trinidad, B. W. I. *Geol. Magazine,* **88**:432–434.

Matthews, R. K. (1974) *Dynamic Stratigraphy.* Englewood Cliffs, N. J., Prentice-Hall, 370 p.

Mayr, E. (1942) *Systematics and the Origin of Species.* Reprint edition. Magnolia, Mass., Peter Smith Publisher, Inc., 334 p.

Mayr, E. (1954) Change of genetic environment and evolution. *In* Huxley, J., Hardy, A. C., and Ford, E. B., eds. *Evolution as a Process.* London, Allen and Unwin, p. 157–180.

Mayr, E. (1963) *Animal Species and Evolution.* Cambridge, Mass., Harvard University Press, 797 p.

Mayr, E. (1969) *Principles of Systematic Zoology.* New York, McGraw-Hill, 428 p.

Mayr, E. (1971) *Populations, Species and Evolution.* Cambridge, Mass., Harvard University Press. 453 p.

Mayr, E., Linsley, E. G., and Usinger, R. L. (1953) *Methods and Principles of Systematic Zoology.* New York, McGraw-Hill, 328 p.

Medawar, P. B. (1945) Size, shape, and age. *In* Le Gros Clark, W. E., ed. *Essays on Growth and Form Presented to D'Arcy Wentworth Thompson.* Oxford, The Clarendon Press, p. 157–187.

Melville, R. (1973) Continental drift and plant distribution. *In* Tarling, D. H., and Runcorn, S. K., eds. *Implications of Continental Drift to the Earth Sciences,* New York, Academic Press, v. 1, p. 439–446.

Middlemiss, F. A., Rawson, P. F., and Newell, G., eds. (1969) *Faunal Provinces in Space and Time.* Liverpool, Seel House Press, 236 p.

Moore, H. B. (1968) *Marine Ecology.* New York, John Wiley & Sons, 493 p.

Moore, R. C., Lalicker, C. G., and Fischer, A. G. (1952) *Invertebrate Fossils.* New York, McGraw-Hill, 766 p.

Muir-Wood, H. M. (1965) Chonetidina. *In* Moore, R. C., ed. *Treatise on Invertebrate Paleontology.* Boulder, Colo., Geological Society of America and University Press of Kansas (joint publication), Part H. Brachiopoda, p. 412–439.

Muus, K. (1973) Settling, growth and mortality of young bivalves in the Øresund. *Ophelia,* **12**:79–116.

Natland, M. L. (1933) The temperature-and-depth distribution of some Recent and fossil Foraminifera in the southern California region. *Scripps Inst. Ocean. Bull. Tech. Ser.,* **3**:225–230.

Nevesskaya, L. A. (1967) Problems of species differentiation in light of paleontological data. *Paleont. Jour.,* **1967**:1–17.

Newell, N. D. (1942) Late Paleozoic pelecypods: Mytilacea. *Geol. Surv. Kansas, Publ. 10, part 1,* 123 p.

Newell, N. D. (1956) Fossil populations. *In* Sylvester-Bradley, P. C., ed. *The Species Concept in Paleontology.* London, Systematics Association. Publication 2, p. 63–82.

Newell, N. D. (1959) Adequacy of the fossil record. *Jour. Paleont.,* **33**:488–499.

Newell, N. D. (1967) Revolutions in the history of life. *Geol. Soc. Amer. Spec. Paper,* **89**:63–91.

Newell, N. D., and Boyd, D. W. 1970. Oyster-like Permian bivalvia. *Bull. Amer. Mus. Nat. Hist.,* **143**:221–281.

Newell, N. D., Rigby, J. K., Fischer, A. G., Whiteman, A. J., Hickox, J. E., and Bradley, J. S. (1953) *The Permian Reef Complex of the Guadalupe Mountains Region, Texas and New Mexico.* San Francisco, W. H. Freeman and Co., 236 p.

Nichols, D. (1959a) Changes in the Chalk heart-urchin *Micraster* interpreted in relation to living forms. *Philos. Trans. Roy. Soc. London,* **B. 242**:347–437.

Nichols, D. (1959b) Mode of life and taxonomy in irregular sea-urchins. *System. Assoc.,* **3**:61–80.

Nicol, D. (1962) The biotic development of some Niagaran reefs—an example of an ecological sucession or sere. *Jour. Paleont.*, **36**:172–176.

Odum, E. P. (1959) *Fundamentals of Ecology*. Philadelphia, Saunders, 546 p.

Odum, H. T., and Odum, E. P. (1955) Trophic structure and productivity of a windward coral reef community on Eniwetok Atoll. *Ecol. Monographs,* **25**:291–320.

Olson, E. C. (1965) *The Evolution of Life*. London, Weidenfeld and Nicolson, 300 p.

Olson. E. C. (1972) *Diplocaulus parvus* n. sp. (Amphibia: Nectridea) from the Chickasha Formation (Permian: Guadalupian) of Oklahoma. *Jour. Paleont.*, **46**:656–659.

Osborn, H. F. (1929) The titanotheres of ancient Wyoming, Dakota, and Nebraska. *U.S. Geol. Surv. Monogr. 55,* 2 vols.

Ovcharenko, V. N. (1969) Transitional forms and species differentiation of brachiopods. *Paleont. Jour.*, **1969**:67–73.

Owen, G., and Williams, A. (1969) The caecum of articulate brachiopods. *Proc. Roy. Soc. London,* **B. 172**:187-201.

Oxnard, C. (1973) *Form and Pattern in Human Evolution*. Chicago, University of Chicago Press, 218 p.

Paine, R. T. (1966) Food web complexity and community stability. *Amer. Naturalist,* **103**:91–93.

Palmer, A. R. (1955) The faunas of the Riley Formation in central Texas. *Jour. Paleont.*, **28**:709–786.

Palmer, A. R. (1957) Ontogenetic development of two olenellid trilobites. *Jour. Paleont.*, **31**:105–128.

Palmer, A. R. (1965) Biomere—a new kind of stratigraphic unit. *Jour. Paleont.*, **39**:149–153.

Pannella, G., and MacClintock, C. (1968) Biological and environmental rhythms reflected in molluscan shell growth. *In* Macurda, D. B., ed., *Paleobiological Aspects of Growth and Development, a Symposium*. Menlo Park, Calif., Paleontological Society. Memoir 2, p. 64–80.

Patterson, B., and Pascual, R. (1972) The fossil mammal fauna of South America. *In* Keast, A., Erk, F. C., and Glass, B., eds. *Evolution, Mammals, and Southern Continents*. Albany, State University of New York Press, p. 247–309.

Perkins, B. F. (1960) *Biostratigraphic Studies in Comanche (Cretaceous) Series of Northern Mexico and Texas*. Boulder, Colo., Geological Society of America. Memoir 83, 138 p.

Peterson, C. G. J. (1913) Valuation of the sea. II. The animal communities of the sea bottom and their importance for marine zoogeography. *Rep. Danish Biol. Sta.*, **21**:1–44.

Pickett, J. W., and Jell, J. S. (1974) The Australian tabulate coral genus *Hattonia*. *Palaeontology,* **17**:715–726.

Pilbeam, D. (1970) *The Evolution of Man*. New York, Funk and Wagnalls, 216 p.

Pilbeam, D., and Gould, S. J. (1974) Size and scaling in human evolution. *Science,* **186**:892–901.

Pielou, E. C. (1975) *Ecological Diversity*. New York, John Wiley & Sons, 165 p.

Porter, J. W. (1974) Community structure of coral reefs on opposite sides of the Isthmus of Panama. *Science,* **186**:543–545.

Rasmussen, H. W. (1976) Function and attachment of the stem in Isocrinidae and Pentacrinitidae: review and interpretation. *Lethaia,* **10**:51–57.

Raup, D. M. (1966) Geometric analysis of shell coiling: general problems. *Jour. Paleont.*, **40**:1178–1190.

Raup, D. M. (1967) Geometric analysis of shell coiling: coiling in ammonoids. *Jour. Paleont.*, **41**:43–65.

Raup, D. M. (1968) Theoretical morphology of echinoid growth. *In* Macurda, D. B., ed., *Paleobiological Aspects of Growth and Development, a Symposium.* Menlo Park, Calif., Paleontological Society, Memoir 2, p. 50–63.

Raup, D. M. (1972) Taxonomic diversity during the Phanerozoic. *Science,* **117**:1065–1071.

Raup, D. M. (1975a) Taxonomic diversity estimation using rarefaction. *Paleobiol.*, **1**:333–342.

Raup, D. M. (1975b) Taxonomic survivorship curves and Van Valen's Law. *Paleobiol.*, **1**:82–96.

Raup, D. M. (1976a) Species diversity in the Phanerozoic: a tabulation. *Paleobiol.*, **2**:279–288.

Raup, D. M. (1976b) Species diversity in the Phareozoic: an interpretation. *Paleobiol.*, **2**:289–297.

Raup, D. M., and Gould, S. J. (1974) Stochastic simulation and evolution of morphology—towards a nomothetic paleontology. *System. Zool,* **23**:305–322.

Raup, D. M., and Michelson, A. (1965) Theoretical morphology of the coiled shell. *Science,* **147**:1294–1295.

Raup, D. M., and Seilacher, A. (1969) Fossil foraging behavior: Computer simulation. *Science,* **166**:994–995.

Raven, H. C. (1935) Wallace's Line and the distribution of the kangaroo family in relation to habitat. *Amer. Mus. Novit.,* **1309**:1–33.

Raven, P. E., and Axelrod, D. I. (1972) Plate tectonics and Australasian paleobiogeography. *Science,* **176**:1379–1386.

Reid, R. E. H. (1968) Bathymetric distribution of Calcarea and Hexactinellida in the past and present. *Geol. Magazine,* **105**:546–559.

Rensch, B. (1960) *Evolution above the Species Level.* New York, Columbia University Press, 419 p.

Rhoads, D. C. (1966) Missing fossils and paleoecology. *Discovery* (Peabody Mus. Nat. Hist., Yale Univ.), **2**:19–22.

Rhoads, D. C. (1967) Biogenic reworking of intertidal and subtidal sediments in Barnstable Harbor and Buzzards Bay, Massachusetts. *Jour. Geol.,* **75**:461–476.

Rhoads, D. C., and Morse, J. W. (1971) Evolutionary and ecologic significance of oxygen-deficient marine basins. *Lethaia,* **4**:413–428.

Rhoads, D. C., Speden, I. G., and Waage, K. M. (1972) Trophic group analysis of Upper Cretaceous (Maestrichtian) bivalve assemblages from South Dakota. *Bull. Amer. Assoc. Petrol. Geol.* **56**:1100–1113.

Rhoads, D. C., and Young, D. K. (1970) The influence of deposit-feeding organisms on sediment stability and community trophic structure. *Jour. Marine Research,* **28**:150–178.

Richards, R. P. (1972) Autecology of Richmondian brachiopods (Late Ordovician of Indiana and Ohio). *Jour. Paleont.,* **46**:386–405.

Richardson, E. S., Jr., ed. (1971) *Extraordinary Fossils.* Proceedings of the North American Paleontological Convention, Part I. Lawrence, Kansas., Allen Press, p. 1153–1269.

Romer, A. S. (1959) *The Vertebrate Story.* Chicago, University of Chicago Press, 437 p.

Romer, A. S. (1966) *Vertebrate Paleontology.* Chicago, University of Chicago Press, 468 p.

Ross, C. A. (1967) Development of fusulinid (Foraminiferida) faunal realms. *Jour. Paleont.,* **41**:1341–1354.

Ross, C. A., and Sabins, F. F. (1965) Early and Middle Pennsylvanian fusulinids from southeast Arizona. *Jour. Paleont.,* **39**:173–209.

Ross, D. A. (1970) *Introduction to Oceanography.* Englewood Cliffs, N.J., Prentice-Hall, Inc. 384 p.

Ross, H. H. (1974) *Biological Systematics.* Reading, Mass., Addison-Wesley, 345 p.

Rowe, A. W. (1899) An analysis of the genus *Micraster,* as determined by rigid zonal collecting from the zone of *Rhynchonella cuvieri* to that of *Micraster coranguinum. Quart. Jour. Geol. Soc. London,* **55**:494–547.

Rowell, A. J. (1967) A numerical taxonomic study of the chonetacean brachiopods. *In* Teichert, C., and Yochelson, E. L., eds. *Essays in Paleontology and Stratigraphy: R. C. Moore Commemorative Volume.* Lawrence, Kans., University of Kansas Department of Geology Special Publication 2, p. 113–140.

Rudwick, M. J. S. (1961) The feeding mechanism of the Permian brachiopod *Prorichthofenia. Palaeontology,* **3**:450–471.

Rudwick, M. J. S. (1964) The inference of function from structure in fossils. *Brit. Jour. Philos. Sci.,* **15**:27–40.

Rudwick, M. J. S. (1970) *Living and Fossil Brachiopods.* London, Hutchinson, 199 p.

Runnegar, B., and Newell, N. D. (1971) Caspian-like relict molluscan fauna in the South American Permian. *Bull. Amer. Mus. Nat. Hist.,* **146**:1–66.

Ruzhentsev, V. Ye. (1964) The problem of transition in paleontology. *Internat. Geol. Rev.,* **6**:2204–2213.

Salfeld, H. (1913) Uber Artbildung bei Ammoniten. *Zeitschr. Deutsch. Geol. Gesell., Monatsber.,* **65**:437–440.

Sanders, H. L. (1956) Oceanography of Long Island Sound, 1952–1954, X. The biology of marine bottom communities. *Yale Bingham Oceanographic Coll.,* **56**:345-414.

Sanders, H. L. (1968) Marine benthic diversity: a comparative study. *Amer. Naturalist,* **102**:243–282.

Sanders, H. L., Hessler, R. R., and Hampson, G. R. (1965) An introduction to the study of deep-sea benthic faunal assemblages along the Gay Head-Bermuda transect. *Deep-Sea Res.,* **12**:845–867.

Savory, T. (1962) *Naming the Living World.* London, English University Press, Ltd., 128 p.

Schaeffer, B., Hecht, M. K., and Eldredge, N. (1972) Phylogeny and paleontology. *Evolutionary Biology,* **6**:31–46.

Schäfer, W. (1972) *Ecology and Paleoecology of Marine Environments.* Chicago, University of Chicago Press, 568 p.

Scheltema, R. S. (1968) Dispersal of larvae by equatorial ocean currents and its importance to the zoogeography of shoal-water tropical species. *Nature,* **217**:1159–1162.

Schenk, E. T., and McMasters, J. H. (1956) *Procedure in Taxonomy,* 3rd ed. Revised by A. M. Keen and S. W. Muller, Stanford, Calif., Stanford University Press, 119 p.

Schopf, T. J. M., ed. (1972) *Models in Paleobiology.* San Francisco, Freeman, Cooper, and Company, 250 p.

Schopf, T. J. M., Raup, D. M., Gould, S. J., and Simberloff, D. S. (1975) Genomic versus morphologic rates of evolution: influence of morphologic complexity. *Paleobiology,* **1**:63–70.

Scott, G. H. (1975) An automated coordinate recorder for biometry. *Lethaia,* **8**:49–52.

Seilacher, A. (1962) Paleontological studies on turbidite sedimentation and erosion. *Jour. Geol.,* **70**:227–234.

Seilacher, A. (1964) Biogenic sedimentary structures. *In* Imbrie, J., and Newell, N. D., eds. *Approaches to Paleoecology.* New York. John Wiley & Sons, p. 296–316.

Seilacher, A. (1967) Fossil behavior. *Scient. Amer.,* **217**(8):72–80.

Seilacher, A., Drozdzewski, G., and Haude, R. (1968) Form and function of the stem in a pseudoplanktonic crinoid (*Seirocrinus*). *Palaeontology,* **11**:275–282.

Selley, R. C. (1970) *Ancient Sedimentary Environments.* Ithaca, N. Y., Cornell University Press, 237 p.

Sepkoski, J. J. (1975) Stratigraphic biases in the analysis of taxonomic survivorship. *Paleobiol.,* **1**:343–355.

Shaw, A. B. (1964) *Time in Stratigraphy.* New York, McGraw-Hill, 365 p.

Shelford, V. E. (1963) *The Ecology of North America.* Urbana, University of Illinois Press, 610 p.

Simpson, G. G. (1940a) Mammals and land bridges. *Jour. Washington Acad. Sci.,* **30**:137–163.

Simpson, G. G. (1940b) Types in modern taxonomy. *Amer. Jour. Sci.,* **238**:413–431.

Simpson, G. G. (1944) *Tempo and Mode in Evolution.* New York, Columbia University Press, 237 p.

Simpson, G. G. (1951) *Horses: The Story of the Horse Family in the Modern World and Through Sixty Million Years of History.* New York, Oxford University Press, 247 p.

Simpson, G. G. (1952) How many species? *Evolution,* **6**:342.

Simpson, G. G. (1953) *The Major Features of Evolution.* New York, Columbia University Press, 434 p.

Simpson, G. G. (1960a) Notes on the measurement of faunal resemblance. *Amer. Jour. Sci.,* **258**:300–311.

Simpson, G. G. (1960b) The history of life. *In* Tax, S., ed. *Evolution after Darwin.* Chicago, University of Chicago Press, vol. 1, p. 117–180.

Simpson, G. G. (1961a) Historical zoogeography of Australian mammals. *Evolution,* **15**:431–446.

Simpson, G. G. (1961b) *Principles of Animal Taxonomy.* New York, Columbia University Press, 247 p.

Simpson, G. G., Roe, A., and Lewontin, R. C. (1960) *Quantitative Zoology.* New York, Harcourt, Brace, and Co., 440 p.

Sloss, L. L. (1963) Sequences in the cratonic interior of North America. *Geol. Soc. Amer. Bull.,* **74**:93–114.

Sneath, P. H. A., and McKenzie, K. G. (1973) Statistical methods for the study of biogeography. *Palaeont. Assoc. Special Papers in Palaeont.,* **12**:45–60.

Sneath, P. H. A., and Sokal, R. R. (1973) *Numerical Taxonomy.* San Francisco, W. H. Freeman and Company, 574 p.

Sohl, N. F. (1960) Archaeogastropoda, Mesogastropoda and Stratigraphy of the Ripley, Owl Creek, and Prairie Bluff Formations. *U. S. Geol. Surv. Prof. Paper 331-A,* 151 p.

Sokal, R. R., and Rohlf, F. J. (1969) *Biometry: The Principles and Practice of Statistics in Biological Research.* San Francisco, W. H. Freeman and Company, 776 p.

Sokal, R. R., and Rohlf, F. J. (1975) *Introduction to Biostatistics.* San Francisco, W. H. Freeman and Company, 368 p.

Sorgenfrei, T. (1958) Molluscan Assemblages from Marine Middle Miocene of South Jutland and Their Environments. *Geol. Surv. Denmark, II Ser. No. 79,* 503 p.

Speden, I. G. (1970) The type Fox Hills Formation, Cretaceous (Maestrichtian), South Dakota, Part 2. Systematics of the Bivalvia. *Bull. Peabody Mus. Nat. Hist.,* **33**:1–222.

Spjeldnaes, N. (1951) Ontogeny of *Beyrichia jonesi* Boll. *Jour. Paleont.,* **25**:745–755.

Stanley, S. M. (1968) Post-Paleozoic adaptive radiation of infaunal bivalve molluscs—a consequence of mantle fusion and siphon formation. *Jour. Paleont.,* **42**:214–229.

Stanley, S. M. (1970) *Relation of Shell Form to Life Habits in the Bivalvia.* Boulder, Colo., Geological Society of America. Memoir 125, 296 p.

Stanley, S. M. (1972) Functional morphology and evolution of byssally attached bivalve mollusks. *Jour. Paleont.,* **46**:165-212.

Stanley, S. M. (1973) An explanation for Cope's Rule. *Evolution,* **27**:1–26.

Stanley, S. M. (1974) Relative growth of the titanothere horn: A new approach to an old problem. *Evolution,* **28**:447–457.

Stanley, S. M. (1975) A theory of evolution above the species level. *Proc. Nat. Acad. Sci. U. S.,* **72**:646–650.

Stanley, S. M. (1976) Stability of species in geologic time. *Science,* **192**:267–269.

Stanley, S. M. (1977) Trends, rates, and patterns of evolution in the Bivalvia. *In* Hallam, A., ed. *Patterns of Evolution.* Amsterdam, Elsevier, 209–250.

Stanley, S. M. (1978) Chronospecies longevity, the origin of genera, and the punctuational model of evolution. *Paleobiology* (In press).

Stanton, R. J., and Dodd, J. R. (1970) Paleoecologic techniques—comparison of faunal and geochemical analyses of Pliocene paleoenvironments, Kettleman Hills, California. *Jour. Paleont.,* **44**:1092–1121.

Stebbins, G. L. (1965) The probable growth habit of the earliest flowering plants. *Ann. Missouri Bot. Garden,* **52**:457–468.

Stehli, F. G. (1966) Taxonomic diversity gradients and pole location: The Recent model. *In* Drake, E. T., ed. *Evolution and Environment.* New Haven, Conn., Yale University Press, p. 163–227.

Stehli, F. G., and Wells, J. W. (1971) Diversity and age patterns in hermatypic corals. *System. Zool.,* **20**:115–126.

Stein, R. S. (1975) Dynamic analysis of *Pteranodon ingens:* a reptilian adaptation to flight. *Jour. Paleont.,* **49**:534–548.

Stephenson, W., Williams, W. T., and Cook, S. D. (1972) Computer analysis of Petersen's original data on bottom communities. *Ecol. Monographs,* **42**:387–415.

Stoll, N. R., et al., eds. (1961) *International Code of Zoological Nomenclature.* London, International Trust for Zoological Nomenclature, 176 p.

Strahler, A. N. (1969) *Physical Geography.* New York, John Wiley & Sons, 733 p.

Strahler, A. N. (1971) *The Earth Sciences.* Second edition. New York, Harper & Row, 824 p.

Strauch, F. (1968) Determination of Cenozoic sea-temperatures using *Hiatella arctica* (Linné). *Palaeogeogr., Palaeoclimat., Palaeoecol.,* **5**:213–233.

Surlyk, F. (1972) Morphological adaptations and population structures of the Danish Chalk brachiopods (Maastrichtian, Upper Cretaceous). *Kongelige Danske Vidensk. Selsk. Biol. Skrifter,* **19**:1–57.

Sutherland, J. P. (1974) Multiple stable points in natural communities. *Amer. Naturalist,* **108**:859–873.

Sverdrup, H. U. (1942) *Oceanography for Meteorologists.* Englewood Cliffs, N. J., Prentice-Hall, 246 p.

Tappan, H. (1968) Primary production, isotopes, extinctions and the atmosphere. *Palaeogeogr., Palaeoclimat., Palaeoecol.,* **4**:187–210.

Teichert, C. (1956) How many fossil species? *Jour. Paleont.,* **30**:967–969.

Teichert, C. (1958) Cold- and deep-water coral banks. *Bull. Amer. Assoc. Petrol. Geol.,* **42**:1064–1082.

Thayer, C. W. (1974) Substrate specificity of Devonian epizoa. *Jour. Paleont.,* **48**:881–894.

Thompson, D'A. W. (1942) *On Growth and Form.* New York, Cambridge University Press, 1116 p.

Thorson, G. (1950) Reproductive and larval ecology of marine bottom invertebrates. *Biol. Rev.,* **25**:1–45.

Thorson, G. (1957) Bottom communities (sublittoral or shallow shelf). *Geol. Soc. Amer. Mem.* **67**(1):461–534.

Thorson, G. (1966) Some factors influencing the recruitment and establishment of marine benthic communities. *Netherl. Jour. Sea Res.,* **3**:267–293.

Valentine, J. W. (1963) Biogeographic units as biostratigraphic units. *Bull. Amer. Assoc. Petrol. Geol.,* **47**:457–466.

Valentine, J. W. (1970) How many marine invertebrate fossil species? *Jour. Paleont.,* **44**:410–415.

Valentine, J. W. (1973) *Evolutionary Paleoecology of the Marine Biosphere.* Englewood Cliffs, N.J., Prentice-Hall, 511 p.

Valentine, J. W., and Moores, E. M. (1970). Plate tectonic regulation of faunal diversity and sea level: a model. *Nature,* **228**:657–659.

Van Valen, L. (1973a) A new evolutionary law. *Evol. Theory,* **1**:1–30.

Van Valen, L. (1973b) Are categories in different phyla comparable? *Taxon,* **22**:333–373.

Van Valen, L. (1975) Group selection, sex, and fossils. *Evolution,* **29**:87–94.

Vermeij, G. J. (1976) Interoceanic differences in vulnerability of shelled prey to crab predation. *Nature,* **260**:135–136.

Vine, F. J., and Matthews, D. H. (1963) Magnetic anomalies over ocean ridges. *Nature,* **199**:947–949.

Vokes, H. E. (1957) Miocene fossils of Maryland. *Maryland Dept. Geol., Mines and Water Res., Bull.,* **20**:1–48.

Wainwright, S. A., Biggs, B. D., Currey, J. D., and Gosline, J. M. (1976) *Mechanical Design in Organisms.* New York, Halsted Press, 423 p.

Walker, K. R., and Laporte, L. F. (1970) Congruent fossil communities from Ordovician and Devonian carbonates of New York. *Jour. Paleont.,* **44**:928–944.

Waller, T. R. (1969) *The Evolution of the* Argopecten gibbus *Stock (Mollusca: Bivalvia), with Emphasis on the Tertiary and Quaternary Species of Eastern North America.* Menlo Park, Calif., Paleontological Society. Memoir 3, 125 p.

Webb, L. J. (1959) Physiognomic classification of Australian rain-forests. *Jour. Ecology,* **47**:551–570.

Weimer, R. J., and Hoyt, J. H. (1964) Burrows of *Callianassa major* Say, geologic indicators of littoral and shallow neritic environments. *Jour. Paleont.,* **38**:761–767.

Wells, J. W. (1956) Scleractinia. *In* Moore, R. C., ed. *Treatise on Invertebrate Paleontology.* Boulder, Colo., Geological Society of America and University Press of Kansas (joint publication), Part F, Coelenterata, p. 328–444.

Wells, J. W. (1957) Coral reefs. *Geol. Soc. Amer. Mem.* **67**(2):609–631.

Weller, J. M. (1960) *Stratigraphic Principles and Practice.* New York, Harper & Brothers, 725 p.

Westoll, T. S. (1949) On the evolution of the Dipnoi. *In* Jepsen, G. L., Simpson, G. G., and Mayr, E., eds. *Genetics, Paleontology,* and *Evolution.* Princeton, N.J., Princeton University Press, p. 121–184.

Whittington, H. B. (1957) The ontogeny of trilobites. *Biol. Rev.,* **32**:421–469.

Williams, A. (1973) Distribution of brachiopod assemblages in relation to Ordovician palaeogeography. *Palaeont. Assoc. Special Papers in Palaeont.,* **12**:241–269.

Wilson, E. O., and Bossert, W. H. (1971) *A Primer of Population Biology.* Stamford, Conn., Sinauer Assoc., 192 p.

Wolfe, J. A., and Hopkins, D. M. (1967) Climatic changes recorded by land floras in northwestern North America. *In* Yawata, I., and Sinoto, Y. H., *11th Pacific Sci. Congr.,* Honolulu, Bishop Museum Press, p. 67–76.

Woodin, S. A. (1974) Polychaete abundance patterns in a marine soft-sediment environment: the importance of biological interactions. *Ecol. Monographs,* **44**:171–187.

Woodring, W. P., Stewart, R., and Richards, R. W. (1940) Geology of the Kettleman Hills Oil Field, California. *U.S. Geol. Surv. Prof. Paper 195,* 170 p.

Yonge, C. M. (1962) On the primitive significance of the byssus in the Bivalvia and its effects in evolution. *Jour. Marine Biol. Assoc. United Kingdom,* **42**:112–125.

Zangerl, R. (1965) Radiographic techniques. *In* Kummel, B., and Raup, D. M., eds. *Handbook of Paleontological Techniques.* San Francisco, W. H. Freeman and Company, p. 305–320.

Ziegler, A. M. (1965) Silurian marine communities and their environmental significance. *Nature,* **207**:270–272.

Ziegler, A. M., Boucot, A. J., and Sheldon, R. P. (1966) Silurian pentameroid brachiopods preserved in position of growth. *Jour. Paleont.,* **40**:1032–1036.

Ziegler, A. M., Cocks, L. R. M., and Bambach, R. K., (1968) The composition and structure of Lower Silurian marine communities. *Lethaia,* **1**:1–27.

Zohary, D., and Feldman, M. (1962) Hybridization between amphidiploids and the evolution of polyploids in the wheat *(Aegilops-Triticum)* group. *Evolution,* **16**:44–61.

Indexes

Author Index

Agassiz, L., 161
Ager, D. V., 302
Alexander, R. M., 188, 195
Alexander, R. R., 264, 265
Andrews, H. E., 97
Arkell, W. J., 206, 366
Arthur, M. A., 446, 447
Asgaard, U., 342
Axelrod, D. I., 431, 434

Bacescu, M., 254
Bailey, I. W., 394
Bakker, R. T., 296, 353, 428, 442, 443, 445, 447
Bambach, R. K., 75, 201, 287, 291
Baron, J. G., 422
Bell, W. C., 224
Berry, W. B. N., 229
Biggs, B. D., 196
Bird, J. M., 437
Blackwelder, R. E., 128
Blatt, H., 9
Boardman, R. S., 66, 68
Bonner, J. T., 68, 195
Bossert, W. H., 410, 412
Boucot, A. J., 78, 277, 330
Bowen, Z. P., 253
Boyd, D. W., 327
Brace, W., 78
Bramwell, C. D., 185, 187, 188, 195
Brasier, M. D., 404
Bridges, P. H., 289
Bromley, R. G., 342
Brooks, J. C., 415
Brown, C. A., 44
Bulman, O. M. B., 355
Buss, L., 281

Callomon, J. H., 73
Camp, C. L., 44
Campbell, K. S. W., 56
Carpenter, F. M., 10
Chaloner, W. G., 395, 428
Chave, K. E., 16, 17, 18
Cheetham, A. H., 68
Cifelli, R., 369, 371
Clarke, G. L., 233, 238
Clarkson, E. N. K., 180, 181, 182, 183, 184
Cloud, P. E., 355, 364
Coates, A. G., 438
Cocks, L. R. M., 291
Colbert, E. H., 143, 314, 315, 358, 359, 390, 424, 425, 426, 428
Connell, J. H., 381
Cook, S. D., 289
Cooper, G. A., 122, 123
Corliss, J. O., 135
Cox, C. B., 397, 448
Craig, G. Y., 71, 80, 81, 298
Creber, G. T., 395, 428
Crompton, A. W., 188
Currey, J. D., 196

Dacqué, E., 195
Darwin, C., 344, 434
Davis, M. B., 397, 399
Dayton, P. K., 280, 287
DeBeer, G. R., 355, 377
deMar, R., 78

Descartes, R., 180, 181
Dewey, J. F., 419, 420, 437
Dietrich, J. W., 311
Dietz, R. S., 417, 427, 428, 429
Dobzhansky, T., 100, 166
Dodd, J. R., 263, 264, 265
Donahue, J., 399
Donovan, D. T., 229
Dorf, E., 235, 393, 395, 399
Doyle, J. A., 361, 362, 363
Driscoll, E. G., 269
Drozdzewski, G., 245
Dunbar, C. O., 224, 229, 294
Durham, J. W., 5, 26, 133
DuToit, A. L., 425

Easton, W. H., 10, 72
Eaton, R. M., 54, 154
Ehrlich, A. H., 383
Ehrlich, P. R., 383
Eicher, D. L., 230
Eldredge, N., 113, 128, 326, 348
Elliott, D. H., 426
Elton, C. S., 203
Eltringham, S. K., 302
Emlen, J. M., 305
Erben, H. K., 372, 373
Erickson, R. O., 70
Ericson, D. B., 215, 402, 403
Ewing, M., 215

Feldman, M., 91
Finks, R. M., 258
Fischer, A. G., 356, 393, 446, 447
Fisher, R. A., 106
Fisher, W. L., 311
Fisk, H. N., 294
Flessa, K. W., 438
Flint, R. F., 398
Florin, R., 431
Forcier, L. K., 280

Galileo, 64
Garrett, P., 257
Gernant, R. E., 260
Gingerich, P. D., 310
Ginsburg, R. H., 15
Gordon, W. A., 442, 443
Goreau, T. F., 375
Gosline, J. M., 196
Gould, S. J., 62, 64, 68, 113, 128, 196, 326, 333, 347, 348, 355
Grant, R. E., 29, 122, 123, 239, 240, 241, 243, 245
Grant, V., 10, 107, 325, 349
Grassle, J. F., 287
Graus, R. R., 400
Gregor, C. B., 9

Haeckel, E., 354, 355
Hallam, A., 71, 80, 81, 298, 377, 448
Hanna, G. D., 44
Harland, W. B., 12, 26, 227, 229
Harper, C. W., 330
Hartman, W. D., 375, 377
Haude, R., 245
Haug, E., 441
Hay, W. W., 33
Hayami, I., 308, 309
Hays, J. D., 441
Healey, I. N., 397, 448
Heaslip, W. G., 132
Hecht, M. K., 330
Hecker, R. F., 302
Hedberg, H. D., 224
Hedgpeth, J. W., 302
Heirtzler, J. R., 422
Hennig, W., 146, 149
Hertel, H., 196
Hess, H. H., 417
Hessler, R. R., 255, 287
Heywood, V. H., 149
Hickey, L. J., 361, 362, 363
Holden, J. C., 427, 428, 429
Holdren, J. P., 383
Holmes, A., 423
Hopkins, D. M., 395, 396
House, M. R., 405
Hoyt, J. H., 259
Hudson, J. D., 33, 265, 266
Hughes, N. F., 448
Hunt, A. S., 50
Hunt, O. D., 239
Huxley, J. S., 62, 68
Huygens, C., 180, 181

Imbrie, J., 87, 88, 90, 100, 302, 438
Israelsky, M., 215

Jackson, J. B. C., 281, 287, 288, 305
Jackson, R. T., 153, 160
Jardine, N., 432, 433
Jeletsky, J. A., 229
Jell, J. S., 120
Jerison, H. J., 345, 351, 352
Johnson, J. G., 441, 442
Johnson, R. G., 298, 301
Jones, R. L., 9
Jurva, R., 262

Kaesler, R. L., 41, 42, 95, 96, 97, 149
Kauffman, E. G., 284, 287, 438
Kaufmann, R., 365, 368
Kermack, K. A., 338
Kessling, R. V., 260
Kier, P. M., 242

King, P. B., 252
Kinne, O., 261, 302
Knight, J. B., 375
Koch, D. L., 240
Koch, K. S., Jr., 94
Kozlowski, R., 369, 371
Kraft, P., 31
Krinsley, D. H., 399
Krumbein, W. C., 230
Kummel, B., 44
Kurtén, B., 83, 328, 329, 343, 344, 434

Ladd, H. S., 302
Lalicker, C. G., 356
Lang, J., 284
Laporte, L. F., 293, 302
Lawrence, D. R., 297
Lawson, D. A., 184
Le Gros Clark, W. E., 68, 347
Leich, H., 22
Lemche, H., 375
Leon, R., 31
LePichon, X., 422
Levinton, J. S., 287
Levi-Setti, R., 180, 181, 182, 183
Lewontin, R. C., 100
Liem, K. F., 415
Linck, O., 48
Link, R. F., 94
Linnaeus, C., 131
Linsley, E. G., 116
Lowenstam, H. A., 284

MacArthur, R. H., 304, 305, 381, 408, 410, 411, 413, 414, 448
MacClintock, C., 47
Macurda, D. B., 68
Makurath, J. H., 292
Martin, P. S., 399
Martin-Kaye, P., 76, 77
Matthews, D. H., 418
Matthews, R. K., 230
Mayr, E., 102, 105, 107, 116, 128, 150, 213, 324, 325, 348, 349, 377
McAlester, A. L., 269, 448
McKenna, M. C., 406, 428, 431
McKenzie, D., 432, 433
McKenzie, K. G., 406
McKinney, F. K., 66
McLean, J. D., Jr., 44
McMasters, J. H., 128
McNeil, J., 149
Medawar, P. B., 60
Melville, R., 428, 431
Michelson, A., 174
Middlemiss, F. A., 448
Moore, H. B., 302
Moore, P. D., 397, 448
Moore, R. C., 356
Moores, E. M., 441
Morse, J. W., 253, 254
Muir-Wood, H. M., 137
Muus, K., 287

Natland, M. L., 258
Nevesskaya, L. A., 113
Newell, G., 448
Newell, N. D., 26, 100, 111, 192, 249, 252, 302, 327, 435, 437, 439
Nichols, D., 338, 339, 340, 342
Nicol, D., 285

Odum, E. P., 234, 238, 283
Odum, H. T., 283
Oliver, W. A., Jr., 68
Olson, E. C., 118, 119, 377
Osborn, H. F., 334
Ostrom, J. H., 247
Ovcharenko, V. N., 113, 324, 325
Owen, G., 272
Oxnard, C., 44, 99, 100
Ozawa, T., 308, 309

Paine, R. T., 280
Palmer, A. R., 59, 209, 212, 221
Pannella, G., 47
Pascual, R., 407
Patterson, B., 407
Pavlow, A. P., 355
Perkins, B. F., 208
Peterson, C. G. J., 289
Pickett, J. W., 120
Pielou, E. C., 235
Pilbeam, D., 347, 348
Pitman, W. C., 441
Pojeta, J., 242
Porter, J. W., 284

Rasmussen, H. W., 243
Raup, D. M., 4, 7, 9, 10, 26, 44, 52, 53, 169, 171, 172, 173, 174, 177, 246, 329, 355, 360
Raven, H. C., 434
Raven, P. E., 431, 434
Rawson, P. F., 448
Reid, R. E. H., 258
Rensch, B., 378
Reynolds, S. H., 334
Rhoads, D. C., 199, 201, 253, 254, 258, 268
Richards, R. P., 242
Richardson, E. S., Jr., 26
Richter, R., 246
Rodda, P. U., 311
Rodgers, J., 224, 229, 294
Roe, A., 100

Rohlf, F. J., 100
Romer, A. S., 334, 354
Ross, C. A., 210
Ross, D. A., 382
Ross, H. H., 150
Rowe, A. W., 338
Rowell, A. J., 137, 147
Rudwick, M. J. S., 179, 196, 239, 357
Runnegar, B., 435, 437
Ruzhentsev, V. Y., 113, 324

Sabins, F. F., 210
Salfeld, H., 365
Sandberg, P. A., 33
Sanders, H. L., 255, 267, 287
Savory, T., 128
Schaeffer, B., 146, 148
Schäfer, W., 26, 302
Scheltema, R. S., 204, 205
Schenk, E. T., 128
Schindewolf, O., 355
Schopf, T. J. M., 304, 378
Scott, G. H., 44
Seilacher, A., 48, 199, 203, 243, 245, 246, 258
Selley, R. C., 295, 302
Sepkoski, J. J., Jr., 329
Shaw, A. B., 216, 217, 218, 219, 220, 221, 227, 229
Sheldon, R. P., 277
Shelford, V. E., 389
Shinn, E. A., 275
Simpson, G. G., 4, 26, 100, 110, 113, 128, 134, 135, 136, 150, 204, 212, 225, 226, 304, 307, 315, 331, 332, 333, 334, 349, 357, 360, 365, 368, 372, 378, 406, 423, 428, 433
Sinnott, E. W., 394
Sloss, L. L., 230, 439, 440
Sneath, P. H. A., 150, 406
Sohl, N. F., 35, 284, 287
Sokal, R. R., 100, 150
Sorgenfrei, T., 6, 8, 262
Speden, I. G., 273
Spjeldnaes, N., 83
Stanley, S. M., 192, 193, 194, 269, 273, 315, 318, 327, 329, 333, 348, 349, 353, 354
Stanton, R. J., 263, 264, 265
Stebbins, G. L., 361
Stehli, F. G., 392, 408, 409, 410
Stein, R. S., 185, 186, 187, 188, 196
Stephenson, W., 289
Stoll, N. R., 128
Strahler, A. N., 382, 384, 386, 388
Strauch, F., 337
Strimple, H. L., 240
Stürmer, W., 23
Surlyk, F., 268, 269, 271, 272, 305, 306
Sutherland, J. P., 281
Sverdrup, H. U., 384

Tappan, H., 448
Teichert, C., 26, 149, 256
Thayer, C. W., 272
Thompson, D'A. W., 58, 62, 63, 68, 196
Thorson, G., 13, 287, 289

Usinger, R. L., 116

Valentine, J. W., 4, 5, 26, 206, 207, 302, 378, 441
Van Valen, L., 10, 303, 304, 331, 352
Vermeij, G. T., 400
Vine, F. J., 418
Vokes, H. E., 38
von Baer, K. E., 354

Wainwright, S. A., 196
Walker, K. R., 293
Wallace, A. R., 434
Waller, T. R., 312
Waters, J. A., 41, 42
Webb, L. J., 394
Wegener, A., 422, 423
Weimer, R. J., 259
Weller, J. M., 224, 226, 230
Wells, J. W., 319, 401, 408, 409, 410
Wenz, W., 375
Westermann, G. E. G., 34
Westoll, T. S., 315
Whitfield, G. F., 185, 187, 188, 195
Whittington, H. B., 49
Williams, A., 272, 435
Williams, W. T., 289
Wilson, E. O., 408, 410, 411, 412, 413, 414, 448
Wingstrand, K. G., 375
Wolfe, J. A., 395, 396
Wollin, G., 215, 402, 403
Woodin, S. A., 288
Woodring, W. P., 18
Wright, S., 166

Yochelson, E. L., 149
Yonge, C. M., 353
Young, D. K., 268

Zangerl, R., 31
Ziegler, A. M., 275, 277, 289, 290, 291, 292
Zohary, D., 91

Subject Index

Page numbers referring to illustrations are given in **boldface** type.

acme, 209
adaptation, 104, 165–196
  See also functional morphology
adaptive landscape, 166, **167,** 169, 170, 304
adaptive peak, 166, **167,** 169, 170
adaptive radiation, 307, 317–324, 328, 357–363, **363, 370**
adaptive valley, 166, **167,** 176
adaptive zone, 365
ahermatypic coral, 256
algae, **245**
  boring, 15
  calcareous, role in reefs, 284
  mechanical destruction by, 16–18
allometric growth, 62, **64,** 68
allometric relationships in evolution, 345
allopatric populations, 102, 108
allopatric speciation, 104–105
amber, Baltic, 21
American Commission on Stratigraphic Nomenclature, 197, 229
ammonoids, **39, 40**
  biogeography of sphenodiscid, **443**
  coiling geometry of, **168, 171**
  evolution of, 167–169, 355, 365, **366, 367, 373**
  sexual dimorphism of, **73**
  sutures, **34**
  theoretical morphology of, 171–175
amphibians, 4, **119**
  biogeography of, 431–432
  species-area relationship, **411**
  species description, 118–119
analogous structures, 178
anisometric growth, 61, **62**
  explanation of, 63–66
Archaeocyatha, 4
Archaeopteryx, 21, **22,** 141, 372
arthropods
  eurypterid, 97
  growth of, 50, 80
  insect, 4
  in Solnhofen Limestone, **22**
  See also crustaceans; ostracodes
assemblage, fossil, 75–90, **298**
  death, 297
  life, 297
  variation between, 87–90
  variation within, 85–86
assemblage zone, 209
Australopithecus, **347,** 348
autecology, 278

bacteria, 16
Baltic amber, 21
bat, anatomy of, **186**
bathyal zone, 236
benthonic organism, 236
Bergmann's Rule, 337, 343
bibliographic sources in paleontology, 161–162
Bibliography of North American Geology, 162
Binomial Theorem, 308
biofacies, 203
Biogenetic Law, 354
biogeography, **70, 205, 331,** 379–448, **389, 390**
  migration, 204, 209, 407
  relicts, 415

biogeographic province, 205, **206,** 207, 387–391
Biological Abstracts, 162
biomass, 234, **238,** 385
biome, 387–391, **389**
biostratigraphy, 197–230
  accuracy of correlation, 225–229, **223**
  acme, 209
  biostratigraphic unit, 203–207
  quantitative correlation, 209–213, 216–222
  See also zone
biostratinomy, **297**
biotic succession, 198
biozone, 203–207
birds, fossil
  Archaeopteryx, 21, **22,** 141, 372
bivalves, **19, 38, 77, 242, 269**
  classification of, 141–142
  coiling geometry of, 175
  electron microscopy of, **32**
  endobyssate, **194,** 353
  epibiont, 242
  epibyssate, **194,** 353
  evolution of, **132, 312, 318,** 337
  evolutionary rates of, 316–324, **318**
  feeding mechanism, related to sediment type, 267–268
  functional morphology of, 192–195, **193, 194**
  growth bands in, **47**
  life habits of, **237, 269, 273, 276**
  mechanical destruction of, **17, 18**
  mortality rates of, 80
  orthogenesis of, **334, 335**
  neoteny in, 353
  post-mortem transport of, **77,** 78
  rudist, 284, **286**
  as salinity indicators, **262, 264,** 265
  shape variation due to deformation, **75,** 76
  shell shape of, **38**
  size variation related to age, **71**
  sorting of, **71, 77**
  survivorship in, **81, 332**
  Venericardia evolution, **132**
bivariate analysis, 86, 89
  of brachiopods, **90**
boring organisms, **15,** 272
bottleneck effect, 344–345
brachiopods, 4, **29, 30, 122, 242, 270, 271, 306, 325**
  biometrics of, 87–90
  classification of, **137,** 141
  coiling geometry of, 175
  dendrogram of, **137**
  distribution related to substratum, 268
  diversity through time, 11, **12**
  ecology of, 268, 305
  evolution of, 324–325, **353,** 357
  growth of, 55, **56, 357**
  life habits of, **241, 245, 270–272, 277, 306**
  Lingula community, 289, **291**
  post-mortem transport of, 78
  numerical taxonomy of, 146, **147**
  as salinity indicators, 265
  speciation in, 324–325
  species description, **123**
  stereophotographs of, **29**
  X-ray photography of, **30**
browser, 273
bryozoans, 4
  abundance of, correlated with abundance of brachiopods, 268–269
  growth of, **67,** 68, **281**
  mechanical destruction of, 16, **17**
  Ordovician, **242**
Burgess Shale, 21
burrows, **200, 202, 246,** 258, **259,** 273
  Skolithos, **288**

Callianassa burrows, 258, **259,** 273
canonical analysis, 98, **99**
carnivore, **233, 234,** 239, 351–352
cephalopods
  coiling geometry of, **39, 40,** 43, **54, 57,** 58, 167–170, **168, 171,** 175–176
  distribution of modern *Nautilus,* 404, **405**
  evolution of, 11, 372
  life habits of, **237**
  as salinity indicators, 265
  sexual dimorphism in, **73**
  Spirula, **168,** 169
  variation in form of, **168**
  X-ray photograph of, **54**
  See also ammonoids
characters
  multi-state, 144
  phenetic, 144
  taxonomic, 124–125, 138–142
  two-state, 144
  weighting of, 138–139
chronospecies, 313, 329
chronostratigraphic unit, 225
clade, 149, **317,** 364
cladistic taxonomy, 146–149
cladogram, **148**
classification. *See* taxonomy
CLIMAP, 403
climate, 380–387
  Bergmann's Rule, 343
  Cenozoic climates, **337, 395, 396, 398, 399**
  climatic zones, 386
climax community, 280, 284
cline, 105–106
Code of Stratigraphic Nomenclature, 197
coiling direction, 173, 213

coiling geometry
  ammonoid, **168,** 169
  bivalve, 175
  brachiopod, 175
  cephalopod, **39, 40,** 43, **54, 57,** 58, 169, 175
  foraminiferan, 176, **214**
  gastropod, 173, **174**
colonial organisms, **281**
  growth, 66–68, **67**
community
  biologic, defined, 232, 278
  climax, 284
  fossil, 278–296, **290**
  Lingula, 289, **290, 291**
  preservation, models of, 299–301
  recurrent, 278, 289
  reef, 282–287
  Silurian, **290, 291**
  soft-bottom, 287–293
  terrestrial, 293–296
computers in paleontology, 29–31, **34,** 43, **53,** 58, 90
computer simulation
  of cephalopod morphology, **168**
  of coiled forms, 175, **177**
  of echinoid morphology, **53**
  of gastropod morphology, **174**
continental drift, 423
continental shelf, **235,** 236, **260**
continental slope, **235,** 236, 255
convergence, evolutionary, 165, 170
  adaptive, 363
  and homeomorphy, 363–364
coordinate transformation, 58, **59, 194**
Cope's Rule, 352
coral reefs, **257,** 282–284, **283**
corals
  ahermatypic, 256
  biogeography of, **409, 410,** 438
  evolution of, **320–321,** 355, **356**
  growth of, 66, **281, 356**
  hermatypic, 256, **257,** 400, 401, 408
  as salinity indicators, 265
  species description, **120, 121**
Coriolis Force, 381–383, **382**
correlation by means of fossils, 207–229
crinoids
  Seirocrinus, 243, **244**
  snails attached to, **242**
  stem growth of, 47–49, **48**
crustaceans
  Callianassa burrows, 258, **259**
  life habits of, **237**
  ultraviolet photograph of, **30**
currents
  ocean, **205,** 380–385, **382, 384**
  sorting by, **77**
  tidal, **283**
data matrix, **92**
death assemblage, 298
decomposers, **233, 234**
deformation of fossils, **75,** 76
deme, 69
dendrogram, **145**
  brachiopod, 137, **147**
  echinoid, **157**
deposit feeder, 239, 268, 287
depth of water, 255–261
diagnosis (taxonomic)
  defined, 117
  in species description, 117–125
diastem, 199
digitizing, 29–31, **34,** 43, 86
dimorphism
  foraminiferan, **72**
  cephalopod, 72, **73**
dinosaurs
  endothermy of, 296
  evolution of size of, **314**
  evolutionary rates of, 313
  phylogeny of, 143
  tracks, **247**
dispersal (biogeographic), **205,** 404–408
divergence, evolutionary, 170
diversity
  gradient, 391, **392, 409, 410**
  taxonomic, 234, 263, 333
diversity. *See* species; taxonomic diversity
doubling time, 316–317, **318**
dynamic method of survivorship analysis, 85

echinoderms, 4
  cystoid life position, **240**
  diversity through time, 11, **12**
  growth of, 47, 51
  mechanical destruction of, **17, 18**
  as salinity indicators, 265
echinoids, **153, 154**
  computer simulation of, **53**
  description of, 152–155
  in faunal list, **159**
  growth of, 47, 51, **52, 53**
  identification of, 152–155
  key for, 155, **156,** 157
  life habits of, **340, 341**
  Micraster, evolution of, 338–342, **339–342**
  phylogeny of, **133**
  synonymy of, **160**
ecologic displacement, 358
ecologic replacement, 365–369
ecologic succession, 280, 284, **285**
ecological niche, 232
ecosystem, 232
  energy pyramid of, **234**
  flow of materials through, **233**

electron microscopy, 28, **32, 33**
endemic fauna, 407
endemic species, 205
energy pyramid, 234
epibiont organism, 241, **242**
epifaunal organism, **193,** 236
epiphytes, 241
epizoans, 241, **242**
euryhaline organism, 261, 265–267
eurypterids, R-mode analysis of, 97
evolution, 303–378
  adaptation, 165–170
  allometric relationships, 62, 345
  and classification, 130
  convergence, 165, 170, 363–365
  of dinosaurs, **143**
  divergence, 170
  explosive, 307
  of horses, **136**
  grade of, 364
  inadaptive, 349
  iterative, 365–368, **366, 367, 368, 370**
  mosaic, 313
  nonadaptive, 349
  parallel, 364
evolutionary rates, 4, 106–107, 307–330, **322, 323**
  doubling time, 316–318
  in lungfish, 327, **336, 337**
  phyletic, 313, 335–355
  phylogenetic, 313, 335
  rate of progressive evolution, 307
  taxonomic frequency rates, 307, 316–324
evolutionary species, 110, 135, 313
evolutionary trends, 307–308, 333–335, **350**
  orthogenesis, 333
  phyletic, 335–348
  phylogenetic, 335, 348–355
extinction, 105, 304–307, 356
  mass, 307, 332, 438–448
  pseudoextinction, 330
  rates, 319–321

facies, 203
facies fossil, 204
factor analysis, 94, 97
faunal list, **159**
feeding mechanisms, 239, 267–268
fishes, **359**
  evolution of, **315,** 359
  geologic history of, **358, 359**
food chain, 233, **238**
food web, 233, 385
foraminiferans
  arenaceous, as salinity indicators, 265
  coiling in, 176, 213, **214**
  as climatic indicators, **402,** 403
  dimorphism of, **72**
  ecological distribution of, **260**
  evolution of, 445–447
  evolutionary lineage within a species, 308–309
  iterative evolution, 369, **370**
  See also fusulinids
formation, stratigraphic, 198
fossilization. *See* preservation
founder effect, 106
Fourier analysis, 40, **41,** 42
frugivore, 273
functional morphology, 163–196, 243–245
  of bivalves, 192–195, **193, 194**
  of flying reptiles, 182–188
  of jaw mechanics, 188–192
  of trilobites, 180–182
fusulinids
  evolution of size in, 308–309, **309**
  factor analysis of, 95–97
  phenogram of, **96**
  principal components analysis of, **95,** 97
  stratigraphic ranges of, **210, 211**

gastropods, **19, 35, 242**
  biogeography of, **392**
  biological destruction of, **15**
  coiling geometry of, **172, 173,** 173–175
  computer simulation of, **174**
  descriptive terminology, 35–36, **37**
  evolution of, **311**
  larvae of, **205**
  life habits of, **237**
  mechanical destruction of, **17, 18**
  as salinity indicators, **262,** 265
  species description, 35–36
  X-ray photography, **15**
gene flow, 69, 105
gene pool, 69
genetic drift, 106
genus, type, 131
geochronologic time scale, 225, **228**
geographic isolation, 104
geographic speciation, 104–106
geologic time scale, 222, **228**
Gondwanaland, **424, 425, 427**
grade (of evolution), 364
gradualistic model of evolution, **326,** 327, 330
granivore, 273
graptolites
  affinities of, **371**
  evolution of, 355, 369
  infrared photograph of, **30**
grazer, 239, 273
group, stratigraphic, 198
growth, 45–68, **60**
  allometry, 62–66, **64**
  anisometric, 61, **62, 173**
  Biogenetic Law, 354

bivalve, **47, 71**
brachiopod, **56, 357**
bryozoan, 67
cephalopod, **54, 57**
colonial organisms, 66–68, **67**
coral, **356**
of crinoid stems, **48**
echinoid, **52, 53**
heterochrony in, 355
isometric, 61, 86
lines, **47,** 400
rates, 59–62
sponge, **65**
trilobite, **49, 50, 59**
types, 46–51
growth bands, **47, 53,** 400
guide fossil, 213

habits, life. *See* life habits
harmonic analysis, **40,** 41
Haug Effect, 441, 445
herbivore, **233, 234,** 351–352
hermatypic coral, 256, **257,** 400, 408
heterochrony, 355
higher category (taxonomic), 129–150
criteria for, 138–142
defined, 129
monotypic, 131
type genus, 131
type species, 131
See also taxonomy
holotype, 113, 125
homeomorphy, 363–364
homeostasis, 324
hominids, 76, 346, 347–348
allometry of, **346**
Australopithecus, **347,** 348
multivariate analysis of, 98–100, **99**
Homo, **346,** 348
homology, 178, 243
homoplasy, 178
homonym, 126
horses, evolution of, **136,** 349–352, **351**
Hunsrück Shale, 21, **23,** 372, **373**
hybridization, 101–103
hypodigm, 113–114

identification of fossils, 151–162
by computer, 158
by keys, 155–157
inadaptive evolution, 349
incertae sedis, 131
index fossil, 213, 225
infaunal organism, 236
infrared photography, 28, **30**
insects, 4, 10–11, 21
internal structures, 44

International Code of Zoological Nomenclature, 114–116, 127, 128, 131
International Commission on Zoological Nomenclature, 114, 117, 125, 127, 131
International Rules of Botanical Nomenclature, 114–115
intertidal zone, 236
Irish "elk," **335**
island arc, 418
island biogeography, 408–413, **411, 412, 414**
isochronous units, 224
isolation
reproductive, 101–103, 105–109
geographic, 104
isometric growth, 61, **62,** 86
isotopic composition, 263, **264,** 403
iterative evolution, **360,** 365–368, **366, 367, 368**

jaw mechanics, 188–192, **188, 189, 191**
JOIDES, 445

key, 155–157
key bed, 226

La Brea tar deposits, 21
Lagerstätten, 21
larvae, **30, 205,** 404
Laurasia, 428
lectotype, 126
level bottom zone, 236
life assemblage, 297
life habits, 239–243
bivalve, **237,** 269, **273, 276**
brachiopod, **241, 245, 270, 271,** 272, **277, 306**
cephalopod, **237**
crinoid, **244**
crustacean, **237**
echinoid, **237**
gastropod, **237**
mollusc, 192–195
sponge, **237**
worm, **237**
life table, 80–85, **82**
limiting factors, 248–274, 278–282
lineage, evolutionary, 103, **110, 111,** 112, **134, 309, 310, 311, 364**
lithofacies, 203
littoral zone, 236
living fossil, 327, 330

macroevolution, 303, **349**
mammals
adaptive radiation of, 318–324, **318,** 328, 365
bat anatomy, **186**
biogeography of, 204, **331, 432,** 434
carnivores, 351–352

mammals *continued*
  diversity through time, 11, **12**
  evolution of size, **343, 344, 345, 351**
  evolutionary rates of, **310, 360**
  horses, evolution of, 136, 349–352, **351**
  Irish "elk," **335**
  jaw mechanics, 188–192, **188**
  migration, 204, 407
  orthogenesis, **334, 335**
  Pleistocene mammoths, 343, **344**
  species duration, 328–330
  survivorship, **332**
  See also hominids
man, fossil, 76, 98–100, **99**
mass extinction, 307, 332, 438–448
matrix, data, **92**
Mazon Creek Shale, 21
measurement of fossils, 37–43
member, stratigraphic, 198
microevolution, 303, **349**
migration
  species dispersal, 404
  "sweepstakes," 204
molluscs, 4, **19**
  evolutionary rates, 316–324, 415
  growth of, 46, **54,** 57–58
  life habits of, 192–195
  Neopilina, 327, 372–375, **374**
  population densities of, **13,** 14
  rarefaction in, 5–7, **6, 8**
  rudist, 284, **286,** 287, **443**
  shell thickness of, 179
molting, **49, 50,** 80, 84–85
monoplacophorans, 327, 372–375, **374**
monotypic taxa, 131
morphocline, 148
morphologic terms, 35–36, **37**
morphospecies, 109
mosaic evolution, 313
multiple-effect factors, 179
multi-state character, 144
multivariate analysis, 58, 86, 90–100
  canonical analysis, 98, **99**
  factor analysis, 94, 97
  graphical method, **91**
  principal components, **95,** 97
  Q-mode, **93**
  R-mode, **93,** 97
Mytilus, 195, 275, **276**

naming of species, 114–117, 126–127
natural selection, 103–104, 304, **349**
Nautilus
  coiling geometry of, **54, 168,** 169
  distribution of, 404, **405**
nektonic organism, 236
Neopilina, 327, 372–375, **374**
neoteny, 353
neotype, 126
niche, ecological, 232
nomenclature (taxonomic)
  International Code, 114–116, 127, 128
  International Commission, 114, 117, 125, 127
  International Rules, 114–115
nonadaptive evolution, 349
norm of reaction, 59, 74
numerical taxonomy, 142–147, **147,** 229

oceans
  currents, **205,** 380–385, **383, 384**
  food cycle, **238**
ontogeny, 45–68, 353–354
  heterochrony, 355
  rates, 59–62
  recapitulation, 354
  types, 46–51
  See also growth
ontogenetic variation, 71–72, **173**
operational taxonomic unit (OTU), 142
opportunistic species, 280
optical scanning, 31
orthogenesis, 333, **334, 335**
ostracodes
  electron microscopy, **33**
  harmonic analysis, **41,** 41–43
  life table, 84–85
  molt stages, **83**
  as salinity indicators, 265
  survivorship curves, **83**
oxygen, 248–253
  in the atmosphere, 448
  in the Black Sea, **254**
oxygen isotopes
  as indicators of paleotemperature, 403
  ratios used to estimate salinity, 263, **264**

paedomorphosis, 353
paleoclimate, 393–403
paleoecology, 231–302
paleospecies, 110, 313
palynology, 397
Pangea, 426
paradigm, 179
parallel evolution, 364
parasite, **233**
paratype, 126
pelagic organism, 236
phenetic classification, 142
phenogram, **96,** 97, 145
phenotype, 103, 142
photic zone, 236
photography of fossils, 28–29
  infrared, 28, **30**
  stereo, **29**
  ultraviolet, 28, **30**
  X-ray, **23,** 28, **30,** 44, **54**

phyletic
  evolution, 103, 307–316
  extinction, 330
  speciation, 103
  transition, 103, 104
phyletic gradualism model, **326,** 327
phylogenetic tree, 131–136, **317, 326**
  bivalve, **132**
  dinosaur, **143**
  echinoid, **133**
  horse, **136**
  protozoan, **135**
phyletic trends, 335–348
phylogenetic trends, 335, 348–355, **350**
phytoplankton, 236, 279
pioneer organism, 275, 280
planktonic organisms, **205,** 236, 279
plants
  biogeography of, **70, 388, 397,** 424–425, 428
  as climatic indicators, 393–395, **396, 398, 399**
  evolution of angiosperm, **362, 363**
  preservation of, 4, 24, 361
  role in ecosystem, **233**
  turtle grass beds, **275**
plate tectonics, 416–417, **417, 419, 420, 421, 422**
pleiotropy, 179
Pleistocene
  hominid skulls, 76
  mammoths, size of, 343
plesiotype, 126
populations, 69–100
  allopatric, 102, 108
  defined, 69, 232
  deme, 69
  spatial distribution, **70,** 274, 275–277
  sympatric, 102
population dynamics, 78–85
Porifera. *See* sponges
post-mortem information loss, 76, **77,** 78, 296–298
post-mortem transport, 24, 25, 76, **77,** 404
Precambrian, 371, **436**
predation, 296
  as cause of extinction, 304, **305**
  role in preserving diversity, 281, 284, 287
preservation, 3–5, 14–23, 113, 360
  biologic aspects of, 14–15
  of bryozoans, 16
  of calcareous algae, 16
  chemical aspects of, 20
  effect of bacteria on, 16
  effect of habitat on, 21–23
  effects on variation on, 20, 75
  evidence of life habits, 239–243
  of insects, 4, 10–11, 21
  mechanical aspects of, 16–18, **17, 18**
  of major biologic groups, 11
  models of, 299–301
  of molluscs, 16
  processes of, 14–23
  of soft parts, 20
primates, 76, 98–100, **99**
principal components analysis, **95,** 97
Principle of Similitude, 63–64
Priority, Law of, 116
producers, **233, 234**
protozoans
  phylogeny, **135**
  preservation of, 4
  See also foraminiferans; fusulinids
province, biogeographic, 205, **206,** 207, 387–391
pseudoextinction, 330
Pteranodon, 182–188, **185, 186, 187**
Pterodactylus, 182
punctuational model of evolution, **326,** 330, 348
punctuated equilibrium model, 326

Q-mode analysis, 92, **93**
quantitative correlation, 216–222

R-mode analysis, 92, **93,** 97
radiography. *See* X-ray photography
radiolarians, electron microscopy of, **32**
rarefaction, **6, 8**
rate of evolution. *See* evolutionary rates
Recapitulation, Principle of, 354
rectangular model of evolution, **326**
reefs, organic, **257,** 282–287, **283**
  Permian, **250**
  rudist, **286**
  Silurian, 284, **285**
refugium, 307
regression, **215**
relicts, 415, 431–432
reproductive isolation, 101–103, 105–109
reptiles
  biogeography of, 425–426, 431–432
  distribution of modern lizards, **392**
  diversity through time, 12, 13
  evolutionary rates of, 313, 365
  growth and allometry of, **64**
  jaw mechanics of, 188–192, **188, 191**
  Pterodactylus, 182
  pterosaurs, functional morphology of, 182–188, **185, 186, 187**
  species-area relationship, **411**
  See also dinosaurs
ring species, 105–106
riparian habitats, 361–363
rock-stratigraphic unit, 198–203
rudists
  biogeography of, **443**
  in reef formation, 284, **286,** 287

salinity, 261–267
  classification, 261
  euryhaline organisms, 261

salinity *continued*
- estimated by oxygen isotope ratios, 263, **264**
- relation to species diversity, **262**
- stenohaline organisms, 261, 265, **266**

scavenger organisms, **233**
sediment-organism relationships, 21, 267–273
sediment survival, 8, **9**
series, defined, 224
sessile organism, 236
sexual dimorphism
- in cephalopods, **73**
- in foraminiferans, **72**

sexual selection, 344
sibling species, 108
Similitude, Principle of, 63–64
size
- Bergmann's Rule, **337,** 343
- Cope's Rule, 352
- evolution of, **309, 311, 314, 336, 337,** 343, **344, 345, 351, 353**

size-frequency distribution, **80, 89**
- bivalve, **71**
- brachiopod, **87, 88,** 90
- ostracode, **83**

Solnhofen Limestone, 21, **22,** 24, **30,** 182
sorting, 76, **80**
speciation, 103
- geographic, 104–106
- rate of, 106–107, 316–324
- sympatric, 106, 435

species, 101–128
- chronospecies, 313, 329
- definition, 102, 108
- description, 35–36, 114–125
- diagnosis, 117–125
- discrimination, 107–110, **110, 111, 112**
- diversity, 3–13, **6, 8, 9,** 234, **254,** 263
- duration, 4, **322, 323**
- endemic, 205
- evolutionary, 110, 135, 313
- evolutionary rates of, 106–107
- morphospecies, 109
- opportunistic, 280
- paleospecies, 110, 313
- pioneer, 280
- ring, 105–106
- sibling, 108
- successional, 110, 313
- type, 131

species-area relationships, 408–413, **411, 412, 414**
species problem in paleontology, 108–114, **110, 111, 112**
species selection, 348, **349, 350**
Spirula, **168,** 169
sponges, 4
- astrorhizae, 377
- *Cliona,* **15**
- growth, **65**
- life habits, **237**
- sclerosponges, 375–377, **376**

stage, defined, 224
statistical significance, tests of, 89
stenohaline organism, 261
stereophotography, **29**
stratigraphic correlation, 207
stratigraphic ranges, 207–209, **211**
stratigraphic zones, 207–209
stratigraphy. *See* biostratigraphy
stromatoporoids, **376**
subduction zone, 418, **419,** 435
sublittoral zone, 236
subspecies, **87, 88, 90,** 105
substratum
- as limiting factor, 267–273, 279–280
- relation to feeding mechanisms, 267–268

succession
- biotic, 198
- ecologic, 280, 284, **285**

successional species, 110, 313
survivorship curve, 79–85, **79, 81, 83, 332**
suspension feeder, 239, 268, 287
symbiotic association, 284
sympatric
- populations, 102
- speciation, 106, 435

synecology, 278
synonym, 127
synonymy, 113–114, 117, **160**
syntype, 125
system, stratigraphic, 222, 224

taphonomy, 296, **297**
taxonomic character, 124–125
taxonomic diversity, **6, 8, 9, 12,** 234, **254,** 263, 333
- gradients in, 391, 392, **409, 410**
- in relation to salinity, **262**

taxonomy
- cladistic, 146–149
- numerical, 142–146
- phenetic, 142

temperature, 248
- Bergmann's Rule, **337,** 343
- oxygen isotope ratios, 403
- relation to size of organism, **336, 337**

theoretical morphology, 170–171
time-rock unit, 224
time scale, geologic
- absolute, 222, **228**
- geochronologic, 225
- relative, 222, **228**

time-specific method of survivorship analysis, 84
topotype, 126
trace fossils, **22,** 199, **200,** 202, 246–248, **246, 247, 288**
tracks and trails, **22,** 246–248
transformers, 233

transgression, **215, 292**
transspecific evolution, 303
Treatise on Invertebrate Paleontology, 161
trilobites, **23**
  Cambrian, **212, 221**
  growth of, **49, 50, 59**
  iterative evolution of, **212,** 368–369, **368**
  lens morphology of, 180–182, **181, 183**
  migration, 209
  speciation, 326
  vision in, **184**
  X-ray photograph, **23**
two-state character, 144
type area, 222
type genus, 131
type section, 222
type species, 131
type specimens, 113, 125–127

ultraviolet photography, 28, **30**
univariate analysis, 85–86, 88
  of brachiopods, **87**

vagile organism, 236
variation
  between assemblages, 87–90
  description of, 85–100
  within assemblages, 85–86

Wallace's Line, 432–434, **433**
weighting of characters, 138–139
worms, 288–289
  boring into brachiopod shells, 272
  life habits of, 237

X-ray photography, 44
  of brachiopod, **30**
  of cephalopod, **54**
  of gastropod, **15**
  of *Nautilus,* **54**
  of trilobite, **23**

zone (biostratigraphic), 207–209
  assemblage zone, 209
  biozone, 203–207
zoogeography. *See* biogeography
Zoological Record, 162
zooplankton, 236, 279, 284
zooxanthellae, 256, 284